History of Mathematics
From a Mathematician's Vantage Point

NICOLAOS K. ARTEMIADIS

Translated by
NIKOLAOS E. SOFRONIDIS

AMERICAN MATHEMATICAL SOCIETY
Providence, Rhode Island

This work was originally published in Greek by the Academy of Athens under the title "Ιστορία των Μαθηματικών" ©Academy of Athens. The present translation was created under license for the American Mathematical Society and is published by permission.

Translated from the Greek by Nikolaos E. Sofronidis

2000 *Mathematics Subject Classification.* Primary 01-01, 01A05.

For additional information and updates on this book, visit
www.ams.org/bookpages/hismat

Library of Congress Cataloging-in-Publication Data
Artémiadis, Nikolaos K., 1917–
  [Istoria ton mathematikon. English]
  History of mathematics : from a mathematician's vantage point / Nikolaos K. Artémiadis ; translated by Nikolaos E. Sofronidis.
    p. cm.
  Includes bibliographical references and index.
  ISBN 0-8218-3403-7 (acid-free paper)
  1. Mathematics–History. I. Title.

AQ21.A7513 2004
510'.9–dc22   2004046178

Copying and reprinting. Individual readers of this publication, and nonprofit libraries acting for them, are permitted to make fair use of the material, such as to copy a chapter for use in teaching or research. Permission is granted to quote brief passages from this publication in reviews, provided the customary acknowledgment of the source is given.

Republication, systematic copying, or multiple reproduction of any material in this publication is permitted only under license from the American Mathematical Society. Requests for such permission should be addressed to the Acquisitions Department, American Mathematical Society, 201 Charles Street, Providence, Rhode Island 02904-2294, USA. Requests can also be made by e-mail to reprint-permission@ams.org.

© 2004 by the American Mathematical Society. All rights reserved.
The American Mathematical Society retains all rights
except those granted to the United States Government.
Printed in the United States of America.

∞ The paper used in this book is acid-free and falls within the guidelines
established to ensure permanence and durability.
Visit the AMS home page at http://www.ams.org/

10 9 8 7 6 5 4 3 2 1   09 08 07 06 05 04

Dedicated to the memory of my Parents
Kyriakos and **Despina**
and of my brother **Artemios**
Dedicated to the memory of the colleague
Athanasios Panagos

CONTENTS

PREFACE vii
ACKNOWLEDGEMENTS FOR THE ENGLISH EDITION ix

PART I

INTRODUCTION 3
Chapter 0: *Remarks on the History and the Philosophy of Mathematics* 5
Chapter 1: *Mathematics of the Egyptians* 11
Chapter 2: *Mathematics of the Babylonians* 13
Chapter 3: *Prime Numbers* 15
Chapter 4: *Mathematics of the Greeks — The beginnings* 19
Chapter 5: *Pythagoras — The Pythagorean School* 21
Chapter 6: *Heraclitus, Parmenides, Zeno, Empedocles, and Democritus* 33
Chapter 7: *The Mathematical School of Athens* 37
Chapter 8: *Plato and Aristotle* 49
Chapter 9: *Groups* 51
Chapter 10: *The Impossibility of Solving the Three Classical Problems of Antiquity* 59
Chapter 11: *Euclid* 67
Chapter 12: *Hilbert's Foundation of Euclidean Geometry* 77
Chapter 13: *Basic Properties of Axiomatic Systems* 83
Chapter 14: *Spaces and Geometries* 89
Chapter 15: *Non-Euclidean Geometries* 103
Chapter 16: *The Geometry of Experience* 111
Chapter 17: *Aristarchus, Archimedes, Apollonius, and Eratosthenes* 123
Chapter 18: *The Period from 200 B.C. to 500 A.D. in Alexandria* .. 133
Chapter 19: *A Brief Review of the History of Greek Mathematics* ... 149
Chapter 20: *Mathematics in China and India* 155
Chapter 21: *Mathematics of the Arabs* 161
Chapter 22: *Europe during the Middle Ages* 165
Chapter 23: *Renaissance (1400-1600)* 171
Chapter 24: *The Seventeenth Century* 183
Chapter 25: *The Eighteenth Century* 205

PART II

INTRODUCTION 247
Chapter 1: *Short Biographies of Mathematicians of the 19th and 20th Centuries* 249
Chapter 2: *Revival of Synthetic Geometry* 279

Chapter 3: *The System of Real Numbers* 283
Chapter 4: *The System of Complex Numbers and the Quaternions* .. 291
Chapter 5: *The Fundamental Theorem of Algebra* 297
Chapter 6: *Set Theory* 299
Chapter 7: *Logic* 313
Chapter 8: *Functional Analysis* 323
Chapter 9: *Topology* 345
Chapter 10: *Functions of Real Variables* 357
Chapter 11: *Abstract Algebra* 367
Chapter 12: *Categories and Functors* 379
Chapter 13: *Recent Discoveries and Achievements* 385
Chapter 14: *Language* 401
Appendix I: *The Fast Fourier Transform (FFT)* 405
Appendix II: *Historical Roots and Basic Notions in the Theory of Distributions* by L. Schwartz 411
Appendix III: *The Theory of Wavelets* 425
CURRICULUM VITAE 431
BIBLIOGRAPHY 433
CHRONOLOGICAL TABLE 437
NAME INDEX 447
SUBJECT INDEX 451

PREFACE

This book is aimed at the young mathematician who is at the end of his/her basic studies or at the beginning of his/her scientific career. The author vividly remembers the longing he and his classmates had during their student years and later as graduate students for a source book where they could find a clear, simple, and yet general description of the science of mathematics, the forest they were about to enter for life. They believed the science of mathematics to be the mother of all science on which one finds the foundation for productive imagination, and of clear and fine thought, as well as criteria for and prototypical examples of objective truth in all intellectual activity. Today, the forest of mathematics has become too dense and too large. It is for this reason that the author, who is now at the end of his scientific career, attempted to write this book, to respond to the desires of present-day students and workers in the science of mathematics, who like him want an authoritative and complete source on the history of the science of mathematics.

The book is aimed at students, teachers, and researchers of mathematics. It is generally agreed that to foresee the future of mathematics one needs to study the history and the current status of the science of mathematics. My intent here is not to guide the reader in his/her historical journey by adopting the classical route (list of names, dates, events, short descriptions of topics, etc.) followed in many important history books. Besides, the author is not a historian, but rather a teacher, who focuses on teaching mathematics within a general historic framework. This justifies the explanatory subtitle of the book, which indicates that the book is about the history of mathematics from the point of view of a mathematician. I hope that the reader will be helped by the book's content and style and that he/she will be oriented toward directions which are appropriate for his/her research and interests.

Having in mind the reader to whom I am referring, I tried to present some of the main discoveries and creations in mathematics since ancient times until the end of the 20th century. The effort is restricted to the presentation of central ideas which are fundamental and which have generally influenced the generation and shaped the course of mathematics.

The book has an additional special value for the student of mathematics. It is well known that the usual university courses convey the impression that the various subjects being taught are unrelated. Contrary to this, the history of mathematics reveals a general picture which not only connects one subject with the rest, but also connects each subject to the main body of mathematical thought. Also, good organization and logical exposition of various topics convey to the student the wrong impression that it is natural for mathematicians to move easily from one theorem to the next overcoming all difficulties and reaching a full understanding of the whole situation. The history of mathematics, on the contrary, teaches the student that progress is achieved step-by-step and results come via many different avenues. Thus, the young scientist learns that often it takes tens or hundreds of years of effort to achieve a few steps of progress, and that even then many gaps are left behind needing subsequent investigation. The student, and the scientist in general, by becoming aware of these difficulties and realities, should recognize that they need to equip themselves with courage, patience, and persistence, in order to face their own problems calmly, unafraid of the difficulties which may arise from their research and may lie ahead in their careers.

Without knowing the concepts, the methods, and the results discovered and developed by previous generations, starting with ancient Greeks, it is not possible to understand the more recent accomplishments, even those of the last one hundred years. Leibniz maintained that whoever understands the work of Archimedes and the work of Apollonius, is less likely to express admiration for the accomplishments of subsequent generations. I am convinced that the material in this book will help the reader appreciate fully the leading role of the Greek spirit in the development and the course of mathematics.

In addition to the main text, which, as I have already pointed out, can be considered as an historical introduction to the science of mathematics, I have also added Chapter 0 to Part I and Chapter 14 to Part II, which, respectively, refer briefly to observations on the history and philosophy of mathematics and the role of language as an instrument for conveying ideas.

For the assimilation of the text, I have assumed that the reader is only familiar with the material which is normally covered by a basic degree in mathematics. There are many topics which I did not address. This is mainly because I wish to keep the book within reasonable size limits for the reader. With regard to the omission of these topics, it is instructive to remember the words of young Evariste Galois: "Unfortunately, it is little recognized that the most important scientific books are those in which the author introduces subjects which he himself does not know. An author who hides from the reader the difficulties he himself faced as an author hurts the reader."

The author thanks the Research Committee of the Academy of Athens for its support of both the writing and the publishing of the book.

He also wishes to thank:

α) Nikolaos K. Stefanidis, Professor Emeritus of the Aristotle University of Thessaloniki, who read the entire manuscript and made useful observations and improvements;

β) Dr. Gregorios Gkizelis, Director of the Center for Research of the Greek Society of the Academy of Athens, who meticulously looked after the entire edition; and

γ) Miss Theodora Gkizelis of the Research Institute of Pure Mathematics, who typed a large part of the manuscript and helped with the usual editorial changes.

I owe special thanks and appreciation to my wife Zafiria for her contribution, help, and continuous support throughout the years it took to complete the writing of this volume.

Last but not least I thank Mr. Stelios Tsapepas for the elegant printing of the volume.

Athens 2000   Nicolaos K. Artemiadis

ACKNOWLEDGEMENTS FOR THE ENGLISH EDITION

I am deeply indebted to Professor Athanasios Fokas for his careful reading of both the Greek and English versions of the book. His corrections and suggestions greatly improved the English edition.

I am sincerely grateful to the Research Committee of the Academy of Athens for its crucial financial support regarding the English edition.

I also wish to express my appreciation to Dr. Nikolaos Sofronidis for the translation of the book into English and for his useful remarks. The help of my colleague in the Academy of Athens, Professor Loucas G. Christophorou, in the English edition of the book is greatly appreciated. Thanks are also due to Professor G. Anastasiou for bringing the Greek edition of the book to the attention of the American Mathematical Society (AMS), and to Dr. Sergei Gelfand of the AMS for making the necessary arrangements for the publication of the English edition and for his excellent collaboration.

PART I

INTRODUCTION

In order to be able to determine when approximately the history of mathematics begins, one must first give the definition of the term "mathematics".

Today what we have at our disposal regarding the birth and the development of the notion of a number in peoples of the prehistoric times, are simply conjectures and hypotheses, which come from observations that were made subsequently.

The prehistoric period in Egypt and Mesopotamia ends around 3000 B.C., while in the large riverside plains of India and China this period ends a little later.

Since in the ancient world agriculture was the basis of civilization in the riverside plains, the primary work of the officials was to control the systems of watering via appropriate works of irrigation, draining, drawing and channeling of water. Secondly, in order to levy the relative taxes, they had to measure the lands and the harvests, and thirdly it was necessary to establish a certain calendar based on observations of the celestial bodies. Consequently, in order to cope with all these tasks, it was necessary for them to have a certain mathematical knowledge. Moreover, in order to build palaces and altars, it was necessary for them to have additional mathematical knowledge. The information at our disposal regarding the development of mathematics during these years comes from primitive objects of art. The possibility of discovering other similar objects is likely to change our knowledge and viewpoints regarding the history of mathematics during the period in question.

The beginning of the history of mathematics cannot be placed with certainty at any school or period before that of the Greeks of Ionia, i.e., around 500 B.C. During that time, mathematical knowledge began to proliferate in Ancient Greece, which contained, in addition to today's Greece, the coasts of Asia Minor (Ionia), the Great Greece (South Italy and Sicily), and later Alexandria (in Egypt).

Around 300 B.C., the center of mathematics was transferred from Athens to Alexandria in Egypt, where it remained for the next 800 years since all the books were kept there.

During 500 A.D., the Mediterranean civilization ceased to develop perhaps due to the epidemic diseases that repeatedly manifested themselves.

Around 800 A.D., mathematics, as it was formulated in Alexandria, was transferred to India, where a long-standing mathematical tradition was created.

Helped by the translations of the Greek texts, the Arabs developed and disseminated again mathematical knowledge in the area of the Mediterranean and finally in the rest of Europe.

During the period of the Renaissance, mathematics flourished in Italy and with the help of typography it was disseminated in Western and Central Europe.

Today mathematics is cultivated in every industrial country of the world.

This book is divided into two parts. Part I is restricted essentially to the History of Mathematics until the end of the 18th century. After this period mathematics tends to be more specialized and thus it is less accessible to the reader. Let us note however that many times we refer also to subjects that took place after the 18th century, either in order to avoid the tedious chronological development, or in order to present simultaneously answers to long-standing questions (provided that this presents no difficulties to the reader).

Part II treats certain more specialized subjects which were developed during the last two centuries and in which spectacular progress was made. Greater emphasis is given to those subjects that concern the foundation of mathematics.

Due to the interest of the author in philosophy, the entire text is very often accompanied by remarks, comments and viewpoints of philosophic content.

For reasons of clarity and due to the peculiar vocabulary used in a mathematical text, it was judged advisable to make sometimes repetitions, lengthy remarks, use of mixed Greek words together with foreign terms, etc. In this sense greater attention is given to the clarity and the exactness of the text presented than to the elegance of writing.

CHAPTER 0

REMARKS ON THE HISTORY AND THE PHILOSOPHY OF MATHEMATICS

History and philosophy of mathematics do not occupy a central place in the work of most mathematicians, whether they are instructors or researchers. A consequence of this fact is the creation in most of us of an incomplete and dogmatic picture of these matters, something that influences the work of all of us and proves the validity of what Henri Poincaré alleged, i.e., that "If we want to predict the future of mathematics, we have to study its history and its present situation". We could perhaps add to these last words that the deeper study of the history and the philosophy of mathematics is important in its own right because, if nothing else, it contributes to the essential improvement of the teaching of mathematics. But let us discuss the subjects above in somewhat more detail.

History

History of mathematics tells us which ideas appeared, which theorems were proved, and when. Sometimes it relates to matters of a different nature, but its main work is to discuss how mathematics were developed and how ideas were combined and gave birth to new ideas. If we take into account this point of view, it follows that the reasons for which we must study the history of mathematics are (1) to understand better and evaluate correctly the results that we already know and (2) to realize that no matter how beautiful and of a general character are the results obtained up to this day, it is possible to advance and improve them even further. It may happen that a day will come when the theory of linear operators in a Hilbert space will be a very ordinary subject, but this theory will not cease to preserve its beauty.

History of mathematics, considered in this light, differs a lot from other histories. This difference can be understood if we take into account the nature of mathematics itself: a Euclidean system of eternal truths. History of mathematics progressively reveals these truths, one after the other, to the human thought.

The most important difference between the history of mathematics and other histories is that the former has no need of interpretation. The importance of a mathematical result is sought only in the relations that exist between this result and the remaining body of mathematics. The set of mathematical knowledge acquired up to this day, is the only measure by which one can properly evaluate the importance of a mathematical result.

The incomplete picture we have formed about the development of mathematics, due to our limited knowledge of the history of mathematics, prevents us from seeing the various critical periods that appeared in relation to the foundations of mathematics; this in its turn prevents us from perceiving the essential changes that have occurred in mathematics and in its role. When essential changes in the role played by mathematics are not perceived, then there is a danger that this science will be considered as being completely isolated from the greater part of human activities. The belief that mathematics is isolated unfortunately exists among the public and among our students, and constitutes a great obstacle to becoming familiar with mathematics. A deeper knowledge of the history of mathematics gives one the opportunity to examine the existing relation between this science and other aspects of culture.

Philosophy

Below we will discuss briefly the "Philosophy of Mathematics", a subject which must be touched upon in a book about the history of mathematics. But what is the meaning of the phrase "philosophy of mathematics"? Its meaning is revealed in two projects which are considered to be the central subjects of the philosophy of mathematics. One is the "**hermeneutical project**", whose study leads to an appropriate interpretation and to a context of the theory of mathematics and its use, while the other is the so-called "**metaphysical project**" which asks the question whether abstract objects exist.

We present some necessary terminology. "Abstract object" is an object which exists outside spacetime, i.e., an object which exists but does not belong to spacetime. Such objects are neither natural nor mental, and do not constitute or generate a cause. The belief in the existence of such objects is called "Platonism" while the non-belief in these is called "Anti-Platonism".

A "mathematical object" is an abstract object which belongs in the area of mathematics as, for example, a number, a function or a set. Finally, "mathematical Platonism" is the viewpoint that mathematical objects exist, while "mathematical Anti-Platonism" is the viewpoint that such objects do not exist. For simplicity we will usually refer to these two notions as "Platonism" and "Anti-Platonism".

The hermeneutical project and the metaphysical project are not completely separate. Philosophers who are mainly interested in the hermeneutical project, very often state views about the metaphysical project, which means that they accept either Platonism or Anti-Platonism. Similarly, philosophers who are interested in the metaphysical project, always state views which concern the hermeneutical project, i.e., they accept some interpretation of mathematical theory and practice.

The difference between these two groups lies in what each one considers to be its main interest.

Hartry Field is an example of a philosopher mainly interested in the metaphysical project. Almost his entire work in the philosophy of mathematics aims to prove that it is not necessary to believe in the existence of mathematical objects. David Hilbert is a good example of a philosopher whose main interest lies in the hermeneutical project.

Mark Balaguer in his book (*Platonism and Anti-Platonism in Mathematics*, *Oxford Univ. Press*, 1998, *pp.* 217), after a detailed and complete discussion of the subject, comes to the conclusion that it is impossible to discover a rational reason to believe or not to believe in the existence of abstract objects by studying some mathematical theory and practice. In other words, Balaguer comes to the conclusion that there exist no powerful arguments in favor of or against Platonism in the philosophy of mathematics.

We will not follow Balaguer in his complicated arguments and thoughts. We will quote a viewpoint which is a favorite of the author of this book and which refers to mathematical Platonism.

According to this viewpoint, human thought seems to be led and directed towards an outside "truth", which is independent of the human being, a truth which has an objective existence and which is partially revealed to only someone or some of us. This viewpoint is supported, for example, by the existence of an object called the "Mandelbrot Set" which we will discuss in detail in Chapter 13 (of Part II) in Chaos Theory. This set was discovered in 1990 by Benoit Mandelbrot and is a fractal which was produced with the aid of a simply programmed computer and which is considered to be the most complicated and the most beautiful figure that ever appeared in mathematics.

This beautiful structure of the Mandelbrot set was neither invented by anyone nor designed by any group of mathematicians. Mandelbrot himself, who first studied the set, did not have the slightest idea about the structure that was concealed in it. Indeed, when the first pictures started to appear in the computer, Mandelbrot thought that these first hazy patterns were due to a malfunctioning of the computer. In addition, no one can completely perceive all the details of the complicated structure of this set. As it appears, this structure of the Mandelbrot set does not constitute part of our thought or part of our reasoning; we had not conceived it intellectually before, but it has its own reality. If any mathematician or computer tries to study this set, his/her or its findings will be approximations of the same fundamental structure. So we can say that the Mandelbrot set does not constitute an invention of the human mind but constitutes a "discovery". Just as Mount Olympus exists somewhere, so does the Mandelbrot set "exist" somewhere and we discover it.

Beyond the Mandelbrot set, there exist other examples of mathematical achievements which are distinguished by the same existential reality, i.e., achievements which do not constitute intellectual constructions of some mathematician, but they are discoveries. Such an example is the discovery of the system of complex numbers. The primary reason for the introduction of these numbers was to make sense of the square roots of negative numbers, and it was thought at the beginning that the complex numbers were just a mathematical invention whose goal was to satisfy some concrete need. However, later it became clear that these objects led to more important results than the ones for which they were originally intended. In the course of time, many "magical", we would say, properties and results were obtained, whose existence was not at all suspected by any of the great mathematicians which had studied the complex numbers. We can say that these results and these properties existed before. The "magic" of these properties existed in the structure of the complex numbers, and it was discovered gradually. We would perhaps be more persuasive if we could comment more on the above properties from a mathematical point of view by referring to Cauchy's Integral or to Riemann's Mapping Theorem which we will examine in subsequent chapters.

The above makes us wonder: Is mathematics an "invention" or a "discovery"? Are mathematical achievements inventions of mathematical science or do they constitute existing objectivities (realities) which are gradually discovered by mathematicians? For this reason, let us discuss the subject in somewhat more detail and more depth.

We mentioned above that there exist mathematical achievements which we characterized as discoveries and not as inventions, and we gave two such examples. These achievements, i.e., these discoveries, have the following fundamental feature: the structure that results in each one of them has consequences which are much more extensive and much more important than the original reasons for which this structure was created. For example, in the case of the complex numbers, the original reason for the discovery, which was to be able to talk about the square root of negative numbers, had consequences which are infinitely disproportionate to it.

One might perhaps claim that in the cases where mathematical achievements are discoveries and not inventions, we find ourselves in front of "divine creations" which already exist and which we gradually discover!

But let's not forget that there exist mathematical structures which are not characterized by the above property. Examples of such structures are the cases where a mathematician, in his or her effort to prove some concrete result, is forced to introduce some devices whose structure is not characterized by the above property. In these cases, the result that occurs (the structure that occurs) no longer has any general consequences; it simply gives the solution to the problem that we are preoccupied with at the moment. The results obtained in these cases must be characterized as inventions rather than as discoveries. They are "human creations"!

It follows from what was presented above that mathematical discoveries are achievements which are much greater and much more important than inventions.

A similar distinction seems to be made also in works of art. The greatest works of art are considered to be closer to "divinity" than the remaining, "less" important ones. Artists believe that the greatest and the most important of their works reveal eternal truths which have a priori an exquisite existence, while their less important works constitute arbitrary constructions, they are works of mortals!

I think, though, that it must be emphasized that what happens in mathematics, at least in the case of fundamental notions, is more powerful and deeper than what the artists believe happens in art. As opposed to what one expects in art, mathematical ideas have a uniqueness and a universality of an imperative nature and of a different order.

The viewpoint that various mathematical notions and mathematical objects exist independently of spacetime is of course known from the days of the ancient Greeks (360 B.C.) and is due to the great Greek philosopher Plato from whom the name "Platonism" derives. After quoting the above viewpoint, which presents an argument that supports Platonism, let's go back to further comments and remarks on philosophy, history and mathematics.

Many people believe that the transition from history to philosophy constitutes a transition from an area which is a bit hazy to an area which is very hazy. Historical remarks may be somewhat relevant to mathematics and may contribute to a better teaching of mathematics, but what can we expect from philosophy? The main interest in philosophy emerged from the crisis that appeared in relation to the foundation of mathematics, which is due to the discovery of existing antinomies (paradoxes) in set theory. The effort to remove the antinomies led to the creation of a powerful "mathematical logic". Mathematics came out of this crisis reinforced, and does not appear to need a "philosophy" in order to rely on it.

The very idea of philosophy of mathematics appears to be reckless. Many wonder whether all this talk, these "babbles" as they say, on Platonism, realism and the like, mean anything. Was Archimedes or Newton or Gauss a philosopher? Philosophers may talk forever as opposed to mathematicians who work.

The differences between philosophy, history, and mathematics are based on the existing difference of clarity between philosophical, historical and mathematical problems. Philosophical problems are never clear and for this reason positivists can "live" without philosophy.

Historic problems vary from being very simple as, for example, the question of when the first Constitution was adopted in Greece, to philosophical problems as, for example, the problem of how the crisis of the 1920s affected the political situation in Greece. If we compare mathematical problems with problems of this kind, we observe that mathematical problems are absolutely clear.

Even though a mathematical problem may be very difficult, we believe that when it is solved, we will conceive that it *is* solved. It appears that there exists a qualitative difference between the complexity of mathematical problems and the hazy situation which is observed in philosophical problems. Mathematical problems demand solutions while philosophical problems demand more thinking.

Despite this, it is believed that an appropriate philosophy of mathematics suggests that the above mentioned difference between philosophy and mathematics is not as fundamental as it appears to be. In order to see this, first of all we need a more careful and more detailed discussion of the subject.

The problems that concern the philosophy of mathematics can be divided into the following subjects: logical foundations—loss of certainty—the nature of proof—the relation that exists between mathematical knowledge and the real world—and finally, what was mentioned above, namely the "ontological position" of mathematical objects such as numbers, functions, sets, etc.

A comprehensive presentation of developments in these subjects obviously goes beyond the scope of the present book. We will simply refer superficially to some of them.

The most extensive activity during the 20th century was the one referring to the foundation of mathematics. The reasons for which mathematicians paid attention to the foundation of mathematics was the discovery of various contradictions (which they euphemistically called "paradoxes") in mathematics and particularly in set theory. One of these paradoxes is the paradox of Burali-Forti (1861-1931): the set $\mathcal{O}$ of all ordinal numbers, which is well-ordered, must be isomorphic to an ordinal number which is greater than any member of $\mathcal{O}$. But then this ordinal number is greater than any ordinal number!

During the first years of the 20th century, other paradoxes were also discovered. Obviously the discovery of these contradictions greatly annoyed the mathematical community. Another problem, whose existence was gradually recognized, was whether mathematics is consistent. Due to the existence of paradoxes in set theory, its consistency or inconsistency had to be clarified.

During the second half of the 19th century some scientists had already begun to occupy themselves with the foundation of mathematics and in particular with the relation that exists between logic and mathematics. Some researchers wanted to lay the foundations of mathematics on logic only. Others questioned the universal application of logic and the soundness of some proofs given in various propositions.

These issues, which were smouldering even before 1900, evolved into a vivid juxtaposition of views, and this resulted in the "foundation of mathematics" becoming a subject of wide interest.

The history of "loss of certainty" is presented to us by Kline in his book: *Mathematics: The Loss of Certainty (Oxford University Press, Oxford,* 1980*)*. In this book he describes the formation (during the 16th century) of the idea that God designed nature mathematically. The emergence of this idea was the result of the tension that was created between the religious doctrines of that time and the trend of returning to classic thought. That time was a golden era for mathematics.

The rest of Kline's book discusses the gradual loss of the "innocence" of "certainty". It was proved that mathematics is subject to the human non-infallibility. This loss of certainty presents serious philosophical problems which are evidently related to mathematics.

The subject of the "ontological position", which is closely related to the question of whether mathematical results are discoveries or inventions, was discussed earlier.

The last subject that we will touch upon is the difference that exists between mathematics and empirical sciences. The object of study of empirical sciences is material reality, while the objects of study in mathematics are "ideas" (we do not examine who conceives the ideas).

This difference between empirical sciences and mathematics is described by various philosophers as follows:

In the traditional (Euclidean) model of mathematics, truth is placed on the top, in the form of unquestionable axioms, and proceeds downwards, through inductive paths, to various theorems. By contrast, empirical sciences do not begin with unquestionable axioms.

What is "unquestionable" in empirical sciences is located at the "bottom" and it is the reality of facts. Conclusions proceed upwards towards theoretical assumptions. If a prediction contradicts the facts, then one comes back to the theory which must be changed.

In the meantime, both the notions of "fact" and "objective reality" started to become problematic. Thus, the general views on mathematics and empirical sciences began to change and move towards each other.

In conclusion, regarding our remarks on history and philosophy of mathematics, we do not want the reader to think that we support the view that history and philosophy must be taught in courses in mathematics. However, we believe that the instructor of these courses must be well informed in these matters in order to make his or her teaching more effective.

When one teaches a course whose content is not determined exactly, it is necessary that he/she knows much more about the subject of the course than what he/she will actually present to the students. It is necessary that the instructor has investigated several different paths and has chosen the one through which he/she will lead the students to the pursued aim. The same of course is true for the research mathematician.

CHAPTER 1

MATHEMATICS OF THE EGYPTIANS

Aristotle had the opinion that mathematics in Egypt were developed by the priests because they had enough free time at their disposal (*Metaphysics* 981 b, 23-24). On the contrary, Herodotus believed that the annual overflows of the Nile River demanded the repeated determination of the boundaries of the fields and that this led to the birth of Geometry.

The main sources about the history of mathematics in Egypt, are two papyri which were discovered in the 19th century and are named "Moscow's Papyrus" and "Rhind's Papyrus" respectively.

"Moscow's Papyrus" dates back to 1850 B.C. The most interesting result included in this papyrus is the calculation of the volume V of a truncated square pyramid. If $b = 0$, then this formula gives the volume of a square pyramid.

"Rhind's Papyrus", which was copied from the original in 1650 B.C. by someone named Ahmose, was acquired by Alexander Henry Rhind in Luxor in Egypt in 1858 and was then bought by the British Museum in 1865. "Rhind's Papyrus" contains the solution of a series of elementary problems such as the following: "how many bricks are needed to build a ramp of certain dimensions?" and other similar problems of practical nature.

The Egyptians used the decimal number system, without however clarifying what is the value that a digit takes depending on its position in the full number. They used fractions, which they (curiously) insisted on representing as sums of other unitary fractions (i.e., fractions with numerator one). Thus, the fraction $\frac{2}{9}$ was written as $\frac{1}{6} + \frac{1}{18}$, $\frac{19}{8}$ was written as $2 + \frac{1}{4} + \frac{1}{8}$ and so on. One easily recognizes that every fraction of the form $\frac{2}{2m+1}$ can be expressed as the sum of two unitary fractions of the form $\frac{1}{m+1}$ and $\frac{1}{(m+1)(2m+1)}$. But the Egyptians did not always follow this last method. For example, Ahmose writes $\frac{2}{45} = \frac{1}{30} + \frac{1}{90}$. The Hungarian mathematician Paul Erdös (1913-1996) suggested the following problem: "If n is an odd integer greater than 4, then the fraction $\frac{4}{n}$ can be written as the sum of three different unitary fractions". This problem still remains unsolved.

CHAPTER 2

MATHEMATICS OF THE BABYLONIANS

When we refer to the mathematics of the Babylonians, we mean the mathematics that was developed in ancient Mesopotamia — the country between the Tigris and Euphrates rivers, which is approximately today Iraq.

Before we proceed, we will remind the reader about a few things (which we will need below) concerning the bases of number systems, the ways of writing numbers in these systems, etc.

We know how a positive integer is written in the decimal number system. For example, we have

$$541 = 5 \cdot 10^2 + 4 \cdot 10^1 + 1 \cdot 10^0,$$
$$3205 = 3 \cdot 10^3 + 2 \cdot 10^2 + 0 \cdot 10^1 + 5 \cdot 10^0.$$

In general, it is known that if b is an integer greater than 1 and if a is an arbitrary positive integer, then a can be written uniquely in the form

$$a = a_n b^n + a_{n-1} b^{n-1} + ... + a_2 b^2 + a_1 b + a_0 \qquad (1)$$

where the a_i are integers, $0 \leq a_i < b$ ($i = 0, 1, 2, ..., n$).

In this case we say that the number a is written in the number system with base b and we write

$$a = (a_n a_{n-1} ... a_2 a_1 a_0)_b$$

where if there is no danger of confusion the subscript b can be omitted.

For example, we have

$$34 = (100010)_2 = 1 \cdot 2^5 + 0 \cdot 2^4 + 0 \cdot 2^3 + 0 \cdot 2^2 + 1 \cdot 2^1 + 0 \cdot 2^0,$$
$$2013 = (5604)_7 = 5 \cdot 7^3 + 6 \cdot 7^2 + 0 \cdot 7^1 + 4 \cdot 7^0.$$

The mechanism of converting the expression of a number from the decimal system to some other system is indicated in (1) and does not present difficulties. We quote the example of conversion of the expression of 643 to the ternary system. Dividing 643 successively by 3, we get

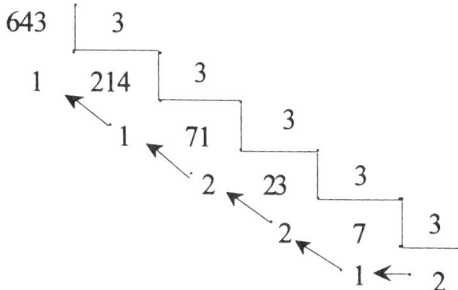

We write the last partial quotient as 2 and to the right of 2 we write the remaining partial quotients that we have obtained in the order shown by the arrows. We have
$$643 = (212211)_3 = 2 \cdot 3^5 + 1 \cdot 3^4 + 2 \cdot 3^3 + 2 \cdot 3^2 + 1 \cdot 3^1 + 1 \cdot 3^0.$$
In addition, if $b = 60$, then the number 5130 is written in the system with base 60 (which was used by the Babylonians) as follows:
$$5130 = (1, 25, 30)_{60} = 1 \cdot 60^2 + 25 \cdot 60^1 + 30 \cdot 60^0.$$
Here we have used commas to separate the positions of the digits.

The number of sources that inform us about the study and the development of the mathematics of the Babylonians is large, and probably this number will increase in the future.

The Babylonians kept precise records of astronomical observations during a long period of time, and in contrast to the Egyptians, the mathematics with which they occupied themselves were not for practical use only. They used the number system with base 60 ($b = 60$), where the value of a digit depended on the position it had in the expression of the number, in the same way as our decimal system uses today. They also used fractions with denominators consisting of powers of 60. For example,
$$(1, 25; 30)_{60} = 1 \cdot 60 + 25 + \frac{30}{60} = 85\frac{1}{2},$$
$$(1; 25, 30)_{60} = 1 + \frac{25}{60} + \frac{30}{60^2} = 1\frac{17}{40},$$
where the symbol (;) is used to distinguish the beginning of the fractional part of the number. They did not have a symbol analogous to the "comma" which we use today for decimal numbers and until the 4th century B.C. there was no symbol for "zero". Thus the exact position of a digit was deduced from the context of the expression considered.

The Babylonians had at their disposal a multiplication matrix, as well as matrices which gave the inverses, the squares and the cubes of numbers. They used these matrices for the solution of even third order equations and of systems of second order equations as well. They gave exact solutions to second order equations whose statements were given in words and not in mathematical expressions. They rejected the negative solutions (when such solutions occurred), but they accepted both solutions of the equation when they were positive. They also occupied themselves with finding integer solutions to the equation $a^2 + b^2 = c^2$. They were able to find solutions which were apparently not obtained by the method of verifying the equation via successive trials, because the corresponding numbers are very large, as for example $a = 12709$, $b = 13500$, $c = 18541$. They even considered the approximate calculation of square roots of integers, which shows that some relation exists between the mathematics of the Babylonians and the mathematics of the Ancient Greeks.

CHAPTER 3

PRIME NUMBERS

The role played by prime numbers in the Science of Mathematics is so fundamental that it is necessary, before we proceed, to present some basic material about this extremely important part of mathematics.

An integer greater than 1 is called **prime** if and only if it can be divided exactly only by itself and by one. Examples of prime numbers are

$$2, 3, 5, 7, 11, 13, 17, 19, 23, 29, 31, 37, \ldots$$

An integer which is not prime is called **composite**. It was known to the ancient people that it is not always possible to use an arbitrary pile of pebbles in such a way that they form a rectangular parallelogram. If for example the pile has 17 pebbles this cannot be done, while if it has 16 it can. This means that the difference between a prime and a composite number was known to ancient people.

The Ancient Greeks were the first to prove that there are infinitely many prime numbers. In the *Elements* of Euclid (300 B.C.) we find the following proposition:

Euclid's Theorem (Book IX, Proposition 20).
Given a finite set of prime numbers $p_1, p_2, \ldots, p_k$, there exists a prime number which does not belong to this set.

Proof. We consider the integer $n = p_1 p_2 \ldots p_k + 1$. It is obvious that the division of n by any of the numbers p_i ($i = 1, 2, \ldots, k$) gives remainder 1, which means that none of the numbers p_i divides n exactly. Now if we consider the set of the divisors of n which are greater than 1 (such divisors exist, for example, n itself) and if p is the smallest of these divisors, then $p \neq p_i$ for every i, and p is a prime number, because if p had some divisor different from 1 and p, then this divisor would also be a divisor of n, which is a contradiction because p is the smallest divisor of n which is greater than 1.

An elegant generalization of the above theorem of Euclid is the following theorem which is due to **Johann Peter Gustav Lejeune Dirichlet** (1805-1859): "Every arithmetic progression $a, a+d, a+2d, a+3d, \ldots$, where the numbers a and d are coprime (i.e., their greatest common divisor is 1), contains infinitely many prime numbers." The proof of this result is far from being simple.

Euclid (*Elements*, Book IX, Proposition 14) proves that if an integer n has no divisors of the form k^2 ($k \neq 1$), then n can be expressed uniquely as a product of prime factors (under the assumption that the prime factors are written in increasing order).

This last proposition, known as the "theorem on the uniqueness of factorization into prime factors", was proved in its full generality, i.e., without the assumption of non-existence of a divisor of the form k^2, by **Carl Friedrich Gauss** (1777-1855).

Theorem. Every integer greater than 1 can be expressed uniquely as a product of prime factors.

Proof. There exist several proofs. One of them is the following. Let us assume that the proposition is not true and that there exist integers greater than 1 that can be expressed as a product of prime factors in more than one way. Let n be the smallest of these numbers, so that n can be factorized in at least two different ways:
$$n = p \cdot q \cdot r ... = p' \cdot q' \cdot r' ...$$
We assume that the prime factors are written in non-decreasing order, i.e.,
$$p \leq q \leq r \leq ... , \quad p' \leq q' \leq r' \leq ...$$
From the assumption about n, it follows that $p \neq p'$. Indeed, if $p = p'$, then the division of both expressions of n by p would give an integer less than n which can be expressed as a product of prime factors in two different ways. Without loss of generality, we can assume that $p' < p$. Consequently, we have:
$$p' < p \leq q \leq r \leq ... \qquad (2)$$
Since n is not prime, we have $n \geq p^2$ and $n > pp'$. From the assumption about n, it follows that the integer $n - pp'$ is expressed uniquely as a product of prime factors.

If in the expression $n - pp'$ we successively replace n with its two different expressions, we find:
$$n - pp' = p \cdot (qr... - p') = p' \cdot (q'r'... - p).$$

This means that the integers p and p' divide $n - pp'$ and consequently there exists some integer z such that $n - pp' = pp'z$. Hence $qr... - p' = p'z$, i.e., p' divides $qr...$. But since $qr... < n$, it follows that $qr...$ is expressed uniquely as a product of prime factors, which means that p' must coincide with some of the integers $q, r, ...$, which contradicts (2). QED

Eratosthenes from Cyrene (born around 284 B.C. in Cyrene, died around 200 B.C. in Alexandria), had invented the following method of constructing prime numbers, known as the "Sieve of Eratosthenes": Write in increasing order all integers greater than 1. Delete all multiples of 2 apart from 2. Delete all multiples of 3 (which are not deleted yet) apart from 3 and so on. The non-deleted numbers constitute the set of prime numbers.

Several mathematicians tried to find a formula which gives all prime numbers. For example, the trinomial $x^2 - x + 41$ is a prime number for $x = 0, 1, 2, ..., 40$ (Euler), but for $x = 41$ it stops being a prime number. It is also known (Legendre) that for $x = 1, 2, ..., 16$ the trinomial $x^2 + x + 17$ is a prime number, while $2x^2 + 29$ is a prime number for $x = 1, 2, ..., 28$. For $x = 17$ and $x = 29$ the above formulae fail.

In 1970 **Yuri Matiyasevic** constructed a polynomial $f(x, y, z, ...)$ in several variables with integer coefficients such that its positive values at positive integer numbers $x, y, z, ...$ are all prime numbers, and only them. Related to the question of finding all prime numbers is the following proposition, where the logarithms considered have base e $\left(e = \lim_{n \to \infty} \left(1 + \frac{1}{n}\right)^n\right)$.

Prime Number Theorem
If $\pi(x)$ is the number of primes which are less than or equal to x, then
$$\lim_{x\to\infty} \frac{\pi(x)\log x}{x} = 1. \tag{3}$$

This proposition was proved almost simultaneously and independently by the French mathematician **Jacques Hadamard** (1865-1963) and the Belgian mathematician **Charles Jean de la Vallée Poussin** (1866-1962).

From this theorem it follows that the nth prime number, say p_n, is approximately equal to $n\log n$. The symbol $\approx$ which we use below means "approximately equal".

Indeed $\pi(p_n) = n$. From (3) we have $p_n \approx n\log p_n$ from which it follows that $\log p_n \approx \log(n\cdot\log p_n) = \log n + \log\log p_n$. Hence $p_n \approx n\log p_n \approx n\cdot\log(n\cdot\log p_n) = n\cdot(\log n + \log\log p_n)$. But the quantity $n\log\log p_n$ is negligible relative to $n\log n$, thus $p_n \approx n\log n$.

The conjecture of **Goldbach** (1742):

Every even integer greater than 2 is the sum of two prime numbers.

Prime numbers have applications in cryptography. The basic idea is the following: Mathematicians can often prove that an integer has factors without being able to find these factors. This has been used as a basis for the construction of cryptographic codes: the designer of the code selects two very large prime numbers, α and β, which usually have from 50 to 80 digits each, and constructs the product $\gamma = \alpha\beta$. The designer of the code knows the factors of γ, but the person who will try to break the code, even though he/she can find out (using theorems of Euclid and Fermat) that γ is composite, cannot find the factors α and β in a reasonable amount of time, and this cannot break the code. Since the number γ is known, these codes are called "Public key cryptographic codes".

CHAPTER 4

MATHEMATICS OF THE GREEKS — THE BEGINNINGS

The intellectual activity in the civilizations developed in the riverside areas of Egypt and Mesopotamia, began to decline many years before the Christian era. During the period of decline, new cultural trends began to appear in the coasts of the Mediterranean, which resulted in the gradual movement of the center of civilization from the valleys of the Nile, the Tigris and the Euphrates, to the coasts of the Mediterranean.

This movement took place approximately during the period between 800 B.C. and 800 A.D. In order to correctly attribute the sources of the new ideas and the new civilization, the first part of the above time period is called the Greek era.

The first Olympic games took place in 776 B.C., a period during which the magnificent Greek Literature had already been developed, and on which Homer and Hesiod had left their mark.

Facts regarding the mathematics of that period appeared about two centuries later. Greek mathematics and Greek philosophy have their origins in Ionia. Later, various political events contributed to the immigration of many Greeks of Ionia to Italy, where the center of spiritual life moved and remained for some period of time.

After the victorious war of the united Greece against the Persians (490 B.C.), mathematics and philosophy flourished in Athens. Later, after the foundation of Alexandria (331 B.C.), the major scientific developments occurred in this Egyptian city and continued until around 500 A.D.

Thales. The first Greek mathematician and philosopher was Thales of Miletus (who was born around 625 B.C. in Miletus, and died around 547 B.C.), one of the Seven Wise Men of Antiquity. According to the philosopher Proclus, Thales visited Egypt from where he brought back to Greece the knowledge of Geometry.

Thales is believed to have predicted the eclipse of the sun that occurred in the Middle East on May of 585 B.C., apparently by using the astronomical observations of the Babylonians.

Plato reports that Thales was so absent-minded that once he fell into a well (Theaetetus 174 a). However, he was able to deal with practical issues: He constructed a calendar, and used similar triangles to calculate the distance of ships from the shore.

Among his results in geometry are the following:
1) The diameter of a circle bisects the circle.
2) The base angles of an isosceles triangle are equal.
3) When two straight lines intersect, the vertically opposed angles are equal.
4) When two angles and a side of a triangle are respectively equal to two angles and a side of another triangle, these triangles are equal.

5) If A, B, C are three distinct points which lie on the circumference of a circle in such a way that AC is a diameter of the circle, then the angle $\angle ABC$ with vertex B is a right angle.

For the proof of the result 5, which is known as the "Theorem of Thales", Thales must have known the following proposition:

6) The sum of the angles of every triangle is equal to two right angles.

The above theorems were known to the Egyptians and the Babylonians as empirical results, but Thales was the one who proved them. This is precisely the difference between the mathematics of the Prehellenic Era and those of the Hellenic Era: The Greeks were the first to give abstract logical proofs of various theorems. The Greeks transformed Geometry from empirical knowledge to Axiomatic Theory, which determined the model of mathematical thought for the next 2000 years. This transformation of Geometry into an axiomatic theory changed the course of History of Mathematics. This creation of mathematics as a science, which surpassed the practical purposes of everyday life, constitutes a fundamental event in the history of human civilization and has had an enormous impact on the development of every science.

In his youth, Thales was a very successful merchant. During the middle of his life he became a distinguished political figure and later he occupied himself with mathematics, astronomy and philosophy. As a philosopher he alleged that everything consists of water, that everything is full of gods and that the soul is the one that gives birth to the motion. Of course one wonders how we should interpret the opinion that everything consists of water and what is its meaning in mathematics.

An important question in physics is: "Is the matter that exists in the universe ultimately countable—i.e., does it consist of distinct and disconnected particles—or is it ultimately connected and continuous?"

In regard to this question, Pythagoras and Democritus believed that "reality" is essentially countable and they tried to interpret the apparently continuous entities with the use of discrete entities. On the contrary, Thales believed that everything was water, i.e., that the material universe becomes more comprehensible if we accept the existence of a single substance, in particular, water. In this sense Thales raised a problem of immense interest concerning the composition of the universe, a problem which still remains unsolved.

Other Ionian philosophers also supported the opinion that there exists a single substance in the universe. Anaximenes of Miletus believed that this substance was the air, while Heraclitus of Ephesus (born around 540 B.C. and died around 480 B.C.) alleged that everything consisted of fire. Also Anaximander believed in the existence of a unique substance which could take the form of earth, water, or fire and which he called Infinity. These notions would correspond today to the notions of solid, liquid, gas, and energy.

CHAPTER 5

PYTHAGORAS — THE PYTHAGOREAN SCHOOL

Pythagoras (who lived around 570-500 B.C.) was born in Samos. Even though he was a contemporary of Thales, the view that he was a student of Thales does not seem to be true (they had an age difference of about fifty years). The similarities that exist between the interests of the two men can be attributed to the fact that, just like Thales, Pythagoras, according to ancient historic sources (Iamblichus (300 A.D.), Porphyrius (232-304 A.D.), Diogenes Laertius (2nd century A.D.)), traveled to Egypt, to Mesopotamia and to India. It is known [15] that the Babylonians knew what we call today "pythagorean theorem", but the proof of this theorem and of its inverse is due to Pythagoras.

Pythagoras also seems to have studied under the priests (Magi) of the Persian jurist Zoroaster and he was a contemporary of Buddha, of Confucius and of the Chinese philosopher Lao Tze.

Around 525 B.C., Pythagoras immigrated to Croton of South Italy, where he founded the noted "Pythagorean Brotherhood", something between a political party and a religious organization. Pythagoras was essentially a philosopher and a teacher of morality, but his philosophy had a mathematical basis.

He separated the people who attended his lectures into two classes: the listeners or "$\Pi v\theta\alpha\gamma\acute{o}\rho\epsilon\iota o\iota$" and the mathematicians or "$\Pi v\theta\alpha\gamma o\rho\iota\kappa o\acute{\iota}$". In general, it was possible for a listener, after passing three years as such, to be initiated into the second class and then the brotherhood would trust him with its main scientific achievements. Now, following modern terminology, we shall confine the use of the word Pythagoreans to the latter class.

The words "philosophy" (love for wisdom) and "mathematics" (what one has learned) were first invented by Pythagoras in order to describe his intellectual activities. The contemporary meaning of the word "mathematics" is due to Aristotle.

Pythagoras taught that "everything is a number". This meant that everything can be understood through the integers and through the ratio (quotient) of integers. Having the science of mathematics as the foundation of his philosophy, he attributed various properties to numbers and to geometric figures: for example, the number five was the cause of colors. The pyramid was the origin of fire. A solid body was analogous to the "tetrad" ($\tau\epsilon\tau\rho\alpha\kappa\tau\acute{v}\varsigma$) which represented matter consisting of the first initial elements: fire, air, earth, water.

The oath of the initiation to the Brotherhood was the following[1]:

> ναὶ μὰ τὸν ἁματέρα ψυχᾶ παραδόντα
> τετρακτύν
> παγὰν αενάου φύσεως ...

The discovery of irrational numbers

In particular, the pythagoreans believed that given two line segments, there exists always a third line segment which can be used as a common measurement unit for both. In other words, by saying that everything is a number, they meant that if a and b are two arbitrary lengths, then there exist integers p and q such that $\frac{a}{b} = \frac{p}{q}$, i.e., that two arbitrary lengths are commensurable.

The pythagoreans knew that if a vibrating chord is divided into two parts a and b in such a way that a melodic tune is produced, then the lengths a and b are commensurable.

The celebrated discovery that: "the diagonal of every square is not commensurable with its side" seems to have been made by the students of Pythagoras without his knowledge. Besides, it is known that at the School of Pythagoreans every discovery was attributed to Pythagoras, and hence we don't know which discoveries are due to him only.

A simple proof of the fact that the diagonal of every square is not commensurable with its side, is found in Aristotle's *Prior Analytics*, 41a, 23-30, and is the following:

Let $ABCD$ be a square of side $AB = 1$. From the pythagorean theorem it follows that the diagonal $AC = \sqrt{2}$. Let's assume that the lengths AC and AB are commensurable, i.e., that $\sqrt{2} = \frac{AC}{AB} = \frac{p}{q}$, where p and q are positive coprime integers (i.e., their greatest common divisor is equal to 1). We have $p^2 = 2q^2$, hence p^2 is even and consequently p is also even (because the pythagoreans knew that the square of an odd number is odd and the square of an even number is even).

We put $p = 2r$, hence $2q^2 = (2r)^2$ or $q^2 = 2r^2$, which means that q is also even, which contradicts the assumption that p and q are coprime. Therefore the assumption that the lengths AC and AB are commensurable is not true.

Today we express this result by saying that $\sqrt{2}$ is an irrational number.

The discovery of the existence of incommensurable quantities constitutes the beginning of the discovery of **irrational numbers** and seemed to bring about a fatal blow to pythagorean philosophy: that everything depends on the integers. The entire pythagorean theory regarding proportions and similar figures, was based on the assumption that two arbitrary line segments are commensurable. The discovery of the existence of incommensurable quantities was inevitably leading to the rejection of a large number of proofs of propositions from the Geometry of Pythagoreans. The scandal was so big that the pythagoreans tried at first to hide the discovery of incommensurable quantities. It is said that the pythagorean **Hippasus of Metapontum** drowned at sea because he disrespectfully gave the secret away; according to another version, he was expelled from the pythagorean brotherhood and a grave was built for him as if he was really dead.

---

[1]**Translator's note:** The meaning of the oath in English is: *Yes upon the one who delivered to our soul the tetrad, a source of eternal nature ...*

We will see below that the Theory of Proportions of the ingenious Greek mathematician **Eudoxus** (who lived around 400-347 B.C.) put an end to this terrible crisis of mathematics.

The pythagoreans had found a method of approximating the number $\sqrt{2}$ by a rational number. If we use modern notation, this method is expressed as follows: The method depends on the possibility of finding positive integers x and y, which satisfy the equations $x^2 - 2y^2 = \pm 1$.

In particular, we have:
$$\left(\frac{x}{y} - \sqrt{2}\right)\left(\frac{x}{y} + \sqrt{2}\right) = \frac{x^2}{y^2} - 2 = \pm\frac{1}{y^2}$$
or
$$\frac{x}{y} - \sqrt{2} = \pm\frac{1}{y^2\left(\frac{x}{y} + \sqrt{2}\right)}.$$

Since $\frac{x}{y} + \sqrt{2} > 1$, we find
$$\left|\frac{x}{y} - \sqrt{2}\right| < \frac{1}{y^2}.$$

From this last inequality it follows that if it is possible to find integer solutions x and y of $x^2 - 2y^2 = \pm 1$, where y is arbitrarily large, then the number $\sqrt{2}$ can be approximated by the rational numbers $\frac{x}{y}$ as long as we want. For finding such integer solutions x and y of $x^2 - 2y^2 = \pm 1$, the pythagoreans used the following construction: We put $a_1 = 1$, $b_1 = 1$ and we define the sequences $\langle a_n \rangle$, $\langle b_n \rangle$ inductively by
$$a_{n+1} = a_n + 2b_n \ , \ b_{n+1} = a_n + b_n.$$

We find
$$a_1 = 1 \quad , \quad b_1 = 1 \quad , \quad \tfrac{a_1}{b_1} = 1$$
$$a_2 = 3 \quad , \quad b_2 = 2 \quad , \quad \tfrac{a_2}{b_2} = \tfrac{3}{2}$$
$$a_3 = 7 \quad , \quad b_3 = 5 \quad , \quad \tfrac{a_3}{b_3} = \tfrac{7}{5}$$
$$a_4 = 17 \quad , \quad b_4 = 12 \quad , \quad \tfrac{a_4}{b_4} = \tfrac{17}{12}$$
$$\vdots \qquad\qquad \vdots \qquad\qquad \vdots$$

We observe that the ratio $\frac{a_n}{b_n}$ approaches $\sqrt{2}$ as n increases at infinity. In order to prove this, due to the arguments mentioned above, it is enough to prove that
$$a_n^2 - 2b_n^2 = (-1)^n.$$

We work inductively. The last relation holds for $n = 1$. We assume that it holds for n and we prove that it holds for $n+1$. We have
$$a_{n+1}^2 - 2b_{n+1}^2 = (a_n + 2b_n)^2 - 2(a_n + b_n)^2$$
$$= a_n^2 + 4a_n b_n + 4b_n^2 - 2a_n^2 - 4a_n b_n - 2b_n^2$$
$$= -a_n^2 + 2b_n^2 = -(-1)^n = (-1)^{n+1}$$

Therefore the relation $a_n^2 - 2b_n^2 = (-1)^n$ holds for $n+1$ and consequently it holds for all n.

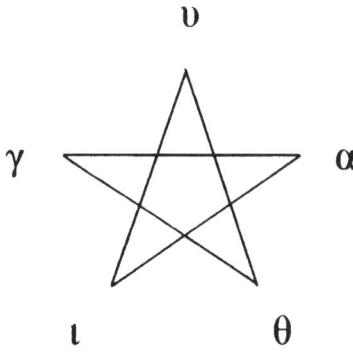

Figure 5.1

One of the characteristic symbols of the pythagoreans was the "triple triangle" (pentagram or star-shaped pentagon or pythagorean star) (see Figure 5.1). This was considered to symbolize health and next to its angles the letters of the word "$υγίεια$[2]" were written (with the difference that the diphthong $ει$ was replaced by the letter $θ$). The number 10 was called "divine number". We note that the figure of the triple triangle contains 10 vertices and that the number 10 is a "triangular" number, i.e., it can be written in the form $n + (n-1) + \ldots + 2 + 1$, where n is a positive integer (in the case of 10, we have $n = 4$).

In general, the discovery of the so-called **polygonal** or **schematic** numbers is due to the pythagoreans. These numbers, considered as the number of points of certain geometric figures (Figure 5.2), constitute the link between geometry and arithmetic.

Triangular numbers are the numbers 1, 3, 6, 10, ..., $\frac{1}{2}(n^2 + n)$, ...
Square numbers are the numbers 1, 4, 9, 16, ..., n^2, ...
Pentagonal numbers are the numbers 1, 5, 12, 22, ..., $\frac{1}{2}(3n^2 - n)$, ...
Hexagonal numbers are the numbers 1, 6, 15, 28, ..., $2n^2 - n$, ...
In general the nth polygonal number of order k is given by the polynomial
$$\frac{1}{2}[(k-2)n^2 - (k-4)n].$$
The number 0 is considered to be a polygonal number.

The Pythagoreans proved various algebraic relations, by studying the polygonal numbers. For example, by examining the sequence of polygonal numbers, they observed that the following relations hold: $n^2 + (2n+1) = (n+1)^2$, $1 + 3 + 5 + \ldots + (2n-1) = n^2$.

By appropriately placing side by side two triangular numbers of the same order, they formed a parallelogram and they observed that the nth triangular number is equal to $\frac{1}{2}n(n+1)$ (see Figure 5.3 for $n = 5$).

Despite the fact that the Pythagorean School constituted mainly a religious and/or political group, its contributions in Arithmetic, Geometry, Astronomy and Music were very important.

Amicable numbers

---

[2]**Translator's note:** $υγίεια$ is the Greek word for health.

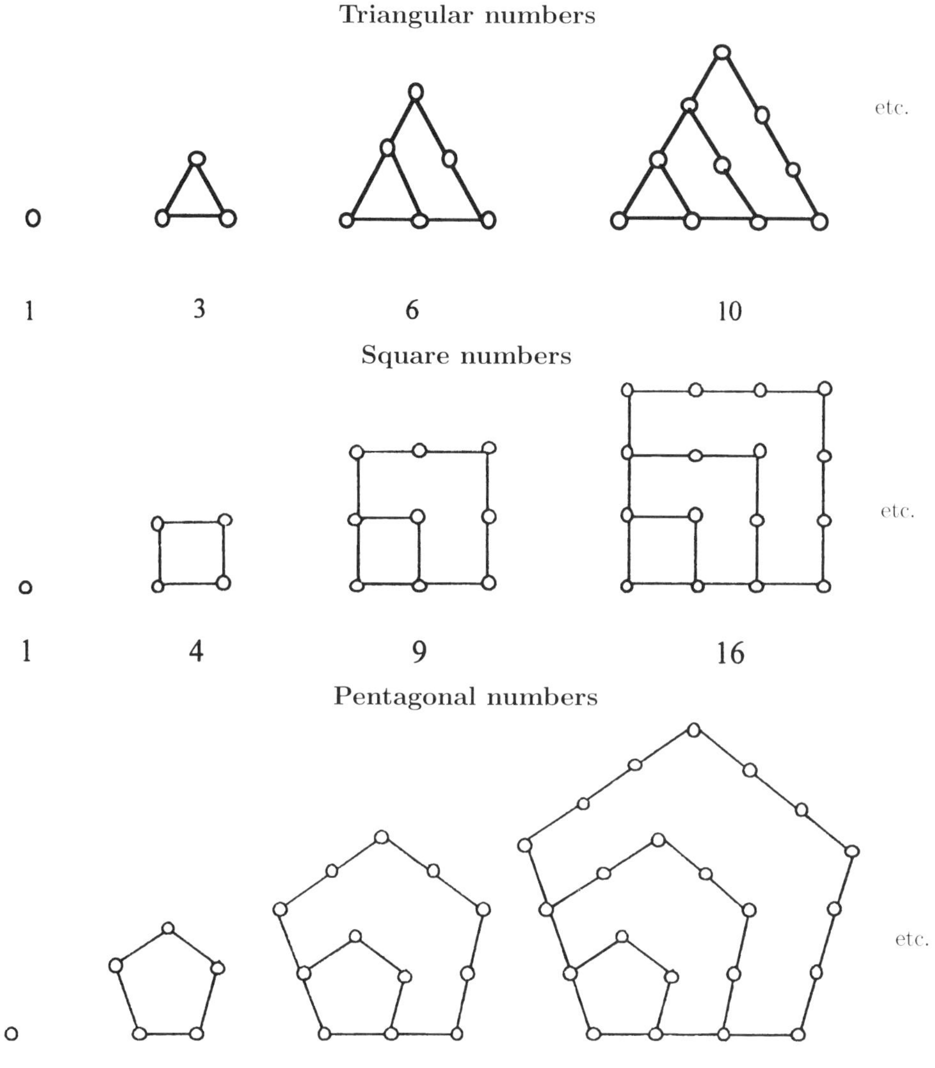

Triangular numbers

o △ △ △ etc.

1   3   6   10

Square numbers

o □ □ □ etc.

1   4   9   16

Pentagonal numbers

o ⬠ ⬠ ⬠ etc.

FIGURE 5.2

Two positive integers are called **amicable**, if each one of them equals the sum of the proper divisors of the other. We remind the reader that proper divisors of an integer n are all the divisors of n apart from n itself.

Pythagoras first discovered that the integers 284 and 220 are amicable numbers, because the proper divisors of 220, i.e., 1, 2, 4, 5, 10, 11, 20, 22, 44, 55 and 110, sum up to 284, while the proper divisors of 284, i.e., 1, 2, 4, 71 and 142, sum up to 220. Once, when Pythagoras was asked "what is friendship", he replied: "friendship is what 220 is with respect to 284".

The discovery of the second pair (17296,18416) of amicable numbers was made many centuries later in 1639 by the French mathematician **Pierre de Fermat**. The third pair was given in 1641 by the French mathematician and philosopher **René Descartes**, while in 1747 the Swiss mathematician **Leonard Euler** gave a

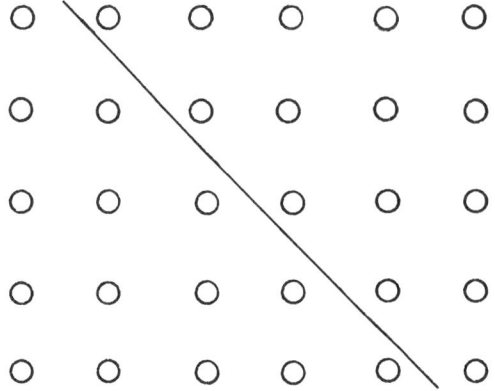

Figure 5.3

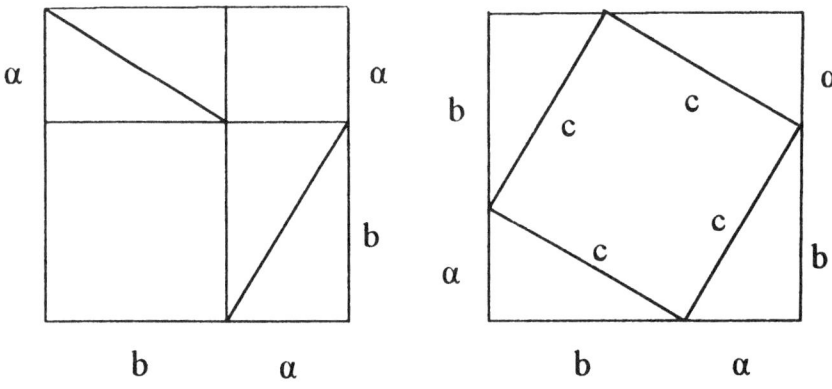

Figure 5.4

list of 30 pairs of amicable numbers, in which he added later another 30 pairs of amicable numbers. Today we know more than 1000 pairs of amicable numbers.

A strange historic phenomenon was the delayed discovery in 1866 of the relatively small pair (1184,1210) of amicable numbers by the sixteen year old Italian **Nicolo Paganini**.

The following result is due to **Thabit ibn Qurra**:
If $p = 3 \cdot 2^{t+1} - 1$, $q = 3 \cdot 2^t - 1$, $r = 9 \cdot 2^{t+1} - 1$ are odd prime numbers, then the numbers $m = 2^{t+1}pq$, $n = 2^{t+1}r$ are amicable.

The pythagorean theorem (*Elements*, Book I, Proposition 47)

The statement of this famous theorem is the following: "**The square of the hypotenuse of every right-angled triangle equals the sum of the squares of its other two sides**". The proof of this theorem and of its converse is due to the pythagoreans and it is conjectured to be similar to the following:

Let a, b be the perpendicular sides and let c be the hypotenuse of the given right-angled triangle, and let us consider the two squares of Figure 5.4, the length of the side of each being $a + b$. The first square consists of six parts, i.e., two squares

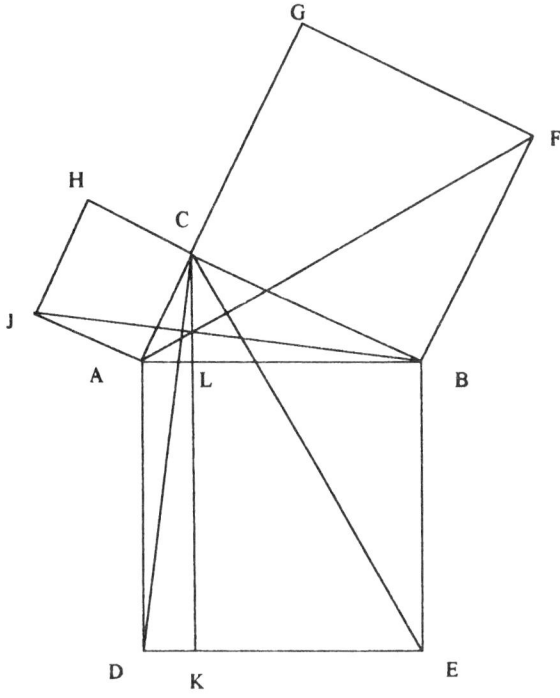

FIGURE 5.5

(with sides the perpendicular sides of the right-angled triangle respectively) and four right-angled triangles equal to the given one. The second square consists of five parts, i.e., a square of side equal to c and four right-angled triangles equal to the given one. By removing four equal right-angled triangles from each square, we find that the square inside the hypotenuse has area equal to the sum of the areas of the squares with sides the other two sides of the right-angled triangle. In order to prove that the central part in the second square is a square of side c, we need to know that the sum of the angles of every triangle equals two right angles, which was known to the pythagoreans.

The proof of the Theorem of Pythagoras, which is given in Euclid's *Elements* (Book I, Proposition 47) is, in general, as follows (Figure 5.5):

$$(AC)^2 = 2 \triangle JAB = 2 \triangle CAD = ADKL,$$
$$(BC)^2 = BEKL.$$

Hence
$$(AC)^2 + (BC)^2 = ADKL + BEKL = (AB)^2.$$

The pythagoreans studied also the relations between the arithmetic mean $\left(\frac{1}{2}(a+b)\right)$, the geometric mean ($\sqrt{ab}$) and the harmonic mean $\left(\frac{2ab}{a+b}\right)$.

Perfect numbers

The discovery of "perfect numbers" is due to the pythagoreans. Let n be a positive integer and let $\sigma(n)$ be the sum of all the divisors of n, including n. The number n is called **perfect** if and only if $\sigma(n) = 2n$. Thus, 6 (with divisors 1, 2, 3, 6) and 28 (with divisors 1, 2, 4, 7, 14, 28) are both perfect numbers.

Until 1952 the number of known perfect numbers was 12 and they were all even. The first three were 6, 28, 496. The last proposition of Book IX of Euclid's *Elements* (300 B.C.) is the following:

Perfect Number Theorem (Proposition IX, 36)
If the number $2^m - 1$ is prime (where $m > 1$ is an integer), then the number $n = 2^{m-1}(2^m - 1)$ is perfect.

Proof. We put $p = 2^m - 1$ and we assume that p is a prime number. Then the divisors of $n = 2^{m-1}p$ are

$$1,\ 2,\ 2^2,\ \ldots,\ 2^{m-1},\ p,\ 2p,\ 2^2 p,\ \ldots,\ 2^{m-1}p$$

whose sum is

$$\sigma(n) = (1 + 2 + 2^2 + \ldots + 2^{m-1})(1 + p) = (2^m - 1)(1 + p) = 2(2^{m-1}(2^m - 1)) = 2n.$$

Therefore n is a perfect number.

Remarks
(α) Despite the fact that n has divisors of the form k^2 (see Chapter 3), Euclid had, in this special case, a rigorous proof of the theorem of unique factorization into prime factors, which he used in order to find the divisors of n. He also had a rigorous proof of the formula which gives the sum of the terms of a geometric progression (IX 35).

(β) In order that the number $2^m - 1$ is prime, it is necessary that m is a prime number. Indeed, if $m = ab$ ($a > 1$, $b > 1$), then the relation $2^m - 1 = (2^a - 1)[(2^a)^{b-1} + (2^a)^{b-2} + \ldots + 2^a + 1]$ implies that $2^m - 1$ is not a prime number.
The fact that the converse of the last proposition is not true follows from the counterexample:

$$2^{11} - 1 = 23 \cdot 89.$$

The following proposition is due to **Francois Édouard Anatole Lucas** (1842-1891): "Let $u_1 = 4$ and $u_{n+1} = u_n^2 - 2$. Let $m > 2$. If $2^m - 1$ divides u_{m-1} exactly, then $2^m - 1$ is a prime number."

Due to computers and the method of Lucas, the following 30 values of m for which $2^m - 1$ is a prime number are known (since 1985):
2, 3, 5, 7, 13, 19, 31, 61, 89, 107, 127, 521, 607, 1279, 2203, 2281, 3217, 4253, 4423, 9689, 9941, 11213, 19937, 21701, 23209, 44497, 86243, 132049, 216091.

The prime numbers of the form $2^m - 1$ are called **Mersenne-primes** in memory of the French monk **Martin Mersenne** (1588-1648) [16].

The Ancient Greeks knew the first four "Mersenne-primes". By our discussion above it follows that the discovery of a new "Mersenne-prime", automatically implies the discovery of a new perfect number. However, it is not known whether there exist infinitely many "Mersenne-primes" or not.

The perfect numbers, which are given by the relative theorem of Euclid, are all even. **It is not known whether there exist odd perfect numbers**. We only know that there do not exist odd perfect numbers less than 10^{100}. We know (Euler) that: "Every even perfect number is of the form $2^{m-1}(2^m - 1)$" [3].

Regular polyhedra

The pythagoreans knew that the plane can be covered (by tiling) with equal regular polygons.

Indeed, if in a polygon with p sides we draw all the diagonals from one of its vertices, then the polygon is partitioned into $p-2$ triangles, thus the sum of its angles is $(p-2)180°$. Hence every angle of a regular polygon with p sides equals $(p-2)180°/p$. If at the attempted tiling, every angle of the p-regular polygon is adjacent to q angles of p-regular polygons, then the sum of these q angles must equal $360°$, i.e., we must have

$$q(p-2)180°/p = 360°$$

or

$$\frac{1}{2} = \frac{1}{p} + \frac{1}{q}.$$

The pairs (p,q) of positive integers which satisfy the last relation are: $(p,q) = (3,6)$, $(p,q) = (4,4)$, $(p,q) = (6,3)$.

The first pair gives a tiling of the plane by equilateral triangles, the second by squares and the third by regular hexagons. These are the only regular polygons which can tile the plane.

A polyhedron is called **regular** when all its faces are equal regular polygons and every vertex has the same number of adjacent faces. We quote five regular polyhedra (Figure 5.6).
(1) The cube: 6 faces (squares), 3 edges at every vertex.
(2) The tetrahedron: 4 faces (equilateral triangles), 3 edges at every vertex.
(3) The octahedron: 8 faces (equilateral triangles), 4 edges at every vertex.
(4) The icosahedron: 20 faces (equilateral triangles), 5 edges at every vertex.
(5) The dodecahedron: 12 faces (regular pentagons), 3 edges at every vertex.

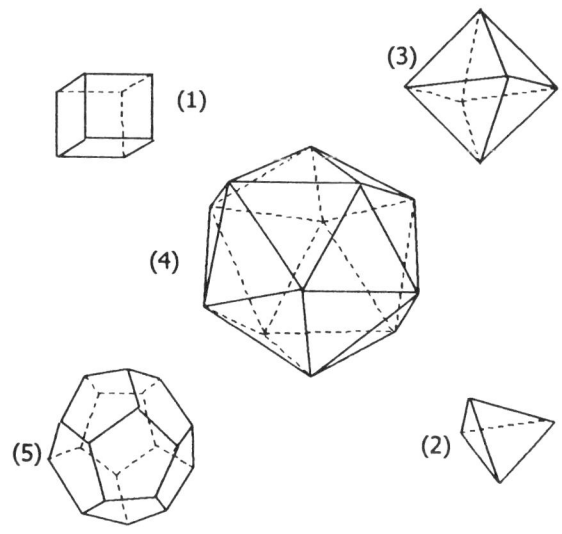

FIGURE 5.6

Plato claimed (*Timaeus* 53-58) that the composition of the universe can be explained by using these five polyhedra. The cube corresponds to earth, the tetrahedron to fire, the octahedron to air, the icosahedron to water and the dodecahedron to the entire universe. Plato explains the boiling of the water by a "chemical equation"

$$\Pi_4 + Y_{20} \to 2A_8 + 2\Pi_4,$$

which means that fire (with 4 faces) combined with water (with 20 faces) produces two atoms of air (with 8 faces each) and two atoms of fire (with 4 faces each).

It is interesting that this viewpoint of Plato has elements of truth: We find the tetrahedron, the cube and the octahedron in nature as crystals. Furthermore, the octahedron, the icosahedron and the dodecahedron are observed in skeletons of a certain microscopic protozoa.

At the end of Euclid's *Elements* (Proposition 21, Book XI) it is proved that the above five regular polyhedra are the only regular polyhedra that exist.

The known formula $F + V - E = 2$ of Euler holds for all simple polyhedra (i.e., polyhedra homeomorphic to a sphere), where F is the number of faces, V is the number of vertices and E is the number of edges of the polyhedron. In the case of a regular polyhedron, the above formula gives the numbers F, V, E, as follows:

Let us assume that every face of the given regular polyhedron has p edges and that every vertex has q adjacent edges. Then

$$p \cdot F = 2 \cdot E \ , \ q \cdot V = 2 \cdot E.$$

By substituting $F = 2E/p$ and $V = 2E/q$ in Euler's formula, we find

$$\frac{1}{p} + \frac{1}{q} - \frac{1}{2} = \frac{1}{E}$$

from which, knowing p and q, we can calculate E and then F and V. Thus, we find the following table:

Tetrahedron	$p = 3,$	$q = 3,$	$E = 6,$	$F = 4,$	$V = 4$
Octahedron	$p = 3,$	$q = 4,$	$E = 12,$	$F = 8,$	$V = 6$
Cube	$p = 4,$	$q = 3,$	$E = 12,$	$F = 6,$	$V = 8$
Icosahedron	$p = 3,$	$q = 5,$	$E = 30,$	$F = 20,$	$V = 12$
Dodecahedron	$p = 5,$	$q = 3,$	$E = 30,$	$F = 12,$	$V = 20$

It is admirable that the Ancient Greeks were able to prove the basic properties of regular polyhedra, without making use of Trigonometry and Differential Calculus. In Book XIII of *Elements* it is proved that for any regular polyhedron, there exists a sphere passing through all its vertices. Furthermore, in the same book, the radius of the circumscribed sphere of every polyhedron is also calculated. For example, the radius of the circumscribed sphere of the icosahedron is calculated to be equal to $\frac{1}{2}\sqrt{\frac{1}{2}(5 + \sqrt{5})}$ if the edge of the icosahedron equals 1.

Music

The pythagoreans motto that "everything is a number" extends to music also. The discovery of some simple laws concerning music is attributed to the pythagoreans. They observed that when the lengths of vibrating chords are expressed as simple ratios of integers, for example two over three (for the fifth) or three over four (for the fourth), then the resulting tones are harmonic. In other words, if a chord gives the sound of the note C, then a similar chord of double length will give the sound of the note C at one octave lower. The tones between these two notes are produced by chords whose lengths are given by the ratios:

$$16:9 \text{ for D}, \ 8:5 \text{ for E}, \ 3:2 \text{ for F},$$
$$4:3 \text{ for G}, \ 6:5 \text{ for A}, \ 16:15 \text{ for B}$$

in ascending order.

The discovery that the fifth and the octave of a note can be produced on the same chord, provided we stop at $\frac{2}{3}$ and $\frac{1}{2}$ of the length of the chord respectively, is also attributed to the pythagoreans. In addition some people have the opinion that the name "harmonic proportion" is due to the harmony produced in this manner because

$$1 : \frac{1}{2} = \left(1 - \frac{2}{3}\right) : \left(\frac{2}{3} - \frac{1}{2}\right).$$

We find ourselves here in front of the most ancient quantitative laws of acoustics and perhaps in front of the most ancient quantitative laws of physics.

The audacious imagination of the pythagoreans led them to the conclusion that celestial bodies emit harmonic tones as they move. This is the so-called "harmony of the spheres".

The viewpoint that the earth is spherical and the idea that the "world" constitutes a harmonically ordered set are often attributed to the pythagoreans. This idea later proved to be very beneficial to the development of astronomy. Even though today these old ideas look perhaps naive, we must not forget that they constituted a motive for the development of mathematics and the exact sciences.

The pythagoreans were among the first who believed that the functions of nature can be perceived via the science of mathematics.

CHAPTER 6

HERACLITUS, PARMENIDES, ZENO, EMPEDOCLES, AND DEMOCRITUS

Aristotle mentions in *Metaphysics* (986 b 4-8) that the pythagoreans had a list of opposite notions: one-many, finite-infinite, male-female, etc. Also there exists a theory that this list led **Heraclitus** (544-484 B.C.), the philosopher from Ephesus, to the philosophic viewpoint that everything that happens, every change, is the result of conflict between opposites.

Heraclitus believed that "everything is in a state of change". He believed that fire was the fundamental substance and that everything could be transformed into fire and vice versa. This viewpoint of Heraclitus brings to mind the modern discovery that mass can be transformed into energy.

The Prussian philosopher **Georg Wilhelm Friedrich Hegel** (1770-1831) apparently was greatly influenced by Heraclitus. Hegel's philosophy is based on the Triptych *Thesis, Antithesis, Synthesis*. Hegel was teaching that in the universe "thesis" and "antithesis" are eternally in conflict in order to produce "synthesis". **Marx** adopted this philosophy and gave to it a materialistic concept. Thus, the viewpoints of Heraclitus ended up being connected with the failed dogmas of Marxist governments.

This retrospection to Heraclitus must not be considered unjustified in a book on the History of Mathematics. We saw above that the viewpoints of Heraclitus led to the Marxist theory, which tries to perceive history not only through the dialectic method but also through mathematics. It is enough to mention that according to Lenin "subtraction" is the antithesis of "addition" and leads to "arithmetic", which is the synthesis of these two; moreover, integration is the "antithesis" of differentiation and their "synthesis" is Calculus. Let it be noted that these last views leave the author of this book indifferent.

Parmenides of Elea (480 B.C.), the philosopher from Elea (South Italy), had a viewpoint opposite to that of Heraclitus and asserted that nothing changes, that change is an illusion, a fallacy.

Zeno of Elea (450 B.C.), who was a student of Parmenides, using various arguments (the so-called "paradoxes") tried to prove that motion cannot exist! What Zeno actually proved was that if one does not allow processes which repeat themselves ad infinitum, i.e., processes which we call "limits", then one cannot make use of mathematics to analyze "motion".

We find Zeno's paradoxes in Aristotle's *Physics* 239b5 - 240a18, 233a21-31, and they are the following:

(1) Let us consider a straight line on which we have marked the integers 0, 1, 2, ..., as well as all rational numbers. A point-particle moving from 0 to 1 must first cover a distance equal to $\frac{1}{2}$, then a distance equal to $\frac{1}{4}$, then a distance equal

to $\frac{1}{8}$, and so on. After n steps the point-particle will have covered a total distance equal to $\frac{1}{2} + \frac{1}{4} + ... + \frac{1}{2^n} = 1 - \frac{1}{2^n}$. Since there exists no n for which $1 - \frac{1}{2^n} = 1$, Zeno concludes that this point-particle will never reach 1!

(2) Both Achilles and a turtle move in the direction from 0 to 1 on the same straight line that was considered in paradox (1). They start simultaneously, the turtle from 1 and Achilles from 0 with twice the speed of the turtle. When Achilles reaches 1, the turtle is at $1+\frac{1}{2}$, when Achilles reaches $1+\frac{1}{2}$, the turtle is at $1+\frac{1}{2}+\frac{1}{4}$, and so on. In general, when Achilles is at point $2 - \frac{1}{2^{n-1}}$, the turtle is at point $2 - \frac{1}{2^n}$. Zeno concludes that Achilles will never reach the turtle. The fact that we see Achilles reaching the turtle simply means that motion is an illusion!

The solution given today to this paradox is similar to the one given to paradox (1):

$$\lim_{n\to\infty}\left(2 - \frac{1}{2^{n-1}}\right) = \lim_{n\to\infty}\left(2 - \frac{1}{2^n}\right).$$

(3) Let us consider a "flying arrow". Since at every moment the arrow is at exactly one place, Zeno deduces that at every moment the arrow is motionless. Thus one comes to the conclusion that the arrow has speed 0 and that in reality it does not move.

This argument of Zeno would hold only if we assumed that time is basically discrete (not continuous), i.e., that there exists a **minimum** time interval, say δ seconds, in other words, no time interval less than δ seconds exists. Then the arrow would be motionless during every time interval δ. Indeed, if the arrow is moving in the direction from point 0 to point 1 during the time interval δ, then a time interval elapses just before the arrow reaches point $\frac{1}{2}$ and another time interval elapses just after the arrow passes point $\frac{1}{2}$. This means that the time interval δ can be divided into two smaller parts, which is a contradiction, since no time interval smaller than δ exists. Hence the arrow is motionless during the time interval δ. In other words, if there existed a positive integer n such that every time interval would be greater than $\frac{1}{2^n}$ seconds, then Zeno would be right: motion would be an illusion.

Note

According to Leibniz, the particle's speed at every moment is given by $\frac{dx}{dt}$, where x is the interval and t the time, while dx and dt are considered as "infinitesimals", The notion of an "infinitesimal" was criticised as "illogical" by **Berkeley**, a philosopher of the 18th century, and the mathematicians of the 19th century agreed with him. Thus the expression $\frac{dx}{dt}$ was redefined as follows:

$$\frac{dx}{dt} = \lim_{\delta t \to 0} \frac{\delta x}{\delta t}$$

where δx is the change of interval that corresponds to the change of time δt, and δx, δt are not necessarily small quantities.

During the middle of the 20th century, **Abraham Robinson** observed that the notion of an "infinitesimal" can be introduced into mathematics "directly". Indeed, let us consider the following infinite set of inequalities:

$$0 < dx, \ dx < 1, \ dx < \frac{1}{2}, \ dx < \frac{1}{3}, \ ... \qquad (4)$$

and let us suppose that this set of inequalities leads to a contradiction. It is generally accepted that, by definition, a mathematical "proof" is carried out following **only** a finite number of steps. This means that **only** a finite number of the inequalities in

(4) is mentioned along the proof of the assumption considered (i.e., that (4) leads to a contradiction). Let $dx < \frac{1}{n}$ be the last of the inequalities mentioned there. Then this finite number of assumptions does not lead to any contradiction since $dx < \frac{1}{n+1}$, combined with $0 < dx$, satisfies all of the inequalities mentioned there.

(4) In the fourth paradox of Zeno, where he tries to prove that motion does not exist, there exist three series of similar objects

$$\begin{array}{cccc} A & A & A & A \\ B & B & B & B \;\rightarrow \\ \leftarrow\; C & C & C & C \end{array}$$

Let us assume, as in the case of paradox (3), that time is discrete and δ is the minimum possible time interval that exists.

The objects A remain motionless. The objects B move uniformly to the right, while the objects C move uniformly to the left with a speed equal to that of the object B. Now if we assume that a time interval equal to δ is needed by an object B in order to pass in front of an object A, then this object B will pass in front of an object C in a time interval equal to half of δ, which is a contradiction, since δ "is" the smallest time interval that exists. This means that no such δ exists and that time is infinitely divisible and not discrete.

The philosopher **Empedocles** (495-435 B.C.) came from Sicily and was a great researcher in cosmology. He believed that matter consists of air, water, fire and earth. He also believed that love and hatred can change the nature of things: as long as love prevails, the universe is contracting, while when hatred prevails, the universe is expanding.

Democritus (420 B.C.), born in Abdera in Thrace, was one of the most important (natural) philosophers of Antiquity. He is known for deep ideas in "cosmology" and for his "atomic" theories. He alleged that the Universe consists of empty space in which microscopic, invisible and indivisible particles exist, which he named "atoms". The atoms are constantly moving. There are infinitely many atoms, but they differ in shape and size, and this is why the various substances differ. Democritus believed that the behaviour of "atoms" is governed by fixed laws and that there exist infinitely many worlds which are in various stages of evolution or decay. His viewpoints regarding the galaxies were analogous to ours today.

Democritus was a determinist. He wrote books on geometry which are not preserved. Apparently he was able to calculate the volume of a pyramid and the volume of a cone by multiplying the area of its base by one third of its height.

CHAPTER 7

THE MATHEMATICAL SCHOOL OF ATHENS

After the victorious war that the united Greek cities conducted against the Persians, in 490 B.C., the city of Athens became a very large and very important cultural center for more than one hundred years. Never in the history of mankind has a civilization achieved such a high standard as the Athenians about 400 B.C. Here we will restrict ourselves only to that which concerns mathematics.

Anaxagoras (500-428 B.C.), who came from Clazomenon, was perhaps the last philosopher of Ionia and is considered to be the forerunner of the Athenian Mathematical School. He settled in Athens in 440 B.C. and taught the results of Ionian Philosophy. He occupied himself with astronomy. He taught that the sun is bigger than Peloponnese. For this viewpoint of his and of others with whom he was trying to explain natural phenomena which were then believed to be due to actions of Gods, he was accused of impiety and was sentenced to imprisonment. During the time of his imprisonment, it is said that he wrote a study regarding the problem of squaring the circle. After he was released from prison, he lived with Pericles and died in Athens in 428 B.C.

The sophist **Hippias of Elis** (420 B.C.) is one of the most ancient mathematicians for whom there is some information in Plato's dialogues.

Hippias discovered the "quadratrix", i.e., the curve used for the trisection of an arbitrary angle and for the construction of the side of a square of area equal to the area of a given circle. The quadratrix is described by Hippias as follows: (Figure 7.1).

Consider the square $OBCA$ (Figure 7.1) of side of length 1, and imagine that the side BC is moving towards OA uniformly and with speed 1 per second, staying parallel to OA, and that the side OB is clockwise and uniformly rotating about point O with angular speed of rotation $\frac{1}{4}$ per second.

So after 1 second BC and OB coincide with OA. At every time t ($0 \leq t \leq 1$) the two moving sides intersect at some point P. The locus of the points P is the "quadratrix".

Using contemporary analytic geometry and trigonometry, it is not difficult to find the equation of the quadratrix.

We consider $\vec{OA}$ and $\vec{OB}$ as the x-axis and the y-axis respectively with origin O. After time t, BC is in position MQ and OB is in position OR. We have

$$1 - t = y = x \cdot \tan(90° - 90°t) = x \cdot \tan(90°(1-t)) = x \cdot \tan(90°y),$$

i.e., the equation of the quadratrix is

$$y = x \cdot \tan\left(\frac{\pi}{2} \cdot y\right), \ 0 \leq y. \tag{5}$$

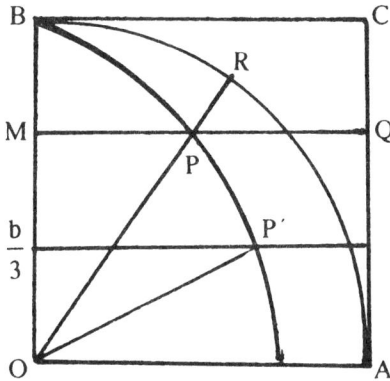

Figure 7.1

At time t the coordinates of P are

$$\left(\frac{1-t}{\tan(90°(1-t))}, 1-t\right).$$

If we set $r = OP$ and $P\hat{O}A = \theta$, then the equation of the quadratrix in polar coordinates is

$$r = \frac{2\theta}{\pi \cdot \sin\theta}.$$

In order to trisect an angle, for example an angle of 60°, it is enough to place it in such a way that its vertex coincides with O, one of its sides lies on the x-axis and the other lies on OR (Figure 7.1). If the side of the angle lying on OR intersects the quadratrix at the point (a, b) and if P' is the point where the straight line $y = \frac{b}{3}$ intersects the quadratrix, then the angle $P'\hat{O}A = 20°$.

If we use the straight line $y = \frac{b}{n}$ instead of the straight line $y = \frac{b}{3}$, then we can divide an arbitrary angle into n equal parts. Moreover, if we assume at (5) that y tends to zero, then x tends to $\frac{2}{\pi}$, which means that the construction of $\frac{2}{\pi}$ implies the squaring of the circle.

The above method of Hippias was criticised by Plato, who believed that in constructions one could only use a ruler (for drawing straight lines) and a compass (for drawing circumferences of circles). However, it was proven by **Pierre Wantzel** (1814-1848) that it is impossible to trisect an arbitrary angle using only a ruler and a compass.

By insisting that geometric constructions can only be done using a ruler and a compass, Plato became the cause of the emergence of many interesting problems. We will come back to this subject later.

The sophist **Antiphon** (425 B.C.) was one of the most ancient Athenian mathematicians who occupied himself with the problem of calculating the area of a circle. Antiphon had the idea that the area of a circle can be calculated by inscribing in the circle regular polygons, the number of the sides of which is constantly doubled. We remind the reader that a regular ν-gon is a polygon with ν equal sides and ν equal angles.

It is straightforward to show that the area of a square inscribed in a circle is larger than half the area of the circle, and that a regular octagon inscribed in a circle has area larger than $\frac{3}{4}$ $\left(=1-\frac{1}{2^2}\right)$ of the area of the circle.

The Greeks knew (*Elements*, XII, 2) that the area of a regular 2^n-gon inscribed in a circle is larger than $1 - \frac{1}{2^{n-1}}$ of the area of the circle. Moreover, it was known that if two triangles ABC and $A'B'C'$ are such that the angles of the vertices $\hat{A}$, $\hat{B}$, $\hat{C}$ are equal to the angles of the vertices $\hat{A}'$, $\hat{B}'$, $\hat{C}'$ and the larger sides in every triangle are the BC and $B'C'$ respectively, then for the areas E and E' of the triangles the following relation holds: $\frac{E'}{E} = \frac{(B'C')^2}{(BC)^2}$. From this last property of similar triangles, it follows that the area of a regular 2^n-gon inscribed in a circle is proportional to the square of the largest diagonal of the regular 2^n-gon, because these largest diagonals are diameters of the circle.

If we (roughly) equate the circle with a regular 2^n-gon (for n very large) inscribed in it, then the above lead us intuitively to the conclusion that the area of the circle "must" be proportional to the square of its diameter. A rigorous mathematical proof of this last result was given by Eudoxus (408-355 B.C.). We present some further information on the subject.

The inscription of a regular polygon in a circle and the ad infinitum duplication of the number of its sides was, of course, the basic idea for treating the problem of calculating the area of a circle. But beyond this, the mathematicians of that time didn't know how to handle the subject further, because the notion of a limit was not known to them.

According to Archimedes, the lemma which today bears his name is due to Eudoxus. It is known as the "axiom of continuity" and constituted the basis of the "method (or axiom) of exhaustion". This method is considered to correspond to what we call today "Integral Calculus".

The axiom of continuity asserts that:

"**If α and β are two quantities and $\alpha \neq 0$, then there exists a natural number n such that $n\alpha > \beta$**".

The "method of exhaustion" follows easily by reductio ad adsurdum from this axiom of Eudoxus (or of Archimedes) and is the following:

"**If from the larger of two given unequal quantities we subtract a portion larger than its half, and then from the remainder we subtract a portion larger than its half, and so on, then at some stage we will obtain a quantity smaller than the smaller of the two quantities given at the beginning**".

This last proposition is given in *Elements* (X, 1), while in all subsequent editions it is found in *Elements* (XII, 2). Let it be also noted that the phrase "method of exhaustion" was not used by the Ancient Greeks. It constitutes a contemporary invention.

Below we make use of contemporary notation in order to prove the following proposition of Eudoxus:

Proposition. If c and C are circles with diameters d and D and with areas a and A respectively, then $\frac{a}{A} = \frac{d^2}{D^2}$.

Proof. (By reductio ad absurdum.)

We assume that $\frac{a}{A} \neq \frac{d^2}{D^2}$. Then either $\frac{a}{A} > \frac{d^2}{D^2}$ or $\frac{a}{A} < \frac{d^2}{D^2}$. Let it be $\frac{a}{A} > \frac{d^2}{D^2}$. Then there exists a quantity $a' < a$ such that $\frac{a'}{A} = \frac{d^2}{D^2}$. We put $a - a' = \epsilon > 0$.

We inscribe in the circles c and C regular polygons of areas p_n and P_n respectively, which have the same number of sides n, and we consider the areas outside the polygons and inside the circles. If the number of sides of the polygons is doubled, then the new polygons will include (as mentioned above) a portion larger than half the areas which are outside p_n and P_n and inside the circles. Consequently it follows from the method of exhaustion that the areas between the polygons and the circles get smaller as the number of sides of the polygons is constantly doubled (i.e., as n increases) and hence at some stage we will have $a - p_n < \epsilon$ or $p_n > a'$. But from what was mentioned above, we have $\frac{p_n}{P_n} = \frac{d^2}{D^2}$ and from the assumption $\frac{a'}{A} = \frac{d^2}{D^2}$ it follows that $\frac{p_n}{P_n} = \frac{a'}{A}$. Consequently the inequality $p_n > a'$, which was obtained above, implies that $P_n > A$, which is a contradiction since P_n is inscribed in the circle with area A. Thus the inequality $\frac{a}{A} > \frac{d^2}{D^2}$ is not true. The proof that the inequality $\frac{a}{A} < \frac{d^2}{D^2}$ is also not true is similar. Therefore $\frac{a}{A} = \frac{d^2}{D^2}$ and the proposition is proved.

Remark. The above takes us back to the question that emerged in the discussion of Zeno's arguments, i.e., whether space is continuous or discrete. If space is discrete, then there exists some minimum area, say e. If n is large enough such that $\frac{1}{2^{n-1}}$ of the area of a circle is less than e, then a regular 2^n-gon inscribed in the circle will cover a portion larger than $1 - \frac{1}{2^{n-1}}$ of the circle's area, i.e., the entire area of the circle.

A bit earlier than Anaxagoras lived the distinguished geometer **Hippocrates of Chios**, who must not be confused with the father of medicine, Hippocrates of Cos. Hippocrates of Chios was born around 470 B.C. and settled in Athens around 430 B.C., initially as a merchant. Later he devoted himself to philosophy and earned a living by teaching geometry.

The largest part of the material contained in Books III and IV of *Elements* is due to Hippocrates of Chios. Hippocrates used the word "power" to express the square of the length of a line segment and since then this terminology became standard in algebra. Hippocrates was one of the first to use in mathematics the method of proof "reductio ad absurdum".

The most important discoveries of Hippocrates are the ones concerning the exact calculation of the area of a curvilinear figure and the known problem of duplication of the cube.

We consider the right-angled triangle ACB ($\hat{C} = 90°$) and with diameters its sides we construct three semicircles, from which the ones with diameters the perpendicular sides of the triangle lie outside the triangle (Figure 7.2).

Let it be noted that due to the converse of the theorem of Thales, (page 20), the semicircle with diameter the hypotenuse AB passes through point C. The areas contained in the two smaller semicircles but not in the semicircle with diameter AB are called **menisci**. We have menisci I and II (Figure 7.2). We will prove that the sum of the areas of menisci I and II equals the area of the triangle ACB.

We know that the area of every circle is proportional to the square of its diameter, i.e., if E and D are the area and the diameter of a circle respectively, then there exists a constant λ (the same for every circle) such that $E = \lambda D^2$. From the pythagorean theorem we have

$$AC^2 + CB^2 = AB^2 \quad \text{or} \quad \frac{\lambda}{2}AC^2 + \frac{\lambda}{2}CB^2 = \frac{\lambda}{2}AB^2.$$

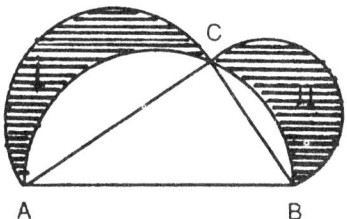

Figure 7.2

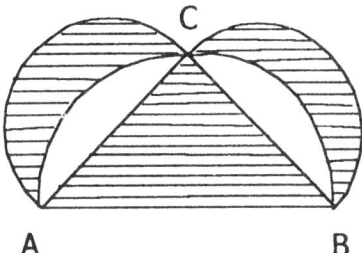

Figure 7.3

In other words:

$$\text{area(semicircle } AC) + \text{area(semicircle } BC) = \text{area(semicircle } AB).$$

By subtracting from both sides of the last equation the non-shaded areas (i.e., the areas where the semicircles overlap), we find

$$\text{area(meniscus I)} + \text{area(meniscus II)} = \text{area}(\triangle ABC) = \frac{1}{2} BC \times AC.$$

Note. Hippocrates proved, in almost the same way, the proposition above in the case where the triangle ACB is isosceles and right-angled (Figure 7.3).

This means that he proved that the area of the shaded meniscus equals half the area of the triangle ACB, i.e., that the squaring of the shaded meniscus is possible.

The story of the problem of duplication of the cube, which occupied Hippocrates' mind, is as follows.

Anaxagoras died in 428 B.C., i.e., the year in which Archytas was born, one year before Plato was born and one year after Pericles died. It is reported that Pericles died of plague which caused the death of one quarter of the population of Athens. The root of the famous problem of the duplication of the cube seems to be found in the deep impression caused by this terrible disaster.

Philoponus (some scholar) reports that the Athenians asked (430 B.C.) the oracle of Apollo in Delos what they should do to stop the plague. The oracle replied that they should double the volume of Apollo's altar, which was made of marble and had the shape of a cube. This was considered to be "easy" and a new altar was built, the length of the edges of which had been doubled. But this didn't stop the plague. The volume of the new altar was eight times and not two times the volume of the old altar.

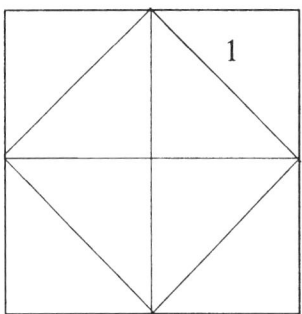

FIGURE 7.4

According to tradition, this was the beginning of the problem of the duplication of the cube which is known as the "Delian Problem":
"Given an edge of a cube construct, using only a ruler and a compass, the edge of a second cube, the volume of which is twice the volume of the first".

Hippocrates reduced the problem to finding numbers x and y such that if α is the edge of the given cube, then the relations

$$\frac{\alpha}{x} = \frac{x}{y} = \frac{y}{2\alpha}$$

hold, from which it follows that the edge of the duplicated cube equals x since $x^3 = 2\alpha^3$. Hippocrates, of course, didn't manage to construct the quantities x and y using only a ruler and a compass.

Socrates (468-399 B.C.) is not considered to be a mathematician, but the inductive method that he used, as well as his insistence that definitions must be exact, justify the reference to him in relation to the development of logic in geometry. As the teacher of Plato, he contributed to the development of this great philosopher. There exists no written work of Socrates. His personality and philosophy are given in Plato's "Dialogues". In the dialogue "Meno", Socrates alleges that every knowledge is a memory. In a conversation with Meno "on the nature of virtue", Socrates claims that he can make his servant "remember" some geometric construction and its proof.

So Socrates asks his servant (who of course has no knowledge of geometry), to construct a square whose area is twice the area of a square of side equal to 1. The first reaction of the servant is to double the side of the given square. Soon though the servant admits his mistake. Then Socrates asks his servant to observe the figure representing a square and the line segments connecting the midpoints of its sides (Figure 7.4), Socrates then convinces the servant to "remember" that the square with side equal to the diagonal of the inner square has area twice the area of the inner square.

Plato (429-349 B.C.), after the "execution" of Socrates in 399 B.C., left Athens and traveled for a few years. During that time he studied mathematics. He visited Egypt (Helioupolis near Cairo), Megara, Cyrene (where he was a student of Theodore, a distinguished pythagorean) and Italy. He remained in Italy for quite a long time and he met there **Archytas** (428-347 B.C.) who was then in charge of the Pythagorean School.

Archytas had also solved the problem of the "duplication of the cube" without of course restricting himself to the use of a ruler and a compass only.

Plato returned to Athens around 380 B.C. where he founded the famous Academy. In Academy's entrance the following phrase[1] was written:

"Μηδείς αγεωμέτρητος εισίτω".

Plutarch (*Symposium Problems* H, 2) informs us that the phrase[2]

"Αεί γεωμετρείν τον Θεόν" or "Αεί ο Θεός γεωμετρεί"

is due to Plato.

Plato, like Pythagoras, was mainly a philosopher whose philosophy, like the pythagorean philosophy, is driven by the idea that the "secret" of the universe is found in "numbers" and "shapes". For this reason, as **Eudemus** of Rhodes (320 B.C.) also remarks, Plato never missed the opportunity to express the existing important relation between mathematics and philosophy. The people who are qualified in these matters agree that, as opposed to what the earlier philosophers did, Plato studied geometry (or some other "exact" science) because he considered it to be a necessary prerequisite for the further study of philosophy.

Even though he was not a mathematician, Plato greatly influenced and contributed to the creation of many important notions in mathematics. He encouraged the study of mathematics because he believed that it helps the person to become wise and consequently righteous.

Plato's objection to the use in geometric constructions of any other instrument apart from a ruler and a compass, became widely accepted and constituted a basic precondition which had to be observed in problems of geometric constructions (*Timaeus* 34a).

One of Plato's dialogues is dedicated to his brilliant student **Theaetetus** who was killed in a battle in 369 B.C. The proof of the following proposition is attributed to Theaetetus: "The square root of a natural number is an irrational number if and only if this number is not the square of some natural number" (Theaetetus 147c - 148b). Theaetetus also occupied himself with the study of regular polyhedra and with the theory of proportions as well.

One of the most important Athenian mathematicians was **Eudoxus** (408-355 B.C.) from Cnidus of Asia Minor. Eudoxus, like Plato, went to Tarantas (South Italy) where he studied mathematics with Archytas. He met Plato in Egypt (in Helionpolis outside Cairo). From there he went to Cyzicus (Sea of Marmaras) where he founded his own school. Finally, he and his students settled in Athens. Eudoxus distinguished himself in astronomy, medicine, geography, philosophy and mathematics. In his youth he studied at Plato's Academy. The contents of Books V and VIII of Euclid's *Elements* are attributed to Eudoxus. Book V contains Eudoxus' Theory of Proportions.

As mentioned earlier, the discovery of irrational numbers caused a major upheaval to the Pythagoreans. First of all this discovery seemed to cause a fatal blow to Pythagorean philosophy that: **everything depends on the integers**. Because how can an irrational number, such as $\sqrt{2}$, depend on the integers, since it

---

[1]**Translator's note:** The meaning of this phrase in English is: *Let no one ignorant of Geometry enter here.*

[2]**Translator's note:** The meaning of both phrases in English is: *God is forever playing the geometer.*

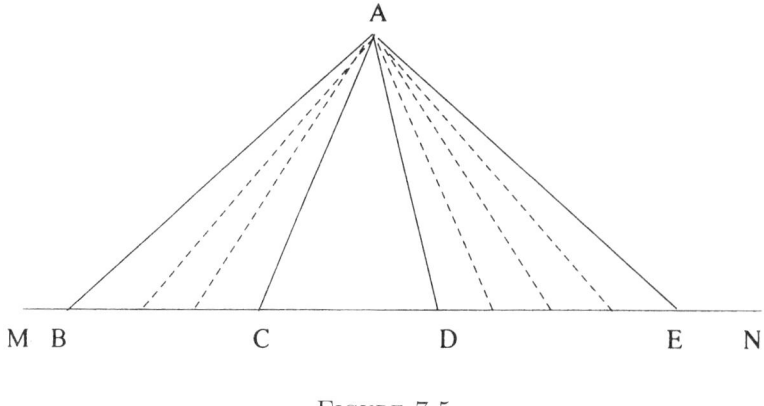

FIGURE 7.5

cannot be expressed as a ratio of two integers? Until then, the entire pythagorean theory of proportions and of similar figures was based on the hypothesis that two arbitrary line segments were always commensurable quantities. Thus, a large part of what the pythagoreans thought they had built, turned out to have precarious bases, because the proofs given were not correct.

This serious crisis was finally resolved in 370 B.C. due to the new definition of proportion between quantities given by Eudoxus. This definition constitutes a crowning moment for mathematics and is as follows:

"If a, b, c, d are four arbitrary quantities, where a and b are of the same kind and c and d are of the same kind as well, then the ratio of a over b equals the ratio of c over d if and only if for any positive integers m and n, we have

$ma\ >$ or $=$ or $<\ nb$ if and only if $mc\ >$ or $=$ or $<\ nd$ respectively.

This definition of proportional quantities given by Eudoxus is completely independent of whether the quantities considered are commensurable or not, and due to this fact the crisis that we mentioned was resolved.

As an example, we will consider a basic proposition of geometry and we will see how the pythagoreans thought they had proved it and what is the correct proof given by Eudoxus.

Proposition. Let ABC and ADE be two triangles whose bases lie on the same straight line MN (Figures 7.5 and 7.6). Then the ratio of the areas of these two triangles equals the ratio of their corresponding bases, i.e.,

$$\triangle ABC : \triangle ADE = BC : DE.$$

Proof (Pythagoreans). Assume (according to the Pythagoreans two **arbitrary** line segments were commensurable) that some common measure of BC and DE is contained p times in BC and q times in DE. We mark the points of partition in BC and DE and we connect them with vertex A. Then the triangles ABC and ADE are partitioned respectively into p and q smaller triangles, which all have the same height and equal bases. Thus, from an earlier known result it follows that these triangles have equal areas. Consequently

$$\triangle ABC : \triangle ADE = p : q = BC : DE.$$

The proposition is "proved".

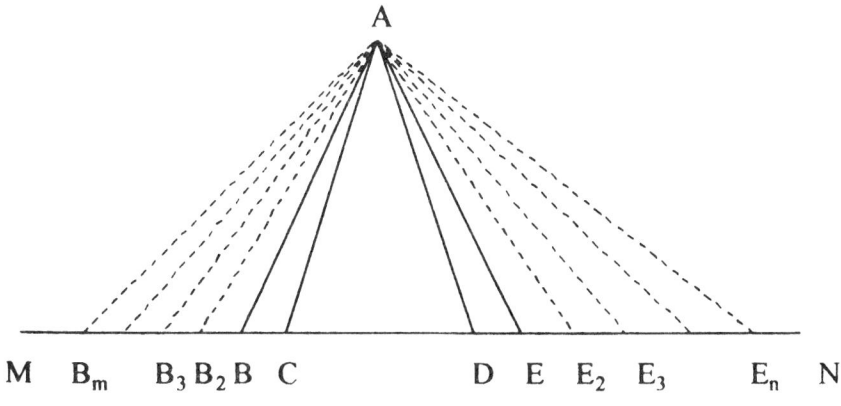

Figure 7.6

Proof (Eudoxus). On the elongation of CB (Figure 7.6), beginning with B, we successively mark $m-1$ line segments equal to CB and we connect the points of partition B_2, B_3, ..., B_m with vertex A. Similarly, on the elongation of DE, beginning with E, we successively mark $n-1$ line segments equal to DE and we connect the points of partition E_2, E_3, ..., E_n with vertex A. Then

$$B_m C = m(BC)\ ,\ \triangle AB_m C = m(\triangle ABC),$$
$$DE_n = n(DE)\ ,\ \triangle ADE_n = n(\triangle ADE).$$

It follows from results proved earlier that:

$$\triangle AB_m C\ >\ \text{or}\ =\ \text{or}\ <\ \triangle ADE_n$$
$$\text{if and only if}$$
$$B_m C\ >\ \text{or}\ =\ \text{or}\ <\ DE_n$$
respectively.

Thus

$$m(\triangle ABC)\ >\ \text{or}\ =\ \text{or}\ <\ n(\triangle ADE)$$
$$\text{if and only if}$$
$$m(BC)\ >\ \text{or}\ =\ \text{or}\ <\ n(DE)$$
respectively.

Consequently, it follows from the definition of proportion given by Eudoxus that

$$\triangle ABC : \triangle ADE = BC : DE.$$

The proposition is proved.

Remark. Nowhere does the last proof mention commensurable or incommensurable quantities and this is because the definition of proportional quantities given by Eudoxus holds in both cases.

Today we express the relation: "a over b is like c over d" with the equality

$$\frac{a}{b} = \frac{c}{d}$$

and we say that this relation is true when $ad = bc$. But such a process presupposes knowledge of the theory of the field of real numbers and knowledge of how we multiply two irrational numbers, something which was of course unknown in the days of Eudoxus. Eudoxus started from the beginning. He could not make use of multiplication to define a proportion, because he defined multiplication through a proportion. Besides, Eudoxus assumed that a, b, c, d were geometric quantities. For example, a and b could be arcs of circles while c and d are angles, hence the equality $ad = bc$ would have no meaning. How does one multiply arcs with circles?

Now if we assume that a/b and c/d are positive real numbers, then from the definition of proportion given by Eudoxus we have that:
• The set of rational numbers greater than a/b = The set of rational numbers greater than c/d
• The set of rational numbers less than a/b = The set of rational numbers less than c/d.

These last equalities are reminiscent of Dedekind's definition of real numbers. The processes that Dedekind and Weierstrass followed, in order to lay the foundations of a rigorously abstract theory of real numbers, look very much like the ones discovered by Eudoxus twenty two centuries earlier.

Eudoxus studied intensely the proportion that resulted from the following problem:
"Divide a given line segment AB in extreme and mean ratio", i.e., find a point H on AB such that
$$\frac{AB}{AH} = \frac{AH}{HB}.$$
Without loss of generality, we assume that $AB = 1$, we set $AH = x$, thus $x^2 + x - 1 = 0$. The positive root $x = \frac{-1+\sqrt{5}}{2}$ defining H is called "golden section".

Eudoxus also proved the following theorem: "the area of a circle is proportional to the square of its diameter" (*Elements* XII, 2). For the proof of this proposition he inscribed in the circle a regular 2^n-gon, where n increases ad infinitum.

Eudoxus is known in the history of mathematics not only for his own work but also for the work of his students. A strong feeling about the importance of tradition existed in Ancient Greece. This feeling was transferred from teacher to student. Thus, Archytas, Theodore and Theaetetus taught Plato. Then the influence of the platonics passed through Eudoxus to the brothers Menaechmus (375-325 B.C.) and Dinostratus.

We know that Hippocrates of Chios had reduced the problem of the duplication of the cube (of edge 1) to the construction of quantities x and y such that $\frac{1}{x} = \frac{x}{y} = \frac{y}{2}$. Menaechmus' discovery of conic sections (ellipse, parabola, hyperbola) came as a natural consequence of Hippocrates' idea and helped Menaechmus solve the Delian Problem as follows: The parabolas $y = x^2$ and $2x = y^2$ intersect at a point whose abscissa x equals the cubic root of 2. These parabolas obviously give the quantities x and y that Hippocrates of Chios was searching for.

Of course, the above solution of the Delian problem is not restricted to the use of a ruler and of a compass only. Besides, as mentioned earlier, **Pierre Wantzel** proved in 1837 that the solution of the Delian problem, using only a ruler and a compass, is impossible.

Note. For historic reasons, we mention that Menaechmus divided conic sections into three categories and examined their properties not by keeping a cone fixed and intersecting it with three different planes, but by keeping a plane fixed and using three different cones to intersect it.

He proved that the intersection of a right circular cone with a plane perpendicular to one of its generators is:

An Ellipse: if the cone is acute-angled (the angle of its vertex is acute),
A Parabola: if the cone is right-angled,
A Hyperbola: if the cone is obtuse-angled.
He also gave methods to construct curves from everyone of the above categories.

Menaechmus was a teacher of Alexander the Great. According to the tradition when Alexander asked Menaechmus if there exist shorter ways of proving the various geometric propositions, he replied: "In the country, there exist private and royal routes, as well as routes for simple civilians. But in geometry, there exists a unique route for everyone". Some writers attribute this anecdote to Euclid.

CHAPTER 8

PLATO AND ARISTOTLE

Plato (427-347 B.C.) believed that the objects of "pure" geometry, i.e., straight lines, circles, triangles, planes, etc., can be perceived only approximately through objects of the real-natural world, while the mathematically precise objects of pure geometry "live" in a different world, in the ideal world of Plato's mathematical notions. Plato's world does not consist of "tangible" objects but of "mathematical entities". We cannot access this world in the common-natural meaning of the term. We can only perceive it. The human mind contacts Plato's world every time it sees a mathematical truth, whose knowledge the mind acquires by visualizing and by using mathematical logic.

Even though this ideal world was considered to be different and more perfect than the material world provided to us by our external experiences, it was also considered to be equally real with the material world. Thus one comes to the conclusion that the objects of pure Euclidean geometry can be studied through the mind, without the need for the "imperfect" natural world of our external experiences to conform to the properties that result from this study.

It is a remarkable fact that Plato's intuition (despite the lack of enough evidence during that time) had predicted this last conclusion, namely:

On the one hand, mathematics must be studied and become understood for its own sake (i.e., it should be considered as an end in itself), without the need to be tested it fully and precisely on objects of our natural experience. On the other hand, the phenomena of the external world can finally be perceived only with the exact use of mathematics — which means that they become perceived through Plato's ideal world, i.e., through the mind.

In other words, according to Plato, in the universe there exist two different categories of beings, the material and the non-material ones. A chair and a sheep belong to the category of material beings, while the soul and a number belong to the category of non-material beings. The figure of a square, which is drawn on a piece of paper, belongs to material things, while the square itself belongs to the space of non-material things.

The category of material things is characterized by variability, uncertainty, ignorance, and imperfection. The drawn figure of a square can be effaced and it is questionable whether everyone of its angles is exactly 90° or whether its sides are completely straight.

The category of non-material beings is characterized by stability, perfection and complete knowledge of them. The sides of a square in this category are absolutely and permanently straight, and we know with absolute certainty that its diagonals are equal.

Plato believed that mathematical beings are not the only non-material beings. Other non-material beings are God, goodness, courage, and the human Soul.

However, the best way for one to begin to understand the non-material world is to study mathematics. "The study of number theory helps the "transition" of the soul from the world of birth to the world of essence and truth (*Republic* 525c). The study of geometry helps us understand the idea of good (*Republic* 525e)."

Plato believed that "mathematical truths" are absolute. By studying these truths we will become "better" ourselves.

Platonism, as a philosophy of mathematics, claims that at least the most basic mathematical objects exist "independently" of the human mind which perceives them. The properties of these objects are **discovered** by the human and they are **not invented**.

Aristotle (384-322 B.C.) was born in Stagira in Chalcidice and died in Chalcis in Euboea. He was a great philosopher—natural philosopher—and showed a great interest in mathematics and mathematical physics, because he believed that these sciences represented "correct reasoning".

Aristotle was a student of Plato, a teacher of Alexander the Great and the founder of the "peripatetic[1] school" (375-323 B.C.). Aristotle disagreed with Plato about the nature of mathematics. In Book XIII of *Metaphysics*, he alleges that when one assumes that the objects of mathematics exist a priori as separate entities, then one comes to conclusions which are contrary to the truth and to the ordinary viewpoints (*Metaphysics* 1077a). Aristotle gives also emphasis to "ideas", but he alleges that they constitute abstractions of material situations. Numbers and geometric shapes are abstract properties of material things.

For Aristotle a word like the word "two" is not a noun referring to an abstract object, but rather an adjective which describes a particular object.

While Plato's viewpoints are compatible with the existence of sets with infinitely many elements, Aristotle is a fanatic supporter of the "finite". He rejected the notion of a set with infinitely many elements and the notion of an infinite magnitude. According to Aristotle, a geometer can have at his/her disposal an arbitrarily large segment of a line, but a geometer cannot have the "whole" line in its infinite entirety.

Under the influence of Plato, Aristotle formulated the "principle" that every (mathematical) proposition is either true or false. But when he tried to apply this principle in matters of everyday life, he doubted its validity.

Aristotle can be compared with the "intuitionists" of the 20th century. His viewpoints are anthropocentric. The existence of numbers is related to the way we describe our world. The notion of infinity must be rejected because the person acts in a "finite" way.

---

[1]**Translator's note:** Aristotle's school came to be called "peripatetic" because he and his students strolled about the *peripatos*, or covered walkway, while he gave his lectures.

CHAPTER 9

GROUPS

The notion of a group is one of the most fruitful acquisitions of contemporary mathematics. Due to the theory of groups, it was revealed that many theories, which appear to be unrelated, are consequences of one theory. This has the tendency to reduce a variety of mathematical subjects to a small number of principles.

But what is a group?

A group is a "set" of objects (or elements) which we can pairwise compose with an "operation" which satisfies a certain number of conditions.

The "set" as well as the "operation" must be such that the composition of two elements of the set always gives an element of the same set. This, for example, happens with the set of integers (negative, positive and zero) and the operation of addition, because the sum of two integers is always an integer. The same thing happens with the set of rational numbers and the operation of multiplication. And the same thing also happens with the set of linear translations of a solid body with operation the succession of two linear translations of this body (because the succession of linear translations is also a linear translation). On the contrary, the same thing does not happen if we consider the set of rotations of a solid body, because the succession of two rotations about two axes which do not intersect is a spiral motion and not a rotation.

The operation of composition must be **"associative"**. This means the following: In the case of addition, $a + (b + c)$ must be equal to $(a + b) + c$. In the case of multiplication, $(a \cdot b) \cdot c$ must be equal to $a \cdot (b \cdot c)$. In the case of the succession of translations, it means that when we execute successively the translations M_1 and M_2 and then the translation M_3, we obtain the same translation with the one obtained if after the translation M_1 we execute the translation that results from the succession of translations M_2 and M_3.

Not every operation is associative. For example, exponentiation is not associative: the number $(a^b)^c$ is in general different from the number a^{b^c}. Indeed, we have $(2^2)^3 = 4^3 = 64$, while $2^{2^3} = 2^8 = 256$.

The original set must satisfy the following two conditions: that the so-called **"unit-element"** exists and that the **"inverse elements"** also exist.

The unit-element e is an element such that when it is composed with any element a of the set, it gives element a itself. Sometimes the unit-element is also called **"zero-element"**. In the case of the first example regarding the integers, the unit-element is the number zero, because $a + 0 = a$. In the case of the second example, the unit-element is 1, because $a \times 1 = a$. In the case of the third example, the unit-element is the zero translation, i.e., a translation of length zero, because the succession of such a translation and of a translation M gives translation M.

The inverse element of an element a is an element such that when it is composed with element a, it gives the unit-element. The inverse of an element a is denoted by a^{-1}. When the unit-element is called zero-element, the inverse of a is called the "opposite" element of a. In the first example regarding the integers, the opposite of a is $-a$, because $a + (-a) = 0$, while in the second example, the inverse of a is a^{-1}, i.e., $\frac{1}{a}$, because $a \times a^{-1} = 1$. In the third example, the opposite translation of a translation M is given by the same line segment, where one considers its origin as its terminus and its terminus as its origin.

The axioms of groups. If we call the operation of composition of two elements a and b **multiplication** and we call **product** the result of this composition, while we denote this product by ab or $a \cdot b$, then we have the following definition of a group.

We call a group a non-empty set of elements which satisfy the following four axioms:

1. Axiom of multiplication: There exists an operation called multiplication which gives an element of the set, when it is applied to two arbitrary elements a and b of the set, considered in this order. The element obtained is denoted by ab or $a \cdot b$ and is called the product of the elements a and b.

Note. The product ab is not necessarily equal to ba. For this reason we mentioned that the elements a, b are considered in the given order, i.e., first a and then b.

2. Axiom of associativity: Multiplication is associative, which means that for every elements a, b and c, we have $(ab)c = a(bc)$.

3. Axiom of the unit-element: There exists an element e called unit such that $ea = ae = a$ for every element a of the set.

4. Axiom of the inverse element: For every element a of the set, there exists an element called "the inverse element of a" and denoted by a^{-1} such that $aa^{-1} = e$.

So the main idea of the notion of a group is the following: a group is a set of elements which we can pairwise compose with an associative operation, which never leads us outside this set, and this operation can be "canceled" if it is appropriately applied once more. Indeed, if we compose a with b, and then we compose their product ab with b^{-1}, we obtain again a, because $(ab)b^{-1} = a(bb^{-1}) = ae = a$.

One can construct a general theory based on the above four axioms, the "Theory of Abstract Groups". But from the moment that the nature of the elements and the operation of composition have been specified, the group is called "concrete". The reader will find a large number of examples in various books on mathematics.

1. Groups of permutations

The strength of group theory as a research tool was first discovered in the theory of algebraic equations, when it allowed the French mathematician **Evariste Galois** to answer a question that remained unsettled for more than two centuries, i.e., that **the general algebraic equation of order greater than 4 has no algebraic solution** (see 9.6).

In the theory of algebraic equations, the notion of a group appears under the form of groups of permutations. We will describe in general the role played by these groups.

Let us consider three letters a, b, c. If we replace these letters with the letters c, a, b respectively, then we say that we performed a "permutation", which we denote

1. GROUPS OF PERMUTATIONS

by $\begin{pmatrix} a & b & c \\ c & a & b \end{pmatrix}$. All possible permutations of these three letters are:

$$\begin{pmatrix} a & b & c \\ a & b & c \end{pmatrix}, \begin{pmatrix} a & b & c \\ b & c & a \end{pmatrix}, \begin{pmatrix} a & b & c \\ c & a & b \end{pmatrix},$$

$$\begin{pmatrix} a & b & c \\ a & c & b \end{pmatrix}, \begin{pmatrix} a & b & c \\ c & b & a \end{pmatrix}, \begin{pmatrix} a & b & c \\ b & a & c \end{pmatrix}.$$

The first of these permutations is not essentially a permutation, because it moves none of the three letters. It is considered though to be a permutation in order to preserve a certain generality in the theorems regarding permutations and it is called the "identity permutation". If we call product of two permutations the succession of two permutations, then we observe that the above six permutations constitute a group.

Indeed, the first axiom is satisfied, because the product of two permutations of the letters a, b, c is obviously a permutation of the letters a, b, c. Since the above six permutations comprise the set of all possible permutations of these letters, every product of permutations is one of these permutations. For example, if we consider the product

$$\begin{pmatrix} a & b & c \\ b & c & a \end{pmatrix} \cdot \begin{pmatrix} a & b & c \\ a & c & b \end{pmatrix},$$

the first permutation replaces letter a with letter b and the second replaces letter b with letter c, so finally letter a is replaced by letter c. In the same way we remark that letter b is replaced by letter b and letter c is replaced by letter a. Hence we have

$$\begin{pmatrix} a & b & c \\ b & c & a \end{pmatrix} \cdot \begin{pmatrix} a & b & c \\ a & c & b \end{pmatrix} = \begin{pmatrix} a & b & c \\ c & b & a \end{pmatrix}.$$

Also it is easily recognized that multiplication of permutations is associative.

The third axiom is satisfied, since it is very easy to recognize that the identity permutation is a unit-element. Finally, for every permutation, as for example for $\begin{pmatrix} a & b & c \\ b & c & a \end{pmatrix}$, we can find an inverse permutation by interchanging its two lines, i.e., the inverse permutation of $\begin{pmatrix} a & b & c \\ b & c & a \end{pmatrix}$ is $\begin{pmatrix} b & c & a \\ a & b & c \end{pmatrix}$ because we have

$$\begin{pmatrix} a & b & c \\ b & c & a \end{pmatrix} \cdot \begin{pmatrix} b & c & a \\ a & b & c \end{pmatrix} = \begin{pmatrix} a & b & c \\ a & b & c \end{pmatrix}$$

where the right-hand side is the unit-element.

Remark. Let it be noted that the order in which the letters of the upper line are written in $\begin{pmatrix} b & c & a \\ a & b & c \end{pmatrix}$ does not matter. This permutation is equal to the permutation $\begin{pmatrix} a & b & c \\ c & a & b \end{pmatrix}$ because the letters, in both permutations $\begin{pmatrix} b & c & a \\ a & b & c \end{pmatrix}$ and $\begin{pmatrix} a & b & c \\ c & a & b \end{pmatrix}$, are written one above the other in the same way.

Thus all possible permutations of three letters form a group. We say that the **order** of this group is equal to 6 because it has six elements.

If we consider the first three permutations of the above group, we remark that they also form a group. Thus we observe the phenomenon that a group contains another group. The order of this subgroup is equal to 3. It is recognized that, apart from the one mentioned, no other system consisting of three permutations of the letters a, b, c constitutes a subgroup.

We ask the reader to form all permutations of four letters as an exercise. The order of the group that results is equal to $4! = 24$. This group contains a subgroup, whose order is equal to 12, and this subgroup contains various other subgroups, which are obviously subgroups of the original group also.

2. The permutations of the roots of an equation

Let us now occupy ourselves with the theory of algebraic equations, i.e., with the equations that we obtain when we set a polynomial of x equal to zero. We know that an algebraic equation of order n with real or complex coefficients has n real or complex roots. The main problem in the theory of equations is the determination of their roots.

Let us first consider the general equation of the fourth order:

$$x^4 + a_1 x^3 + a_2 x^2 + a_3 x + a_4 = 0,$$

i.e., an equation of fourth order whose coefficients are quantities independent of each other (see the Note below), and then the biquadratic equation

$$x^4 + px^2 + q = 0.$$

Note. Quantities "independent of one another" are quantities between which there exists no relation of the form of a polynomial equal to zero, whose coefficients are elements of a given "field" which is considered as the base field. "Field" is every system in which the operations of addition, subtraction, multiplication and division are possible, apart from division by the zero element.

We know that the solution of the biquadratic equation is a lot easier than the solution of the original equation. But what is the reason for that? It is due to the fact that, before attempting any solution process, the roots of the biquadratic equation are less unknown than the roots of the general equation.

Even though we don't know the roots of these equations, we can write for both equations certain relations between the roots and the coefficients. If we call ξ_1, ξ_2, ξ_3, ξ_4 the roots, for the first equation we have the following relations:

$$\xi_1 + \xi_2 + \xi_3 + \xi_4 = -a_1$$
$$\xi_1\xi_2 + \xi_1\xi_3 + \xi_1\xi_4 + \xi_2\xi_3 + \xi_2\xi_4 + \xi_3\xi_4 = a_2$$
$$\xi_1\xi_2\xi_3 + \xi_1\xi_2\xi_4 + \xi_1\xi_3\xi_4 + \xi_2\xi_3\xi_4 = -a_3$$
$$\xi_1\xi_2\xi_3\xi_4 = a_4$$

as well as the relations that result if we combine the relations above. It is proved that, apart from the above, no other relations exist between the roots and the coefficients.

We remark that all these relations are symmetric with respect to ξ_1, ξ_2, ξ_3, ξ_4, which means that they remain unchanged (invariant) **and hence exact**, if we perform an arbitrary permutation of these letters. For example, by applying the permutation $\begin{pmatrix} \xi_1 & \xi_2 & \xi_3 & \xi_4 \\ \xi_2 & \xi_4 & \xi_1 & \xi_3 \end{pmatrix}$, the last relation takes the form $\xi_2\xi_4\xi_1\xi_3 = a_4$, which can be written as $\xi_1\xi_2\xi_3\xi_4 = a_4$.

We can write analogous symmetric relations for the biquadratic equation, but in this case we can also write other non-symmetric relations. From the fact that this equation does not include odd powers of x, it follows that its roots are two pairs of equal in absolute value numbers of opposite sign. If we call (ξ_1, ξ_2) and (ξ_3, ξ_4) the pairs of opposite roots, then we can write the relations

$$\xi_1 + \xi_2 = 0, \quad \xi_3 + \xi_4 = 0.$$

These relations permit us to make a distinction between the roots. For example, we can say for ξ_1 and ξ_2 that their sum is equal to zero, something that we cannot say in the general case of the first equation. Consequently, the relation $\xi_1 + \xi_2 = 0$ does not remain invariant when we substitute ξ_3 for ξ_2, which means that the relation $\xi_1 + \xi_2 = 0$ does not remain unchanged if we perform the permutation $\begin{pmatrix} \xi_1 & \xi_2 & \xi_3 & \xi_4 \\ \xi_1 & \xi_3 & \xi_2 & \xi_4 \end{pmatrix}$ or the permutation $\begin{pmatrix} \xi_1 & \xi_2 & \xi_3 & \xi_4 \\ \xi_1 & \xi_3 & \xi_4 & \xi_2 \end{pmatrix}$.

3. The Galois group of an equation

The fact that certain relations between the roots of the biquadratic equation become invalid, when we perform certain permutations, is due to the fact that we can say certain things for certain roots that we cannot say for other roots, i.e., we can make a distinction between the roots. This means that we are less ignorant regarding the roots of the biquadratic equation than of the roots of the general equation, where all permutations of the roots are permitted to be performed in the relations by the roots.

In the case of the general equation of fourth order we can perform $4! = 24$ permutations, without changing any of the exact relations between the roots to a non-valid one. But which are the permutations that we can perform in the case of the biquadratic equation?

We can interchange ξ_1 and ξ_2, because then $\xi_1 + \xi_2 = 0$ becomes $\xi_2 + \xi_1 = 0$, which is again exact, while $\xi_3 + \xi_4 = 0$ remains invariant. The interchanging of ξ_1 and ξ_2 corresponds to the permutation $\begin{pmatrix} \xi_1 & \xi_2 & \xi_3 & \xi_4 \\ \xi_2 & \xi_1 & \xi_3 & \xi_4 \end{pmatrix}$, which we can write as $\begin{pmatrix} \xi_1 & \xi_2 \\ \xi_2 & \xi_1 \end{pmatrix}$, where the roots not written remain unchanged. Similarly, we can perform the permutation $\begin{pmatrix} \xi_3 & \xi_4 \\ \xi_4 & \xi_3 \end{pmatrix}$, and we can also perform the last two permutations simultaneously, i.e., we can perform the permutation $\begin{pmatrix} \xi_1 & \xi_2 & \xi_3 & \xi_4 \\ \xi_2 & \xi_1 & \xi_4 & \xi_3 \end{pmatrix}$.

But we can also interchange ξ_1 and ξ_3 provided that we interchange ξ_2 and ξ_4 at the same time, i.e., we can perform the permutation $\begin{pmatrix} \xi_1 & \xi_2 & \xi_3 & \xi_4 \\ \xi_3 & \xi_4 & \xi_1 & \xi_2 \end{pmatrix}$, because then the first relation transforms into the second and the second into the first. In addition, we can interchange ξ_1 and ξ_4 provided that we interchange ξ_2 and ξ_3 at the same time, because then the relations in question transform into each other; but this is equivalent to performing the permutation $\begin{pmatrix} \xi_1 & \xi_2 & \xi_3 & \xi_4 \\ \xi_4 & \xi_3 & \xi_2 & \xi_1 \end{pmatrix}$. Finally, this last transformation results also from the following less simple permutations $\begin{pmatrix} \xi_1 & \xi_2 & \xi_3 & \xi_4 \\ \xi_3 & \xi_4 & \xi_2 & \xi_1 \end{pmatrix}$ and $\begin{pmatrix} \xi_1 & \xi_2 & \xi_3 & \xi_4 \\ \xi_4 & \xi_3 & \xi_1 & \xi_2 \end{pmatrix}$.

If to all these permutations we add the identity permutation, which leaves every relation invariant, we observe that only eight permutations are "admissible" (i.e., we can perform them without changing the existing relations between the roots). It is also easily recognized that every other permutation, from the 24 possible permutations, changes at least one of the relations between the roots. So because of the two relations that we can write in the case of the biquadratic equation, sixteen out of twenty-four permutations become unacceptable.

If we are studying an equation of fourth order for which it is possible to write other additional relations, apart from the ones written for the biquadratic equation, then the number of admissible permutations would have been perhaps even smaller.

Consequently: the more we know about the roots of an equation, the smaller the number of admissible permutations.

Conversely, the larger the number of admissible permutations, the less we know about the roots.

It is of utmost importance to note that when an algebraic equation is given to us **the admissible permutations always constitute a group**. This group is called the **Galois group** of the equation considered or simply "the group of this equation".

Consequently the Galois group of a given equation is the set of those permutations of its roots, which when applied to all the relations satisfied by the roots, yield relations which are also valid.

Note. The definition of the Galois group given above is not complete. After denoting the roots by $\xi_1, \xi_2, ..., \xi_n$, we should replace the phrase: "when they are applied to all the relations that we can write satisfied by the roots" with the phrase: "when they are applied to all the relations satisfied by the roots, which have the form of a polynomial of $\xi_1, \xi_2, ..., \xi_n$ equal to zero, whose coefficients are contained in a given field".

For the definition of a field the reader is referred to [7], [8]. Examples of fields are: the set of rational numbers — the set of real numbers — the set of complex numbers. The Galois group differs according to the field we have chosen.

So, as we saw, the number of permutations comprising the group of an equation (we called this number the "order" of the group) gives a measure of our ignorance regarding the roots. In general, the larger the order of its group, the more difficult to solve an equation.

The fact that the admissible permutations constitute a group permits the use of the results of group theory. Hence it makes us able, without solving the equation and without writing any relation regarding the roots, to construct, only with the help of the coefficients, the group of this equation and to use this group in order to solve the equation. The mechanism of this solution depends on the composition of the group of the equation, as well as on the order of this group. The composition of the group of an equation permits us, in some cases, to recognize (before starting the solution process) the most important properties of the roots.

4. The algebraic solution of an equation

In the framework of the present short study it is not possible to address the method of solution of an algebraic equation by using its group. We will limit ourselves to some generalities only.

The second order equation was already solved by the Ancient Greeks. During the 16th century, the third order equation was solved by **Scipione del Ferro**, who never published his solution which was thus lost. The third order equation was solved again by **Girolamo Cardano**. A bit later **Ferrari** gave a solution to the fourth order equation. The solution of an equation of order greater than four presented great difficulties.

But let us examine the problem in more detail.

An algebraic equation of order n has the form

$$x^n + a_1 x^{n-1} + ... + a_{n-1} x + a_n = 0.$$

This equation is called an equation of general form if the coefficients a_1, a_2, ..., a_n are quantities independent of one another, as mentioned earlier. To solve "algebraically" an equation means to find a formula which gives directly the expression of the roots via the coefficients, the symbols $+$, $-$, $\times$, $:$ (i.e., the four rational operations), and via radicals of any order (not only symbols of square roots).

In the solution formula of the general second order equation, a square root exists. In the case of the third order equation, there exist square and cubic roots. The formula that gives the solution of the general fourth order equation contains in addition symbols of roots of fourth order.

Some fifth order equations of a particular form are solved easily, as for example is the equation $x^5 - a = 0$ whose roots can be expressed via radicals. All attempts to solve the general fifth order equation were unsuccessful.

Near the end of 1824, the Norwegian mathematician **Abel**, who died at 27, proved that the general fifth order equation cannot be solved algebraically.

5. The resolvents of an equation

Let us examine more closely the mechanism of solution of an equation. The equation $x^2 + px + q = 0$ has solution $x = \frac{-p \pm \sqrt{p^2-4q}}{2}$. This solution can be written as $x = \frac{-p \pm y}{2}$, where y is the root of the binomial $y^2 - (p^2 - 4q) = 0$.

We observe that we can explicitly express (using all four operations) the roots of a second order equation via the coefficients and via another quantity which is given as the root of a binomial second order equation. This binomial equation is called the **resolvent** of the given equation. Thus, the roots of the given second order equation can be explicitly expressed via the coefficients of the equation and via the roots of a resolvent.

Since the resolvent is a binomial equation, its roots are obtained via radicals. Therefore, the roots of the second-order equation can be expressed via the symbols $+$, $-$, $\times$, $:$ and via radicals.

The entire theory of solutions of an algebraic equation consists of trying to explicitly express the roots of a given equation via the coefficients and via other quantities given as roots of one or more successive resolvents, which are simpler than the equation considered. Then, in the same way, we try to reduce the resolvents of the above resolvents, to simpler resolvents and so on, until we reach a finite sequence of resolvents which cannot be reduced to simpler resolvents. If all these final resolvents are of binomial type, then everything can be expressed via radicals. On the contrary, if it is impossible to reach a finite sequence of resolvents which are all binomials, then the given equation cannot be solved algebraically, i.e., via the extraction of roots. Of course, this does not mean that the roots cannot be expressed via the coefficients and via certain other mathematical symbols, but in order to do this the use of other symbols, apart from +, −, ×, : and radicals, is also required (as, for example, sin, cos, ... or other symbols of the theory of functions).

6. The research of Galois

The determination of the resolvents of a given equation was the main objective of the algebraists for more than two centuries. Despite the remarkable efforts of eminent mathematicians of the 18th century and of the beginnings of the 19th century, the complete unravelling of this difficult problem is due to the ingenuity of a young mathematician, who was killed in a duel at the age of 20. His name is **Evariste Galois** (October 25, 1811 - May 31, 1832).

Even though he was not the one who discovered group theory, Galois was the one who applied the notion of a group to the study of equations and he introduced into group theory new notions, which were so important that he can be considered as the founder of this theory.

As mentioned above, Galois assigned a group of permutations to every equation and proved that certain fundamental properties of the equation are reflected in its group. Galois did not try, as the mathematicians before him, to obtain the resolvents directly from the given equation. His ingenious idea was the discovery of a parallel and easier path which starts not from the equation itself but from the group of the equation. This path gives at every step one of the resolvents, without any trouble. The following simple criterion permits us to recognize whether all these resolvents are of binomial type or whether they can be reduced to equations of this type:

We express the order of the group as a product of factors **following a certain rule**, which depends on the composition of the group. The sequence of these factors is called **composition series of the group** and every factor is called a **composition factor**. A resolvent corresponds to every composition factor. In case one of the factors is a prime number (and in this case only), the corresponding resolvent either is a binomial equation or it can be reduced to a binomial equation. From this, it follows that **"a necessary and sufficient condition for a given algebraic equation to be soluble by radicals is that the composition series of its group consists of prime numbers only"**.

Even though the rule according to which we form the composition series of a group is simple, it is not possible to analyze it here. We will limit ourselves to giving the composition series of the groups of general equations of various orders: The composition series of the group of the general second order equation contains only the factor 2, and hence this equation is solvable by radicals. The same thing happens with the general equations of third and fourth order, whose composition series are 2, 3 and 2, 3, 2, 2 respectively. In the case of the general equation of order n, where $n > 4$, the composition series of its group is always equal to 2, $\frac{n!}{2}$. Since $\frac{n!}{2}$ is never a prime number when $n > 4$, it is impossible to solve by radicals the general equation of order greater than 4.

CHAPTER 10

THE IMPOSSIBILITY OF SOLVING THE THREE CLASSICAL PROBLEMS OF ANTIQUITY

Group Theory has many successful applications to all branches of mathematics. We will limit ourselves to the presentation of an application in the area of elementary mathematics. It is an application of great historic significance, since it refers to the criterion of whether a geometric construction can be carried out using a ruler and a compass only.

Since the days of the Ancient Greeks, a number of problems vividly occupied the minds of the geometers:

The squaring of the circle
The duplication of the cube
The trisection of an angle

The solution of these problems was not achieved for the simple reason that it is impossible to solve them by using a ruler and a compass only.

Despite the fact that the impossibility of their solution was proved a long time ago and there exist many relative publications, I believe that the present, simple and (on purpose) narrative presentation of the subject will help the reader to perceive why it is POINTLESS to try to give a constructive solution to these problems. Of course, people who try to solve them have not ceased to exist even today, but their overwhelming majority are not "professional mathematicians". Perhaps the analysis that follows will contribute to the reduction of the number of people who try to solve these problems in vain.

It is essential to point out again that the problem consists of carrying out the above constructions **using a ruler and a compass only**.

Of course the following question arises: What is the reason for this restriction? The answer is that it is a matter of tradition. Various solutions of these problems were suggested from time to time, but they were not restricted to the use of a ruler and a compass only, i.e., they didn't comply with the terms that seem to have been set by Plato's Academy: **The only admissible constructions are the ones using a ruler and a compass ONLY**.

The problem of squaring the circle consists of constructing the side of a square with area equal to the area of a circle of a given radius, with the help of a ruler and a compass only.

We mentioned before (page 42) how the problem of the Duplication of the Cube arose according to tradition. We repeat its statement here: "Given an edge of a cube, construct the edge of a cube of double volume with the help of a ruler and a compass only".

The problem of trisecting an angle consists of "dividing a given angle into three equal parts with the help of a ruler and a compass only".

At this point we must make the following clarifications:

There exist infinitely many angles that we can trisect using a ruler and a compass only. But, as we will see, there exist also angles, whose trisection with the help of a ruler and a compass only is impossible.

For example, it is easy to trisect a right angle, because its one third corresponds to one twelfth of the circle, which is constructed by inscribing a regular dodecagon in the circle. The latter can be carried out using a ruler and a compass only. On the contrary, the trisection of an angle of $60°$ is impossible.

1. Geometric constructions

The question that we want to answer is the following:

"Which geometric problems can be solved with the help of a ruler and a compass only?"

In order to answer this question, let us examine first all the operations that can be made with the help of a ruler and a compass.

Given a line segment of length a, we can successively construct segments of length $2a$, $3a$, etc. We can also divide a given line segment (or a line segment that we have already constructed) into equal parts. So starting from the segment a, we can construct, for example, the segment $3a$ and then we can divide the segment $3a$ into four equal parts, i.e., we can construct the segment $\frac{3a}{4}$ or $(3/4) \times a$. We will call such a line segment a **rational multiple** of the segment a.

Given two line segments a and b, we can construct the segments $a+b$ and $a-b$, and in general we can construct the sum and the difference of rational multiples of two given segments.

Furthermore, given two line segments a and b, and given a line segment which we consider as the measurement unit of length, we can construct a segment of length x such that the rectangles with sides x, 1 and a, b have equal area, i.e., $x \times 1 = a \times b$ or $x = a \times b$.

By analogy, we can construct the segment $x = \frac{a}{b}$ or the segment $x = \sqrt{a}$.

So we can perform on given line segments (or on already constructed segments) the operations of addition, subtraction, multiplication, division and extraction of square roots.

It can be proved, using analytic geometry, that it is impossible to perform other operations apart from the above with the help of a ruler and a compass only. This follows from the fact that a straight line and a circle, i.e., the curves that can be drawn using the instruments at our disposal, have equations of first and second order respectively. Thus, if for example a line segment of length a is given (as well as a line segment of length 1), it is impossible in general to construct segments of length $\sqrt[3]{a}$, $\sin a$, $\log a$, ..., using a ruler and a compass only.

Note. We said that the construction of $\sqrt[3]{a}$, etc. is impossible "in general", because in some cases, as for example when $a = 8$, it is possible.

Consequently, now we have at our disposal the following answer to the question posed above: "Given one or finitely many line segments, among which one is considered as the measurement unit of length and the rest (if there are any left) have lengths a, b, c, ..., a necessary and sufficient condition for a segment of length x to be constructible with the help of a ruler and a compass only, is that x can be expressed with the help of rational multiples of the quantities 1, a, b, c, ... or combinations of them, using finitely many of the symbols $+$, $-$, $\times$, $:$, $\sqrt[2]{}$".

We will call these five symbols **elementary symbols**.

This answer to the question posed gives a criterion, which at first seems quite simple. But from the moment that one tries to apply it, one is met with serious difficulties.

In order to know whether the length x of the requested line segment can be expressed via rational multiples of the data using the five elementary symbols mentioned above, the simplest procedure, in general, is to treat the problem using analytic geometry. This, in general, leads us to an equation whose root is x. Such a case is, for example, the problem of trisecting an angle: if the angle to be trisected is given via its cosine, which we denote by a, then the cosine of one third of this angle is a root of the equation

$$4x^3 - 3x - a = 0.$$

Sometimes the requested line segment is given **directly**, as it happens in the case of the squaring of the circle

$$x^2 = \pi \rho^2 \quad \text{or} \quad x = \rho\sqrt{\pi}.$$

We note that although x is given directly via π and ρ, π is not a given quantity. So how can we know whether π (and consequently x) can be expressed using only the five elementary symbols? This means that the criterion given above does not satisfy us completely and hence we will try to formulate it in a more practical form.

2. The search for a criterion of geometric construction

In order to find the requested criterion we must first give an answer to the following question:

"If a quantity is given to us either indirectly as a root of a given equation or directly as a function of other quantities, how can we recognize whether this quantity can be expressed via rational multiples of the data using the five elementary symbols?"

Let us consider, for example, the case of the squaring of the circle.

We can always take the radius ρ of the circle to be equal to one and hence the requested segment is given by the relation $x = \sqrt{\pi}$.

So the matter is to verify whether the quantity $\sqrt{\pi}$ or equivalently (because if $\sqrt{\pi}$ is constructible, then so is π and vice versa) whether the quantity π can be expressed with the help of rational multiples of one using the five elementary symbols or equivalently with the help of the integers using the five elementary symbols. So how can we know that there exists no expression, as for example $\frac{5+\sqrt{7}+3\sqrt{11}}{\sqrt{25}}$, which is equal to π?

Let us examine now also the case of the problem of the duplication of the cube, where the requested line segment is given directly. If we take the edge of the given cube to be equal to one, then the volume of this cube is equal to $1^3 = 1$. Let x be the edge of the cube of double volume. Then the volume of the new cube is x^3. So we must have $x^3 = 2 \times 1^3$ or $x = \sqrt[3]{2}$. Since the requested edge is expressed as a cubic root, we could at first (frivolously) conclude that the edge in question is not constructible, because the cubic root is not included in the five elementary symbols. But such a conclusion would be premature, because in order to say that the construction is impossible, we must first prove that the cubic root of 2 cannot be expressed with the help of the five elementary symbols; i.e., we must prove that it is impossible to find an expression, as for example $\frac{5-\sqrt{3}}{2}$, equal to $\sqrt[3]{2}$. It is not obvious a priori that such an expression does not exist, because there exist cases in which this is possible, such as

$$\sqrt[3]{1} = 1, \quad \sqrt[3]{8} = 2, \quad \sqrt[4]{25} = \sqrt{5}.$$

When the length of the requested line segment is given as a root of an equation, it is much harder to give an answer to our question.

But all these cases can be reduced to the following criterion given to us by Group Theory:

"**Given one or finitely many line segments, where one of them is considered as the measurement unit of length, while the others (if there are any left) have lengths a, b, c, ..., a necessary and sufficient condition for a line segment of length x to be constructible using a ruler and a compass only, is that x is the root of an irreducible algebraic equation, whose coefficients are rational functions of the lengths 1, a, b, c, ... of the given line segments, and whose group has order a power of 2**".

Note: The expression "irreducible equation" means that the first part of the equation cannot be expressed as a product of polynomial factors, whose coefficients are also rational functions of 1, a, b, c, ...

It is impossible to give here even a clue about how the above proposition can be proved. The proof is obtained after a long series of propositions of group theory.

Now that we possess a criterion of constructibility using a ruler and a compass only, let us apply it to the three famous problems of antiquity.

3. Application of the criterion of constructibility

The squaring of the circle. We saw that the solution of this problem is reduced to the construction of a segment of length π, where the unit of length is given.

In principle, the above criterion demands that π is a root of an algebraic equation whose coefficients are rational functions of the data. Since the only data here is the line segment of length 1, it follows that the coefficients must be rational numbers. Consequently, π must be first of all a root of an equation with rational coefficients, which means that it must be a root of an equation with integer coefficients.

In 1844, **Liouville** proved that there exist numbers which are not roots of any algebraic equation with integer coefficients. These numbers are called **transcendental**, while the remaining ones are called **algebraic**. The proof given by Liouville simply consists of constructing transcendental numbers. It does not give a way (a method) through which one can recognize whether a number, given a priori, is transcendental or not. The latter problem for a particular number was solved thirty years later, in 1873, when **Hermite** proved for the first time the transcendental nature of the number e, (this number is the base of the system of natural logarithms). Based on this result, **Lindemann** in 1882 proved that the number π is transcendental, thus giving a final answer to the problem of squaring the circle:
It is impossible to square the circle.

The duplication of the cube. We saw before that if we take the edge of a given cube to be equal to one, then the length of the edge of the cube, which has twice the volume of the given one, satisfies the equation $x^3 - 2 = 0$. This equation has integer coefficients, i.e., it satisfies the first condition of the criterion, and it is easily recognized that it is irreducible. But if we consider its group, we observe that its order is equal to 3, i.e., it is not equal to some power of two. Furthermore, there can be no other irreducible equation with integer coefficients, whose root is the requested quantity and whose group has order some power of 2. This follows from the fact that a given quantity cannot satisfy more than one irreducible equation, whose coefficients belong to a given system.

Note. To be precise we must say that "a given quantity cannot satisfy more than one irreducible equation, whose coefficients belong to a given commutative field".

The above imply the conclusion that: **the duplication of the cube is impossible**.

The trisection of an angle. We know from trigonometry the relation:
$$\cos(3\phi) = 4\cos^3\phi - 3\cos\phi,$$
which gives the cosine of the angle 3ϕ when the cosine of the angle ϕ is known. We put $\theta = 3\phi$, $\cos\theta = a$ and we have
$$4\cos^3\frac{\theta}{3} - 3\cos\frac{\theta}{3} - \cos\theta = 0.$$
Thus the cosine of one third of the angle θ is given by the equation
$$4x^3 - 3x - a = 0.$$
The only given quantity in the problem is the angle θ, which is given to us via its cosine which is equal to a. But a cosine equal to a segment a has a meaning only if the line segment, which is considered as a unit of length, is also given to us simultaneously.

So our data are the segment 1 and the segment a, and we observe that the coefficients of the last equation are of the form that the criterion demands. It can be shown that for almost all values of a this equation is irreducible, and furthermore the order of the group of this equation is equal to 6. **Therefore, the trisection of an arbitrary angle is impossible**.

On the contrary, if we give to a certain specific values, then the first part of the equation can be expressed either as a product of two polynomials, one of degree one and the other of degree two, or as a product of three polynomials of degree one. In these cases, the quantity to be constructed will be a root of an irreducible first or second order equation. But the order of the group of a first order equation is equal to 1 (or 2^0) and the order of the group of an irreducible second order equation is equal to 2. Consequently, for the values of a which make the above equation reducible, the corresponding angle can be trisected. Such a case is the case of a right angle. We have $\theta = \frac{\pi}{2}$, hence $a = \cos\frac{\pi}{2} = 0$ and $4x^3 - 3x - a = 4x^3 - 3x = x(4x^2 - 3)$.

4. Construction of regular polygons

Before reaching the age of 19, **C. F. Gauss** (1777-1855) proved that the regular polygon of 17 sides can be constructed using a ruler and a compass only. His method, when presented today, can be considered as an application of the theory of Galois, even though Galois was not born at the time that Gauss made his discovery. Also, apparently Galois did not consider the method of Gauss, when he was elaborating on his theory of equations.

The result proved by Gauss is merely a particular case of a general theorem. Indeed, it follows from the criterion given above that: **"A necessary and sufficient condition for a regular polygon with a prime number of sides to be constructible using a ruler and a compass only is that this prime number is of the form $2^{2^{\mu}} + 1$, where μ is an integer, positive or zero"**. For $\mu = 0, 1, 2, 3, 4$, the expression $2^{2^{\mu}} + 1$ gives the prime numbers 3, 5, 17, 257, 65537. For $\mu = 5$ we obtain a number which is not prime.

When μ ranges over the set of non-negative integers, it is in general not known how the prime numbers of the form $2^{2^{\mu}} + 1$ are distributed, i.e., which ones are primes and which ones are not primes.

It follows also from the same criterion that for a regular polygon of n sides to be constructible with a ruler and a compass only, it is necessary and sufficient that either $n = 2^{\lambda}$ or $n = 2^{\lambda}(2^{2^{\mu}} + 1)(2^{2^{\nu}} + 1)...(2^{2^{\rho}} + 1)$, where $\lambda, \mu, \nu, ..., \rho$ are integers, positive or zero, the numbers $\mu, \nu, ..., \rho$ are pairwise distinct and every expression in parenthesis is a prime number.

Despite the fact that Gauss' career was long and brilliant, he considered his discovery of the regular polygon of seventeen sides one of his most beautiful titles of glory. He even expressed the desire that his grave be inscribed with a regular polygon of seventeen sides, like the grave of Archimedes was inscribed with a sphere inscribed in a cylinder. Even though this desire of Gauss was not fulfilled, his statue in his birthplace Braunschweig is placed on a pedestal of seventeen sides.

5. The geometry of a compass

Let us go back for a while to the proverbial "impossibility of the squaring of the circle". The fact that a mathematician cannot solve this problem is not due to the fact that the methods at his/her disposal are not powerful enough, but to the fact that he/she is constrained to use only two instruments, a ruler and a compass, and to the fact that these instruments are not "capable" of carrying out this construction.

Who would ever think to carry out the excavation of the Isthmus of Corinth using only a shovel made of wax? The lack of tools in this last case is obvious, while in the case of the three problems of Antiquity the lack of proper tools at a mathematician's disposal was revealed only after centuries of efforts.

The mathematicians of Ancient Greece had never suspected that, by restricting the instruments permitted to be used by the geometers to be a ruler and a compass only, they were allowing simultaneously more and less than what was needed. They were allowing less than what was needed for the reasons explained above. But simultaneously they were allowing more than what was needed, because the use of a ruler becomes superfluous from the moment that one has a compass at his/her disposal.

In particular, it can be proved that the problems that can be solved with a ruler and a compass only can also be solved as follows:

1st With the help of a ruler and a compass of fixed (unchangeable) opening.
2nd With the help of a ruler and a fixed circle of a given center drawn forever on a plane.
3rd With the help of a ruler with parallel edges.
4th With the help of a goniometer.
5th With the help of a goniometer with mobile arms.

6th With the help of a compass and without a ruler.

This last result is particularly interesting. For example, it is possible, using only a compass, to determine the point of intersection of two straight lines, each of which is given by two of its points. This discovery is due to the Italian mathematician **L. Mascheroni** (1750-1800) and is found in his book entitled: *La Geometria del compasso* and published in 1797.

We will close this subject with the following anecdote: Mascheroni's book attracted the interest of Napoleon Bonaparte, when Bonaparte was in Italy. After returning to France, Bonaparte attended a meeting of the French Academy, where he directed the discussion to the subject in question and presented some constructions that particularly impressed him. This presentation vividly surprised the audience, among whom were the famous mathematicians Lagrange and Laplace; Laplace, who was a professor of Bonaparte at the Military School of Brienne, shouted in front of everyone present: "We expected everything from you General, except for lessons in mathematics".

Note. As mentioned before, the proof that π is transcendental given by Lindemann in 1882 put an end to the mathematicians' efforts to solve the problem of squaring the circle. Despite this fact, even today there exist people who try to square the circle (or try to solve the remaining classic problems). Up to now, more than ten persons, who "solved" this problem, gave me their "solutions" and asked me (sometimes in the presence of their lawyer) to verify them in order to publish them in scientific journals or to present them at the Academy of Athens! It is a comfort that the overwhelming majority of these people, are not professional mathematicians. But what drives these people to find the solution to these problems although it is proved that their solution is impossible? I couldn't give a complete answer to this question.

One of the main characteristics of these people is that they ignore the meaning of the word "impossible" in mathematics. The meaning of this word in mathematics is not the same with its meaning in ordinary everyday language, where it may be interpreted as: "it is very difficult to achieve it now". For example, the sentence "it is impossible to land a human on any of the satellites of Saturn" is today true, but in the future it may cease to be true. In mathematics though, what is impossible today remains impossible forever. The sum of two even numbers was, is and will be an even number for ever. This essential difference between the two interpretations of the word "impossible" must be completely clear to students of mathematics.

The last person, who "solved" the problem of squaring the circle and visited the author (in 1996), presented a "proof" where the **exact** value of π was calculated to be $\pi = 3.1416$. In that case I thought that it would be much easier to explain the mistake, because it would be enough to show to that person the following (known) proof that π is irrational, hence by-passing the proof that π is transcendental. Unfortunately, even this effort was fruitless.

Here is a sketch of a proof that π is irrational.
We consider the function $f(x) = \frac{x^n(1-x)^n}{n!}$.
It is proved that:

(a) $0 < f(x) < \frac{1}{n!}$ if $0 < x < 1$.

(b) For any non-negative integer k, the numbers $f^{(k)}(0)$ and $f^{(k)}(1)$ are integers.

We assume that the relation $\pi^2 = \frac{a}{b}$ is true (i.e., that π^2 is a rational number), where a and b are positive integers, and we consider the function

$$F(x) = b^n \sum_{k=0}^{n} (-1)^k f^{(2k)}(x) \pi^{2n-2k}.$$

It is proved that:
- **(c)** $F(0)$ and $F(1)$ are integers.
- **(d)** $\pi^2 a^n f(x) \sin \pi x = \frac{d}{dx}\left(F'(x) \sin \pi x - \pi F(x) \cos \pi x\right)$
- **(e)** $F(1) + F(0) = \pi a^n \int_0^1 f(x) \sin \pi x\, dx$
- **(f)** Parts **(a)** and **(b)** imply the inequality $0 < F(1) + F(0) < 1$ when n is large enough.

The inequality $0 < F(1) + F(0) < 1$ contradicts **(c)**. Therefore, π is an irrational number.

CHAPTER 11

EUCLID

The most ancient effort to establish a university (with the contemporary meaning of the word) took place in Alexandria. The foundation, benefiting from a rich state funding, had teaching rooms, libraries, museums, laboratories, gardens, plants and machinery of every kind.

The University of Alexandria very quickly became the intellectual metropolis of the Greeks and of many other peoples for about 1000 years. There, among others, three great mathematical figures appeared, **Euclid**, **Archimedes** and **Apollonius** (during the 3rd century B.C.), who set the directions that mathematics followed until the destruction of the city of Alexandria by the Arab invaders in 641 A.D.

The city and the University of Alexandria were created under the following conditions.

Alexander the Great became king of Macedonia in 336 B.C., at the age of only 20, and by 322 B.C. he had already conquered Asia Minor and Egypt. In 332 B.C. he founded Alexandria on the Mediterranean coast of Egypt, near one of the mouths of the Nile River. A desire of Alexander the Great was to make this city the most magnificent city of the entire world. To this end, the supervision of the entire work was assigned to **Dinocrates**, the architect of the Temple of Diana in Ephesus.

After the death of Alexander the Great in 323 B.C., his entire state was divided into three parts, and Egypt devolved to **Ptolemy I**, one of the most competent generals of Alexander. Ptolemy I was followed by the long dynasty of the Ptolemies which ended with the reign of the famous queen **Cleopatra**.

Ptolemy I made Alexandria the capital of his kingdom. Then a short period of confusion followed, but immediately after the enthroning of Ptolemy I became definite, around 306 B.C., the erection of the buildings of the University began at a location very close to the palace. The University opened around 300 B.C. The supervision of its large library was assigned to the former "superintendent" of Athens, **Demetrius of Phaleron**. The library developed so rapidly that in 40 years it acquired about 600,000 volumes (of papyri). After the end of the dynasty of the Ptolemies, Alexandria devolved to the Romans and was finally occupied by the Arabs in 641 A.D., as mentioned above.

The mathematics department was under the direction of **Euclid** (330-275 B.C.), who was thus the first and one of the most distinguished mathematicians of the University of Alexandria. **Euclid of Alexandria**, the mathematician, must not be confused with **Euclid of Megara** (450-380 B.C.), who was a philosopher and a student of Socrates.

Apart from some anecdotes, we know very few things about the life of Euclid. He was Greek and was probably born in Tyrus. He knew the Geometry taught at Plato's Academy, which suggests evidence that he had studied in Athens. He was an excellent teacher. He wrote books on Optics, Music, Astronomy, etc. But his fame is due to his monumental work *Elements*, which consists of 13 Books (today we would call them chapters) and contains the foundations and the knowledge of all mathematics up to that time. There has been no attempt to write anything similar to Euclid's *Elements*, up to this day. Only the French book series *Éléments de Mathématique* under the name **Nicolas Bourbaki** exists. This series which was first published during the middle of the 20th century (1939), tries to cover all contemporary branches of mathematics.

It is not possible to verify which of the theorems contained in the 13 Books of *Elements* are due to Euclid himself. It is believed that the largest part of the material contained in Books I, II, VI, VIII, IX and XI is due to the **Pythagoreans** (including Archytas). Books III and IV are dominated by the work of **Hippocrates of Chios**. The content of Books V and XII is due to **Eudoxus**, while Books X and XIII are based on the work of **Theaetetus**.

We present a brief summary of the contents of the 13 Books of *Elements*.

Book I: Elementary constructions, equality theorems, areas of polygons, the Pythagorean Theorem.

Book II: Geometric Algebra.

Book III: Geometry of the Circle.

Book IV: Construction of certain regular polygons.

Book V: Eudoxus' "Theory of Proportions".

Book VI: Similar figures.

Books VII, VIII, IX: Number Theory.

Book X: Classification of certain irrational numbers (Theaetetus).

Book XI: Stereometry, volume of simple solids.

Book XII: Areas and volumes calculated with Eudoxus' "method of exhaustion" (integration).

Book XIII: Construction of the five regular polyhedra.

Regarding the organization and the logical structure of *Elements*, it is absolutely certain that they are due to Euclid exclusively.

The unprecedented enormous success of this great work is measured also by the fact that for 2000 years it was used and it is still used as a textbook in Secondary Education. For many centuries the structure of *Elements* set an example for imitation. In his important work **Summa**, **Thomas Aquinas** (1225-1274) used an axiomatic presentation similar to the one in *Elements*. The famous book *Philosophiae Naturalis Principia Mathematica* of **Isaac Newton** is written in the style of *Elements*. Similarly, the Dutch philosopher **Benedikt Spinoza** (1632-1677) follows faithfully the logical structure of *Elements* in his work **Ethica**. It is clear from its organization and presentation that Thomas Jefferson and the other writers of The Declaration of Independence had studied the *Elements*. Abraham Lincoln in his short autobiography that he wrote for the *Chicago Press & Tribune* when he ran for President of the U.S.A. in 1860, stated that "He studied and nearly mastered the six books of the *Elements* when he was a member of the U.S. Congress". Undoubtedly this work of Euclid is the most important scientific book in History, which influenced enormously human thought, and Euclid is rightly considered as one of the great mathematicians of all times [34].

Euclid's grandiose plan was to deduce all mathematical conclusions and all results from a small number of initial definitions and assumptions. The assumptions from which he started are subdivided into the **axioms**, which are used in mathematics in general, and the **postulates**, which concern geometry in particular. We will examine in somewhat more detail the definitions, the axioms and the postulates.

Elements begins with 23 definitions from which we present only the following ten:

1. A **point** is that which has no "part".
2. A **line** is that which has only length and no breadth.
   (The word line means curve.)
3. **The extremities of a line are points**.
   (It follows from this definition that a line or a curve has always finite length. The case of a curve which extends itself to infinity is not presented in *Elements*.)
4. A **straight line** is a line which lies evenly with the points on itself ($\epsilon\xi$ $\acute{\iota}\sigma o\upsilon$ $\tau o\iota\varsigma$ $\epsilon\pi$ $\alpha\upsilon\tau\acute{\eta}\varsigma$ $\sigma\eta\mu\epsilon\acute{\iota}o\iota\varsigma$ $\kappa\epsilon\acute{\iota}\tau\alpha\iota$).
   (It follows from definition 3 that, according to Euclid, a straight line is what we call today a line segment.)
5. A **surface** is that which has only length and breadth.
6. **The extremities of a surface are lines**.
   (Consequently, a surface is also a bounded figure.)
7. A **plane surface** is a surface which lies evenly with the straight lines on itself.
8. A **circle** is a plane figure enclosed by one line called **"circumference"** such that all the straight lines falling on it from one point among those lying within the figure are equal to each other, and this point is called the **centre** of the circle.
9. A **diameter** of the circle is any straight line drawn through the center and terminated in both directions by the circumference of the circle, and such a straight line bisects the circle.
10. **Parallel straight lines** are straight lines which, being in the same plane and extended indefinitely in both directions, do not meet each other in either direction.

The above sentences are not considered to be definitions in the contemporary meaning of the term.

Today points and straight lines are considered to be undefinable primitive notions, based on which are the definitions of curves, etc., in more advanced texts in mathematics.

Later, Euclid gives five postulates and five axioms. He accepts the distinction between these notions already made by Aristotle, i.e., that: "the axioms are truths which are applied to all sciences, while the postulates are applied only to geometry". ($T\alpha$ $\mu\epsilon\nu$ $\alpha\iota\tau\acute{\eta}\mu\alpha\tau\alpha$ $\sigma\upsilon\nu\tau\epsilon\lambda\epsilon\acute{\iota}$ $\tau\alpha\iota\varsigma$ $\kappa\alpha\tau\alpha\sigma\kappa\epsilon\upsilon\alpha\acute{\iota}\varsigma$ $\tau\alpha$ $\delta\epsilon$ $\alpha\xi\iota\acute{\omega}\mu\alpha\tau\alpha$ $\tau\alpha\iota\varsigma$ $\alpha\pi o\delta\epsilon\acute{\iota}\xi\epsilon\sigma\iota$.) Aristotle alleged that the postulates' truth is not necessarily known a priori, but it follows from whether the results of their application agree or disagree with reality. It seems that Euclid had accepted this viewpoint of Aristotle regarding the truth of the postulates. Anyway, in the history of mathematics that followed, both postulates and axioms became accepted as unquestionable truths at least until the time that non-Euclidean Geometry made its appearance.

The **axioms** posed by Euclid are the following:

1. Things which are equal to the same thing are equal to each another.
2. If equals are added to equals, the wholes are equal.
3. If equals are subtracted from equals, the remainders are equal.
4. Things which coincide with each other are equal to each other.
5. The whole is greater than the part.

The **five postulates** posed by Euclid are the following:

1. [It is possible] to draw a straight line from any point to any point.

This means that there exists one and only one straight line which connects two given distinct points. Thus, when we say "straight line", we cannot refer to a great circle of a sphere, because there exist infinitely many great circles connecting two antipodal points. A way to bypass this argument is to identify the antipodal points, and then we go to "elliptic geometry" which also satisfies Postulate 1.

2. It is possible to extend a finite straight line continuously in a straight line.
3. It is possible to draw a circle with any center and diameter.
4. All right angles are equal to one another.

It was argued already in Antiquity that this proposition should be considered as an axiom and not as a postulate.

5. If a straight line intersecting two straight lines makes the interior angles on the same side less than two right angles, the two straight lines, if extended indefinitely, meet on that side on which the angles are less than two right angles.

Euclid's Postulate 5 presents evidence of the ingenuity of the man, because he saw the importance and the necessity of the existence of such a proposition.

There existed many who opposed this proposition and alleged that it is not obvious by itself, as with the remaining postulates. Efforts to give a "proof" of the above proposition began around the time of Euclid, but they all failed. We will come back to this subject when we discuss non-Euclidean Geometries.

Euclid tried to use the above definitions and assumptions as foundations for building the logical conclusions that would follow. He started with the following proposition.

Proposition 1. Construct an equilateral triangle on a given straight line.

Euclid starts with a line segment AB and constructs two circles with centers the points A and B and with radius AB. Then he considers the point Γ where the two circles intersect and proves that the triangle $AB\Gamma$ is equilateral.

The above proof is not rigorous: In general, two circles either intersect at two points or are tangent at one point or have no points in common. In this case the circles indeed intersect at two points, but this fact does not follow from the initial assumptions mentioned above. The figure suggests that they intersect, but although figures are useful and help to illustrate an argument, **they must never form part of our reasoning or part of the proof of a proposition**.

In Euclid's *Elements* many "tacit assumptions" are made, as for example that the circles in the proof of Proposition 1 intersect. Of course, the existence of tacit assumptions has a drawback. We know of course how difficult it is to pin-point the existence of problems when habit dulls the keenness of our vision.

11. EUCLID

David Hilbert gave in 1899 a complete set of axioms for Euclidean Geometry in his famous book *Grundlagen der Geometrie* (Foundations of Geometry) [38].

Book I ends with the proof of the Pythagorean Theorem and of its converse. Before treating the properties of this theorem, Euclid proves first of all, that a square with side the hypotenuse exists. Let it be noted that **Legendre** proved later that the existence of this square implies Postulate 5.

The proof of the Pythagorean Theorem given by Euclid makes use of a theory regarding areas. Today the area of a parallelogram is defined as "length times the width". This definition presupposes knowledge of a certain theory which explains what it means "to multiply two irrational numbers".

Euclid approaches the subject of the area from a more elementary vantage point. He starts from the idea that two polygons have the same area if one of them can be partitioned into triangles, which form the other polygon when they are appropriately pieced together.

The formula that gives the area of the parallelogram as the product of the length times the width appears in Book VI, after Euclid first presented Eudoxus' theory regarding irrational numbers.

In Book II Euclid analyzes in a geometric way certain basic algebraic identities, such as $a(b+c) = ab+ac$, using areas of rectangles to express products of numbers.

In Book III the basic properties of the circle are examined. Euclid prefers to give lengthy proofs in order to achieve mathematical rigor. For example, despite the fact that it is "obvious from the figure" that the points of a chord of a circle lie in the interior of the circle, Euclid gives a proof of this proposition. As mentioned before, Euclid does not always succeed in giving "rigorous" mathematical proofs, but it is obvious that he understood the necessity of such proofs.

In Book IV the constructions of various regular polygons are given.

One of these constructions is that of the regular pentagon, i.e., of an angle equal to $\frac{360°}{5} = 72°$. Today we would try to solve this problem with the help of trigonometry. For a more elegant solution, we can use complex numbers. Let $\theta = 72°$, hence $5\theta = 360°$ and from **de Moivre**'s formula we have:

$$\begin{aligned}(\cos\theta + i\sin\theta)^5 &= \cos 5\theta + i\sin 5\theta \\ &= \cos 360° + i\sin 360° \\ &= 1 + i0 \\ &= 1\end{aligned}$$

Consequently, we want to solve the equation $z^5 = 1$, i.e.,

$$(z-1)(z^4 + z^3 + z^2 + z + 1) = 0,$$

where $z = \cos\theta + i\sin\theta$. We observe that $z^{-1} = \cos\theta - i\sin\theta$, hence $2\cos\theta = z + z^{-1}$. The solution $z = 1$ does not satisfy us, which means that we must solve the equation

$$z^4 + z^3 + z^2 + z + 1 = 0$$

or the equivalent equation

$$z^2 + z + 1 + z^{-1} + z^{-2} = 0.$$

We put $z + z^{-1} = u$, hence $z^2 + 2 + z^{-2} = u^2$ and $z^2 + z^{-2} = u^2 - 2$. Then the equation to be solved is

$$u^2 + u - 1 = 0.$$

If we exclude the negative solution, we find
$$u = \frac{-1+\sqrt{5}}{2}.$$
This number defines the **"golden section"** of a segment of length 1 (page 46). Of course, the ancient Greeks did not follow the above solution, because they did not know the complex numbers and trigonometry was not discovered yet in the days of Euclid. Anyway, the above solution shows that they could construct the angle θ, since the relation $2\cos\theta = \frac{-1+\sqrt{5}}{2}$ demands only rational operations and the extraction of a square root. Indeed, when Euclid constructs in Book IV (Proposition 11) the regular pentagon with some method which seems to us somewhat complicated, he uses (with Proposition 10) the construction of the golden section (Book II, Proposition 11).

The ancient Greeks knew also how to construct squares, regular hexagons, etc., but they did not know how to construct a regular polygon of seven sides.

With the same method used above for the construction of a regular pentagon, we can verify that the construction of a regular polygon of seven sides is reduced to the equation
$$u^3 + u^2 - 2u - 1 = 0.$$
The ancient Greeks did not know of course that the only numerical operations that one can perform with the help of a ruler and a compass, are the rational operations, the extraction of square roots, and combinations of them.

In Book IV the constructions of various regular polygons are given. At the end of Book IV the construction of a regular polygon of fifteen sides is given (Figure 11.1). In order to convince ourselves of the correctness of this construction, it is enough to notice that:
$$\frac{2}{5}360° - \frac{1}{3}360° = \frac{1}{15}360°.$$
This last achievement remained sublime until 1796, when **Carl Friedrich Gauss** (1777-1855) found a way to construct a regular polygon of seventeen sides.

In Book VI Euclid studies similar figures and makes use of the theory of analogies of Book V. Book VI ends with the following theorem: "The length of an arc of the circumference of a circle is proportional to the measure of the corresponding central angle." Here, when Euclid speaks of "arc length", he presupposes implicitly the "completeness" of the plane as a metric space.

It is often reported incorrectly by various people that Euclid's *Elements* is a book of pure geometric content. We have already described two Books (II, V) which treat algebra almost exclusively.

Books VII, VIII and IX treat theorems of elementary number theory. We present some of them. In Book (VII, 2) we find the proof of the "Euclidean Algorithm", i.e., of the procedure for finding the "greatest common divisor" of two positive integers. In Book (IX, 14) we find 1) the "theorem of unique factorization into prime factors" (see page 15) for integers which have no factors of the form k^2, 2) the proof of the proposition that there exist infinitely many prime numbers (IX, 20), 3) the formula giving the sum of the terms of a geometric progression (IX, 35), and 4) the following theorem giving the even perfect numbers:

If the sum $1 + 2 + 2^2 + ... + 2^{n-1}$ of the terms of the geometric progression 1, 2, 2^2, ..., 2^{n-1} is a prime number, then the number $(1 + 2 + 2^2 + ... + 2^{n-1})2^{n-1} = (2^n - 1)2^{n-1}$ is perfect.

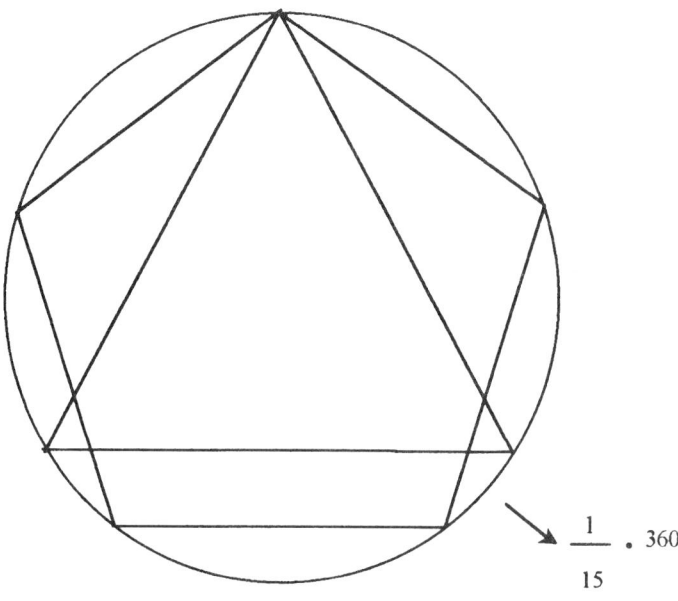

FIGURE 11.1

Book X treats what we call today "extension of degree four" of the field of rational numbers. In particular, we recall the following:

Let $\mathbf{Q}$ be the field of rational numbers. To "adjunct" the quantity $\sqrt{2}$ to the field $\mathbf{Q}$ means to form a field which contains the quantity $\sqrt{2}$ and is the smallest field which contains $\mathbf{Q}$ as a subfield. It is easily recognized that the requested field, which we denote by $\mathbf{Q}(\sqrt{2})$, consists of all rational functions of $\sqrt{2}$ which have as coefficients elements of the field $\mathbf{Q}$. After adjuncting $\sqrt{2}$ to $\mathbf{Q}$, one can continue and "adjunct" to the new field $\mathbf{Q}(\sqrt{2})$ some other element, which does not belong to $\mathbf{Q}(\sqrt{2})$. Hence one obtains a field which is wider than $\mathbf{Q}(\sqrt{2})$, and so on. The above process plays a fundamental role in Galois' theory of equations. Indeed, let us consider, for example, the polynomial $x^2 - 2$ over the field $\mathbf{Q}$. This polynomial is "irreducible" over $\mathbf{Q}$, i.e., it cannot be factored into a product of polynomials of degree less than 2, which have coefficients in $\mathbf{Q}$. But if we adjunct to $\mathbf{Q}$ the quantity $\sqrt{2}$ and we form the field $\mathbf{Q}(\sqrt{2})$, then the polynomial in question is written as $x^2 - 2 = (x - \sqrt{2})(x + \sqrt{2})$ over $\mathbf{Q}(\sqrt{2})$, i.e., it ceases to be "irreducible". In Book X Euclid classifies systematically the irrational line segments of the form $a \pm \sqrt{b}$, $\sqrt{a} \pm \sqrt{b}$, $\sqrt{\sqrt{a} \pm \sqrt{b}}$, $\sqrt{\sqrt{a} \pm \sqrt{b}}$, where a and b are rational quantities. Today we would say that Book X treats irrational numbers of the above form, where a and b are rational numbers. Euclid wants, for example, to know when an expression as $\sqrt{7 + 2\sqrt{6}}$ is equal to an expression as $1 + \sqrt{6}$, where only one symbol of square root appears.

Book XI gives the basic theorems of stereometry. A "cone" is defined as the solid that results from the revolution of a right-angled triangle. A "cube" is the solid that is embodied by six equal squares. Proposition (XI, 21) asserts that every polyhedral solid angle is enclosed by plane angles whose sum is less than four right angles. This proposition is used in Book XIII in order to prove that there exist at most 5 regular polyhedra.

Book XII presents Eudoxus' masterpiece. The dominant idea in it is the "method of exhaustion". Eudoxus gives rigorous proofs of the calculation of the volume of a pyramid, a cone and a sphere, without the help of Differential Calculus.

In Book XIII, for each of the five regular polyhedra, the ratio of its edge over the radius of the circumscribed sphere is calculated. This Book gives a complete theory of regular polyhedra.

The thirteen Books of *Elements* contain a total of 467 propositions. Some old versions include two more Books, which contain additional results regarding regular polyhedra. Both these Books are dated after the time of Euclid. Book XIV is due to **Hypsicles** (150 B.C.). Book XV contains vague and incorrect results, and some of its parts were probably written during the 6th century A.D.

As mentioned before, the *Elements* were used for more than 2000 years for teaching not only geometry but also rigorous logical reasoning in universities and Secondary Education. Despite the existing imperfections, some of which we have already mentioned, the *Elements*, which are called by many people the Bible of mathematics, constitute a crowning achievement in the History of Mathematics. More than 1000 publications of this work followed its first printed version which was published in 1482. No copy of *Elements* from the time of Euclid was preserved. All contemporary versions are based on the revised presentation of this work by **Theon** of Alexandria, a Greek commentator, who lived about seven hundred years after the time of Euclid. The oldest copy of *Elements* was discovered in the beginning of the nineteenth century. Indeed, in 1808, when Napoleon Bonaparte ordered the transportation of the valuable manuscripts from the Italian libraries to Paris, **F. Peyrard** discovered in the Vatican's library a version of *Elements* which was older than the one of Theon. But the study of this older version revealed only minor differences with the version of Theon.

The first complete translation of *Elements* into Latin was made from an Arabic text and not from a Greek one. During the 8th century, a number of Byzantine manuscripts of Greek works were translated into Arabic, and in 1120 the Englishman **Adelard of Bath** translated into Latin one of the above translations of *Elements*.

Other translations into Latin from Arabic texts were made by **Gherard of Cremona** (1114-1187), and similar translations were made by **Johannes Campanus** 150 years after the time of Adelard. The first printed version of *Elements* appeared in Venice and was based on the translation of Campanus.

An important translation of *Elements* from Greek into Latin was made in 1572 by **Commandino**. Commandino's translation served as a basis for many other later translations, one of which is that of **Robert Simson**. Later, many English versions of *Elements* were produced from Simson's translation.

Shortly after the end of the second World War, a reaction emerged against the educational programs which were using *Elements* for teaching geometry as well as rigorous logical reasoning in general. The viewpoint prevailed that geometry is not the most suitable medium for applying Logic. The argument was that on the one hand Euclid's *Elements* were not "rigorous" enough from a mathematical point of view, and on the other hand, Hilbert's treatment of this subject was very hard to use. Thus, it was decided that geometry's place is to be taken by the so-called

"New Mathematics". The French mathematician **Jean Dieudonné**, one of the founding members of the Bourbaki team, suggested that geometry (which he called "Theory of the Triangle" in contempt) should be replaced by Linear Algebra. His phrase "A bas Euclide" (Down with Euclid) is characteristic of his attitude.

Dieudonné's suggestion took a concrete form with the acceptance of a completely new set of programs, which were elaborated by a special governmental committee in France. The aim of these programs was to introduce into the teaching of mathematics the ideas that formed the basis of the work of the Bourbaki team: Sets, relations, mappings, algebra and linear algebra. Despite the fact that the elaboration and formation of these programs occupied the official quarters for a rather long time, a most important fact was overlooked, namely that "teaching" means something beyond introducing new programs. Things got worse when various authors of new books, under the direction of these new programs, "broadened" further the formalist directions. As a result the instructors (after their initial shock) in their effort to assimilate this wave of new notions, ended up either becoming "more than formalists" themselves or passively repeating the "new mathematics", to which the various published books did not cease to add more mistakes. The final result was destructive. Apart from very few cases, pupils and parents considered the "New Mathematics" as some new kind of "plague". Due to the international character of mathematics, this situation spread to many other countries beyond France.

The need for a new change became obvious. Of course, the further analysis of this interesting subject goes beyond the scope of the present book. However, we note that in this adventure the "official quarters" did not think about the danger for Education, when radical changes are attempted without prior experimental testing. In this case, it was assumed that the introduction of new unifying notions, such as algebra and linear algebra, will imply the success of the pursued aim. But this proved to be completely wrong, because the basic role played in mathematics by intuition in general and by geometric intuition in particular were completely ignored.

CHAPTER 12

HILBERT'S FOUNDATION OF EUCLIDEAN GEOMETRY

It was mentioned before that the development of Euclidean Geometry in *Elements* contains many imperfections. First of all, the tacit assumptions made, must not exist in a logical system. For example, in the proof of certain propositions, use is made of the assumption that a straight line has infinite length. This assumption is neither a stated axiom nor a consequence of the initially stated axioms. A straight line can be extended ad infinitum in both of its directions, but this does not mean that this line has infinite length. This, for example, happens in "Elliptic Geometry", where even though a line can be extended ad infinitum, its length is finite.

Use of assumptions, which are neither stated axioms nor consequences of the stated axioms, is made also in the proof of propositions where "superposition of figures" is used. These assumptions constitute an important part of every system of axioms for geometry, since they refer to congruence of figures. The incorrect evaluation of the seriousness of this fact is due to the difficulties encountered, when superposition of figures is used as a method of proof. For example, in order to prove the equality of two triangles, where two sides and the angle defined by them in one triangle are respectively equal to two sides and the angle defined by them in the other triangle, we move the first triangle and we place it over the second in such a way that the two triangles coincide. Here we tacitly assume that the shape of figures does not change when they are moved, thus forgetting that points are "undefinable" elements of our system. Now if we consider geometry as an applied science, where figures can either change place or move, then we cannot ignore the fact that, according to contemporary physics, the dimensions of moving bodies are not equal to their dimensions at rest. The theory of relativity has established that there is an inseparable link between the notions of space and time.

When Euclid constructs lines and circles in order to prove the existence of certain figures, he tacitly assumes that points of intersection occur during these constructions. The construction of an equilateral triangle whose side is given (see page 70) constitutes an example of the use of such an assumption. A logically rigorous theory demands that the existence of these points of intersection, either is proved or it follows directly from the initially stated axioms of the theory. The above defect was corrected by the so-called "axiom of continuity".

Another defect of the Euclidean system is that certain notions are almost never clarified, for example, the two sides of a straight line or the interior of an angle. If one does not clarify these notions, there is a danger of obtaining contradictory conclusions, as is the following:

Paradox: All triangles are isosceles.

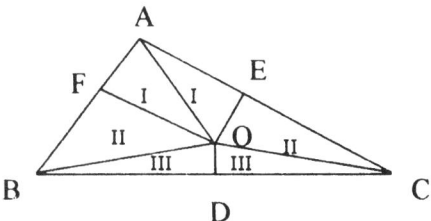

Figure 12.1

Let ABC be an arbitrary triangle. We construct the bisector of the angle A and the normal to the midpoint D of side BC (Figure 12.1). If these straight lines are parallel, then the bisector is perpendicular to BC and the triangle ABC is isosceles. So we assume that these straight lines intersect at point O. We draw the normal OF to AB and the normal OE to AC. Then the triangles marked with I are congruent and $OF = OE$. Also the triangles marked with III are congruent and $OB = OC$. Consequently, the triangles marked with II are congruent and $FB = EC$. From the triangles marked with I, we have $AF = AE$. Therefore, $AB = AC$ and the triangle ABC is isosceles!

In order to explain the above paradox, try to determine the position of O relative to the triangle ABC, by constructing your own figure. Is point O inside or outside the triangle ABC?

The axiomatic foundation of Euclidean Geometry, given by the German mathematician **David Hilbert** (1862-1943) in his book *Grundlagen der Geometrie* (Foundations of Geometry), was able to eliminate from Euclid's foundation of Euclidean Geometry the imperfections discussed above. Hilbert's book was published in 1899 and greatly influenced the development of the science of mathematics in the 20th century. Hilbert understood that it is impossible to define in mathematics all notions and terms used, and hence he initiated the rigorous study of geometry by introducing a number of undefinable notions. The scope of the present book permits only a short description of Hilbert's foundation of Euclidean Geometry.

Hilbert's axioms

1. Undefined notions

- **1.1** A class of undefined elements called "points" which we denote by capital letters $A, B, C, ...$ of the Latin alphabet.
- **1.2** A class of undefined elements called "straight lines" (briefly "lines") which we denote by lower case letters $a, b, c, ...$ of the Latin alphabet.
- **1.3** A class of undefined elements called "planes" which we denote by lower case letters $\alpha, \beta, \gamma, ...$ of the Greek alphabet.
- **1.4** Undefined relations: "Incidence", "Membership", "Betweenness", "Congruence", "Parallelism", "Continuity".

2. Axioms of Incidence

With the term "incidence" we mean a symmetric relation between the elements of the various classes (points, lines, planes) with the following properties:

- **2.1** Given two arbitrary points A and B, there exists a line a which contains A and B.

2.2 Given two arbitrary points A and B, there exists at most one line which contains A and B.

2.3 At least two points lie on every line. There exist at least three points which do not lie on the same line.

2.4 If three points A, B and C don't lie on the same line, then there exists a plane α which contains A, B and C. Given any plane α, there exists a point which lies on α.

2.5 If three points A, B and C don't lie on the same line, then there exists at most one plane which contains A, B and C.

2.6 If two points of a line a lie on a plane α, then every point that belongs to a also belongs to α.

2.7 If two planes have a point in common, then they have at least one more point in common.

2.8 There exist at least four points which do not lie on the same plane.

The next group of axioms contains the axioms that were not included in the group of axioms posed by Euclid and concern the relative ordering between points and between lines.

3. Axioms of Betweenness

3.1 If point B is between points A and C, then A, B and C are three distinct points of the same line and B lies between C and A.

Note. It follows from axiom 3.1 that the term "between" is used only for points that lie on the same line and asserts that the relative position of A and C does not affect B's property to lie between A and C.

3.2 Let A and C be two points and let AC be the line to which they belong. Then there exists at least one point B on AC such that C is between A and B.

Note. Axiom 3.2 asserts the existence of at least three points on every line, something that permits us not to add more powerful postulates to the group of axioms of incidence. Let it be noted that axiom 2.3 also asserts the existence of three points, but they don't lie on the same line.

3.3 Let A, B and C be three points that lie on the same line. Then at most one of them lies between the other two.

Note. Axiom 3.3 in combination with axioms 3.1 and 3.2 permits us to give the following definition: "Let A and B be two points that lie on the same line a. The pair of points A, B or B, A is called the "segment AB" (or the "interval AB"). The points between A and B are called points (or interior points) of the segment AB. A and B are called extremities of the segment AB. All other points of a are called exterior to the segment AB".

3.4 (**Pasch**'s axiom) Let A, B, C be three points which do not lie on the same line. Let b be a line on the plane defined by A, B, C such that none of the points A, B, C lies on b. Let b intersect the line AB at a point D that lies between A, B. Then there exists either a point X on b that lies between A, C or a point Y on b that lies between B, C.

Let it be noted that axioms 3.2 and 3.3 assert that a line is an infinite set.

4. Axioms of Congruence

The axioms of incidence and the axioms of betweenness allow us to define the notions "interval", "side of a line or of a plane", "half line" and "angle".

4.1 If A and B are two points of a line a, and A' is a point of a line a', which might coincide with a, then at a given (a priori determined) side of A' over a' we can find a point B' such that the segment AB is congruent to the segment $A'B'$; in symbols $A'B' \equiv AB$.

4.2 If $A'B' \equiv AB$ and $A''B'' \equiv AB$, then $A'B' \equiv A''B''$.

Note. Axiom 4.2, which refers to "segments", expresses Euclid's axiom: "Things which are equal to the same thing are equal to each other".

4.3 Let AB and BC be segments of a line a, which have no interior points in common, and let $A'B'$ and $B'C'$ be segments of a line a', which also have no interior points in common. If $A'B' \equiv AB$ and $B'C' \equiv BC$, then $A'C' \equiv AC$.

Note. Axiom 4.3, which refers to line segments, expresses the Euclidean Axiom: "If equals are added to equals, the wholes are equal."

4.4 Let $\angle(h,k)$ be the angle formed by two half lines h and k with common vertex, which lie on a plane α. Let a' be a line which lies on a plane α' and let a side of a' be given on α'. Let h' be a half line of a' emanating from a point O' in a'. Then there exists exactly one half line k' in α' such that $\angle(h,k) \equiv \angle(h',k')$ and all interior points of $\angle(h',k')$ lie on the given side of a'. Every angle is congruent to itself.

4.5 If two triangles ABC and $A'B'C'$ satisfy the relations $AB \equiv A'B'$, $AC \equiv A'C'$ and $\angle BAC \equiv \angle B'A'C'$, then $\angle ABC \equiv \angle A'B'C'$.

Note. Axiom 4.5 can be used to prove that: $\angle ACB \equiv \angle A'C'B'$.
Indeed, let us consider the triangles in 4.5 under the assumptions there. We already have that $\angle ABC \equiv \angle A'B'C'$. By applying 4.5 once more to the data
$$AC \equiv A'C', \ AB \equiv A'B', \ \angle ABC \equiv \angle A'B'C',$$
we obtain that $\angle ACB \equiv \angle A'C'B'$.

5. The Axiom of Parallelism

Let a be a line and let A be a point which does not lie on a. Then in the plane defined by a and A, there exists exactly one line which contains A and does not intersect a.

Note. The existence of at least one line passing through A, that does not intersect a, can be proved from the preceding axioms, and hence the above axiom could be stated as follows:

"Let a be a line and let A be a point which does not lie on a. Then in the plane defined by a and A, there exists at most one line which contains A and does not intersect a".

6. Axioms of Continuity

6.1 (Archimedes' Axiom) If AB and CD are two arbitrary segments, then on the line AB there is a finite number of points $A_1, A_2, ..., A_n$ which are so situated that A_1 lies between A and A_2, A_2 lies between A_1 and A_3, and so on, and which satisfy the following conditions:
 (α) $AA_1 \equiv A_1A_2 \equiv ... \equiv A_{n-1}A_n \equiv CD$
 (β) Point B lies between points A and A_n.

6.2 (**Axiom of Linear Completeness**) The points of a line a constitute such a system that when the axioms 2.1–2.3, 3.1–3.4, 4.1–4.5 and 6.1 are satisfied, it cannot be extended further, i.e., it is impossible to add to a more points in such a way that the system of points that results still satisfies the axioms 2.1–2.3, 3.1–3.4, 4.1–4.5 and 6.1.

Note. Axioms 6.1 and 6.2 imply that there exists a one-to-one correspondence between the points of a line and the real numbers.

By using the above axioms, Hilbert proved some of the basic theorems of Euclidean Geometry. Later, others complemented this effort and proved all the remaining propositions of Euclidean Geometry.

The arbitrary character of the above axioms, i.e., their independence from natural reality, brought up the problem of whether this geometry is "consistent", i.e., whether it contains propositions which contradict each other, or whether there exist two propositions which contradict each other (then one of them cannot be proved inside the system). Since Euclidean Geometry was considered to express the truth about the natural space, no one posed the question of its consistency. But the introduction of new axioms and of undefined notions, created a new viewpoint which necessarily demanded the verification of the consistency of the new axiomatic theory.

This problem became even more vital, because the question of the consistency of non-Euclidean Geometries (which will be treated later) was reduced to the question of the consistency of Euclidean Geometry.

Henri Poincaré (1854-1912) raised this issue in 1898 and said that we could accept that an axiomatically structured logical system is "consistent", if one could give a "numerical" interpretation of it. Hilbert went on to prove that Euclidean Geometry is consistent by giving such an interpretation:

In the case of the geometry of the plane, Hilbert identifies a point with an ordered pair (a,b) of real numbers, and a line with a ratio $(u:v:w)$, where u and v are not both zero. The point (a,b) lies on the line $(u:v:w)$ if

$$ua + vb + w = 0.$$

The equivalence is interpreted algebraically with the use of the expressions "parallel translation" and "rotation" of analytic geometry. This means that two figures are congruent, if one of them can be transformed into the other via "parallel translations", "reflections" with respect to an axis and "rotations".

Since every notion can be interpreted numerically, and since it can be made clear that this interpretation satisfies the axioms, Hilbert's argument is that the theorems must also be valid in this interpretation, because they are logical consequences of the axioms. If some contradiction existed in Euclidean Geometry, then a contradiction would also exist in its algebraic interpretation, which is an extension of arithmetic. Consequently, if "arithmetic is consistent", then Euclidean Geometry is also consistent.

Regarding the general question of whether the entire building of mathematics is consistent or not, the reader is referred to [11], where the following conclusion is stated: "If the system of natural numbers is consistent, then the entire mathematical building is consistent".

In an axiomatically structured logical system, it is always desirable to prove the independence of its axioms, i.e., to prove that none of the axioms of the system follows from all or from some of the remaining axioms of the system. Because if the latter happens for a particular axiom of the system, then this is no longer an axiom and it must not be included in the axioms of the system. Hilbert occupied himself with proving the independence of his axioms. It is impossible to prove for any particular axiom in this system that it is independent from the remaining ones, since the meaning of certain axioms depends on some of the axioms preceding them. Thus, what Hilbert managed to prove is that the axioms of anyone of the five groups of his axioms cannot follow from the axioms of the remaining four groups. Hilbert's method consists of giving consistent interpretations or models which do not satisfy any a priori given group of axioms and satisfy the remaining four groups of axioms.

The proofs of the independence of Hilbert's axioms are directly related to the existence of non-Euclidean Geometries. In order to prove the independence of the axiom of parallelism, Hilbert constructed a model of a logical system which does not satisfy the Euclidean axiom of parellelism, but which satisfies the four remaining groups of axioms.

CHAPTER 13

BASIC PROPERTIES OF AXIOMATIC SYSTEMS

One might perhaps think that the structure of a branch of pure mathematics, i.e., of a logical system, simply consists of selecting an arbitrary set of undefined elements, of stating certain axioms regarding the undefined elements, and of constructing a set of conclusions which follow by logical reasoning from the stated axioms and the relative definitions. But this viewpoint is incorrect, because there exist certain basic properties that the system must also satisfy.

We will try to analyze below systematically these properties.

We distinguish three different levels in the study of an axiomatic system.

The first level is the development of certain concrete areas of knowledge. These areas provide examples of the so-called applied axiomatic systems.

The second level is the development of abstract axiomatic systems, such as the particular mathematical models mentioned earlier. The abstract systems provide examples of formal axiomatic systems.

The third level is the theory that studies the properties of formal abstract axiomatic systems. Hilbert called this level of study "Metamathematics".

The main characteristic of an applied axiomatic system is that the primitive data of the system is a (finite and non-empty) set whose elements are defined, i.e., they are entities of the natural world. For example, it can be a number of beads ([21]), or a set of persons belonging to a certain organization, etc. On the contrary, in the case of a formal abstract axiomatic system, the primitive data are terms, or undefined notions as it happens, for example, with Hilbert's foundation of Euclidean Geometry.

We now come to the basic properties of a formal abstract axiomatic system. Among the many properties of such a system, we distinguish the most basic ones, which are the following:

Equivalence, Consistency, Categoricity, and Independence

The property of "Equivalence" refers to pairs of axiomatic systems, while the remaining three properties always refer to a particular axiomatic system.

Equivalence

Two axiomatic systems are called "equivalent", if each one of them implies the other, i.e., if the primitive terms of the first can be defined via the primitive terms of the second and the axioms of the first follow from the axioms of the second, and vice versa. If two axiomatic systems are equivalent, the study of anyone of them does not differ from the study of the other. It is like saying the same thing in two different ways. The idea of equivalent axiomatic systems came up a long time ago, when the geometers were not satisfied with the Euclidean postulate of parallelism and were trying to replace it with some other postulate, equivalent to it but more acceptable.

The contemporary studies of Euclidean Geometry with their various axiomatic bases, clarify the fact that an axiomatic system is not determined uniquely by a certain concrete way of studying it. It depends on which technical terms were selected as being undefined and on which propositions were selected as being unprovable, i.e., its axioms.

Consistency

As mentioned before, an axiomatic system is called "consistent", if no contradictory propositions occur inside the system. "Consistency" is the most important and the most fundamental property of an axiomatic system, because an axiomatic system, which does not have this property, has essentially no value. This is due to the fact that if some proposition A and its negation "not A" can both be proved inside the system, then any other proposition can also be proved inside the system.

Until now, the most successful method of proving the consistency of an axiomatic system is the method of constructing "models" of this system. A "model" of an axiomatic system is obtained by giving concrete interpretations to its primitive terms in such a way that the axioms of the system are transformed into propositions, which are true in relation to certain notions. There exist two types of models: the "concrete models" and the "abstract models". A model is called "concrete" when the interpretations given to the primitive terms of the system are objects or relations taken from the surrounding natural world. While a model is called "abstract" if the interpretations given to the primitive terms of the system are objects and relations that are selected from some other axiomatic system.

When we manage to form a "concrete" model of our system, we have the feeling that we have secured its absolute consistency. Because if contradictory propositions followed from the axioms of our system, then these propositions would also exist in the concrete model of our system, which is impossible since it is accepted that no contradictory propositions exist in the natural world. However, it is not possible to construct a "concrete" model for every axiomatic system that we study. For example, if the system that we study contains infinitely many primitive elements, then it is obvious that it is impossible to find a "concrete" model of our system, because, as it appears, the natural world does not contain infinitely many objects. In the cases that the construction of a "concrete" model is not feasible, we try to construct an "abstract" model by interpreting the primitive terms of our system, say A, through the primitive terms of some other system, say B, in such a way that the interpretations of the axioms of A are logical consequences of the axioms of B. This relative consistency is the best that we can hope for when we apply the method of constructing models to various branches of mathematics, because the branches of mathematics that contain infinitely many primitive terms are indeed many. This, for example, happens in the case of the non-Euclidean Plane Geometry of Lobachevsky: it is possible to construct inside Euclidean Geometry a model of Lobachevsky's Geometry and hence to prove that Lobachevsky's Geometry is consistent if Euclidean Geometry is consistent.

Categoricity

It is more difficult to understand this property since it presupposes the introduction of the notion of "isomorphic interpretations" of an axiomatic system.

As mentioned above, in order to construct a model of an axiomatic system, we give concrete interpretations to the primitive terms of the system in such a way that the axioms of the system are transformed into true propositions. The set of concrete interpretations that we give to the primitive terms is called an "interpretation" of the system. It is obvious that an axiomatic system can have more than one interpretation.

Given any interpretation I of a system Σ, let e be the set of interpretations of the elements of Σ. Also let $\tau_1, \tau_2, \ldots$ be the interpretations of the relations in Σ and let $o_1, o_2, \ldots$ be the interpretations of the operations in Σ.

Now let I' be another interpretation of the same system Σ and let

$$e', \ \tau_1', \ \tau_2', \ \ldots, \ o_1', \ o_2', \ \ldots$$

be the terms of I' that correspond to the terms of I.

If it is possible to establish a one-to-one correspondence between e and e', in such a way that for any index i, between two or more elements of e the relation τ_i holds if and only if the relation τ_i' holds between the corresponding elements of e', and for any index j, the operation o_j acting on one or more elements of e gives an element σ if and only if the operation o_j' acting on the corresponding elements of e' gives the element σ' that corresponds to σ, then we say that the interpretations I and I' of Σ are "isomorphic". The above definition is often formulated briefly as follows:

Two interpretations I and I' of an axiomatic system Σ are called "isomorphic" if and only if it is possible to establish a one-to-one correspondence between the elements of I and I', which is preserved (respected) by the relations and the operations of Σ.

In other words, two isomorphic interpretations of an axiomatic system are absolutely identical (apart from superficial differences in terminology and notation).

After giving the definition of "isomorphic interpretations" of an axiomatic system, we can now define the property of "categoricity" of an axiomatic system.

Definition. An axiomatic system Σ, and hence the corresponding branch of mathematics as well, is said to have the property of "categoricity" (or it is said that it is categorical), if two arbitrary interpretations of Σ are isomorphic.

One usually verifies that a system is categorical by proving that every interpretation of this system is isomorphic to a given interpretation of it. This procedure was actually applied to Hilbert's axiomatic foundation of Plane Euclidean Geometry. That is, it was proved that every interpretation of **Hilbert**'s axioms is isomorphic to the algebraic interpretation given by **Descartes**'s Analytic Geometry. **Lobachevsky**'s non-Euclidean Geometry was also proved to be a categorical system.

A categorical system has both advantages and disadvantages. The most characteristic advantage of a non-categorical system is that it has a wide range of applications, and this is because essentially there exist more than one model of such a system. For example, if we consider the theorems of the so-called "absolute plane Euclidean geometry", that is, the geometry that follows from the first four postulates of Euclid (i.e., without the postulate of parallelism), these theorems (which obviously belong to Lobachevsky's plane geometry as well) also follow from the axioms of Hilbert if we subtract from them the axiom of parallelism. This means that if we drop the axiom of parallelism from Hilbert's axiomatic system, we obtain a non-categorical system. Indeed, if one gives to it the interpretation of Euclidean Geometry and the interpretation of Lobachevsky's Geometry, then these interpretations are not isomorphic.

The structure of a "group" also constitutes an example of a non-categorical axiomatic system. This is because its axioms are satisfied both by interpretations having infinitely many elements and by interpretations having finitely many elements, which means that these axioms are satisfied by non-isomorphic interpretations.

An advantage of categorical systems is that it is often easier to prove various theorems in such a system, because their proof is achieved in models of the system which are more familiar to us. For example, there exist models of Lobachevsky's Plane Geometry inside Plane Euclidean Geometry. Now given that Lobachevsky's Geometry is a categorical axiomatic system and that Euclidean Geometry is more familiar to us than Lobachevsky's Geometry, we can prove theorems of Lobachevsky's Geometry by proving their corresponding ones in Euclidean Geometry.

The categoricity of a system of axioms implies another property of the system called **completeness**.

A system of axioms is called **complete** if it is impossible to add to it one more axiom which is independent of the already existing axioms and consistent with respect to them, without introducing new primitive notions to the system.

Categoricity implies completeness, because if we assume that a set A of axioms is categorical but incomplete, then it is possible to introduce to A an axiom S such that both S and "not S" are consistent together with A. Hence, since the initial set A is categorical, there exist an interpretation of the system consisting of A and S and an interpretation of the system consisting of A and "not S", and the two interpretations are isomorphic. But this is impossible, since these interpretations must satisfy the same propositions, and this does not happen since in one interpretation S is true and in the other "not S" is true.

Independence

We know that an axiom of an axiomatic system is called "independent" if it is not a logical consequence of the remaining axioms of the system.

It is known that Euclid's "postulate of parallelism" has been studied extensively in relation to the notion of independence of an axiom.

For centuries, mathematicians found it difficult to accept that this postulate of Euclid is independent of the remaining four postulates, until the discovery of Lobachevsky's non-Euclidean Geometry revealed the independence of this postulate. From a historical viewpoint, it would not be an exaggeration to say that the beginning of the general study of the properties of axiomatic systems is due to the proof of the independence of the postulate of parallelism.

A way of checking whether an axiom is independent consists of finding an interpretation of the primitive terms of the system which makes this axiom false but which satisfies the remaining axioms of the system. If it is possible to find such an interpretation, then the axiom under consideration cannot be a logical consequence of the remaining axioms. Indeed, if this happened, then the interpretation that transforms the remaining axioms into true propositions would also transform this axiom into a true proposition.

Let it be noted that it is not necessary for the axioms of a system to be independent. In other words, one cannot reject an axiomatic system because some of its "axioms" are not independent. The reason for which a mathematician wants the axioms of his/her system to be independent is his/her desire to base his/her theory on the minimum possible set of assumptions.

It should be noted that sometimes it is advisable for pedagogical reasons to develop a subject starting from an axiomatic system, where there exist "axioms" which are not independent. For example, when one teaches plane geometry in Secondary Education, the proof of a certain theorem in the beginning of the course may present difficulties and hence this theorem can be considered as one of the axioms of the system. Later, when the pupils have acquired the necessary maturity and familiarity with the subject, one can explain to them that what was given initially as an axiom, is not independent of the remaining axioms, and then one can prove this proposition by using the other axioms.

There exist known axiomatic systems that when they were first published contained (to their author's ignorance) axioms which were not independent. An example of such a case is Hilbert's first publication of his axioms of Euclidean Geometry. Indeed, it was proved later that Hilbert's system contained "axioms" that were not independent. Of course this does not reduce the importance of the system of Hilbert, who transformed these "axioms" into theorems by supplying their proofs in a later publication.

CHAPTER 14

SPACES AND GEOMETRIES

For the general public Geometry is probably the art of performing measurements on various figures and determining the relative size of the various parts of a given figure. Besides, this is indeed the etymological meaning of the word geometry.

However, for a contemporary mathematician, Geometry means something different. The art of measuring figures constitutes a very small and perhaps less interesting part of geometry.

The study of the properties of figures which are independent of the notion of measure (i.e., the notion of distance) constitutes a wide field. Of course, it is possible to introduce the notion of measure into the basic notions of this field in order to extend it further, however, even without the notion of measure, there still remains fundamental material to be studied. The additional notion of measure is nothing more than an accessory, which up to a point depends on the "will" of the geometer.

Indeed, no one doubts that a unique line passes through two distinct points. On the other hand, people might give different answers to the question of whether there exist two points on a line whose distance is greater than a given limit of size. As we will see, the notion of measure that we introduce to our system depends on whether or not we accept the existence of such points.

The clear distinction between "metric properties" and "non-metric" properties began in the 19th century. Almost all geometric properties that are studied in the geometry taught in Secondary Education have a metric character. As examples, we mention the following properties:

- The diagonals of every parallelogram intersect at their midpoints.
- Parallel straight lines which intersect given straight lines, define on the given straight lines proportional line segments.

An example of a non-metric property is given by **Brianchon**'s theorem: **"The three diagonals connecting the opposite vertices of a hexagon inscribed in a conic section, pass through the same point"**.

The systematic study of the non-metric properties of figures began under very strange circumstances. The young French second lieutenant **Poncelet** (1788-1867) was imprisoned in Saratov in Russia in 1812 during the Russian-French war. During his imprisonment which lasted for quite a long time, in order to fight his boredom he tried to remember the geometric theories that he was taught at the École Polytechnique (France). However, he realized that he had completely forgotten them! This forced him to follow a new road, entirely on his own, which led him to the discovery of a new branch of mathematics, which we call today Projective Geometry (PG).

PG lacks all considerations which are based on the notions of measure and parallelism, and hence on the notion of distance.

In PG we talk about straight lines passing through the same point (see Brianchon's theorem), about points lying on the same straight line as in **Pascal**'s theorem (**If a hexagon is inscribed in a conic section, then the points of intersection of the straight lines defined by opposite sides (which are assumed to be non-parallel) lie on the same straight line**), about mutually separated pairs (the case of harmonic quadruples), etc. But in PG we do not talk about equal or proportional line segments, and we talk neither about bisectors of angles nor about the area of a surface.

PG in its present form was not entirely conceived by Poncelet. Only in the course of time, due to the work of Poncelet himself, of the French mathematician **Chasles,** of the Swiss geometer **Steiner**, and finally due to the systematic clarification carried out by the German mathematicians **Von Staudt** and **Felix Klein**, did it become possible to eliminate all metric notions and to make PG an autonomous geometry.

1. The elimination of metric properties

It is natural for the inexperienced reader to wonder: What will remain from geometry, if we eliminate everything that is based on the notion of measure? Even the experienced reader will be surprised by the list of notions which as a matter of principle cannot be considered in PG.

In PG one cannot talk about "equal line segments" and one can talk neither about the midpoint of a line segment nor about proportional line segments. In PG one never talks about parallel straight lines, because this leads implicitly to the notion of measure. Indeed, two parallel straight lines which intersect two other parallel straight lines, define on the latter lines equal line segments. Consequently, one cannot talk about a parallelogram and one can talk neither about a rectangle nor about a square. Also the notion of a circle does not exist in PG, because it is impossible to define a circle without using the notion of distance.

This elimination proceeds even further since one ignores the measure of an angle. Hence it is impossible to talk about equal angles or right angles (because a right angle is defined as being equal to its supplementary angle) and hence the notion of being perpendicular is also eliminated. Finally, let it be noted that in PG one can talk neither about the measurement of an area, nor about the measurement of a volume, nor about the translation of a figure (because a translation preserves the measure of various parts of a figure).

At this point it is necessary to make the following observation. Sometimes it happens that non-metric notions give the impression of being metric when they can be defined using metric notions, neglecting the fact that it is also possible to define them without using metric notions. Examples of such cases are the notion of harmonic quadruple, the notion of anharmonic ratio, and the notion of a conic section (i.e., of the intersection of a circular cone and a plane). We will see below that these last notions can be defined without using the notion of measure, which means that their study falls into the scope of PG.

In any case, the elimination of metric properties is quite drastic, and hence one has the right to wonder whether what is left after this elimination is worth being considered as an autonomous field. We will come to this subject later. For the time being we will restrict ourselves to examining the possibility of eliminating any notion of "measure".

2. The points at infinity and the straight line at infinity

In order to make things simpler, we will restrict ourselves to the geometry of the plane. As mentioned above, in PG one cannot talk about parallel straight lines. But is this practically possible? Parallel straight lines in a plane are straight lines that do not intersect. Consequently, if we drop the notion of parallelism, we can only talk about straight lines that intersect. Thus each time we have to deal with a proof of say a theorem, with straight lines which either intersect or do not intersect, we can only consider those that intersect. But such a situation would impose such limitations on the generality of the theorems, that the conclusions of these theorems would cease to be interesting. This obstacle is by-passed as follows:

Let us define the notion of direction using the expression "point at infinity". Of course, this is a matter of vocabulary. Then we can say that two straight lines always have a common point, either a "proper" point (i.e., one of the already existing points) or a point at infinity. In both cases, we call this common point **the common point of the two straight lines**.

Now let us consider the set of all points at infinity (or directions) of the plane. We call this set the **straight line at infinity**. Every proper straight line (i.e., one of the already existing straight lines) has a direction, i.e., it has a point at infinity. We will say that this proper line passes through its point at infinity, and since it has a common point with the straight line at infinity, we say that it intersects the straight line at infinity at this point at infinity.

Up to now we have introduced only verbal expressions. But here is an essential new element of PG: "It makes no distinction between the proper (the existing) points and the points at infinity, and calls all of them simply points. **Consequently, two straight lines will always have a common point**. Also PG makes no distinction between the proper (the existing) straight lines and the straight line at infinity. It calls all of them simply straight lines."

In order to illustrate the importance of the situation created by the introduction of the above element, we will study **Desargues**' theorem, which plays a crucial role in the study of the foundations of geometry.

Let ABC and $A'B'C'$ be two triangles (Figure 14.1) and let us call the pairs of vertices A, A', B, B', and C, C' homologous. Also let us call homologous the pairs of sides $BC, B'C'$, $CA, C'A'$, and $AB, A'B'$. Then Desargues' theorem is related to the following proposition:

"If the straight lines AA', BB' and CC' which connect the homologous vertices of the triangles ABC and $A'B'C'$ either pass through the same point or are parallel, then either the straight lines defined by every pair of homologous sides intersect and the three points of intersection lie on the same straight line, or two homologous sides are parallel to the straight line connecting the points of intersection of the other two pairs of straight lines defined by homologous sides, or the homologous sides are pairwise parallel."

Triangles of this kind are called **homologous triangles**.

The above proposition can also be formulated as follows:

"If the straight lines connecting the homologous vertices of the two triangles either pass through the same point or have the same direction, then either the

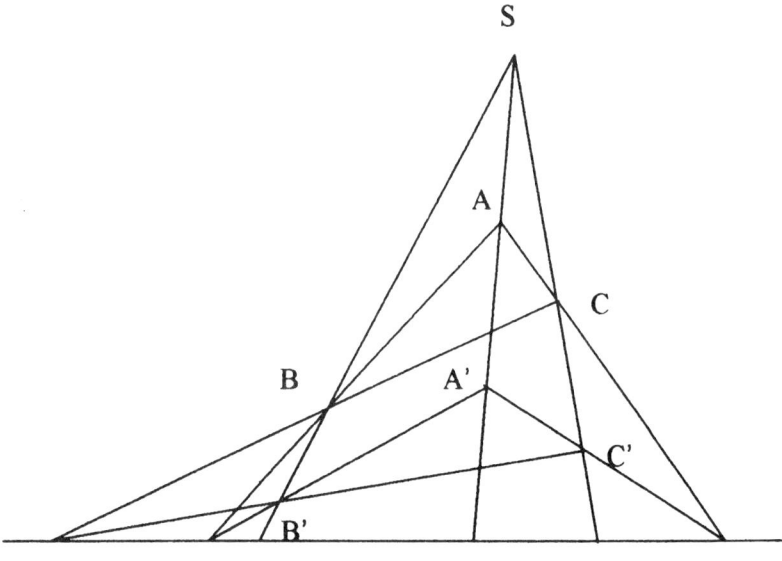

Figure 14.1

straight lines defined by every pair of homologous sides intersect and the three points of intersection lie on the same straight line, or two homologous sides have the same direction with the straight line connecting the points of intersection of the other two pairs of straight lines defined by homologous sides, or the homologous sides have pairwise the same direction."

By using the new verbal expressions introduced above, we could also state this proposition as follows:

"If the straight lines connecting the homologous vertices of the two triangles pass through the same point (either proper point or point at infinity), then either the straight lines defined by every pair of homologous sides intersect at proper points and the three points of intersection lie on the same proper straight line, or the straight lines defined by one pair of homologous sides intersect at a point at infinity which lies on the proper straight line connecting the proper points of intersection of the other two pairs of straight lines defined by homologous sides, or the straight lines defined by homologous sides pairwise intersect at points at infinity (i.e., the three points of intersection lie on the straight line at infinity)."

PG makes no distinction between proper elements and elements at infinity and gives the following simplified statement covering all the previous ones.

Theorem (Desargues)

"If the straight lines connecting the homologous vertices of the two triangles pass through the same point, then the points of intersection of the straight lines defined by homologous sides lie on the same straight line."

This last statement avoids completely the notion of parallelism and includes the case in which certain straight lines are parallel. In order to obtain the various special cases, it is enough to distinguish between proper elements and elements at infinity, but this does not fall into the scope of PG.

It is important to emphasize that in PG not only the various special cases of a theorem are summarized into a single statement, but also in the proof of the theorem no distinction is made between proper elements and elements at infinity so that the same proof holds in all cases.

Therefore PG is not an artificial construction. On the contrary, from a certain point of view it is richer in content than classical geometry.

3. The laws of duality

If we look back at our school days and at our experience as educators, we might perhaps remember that many pupils found the course of classical geometry quite complicated. Perhaps this was due mainly to the lack of guidelines in the succession of theorems and proofs. The solution of almost every problem seemed to demand from us to find some trick.

This phenomenon is not observed in PG. Based on axioms which have a remarkable symmetry, PG first studies the properties of very simple figures. These figures consist of 1) points lying on the same straight line, 2) straight lines passing through the same point and lying on the same plane, and 3) planes passing through the same straight line. Then PG progressively proceeds to the study of more complex figures. The guideline of proofs always remains the same, namely the preservation of a certain correspondence between the elements of the figures considered.

In plane PG the symmetry of the basic propositions is so remarkable, that it is possible to interchange the word "point" and the words "straight line" without changing the set of the relevant propositions. The propositions simply alternate. Examples of such propositions are the following: "Two points determine a straight line" and "Two straight lines determine a point". Consequently, every theorem of plane PG remains true if the word "point" is interchanged with the words "straight line", and the new proposition obtained in this way needs no new proof. We note that this symmetry does not exist in classical geometry, because, for example, the proposition "Two straight lines determine a point" is not true in the case of parallel straight lines.

In 3-dimensional PG, the word "point" can be interchanged with the word "plane". Thus, the proposition "Three points not lying on the same straight line determine a plane" is transformed into the proposition "Three planes not passing through the same straight line determine a point". This last proposition is not true in classical geometry, because, for example, one of the planes can be parallel to the straight line of intersection of the other two (in PG in the lateral case these planes determine a point at infinity).

Note. As in the case of the plane, in 3-dimensional space the direction of a straight line is called "the point at infinity of the straight line". When a plane is parallel to a straight line (or contains a straight line), we say that the direction (i.e., the point at infinity) of this straight line belongs to this plane. If a plane is parallel to a straight line along which two other planes intersect, the direction of this straight line belongs to these three planes (all other directions excluded), which means that the three planes determine this direction, i.e., they determine a point at infinity.

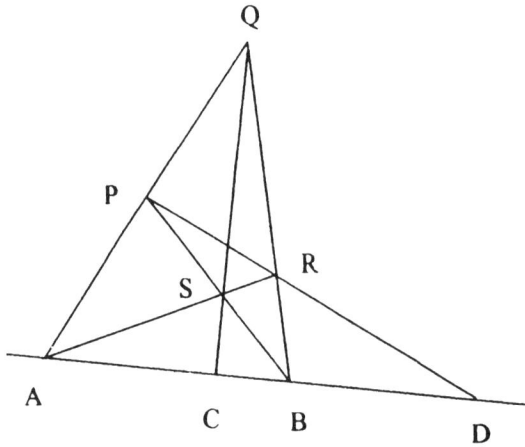

Figure 14.2

The possibility of interchanging the word "point" with the word "straight line" in plane PG is an example of the law of duality. Similarly, the possibility of interchanging the words "point" and "plane" in 3-dimensional geometry is another example of the law of duality. This law doubles the number of propositions valid in PG, without the need for new proofs.

4. The harmonic tetrads

We will show that certain notions which can be defined using the notion of measure (distance), can also be defined without using metric notions. Such a case is the case of harmonic tetrads.

Let A, B, C, D be four points lying on the same straight line (a tetrad of points). From the "metric" point of view, the expression $\frac{(CA)}{(CB)} : \frac{(DA)}{(DB)}$ is called the anharmonic ratio of this tetrad and is denoted by $(ABCD)$; in this expression (CA), (CB), (DA), (DB) denote the measures, in quantity and sign, of the line segments CA, CB, DA, DB respectively. If these points are ordered in such a way that $(ABCD) = -1$, then we say that the tetrad is **harmonic** and that the pairs A, B and C, D are **harmonically separated**; we also say that the points of the pair A, B are **harmonic conjugate** with respect to the points of the pair C, D and vice versa.

It can be shown that a necessary and sufficient condition for a tetrad A, B, C, D to be harmonic is the existence of a "complete tetrad" (i.e., of a system consisting of four points or vertices P, Q, R, S, three of which never lie on the same straight line, and six straight lines or sides which connect these points in pairs), two opposite sides of which pass through point A, another two opposite sides of which pass through point B, the fifth side of which passes through point C, and the sixth side of which lies opposite point C and passes through point D (Figure 14.2). Hence, this property can be used as the definition of a harmonic tetrad of points, and this new definition has the remarkable characteristic that it does not refer to any metric notion.

Furthermore, if we consider a **tetrad of rays** a, b, c, d (i.e., four straight lines lying on the same plane and passing through the same point), then we will say that this tetrad is **anharmonic**, if there exists a "complete quadrilateral" (i.e., a system consisting of four straight lines or sides p, q, r, s, of which no three pass through the same point, and of six points or vertices, at which these straight lines pairwise intersect), with two opposite vertices lying on ray a, with another two opposite vertices lying on ray b, with the fifth vertex lying on ray c, and with the sixth vertex lying on ray d.

We note that the last two definitions follow from each other via the law of duality.

Now let us come back to the harmonic tetrad A, B, C, D. It is easy to prove that if the point D is a point at infinity (this is the case when the straight line PR is parallel to the straight line AB), then the point C is the midpoint of the line segment AB, i.e., $AC = CB$. Consequently, as soon as we are allowed to talk about the point at infinity of a straight line, a metric notion is introduced; indeed, the midpoint of a line segment AB can then be defined as the harmonic conjugate point of the point at infinity of the straight line AB with respect to the extremities of the line segment AB.

Before showing how the notion of anharmonic ratio can also be introduced without using the notion of measure, we must define the so-called projective coordinates.

5. The projective coordinates

We will now show that by taking advantage of the rich structures of analytic geometry, we can introduce in PG coordinates which are completely independent of the notion of measure.

On a straight line d we consider three arbitrary points A_0, A_1 and T as fundamental points, which we call the origin, the unitary point, and the limit point respectively. Let A_2 be the harmonic conjugate of A_0 with respect to the points A_1 and T. Then, let A_3 be the harmonic conjugate of A_1 with respect to the points A_2 and T, and so on. In this way we form a sequence of points such that if A_{i-1}, A_i, A_{i+1} are three successive points of the sequence, then the points A_{i-1}, A_{i+1} separate harmonically conjugate points A_i, T. Such a sequence of points is called a **harmonic sequence**. The index k of a point A_k of this sequence is called the **projective coordinate** of this point. We remark that this definition is not based on any metric notion.

The Cartesian abscissae of the points of an axis are merely special cases of projective coordinates. Indeed, let us assume that the limit point T is the point at infinity of d. Then since the quadruple A_0, A_2, A_1, T is harmonic, we have that $A_0A_1 = A_1A_2$. And since the quadruple A_1, A_3, A_2, T is harmonic, we have that $A_1A_2 = A_2A_3$, and so on. The harmonic sequence in this case is a sequence of equidistant points, and if we take point A_0 as the origin and segment A_0A_1 as the unit segment, then the Cartesian abscissa of point A_i is equal to i, which means that the Cartesian abscissa of a point is equal to its projective coordinate.

It follows from the above that as soon as we can talk about the point at infinity of a straight line, the notion of measure on this straight line is implicitly introduced.

It is interesting to see how we can construct in a simple way a harmonic sequence using only a ruler.

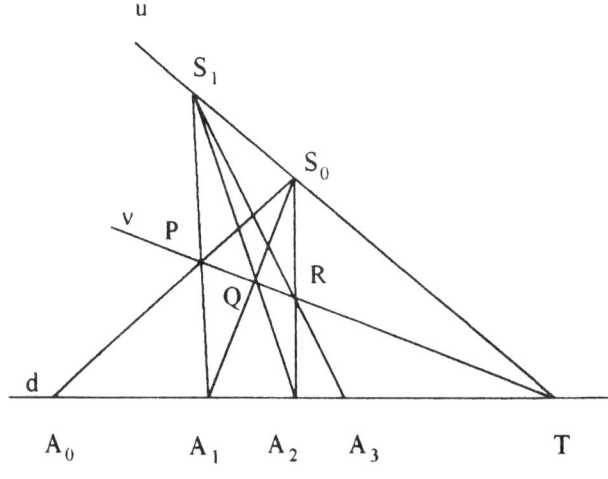

FIGURE 14.3

We take on a straight line d three arbitrary points A_0, A_1 and T, and we draw from point T two arbitrary straight lines u and v (Figure 14.3). We draw two straight lines passing through A_0 and A_1 respectively, which intersect at a point P of v. These straight lines intersect u at points S_0 and S_1 respectively. We will use successively these two points as centers of projection. From center S_0 we project A_1 on v and let Q be this projection. Also from center S_1 we project Q on d. We will prove that this projection is the point A_2 of the sequence. Indeed, the complete quadrangle PQS_0S_1 shows that the pairs of points A_0, A_2 and A_1, T are harmonically separated. Now in order to construct the point A_3 of the sequence, from center S_0 we project A_2 on v and let R be this projection. Then from the center S_1 we project R on d and we obtain the point A_3 of the sequence. Indeed, the complete quadrangle QRS_0S_1 shows that the pairs of points A_1, A_3 and A_2, T are harmonically separated. The construction proceeds always in the same way.

If we chose the point T to be the point at infinity of d and if, for example, apart from points A_0 and A_1, we took the straight line u to be parallel to d, then we could make the same construction using only a ruler, and we would obtain a sequence of equidistant points on d, i.e., a metric sequence.

Working in an analogous manner it is possible to insert between the points with integral indices, points with rational indices, and even points with irrational indices. Moreover, we can also extend this sequence in the opposite direction and obtain points with negative indices. In this way every point of d would have an index, i.e., a projective coordinate.

Next we study **projective coordinates in the plane**.

From a point O (the origin) we draw two straight lines Ox and Oy (the axes). On each of these straight lines we construct a harmonic sequence by taking each time the point O as the origin, and by selecting arbitrarily the unit points A_1, B_1 and the limit points T, T' (Figure 14.4).

In order to determine the coordinates of an arbitrary point P of the plane, we project P on the x-axis from the limit point T' and we project P on the y-axis from

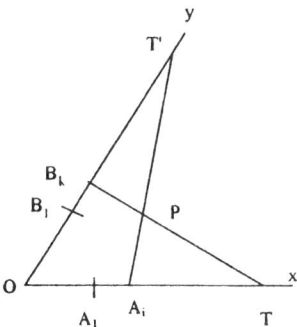

FIGURE 14.4

the limit point T. Let A_i and B_k be these projections. We will call the indices i and k the **projective coordinates** of the point P.

It follows from the above that if T, T' are the points at infinity of the axes Ox, Oy respectively, then the projective coordinates of the point P coincide with its Cartesian coordinates with respect to these axes.

It is often convenient to replace the projective coordinates with the so-called homogeneous coordinates. To this end, we set $x = \frac{x_1}{x_3}$ and $y = \frac{x_2}{x_3}$, and we call **homogeneous projective coordinates** of the point P every triple of real numbers proportional to x_1, x_2, x_3 and different from $(0,0,0)$.

Finally, we mention without proof that the projective coordinates of a point which ranges over a straight line, satisfy a first order equation (just as it happens with the Cartesian coordinates of this point). This equation has the form $u_1 x_1 + u_2 x_2 + u_3 x_3 = 0$. The coordinates u_1, u_2, u_3 or, to be more precise, every triple of real numbers proportional to u_1, u_2, u_3 are called **tangential projective coordinates** (or line-coordinates) of this straight line (and x_1, x_2, x_3 are called **point-coordinates**).

It can be shown that there exists a complete **duality** between point-coordinates and line-coordinates, thus the definition of point-coordinates or line-coordinates can be transformed into the definition of line-coordinates or point-coordinates respectively by interchanging the word "point" with the phrase "straight line".

Remark. It is useful to give a clearer picture of certain fundamental notions. If x_1, x_2, x_3 are the homogeneous coordinates of a point (point-coordinates) and if u_1, u_2, u_3 are the homogeneous coordinates of a straight line (line-coordinates), then the relation $u_1 x_1 + u_2 x_2 + u_3 x_3 = 0$ expresses the fact that this point belongs to this straight line or that this straight line passes through this point. Indeed, for example, if we consider a given straight line d with coordinates 6, -4, 5 by giving to the coordinates x_1, x_2, x_3 all values satisfying the relation $6x_1 - 4x_2 + 5x_3 = 0$, we obtain the coordinates of all the points that belong to the straight line d. This relation called "the equation of the straight line d" represents the so-called "range" (i.e., the set of points lying on a straight line which is called the "carrier") whose carrier is the straight line d. An equation of x_1, x_2, x_3 is called a **point-equation**. Now according to the notion of duality, if we consider a point P with coordinates 2, 5, 3 and we give to the coordinates u_1, u_2, u_3 all values satisfying the relation $u_1 \cdot 2 + u_2 \cdot 5 + u_3 \cdot 3 = 0$, i.e. $2u_1 + 5u_2 + 3u_3 = 0$, then we obtain the coordinates of all the straight lines passing through the point P. The last relation, called "the equation of the point P", represents a **bundle of rays** (i.e., the set of all straight lines or rays in a plane passing through the same point called its "carrier" or its "vertex") whose carrier (or vertex) is the point P. An equation of u_1, u_2, u_3 is called a **tangential equation**.

6. The anharmonic ratio

We can now show that the notion of anharmonic ratio can be introduced without using metric notions, which means that the anharmonic ratio is a projective notion.

For simplicity, we assume that the four points A, B, C, D lie on the x-axis. Let α, β, γ, δ be the projective coordinates of the points A, B, C, D respectively, with respect to a certain harmonic sequence. We call **anharmonic ratio** of the tetrad A, B, C, D the expression $\frac{\alpha-\gamma}{\beta-\gamma} : \frac{\alpha-\delta}{\beta-\delta}$.

Since α, β, γ, δ are numbers, this expression is also a number. It can be shown that if we consider the points A, B, C, D with respect to another harmonic sequence, then the same expression formed by the new projective coordinates of these points has the same numerical value with the first expression. In particular, if we consider a harmonic sequence whose limit point is the point at infinity, then the projective coordinates are transformed into Cartesian coordinates and the above expression coincides, in this case, with the classical expression of the anharmonic ratio. Thus, the above definition of the anharmonic ratio of a tetrad of points, gives the same value with the one given by the classical definition. This means that this definition can replace the classical definition. Moreover, the new definition has the property that it makes no use of metric notions.

It is interesting to note that, as in Cartesian geometry, we can extend the domain of PG by giving to the coordinates complex values. Thus we will refer to the above expression as the anharmonic ratio of four points, even if some of the coordinates α, β, γ, δ are complex numbers.

Let us pass from tetrads of points, to tetrads of rays. By transferring, via the laws of duality, the notion of projective coordinate given above, and by starting from three rays a_0, a_1, t passing through a point S, we can assign an index to every ray passing the point S. If the indices of four rays a, b, c, d passing through S are α, β, γ, δ respectively, then we call the expression $\frac{\alpha-\gamma}{\beta-\gamma} : \frac{\alpha-\delta}{\beta-\delta}$ the anharmonic ratio of these four rays. It can be shown that the numerical value of this expression coincides with the value of the anharmonic ratio obtained by the classical definition. As in the case of tetrads of points, the above definition is also valid if some rays are complex.

7. The introduction of the notion of measure

No matter how interesting PG is, when one studies geometry in general, one must be able to consider metric notions. In classical geometry metric notions are introduced as soon as together with the axioms expressing projective properties (as, for example, is the axiom: "two points determine a straight line"), other axioms are given, which express metric properties (as, for example, is the axiom: "if two line segments are equal to a third line segment, then the two line segments are equal").

In order to build PG, we preserve only those axioms expressing projective properties, and we slightly modify the form of these axioms in order to include in our theory elements at infinity.

In order to be able to pass from PG to a more complete geometry, it would be enough to introduce to our theory the metric axioms that were excluded at the beginning. However, there exists a far more interesting approach which connects directly the metric notions with PG.

We recall that the basic property of PG is that it makes no distinction between the straight line at infinity and the other straight lines of the plane, and that it makes no distinction between the points at infinity and the other points of the plane.

We also recall that as soon as we can talk about the straight line at infinity, we can introduce with the help of projective notions (harmonic tetrads) the measure on this line. If we distinguish between the straight line at infinity and the remaining straight lines of the plane, we can talk about the point of intersection of an arbitrary straight line d with the straight line at infinity, i.e., we can talk about the point at infinity of the straight line d (and then we can introduce the measure on d).

So instead of introducing the measure via metric axioms, we can simply make the following statement: "Among all the straight lines of the plane, there exists one line which we distinguish from the others and which we call straight line at infinity".

We note that the simple distinction between the straight line at infinity and the remaining straight lines is not sufficient in order to introduce the measure of angles and the notion of equality between line segments lying on different straight lines. In order to introduce these last two notions, we must also distinguish, among all points in the plane, two specific points known as **circular points**.

8. Laguerre's formula

In order to formulate clear ideas on this subject, it is necessary to give certain formulas. In 1853, the young **Laguerre**, who was still studying at the École Polytechnique (France), discovered a remarkable expression of the measure of an angle ab:

$$\text{Measure of the angle } ab = \frac{i}{2}\log(abmn).$$

In this formula i is the imaginary quantity $\sqrt{-1}$ and $(abmn)$ denotes the anharmonic ratio formed by the sides of the angle ab and by the straight lines m, n connecting the vertex of the angle ab with the two circular points of the plane mentioned above. These circular points are two imaginary points (i.e., they have imaginary coordinates) which lie on the straight line at infinity and play an important role in geometry. It is remarkable that the elements of the plane which play the most important role in PG, i.e., the straight line at infinity and the circular points, are not proper elements but elements that we have created artificially.

It was mentioned earlier that the notion of the anharmonic ratio is a projective notion. It follows from Laguerre's formula that the measure of an angle can be defined with the help of projective notions only, provided that we distinguish two specific points of the plane.

Remark. The two circular points are two imaginary points which belong to every circle of the plane. The equation of an arbitrary circle in homogeneous projective

coordinates x, y, t is
$$x^2 + y^2 + axt + byt + ct^2 = 0.$$

We remark that the points with homogeneous projective coordinates $(1, i, 0)$ and $(i, 1, 0)$ are the two circular points, because they satisfy the above equation of a circle no matter what are the values of a, b, c.

9. The Euclidean measure of a line segment

For the measure of line segments there exists a formula similar to Laguerre's formula.

We consider a line segment AB and an ellipse whose axes are assumed to be extended indefinitely in such a way that the ellipse constantly dilates in every direction. The straight line on which the segment AB lies, intersects the ellipse at the points M, N which are also moving away ad infinitum. The anharmonic ratio $(ABMN)$ is changing, but it tends to a certain limit. It can be shown that the measure of the segment AB is equal to the limit of the expression $\log(ABMN)$ times a factor c, which tends to infinity as the axes of the ellipse are extended indefinitely:

Measure of the segment $AB = \lim(c \cdot \log(ABMN))$.

But let us for the moment consider the straight line at infinity. When a straight line a is rotating about one of its points, say A, in such a way that it tends to become parallel to a straight line b, then its point of intersection with b moves away ad infinitum and passes through any given point of b, no matter how far the latter is located. We express this fact by saying that this point of intersection tends to infinity. In this way we are led "intuitively" to the notion of the point at infinity (where two straight lines intersect), a notion that was introduced earlier only formally.

Now if we consider the set of all points at infinity in the plane, every straight line in the plane has one common point with this set, and thus we can say that it **intersects** this set at one point. So it is quite natural "intuitively" to identify this set of points at infinity with a straight line. Indeed we called this set earlier the straight line at infinity, but in a completely formal way! We meet this straight line at infinity in all directions, because every straight line has a point at infinity. So we can "intuitively" see that the straight line at infinity surrounds the plane and lies away from all proper points of the plane. Also the ellipse, whose axes are extended indefinitely tends to surround the plane, thus "intuitively" we see that it tends to coincide with the straight line at infinity. Hence the straight line at infinity can be considered as the limit of this dilating ellipse. The points of intersection M, N of the straight line AB with this ellipse, both tend to infinity during the dilation even though they move in opposite directions; or, to be more precise, they tend to the point of intersection of the straight line AB with the straight line at infinity. Hence the measure of a line segment can be defined with the help of projective notions only, provided that we distinguish a specific straight line in the plane, i.e., the straight line at infinity.

It follows from the above considerations that the metric properties of a figure can indeed be reduced to projective properties, but not to projective properties of the figure itself. They are reduced to projective relations between the figure and certain distinguished elements of the plane.

Conclusion. We go from PG to metric geometry by distinguishing certain elements from the remaining ones. We consider as the measure of a segment (or of an angle) a certain expression of projective nature which depends on the distinguished elements and on the extremities of the segment (or on the sides of the angle respectively).

Remark. All the above intuitive considerations regarding the elements at infinity become positive mathematical notions, when we express them through analytic geometry by making use of the homogeneous coordinates x, y, z. The points at infinity are the points whose coordinate z is zero. The relation $z = 0$ expresses the fact that the coordinates of these points satisfy a first-order equation, which we consider as the equation of a straight line, i.e., of the straight line at infinity.

10. The conic sections

Before we proceed to the treatment of non-Euclidean geometries, it would be useful to show how we can define conic sections by using projective notions only.

We will assign each point of the plane to a straight line of the plane in such a way that we form pairs consisting of a point and a straight line. This correspondence can be arbitrary, but it must satisfy the following condition: If a point P and a straight line q are such that P lies on q, then the straight line p that corresponds to P and the point Q that corresponds to q are such that Q belongs to p. Such a correspondence is called a **polarity**. The point P is called the **pole** of the straight line p, which corresponds to P, and the straight line p is called the **polar** of the point P.

A pole does not lie, in general, on its polar. The set of all poles that actually lie on their polars is called a **conic-locus**. Also a polar does not pass, in general, through its pole. The set of all polars that actually pass through their poles is called a **conic-envelope**. The set consisting of the points of a conic-locus and of the straight lines of a conic-envelope is called a **conic**.

It can be shown that this notion of a conic is equivalent to the corresponding classical notion of a conic.

If we express the relations of a polarity with the help of projective coordinates, we find that the coordinates x, y, z of the points that form a conic-locus satisfy a second-order equation called the **point-equation of the conic**. Similarly, we find that the coordinates u, v, w of the straight lines that form a conic-envelope satisfy a second-order equation called the **tangential equation of the conic**. When one of these two equations is known to us, we can find the other.

We note that for simplicity we have denoted here by x, y, z, u, v, w the projective coordinates that were denoted earlier by x_1, x_2, x_3, u_1, u_2, u_3 respectively.

We will not go any further in the study of conic sections from the vantage point of PG. We will only add that, although PG is able to define the conic sections, it is not able to distinguish between the three kinds of curves: ellipse, parabola and hyperbola. This distinction is based on the relation between the conic and the straight line at infinity (which is not intersecting, is tangent to and intersecting the conic respectively).

CHAPTER 15

NON-EUCLIDEAN GEOMETRIES

The interest in Projective Geometry is not only due to the remarkable symmetry that characterizes this geometry, but also due to the fact that all the theorems of Projective Geometry are valid both in Euclidean geometry as well as in the so-called non-Euclidean geometries. This observation leads us naturally to discuss these latter geometries.

It has already been stated several times in this book that the Euclidean postulate of parallelism is less obvious than the other four postulates of Euclidean geometry. This famous postulate, is equivalent to the statement below (although it is formulated differently by Euclid).

"From a given point only one straight line parallel to a given straight line can be drawn".

Indeed, this axiom is less obvious than the following, for example:

"Exactly one straight line passes from two points".

"A straight line can be extended indefinitely".

As mentioned earlier, even in the days of Euclid it was thought that the "postulate of parallelism" can be "proved" by using the remaining axioms, i.e., that it is not an axiom, but a theorem. Remarkable efforts to prove the axiom of parallelism were made by

Proclus (412-485 A.D.)
Saccheri (1667-1733 A.D.)
Thibault (1775-1822 A.D.)

as well as by many others. We know that all these efforts failed. We also know today that the axiom of parallelism is independent of the remaining four axioms. **The fact that Euclid conceived its independence and added the axiom of parallelism to the remaining four axioms constitutes one of the most important of his contributions.**

If we consider the geometry that is based only on the first four axioms of Euclid, then there exist certain important propositions which are equivalent to the axiom of parallelism:

- From a point which does not lie on a straight line a, exactly one straight line can be drawn that is parallel to a (**Proclus**).
- Every line segment is the side of a square (**Legendre**).
- Not all similar triangles are congruent (**Wallis**).
- There exists at least one triangle which has the property that the sum of its angles is 180° (**Legendre**).
- Every triangle has a circumscribed circle (**Legendre**).

But none of the above propositions is more obvious than the one given by Euclid (i.e., the axiom of parallelism).

The failure to prove the axiom of parallelism by using the other four axioms, gradually strengthened the point of view that this axiom is independent of the remaining axioms. It seems that **Gauss** was the first to suggest the correctness of this point of view. In a letter to **Franz Taurinus**, in 1824, Gauss mentions that he is certain that the axiom of parallelism cannot be proved.

In the beginning of the 19th century, in order to prove the independence of the axiom of parallelism, certain mathematicians had the idea to construct a geometry which preserves all the principles of Euclidean geometry, with the exception that the axiom of parallelism be replaced by the following:

"In a plane that contains a given straight line and a given point, it is possible from this point to draw more than one straight line that does not intersect the given straight line". The geometry that is based on these principles is called **Lobachevsky's Geometry** or **Hyperbolic Geometry**.

N. I. Lobachevsky (1792-1856) was a Russian mathematician and made this discovery at the University of Kasan in 1829. About the same time and independently from Lobachevsky, the Hungarian mathematician **J. Bolyai** discovered essentially the same results.

If the axiom of parallelism could be proved by using the other four axioms, then Hyperbolic Geometry would contain a contradiction. Lobachevsky did not encounter any such contradiction during the development of this geometry, but this does not prove of course that no such contradiction exists.

The non-existence of a contradiction was actually proved many years later. In 1868, **E. Beltrami** (1835-1900) constructed a Euclidean model of hyperbolic geometry. This construction proved that if hyperbolic geometry contained a logical contradiction (for example, that the axiom of parallelism and the axiom introduced by Lobachevsky are both valid), then this contradiction could be transformed into a contradiction in Euclidean geometry (or Parabolic Geometry, as it is also called). But since we accept that no contradiction exists in Euclidean geometry, it follows that no contradiction exists in hyperbolic geometry. This proves that these two geometries have exactly the same logical value and gives a final answer to the famous problem of giving a "proof" of the postulate of Parallelism.

In 1882, in the first article published by the journal *Acta Mathematica*, the French mathematician **Henri Poincaré** gave a second model of hyperbolic geometry, which is the following:

We call "point" every point of the Cartesian plane which is located in the interior of the unit circle $x^2 + y^2 = 1$. We call "straight line" every diameter of the unit circle (with its extremities excluded), as well as every circular arc which is located in the interior of the unit circle and is orthogonal to the unit circle (two arcs are orthogonal to one another if they intersect at right angles).

The notion of "between" is defined in the obvious way. The equality between two line segments is defined as follows: Let AB be a line segment, which means that AB is either a portion of a certain diameter of the unit circle or a portion of a circular arc which is orthogonal to the unit circle. Let A^* be the extremity of the diameter or of the circular arc, which lies on that part of the diameter, or of the circular arc respectively, that is in the side of A. We define point B^* in a similar way. We put

$$d(AB) = (AB^*/BB^*)(BA^*/AA^*),$$

where the segments on the right side of the equality are usual Euclidean segments.

The segments AB and CD are called equal, in the sense of Poincaré, if and only if $d(AB) = d(CD)$.

The "angle", according to Poincaré, between two "straight lines" is the Euclidean angle formed by the tangents at the point where these "straight lines" intersect. The equality between angles is defined in the usual way. We will not comment any further on this model. For more details the reader may consult the book: *A Survey of Geometry*, H. Eves (Boston: Allyn and Bacon, 1963).

After Lobachevsky, in 1854, **G. F. B. Riemann** (1826-1866) founded a new geometry by replacing the axiom of parallelism with the following: "Two straight lines lying on the same plane always have a point in common". This axiom is equivalent to the axiom: "From a given point in a plane α not lying on a given straight line a in the plane α, no straight line can be drawn that is parallel to a". But, since the existence of at least one parallel line follows from the remaining axioms, such a geometry would have a priori an obvious contradiction, unless simultaneously the axioms that imply the existence of this parallel were not abandoned or modified appropriately. To this end, depending on which axioms of Euclidean geometry one abandons, one obtains different geometries, the so-called **Riemannian geometries**. Among these geometries the most interesting one is called **Elliptic Geometry**, and we will examine it later.

All Riemannian geometries as well as Lobachevsky's geometry, have no contradictions.

We are as confident about the logical structure of all these new geometries (the so-called non-Euclidean geometries), as we are with the classical Euclidean geometry. The question of whether these new geometries are compatible with the natural phenomena that we observe and encounter in our daily experiences will be treated later.

In the beginning of the 19th century, the study of non-Euclidean geometry was considered by many mathematicians as an occupation that lacks scientific interest. But during the second half of the 19th century, a deeper study of non-Euclidean geometry was made using some contemporary powerful tools (differential geometry and group theory). The interest for non-Euclidean geometry rose rapidly in the beginning of the 20th century, when progress in astronomy revealed the fact that elliptic geometry is perhaps more appropriate than Euclidean geometry for interpreting the observations made in astronomy.

In addition, the following fact was established, which is of great theoretical interest: The entire Projective Geometry is valid in non-Euclidean geometry, as it is valid in Euclidean geometry. The difference between these geometries depends on how one introduces the notion of measure. Depending on how one defines the measure of a line segment and the measure of an angle, one obtains accordingly either Euclidean or elliptic or hyperbolic geometry. We will give an outline of these differences.

1. The non-Euclidean measure of line segments

We recall that in order to introduce the notion of measure in Projective Geometry, it is enough to distinguish (to separate) among all straight lines the straight line at infinity, and among all points the circular points. One might wonder what

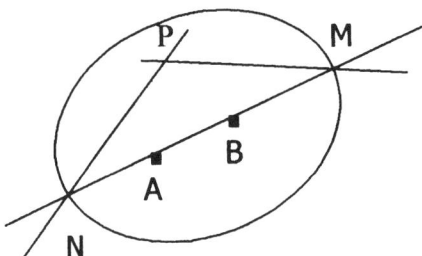

Figure 15.1

would happen to the metric aspects of a geometry if one (without changing anything else) distinguishes other elements (as opposed to the straight line at infinity and the circular points).

At first this question might appear naive. However, after examining the matter carefully, one understands that this question leads to other geometries, the so-called **non-Euclidean geometries**.

We recall that the straight line at infinity can be considered as the limit of an ellipse whose axes are extended indefinitely. In order to simplify the resulting relative equations, let us assume, without changing anything essential in the presentation of the subject, that this ellipse is a circle. Let

$$\epsilon(X^2 + Y^2) + Z^2 = 0$$

be the equation of the circle in homogeneous Cartesian orthogonal coordinates, where ϵ is a negative number. If we let ϵ tend to zero, then the radius of the circle, which is equal to $\frac{1}{\sqrt{-\epsilon}}$, tends to infinity. For $\epsilon = 0$ this equation takes the form $Z^2 = 0$ and represents the straight line at infinity, which we consider as a double straight line (see the remark below), and in this case we say that the circle has degenerated.

If ϵ takes positive values, then no real value of X, Y, Z satisfies the given equation, which represents then an imaginary circle. Thus the circle "goes" from the real to the imaginary domain by "passing" through the straight line at infinity.

We have introduced the notion of the metric in geometry by making use of the above circle only when it degenerates to the straight line at infinity. What will happen to our geometry if, in order to define the notion of measure, we make use of this circle either before it meets the straight line at infinity or after it meets the straight line at infinity when it becomes complex?

Let us examine the first case first, and let us consider a point A in the interior of the circle or, in general, of the ellipse, which was assumed earlier to be a circle in order to simplify the equations (Figure 15.1). We call distance of two points A and B the expression $c \cdot \log(ABMN)$, where M and N are the points of intersection of the straight line AB and the ellipse, and c is a constant.

Now we can find from the last formula the distance AM (under the assumption that we assigned to c a positive value),

$$c \cdot \log(ABMN) = c \cdot \log(\infty) = +\infty.$$

We also find that the distance AN is equal to $c \cdot \log(0) = -\infty$.

So on every straight line passing through the point A, there exist two points which lie infinitely far from the point A. An observer that moves in the interior of the ellipse and calculates distances according to the metric described above, will have the impression that the ellipse lies infinitely far in all directions and consequently the interior of the ellipse extends indefinitely in all directions. Now if we call "parallel" two straight lines that intersect at a point that lies infinitely far, then from any point P we can draw two straight lines parallel to a given straight line AB, namely the straight lines PM and PN (Figure 15.1).

The geometry obtained in this way coincides with hyperbolic geometry. It can be shown that in this geometry the sum of the angles of a triangle is always less than π, while the measure of an angle ab, as it is known, is by definition equal to $\frac{i}{2}\log(abmn)$, where m and n are the (imaginary) tangents of the ellipse, which eminate from the vertex of the angle.

Here we cannot discuss this further, and in particular, we cannot discuss the role played by the points that lie outside the ellipse.

The second case is also interesting: the ellipse that is used as a basis for the definition of measure is imaginary.

In the expression $c \cdot \log(ABMN)$ the points M and N are the points of intersection of the straight line AB with an imaginary curve, which means that these points are also imaginary.

Let us examine how many points a given straight line has at infinity. Let us assume that the point B ranges over the straight line MAN. The distance AB, which is equal to $c \cdot \log(ABMN)$, becomes infinite only if the anharmonic ratio $(ABMN)$ is either infinity or zero. But this happens only if the point B coincides with either point M or point N. Since these two points are imaginary, no real point can coincide with them. Thus a straight line has no real point that lies infinitely far. Consequently, two straight lines never intersect at infinity, which means that no parallel straight lines exist.

The geometry that we obtain in this way coincides with elliptic geometry. In this geometry (where the measure of angles is always given by the expression mentioned above) the sum of the angles of a triangle is always larger than π. We note that since the points M and N are imaginary, the logarithm of the anharmonic ratio $(ABMN)$ is an imaginary number (it can be shown that its real part is equal to zero). Consequently, if we want a real line segment to have real measure, we must give to the constant c an appropriate imaginary value. If we put $c = c_e i$, then we have:

$$\text{Measure of the line segment } AB = c_e \cdot i \cdot \log(ABMN),$$

where c_e is a real constant.

We will return to elliptic geometry later in order to give an intuitive picture of it.

Before we close this section, we make a few clarifying remarks. The reader might object to the above considerations by claiming that they are based on an equation given in Cartesian coordinates, which means that before we proceed to the definition of the various other metrics, it is already assumed that the Euclidean metric is known.

This objection is justified, but we could have avoided this problem by making use of projective coordinates. The only reason for using Cartesian coordinates and for assuming that the dilating ellipse is a circle is to make the text simpler and clearer. As we will see immediately below, these assumptions do not affect the essence of the matter.

Let us replace the homogeneous Cartesian coordinates X, Y, Z with the projective homogeneous coordinates x, y, z. Then the equation

$$\epsilon(x^2 + y^2) + z^2 = 0 \quad (\epsilon < 0)$$

represents a certain conic section. However, we do not know the type of this conic section because, as we mentioned before, Projective Geometry is not capable of making a distinction between the various conics, and furthermore, it cannot distinguish between an arbitrary conic and a circle. When ϵ becomes zero, the above equation takes the form $z^2 = 0$, and now it does not represent (as it did before) the straight line at infinity, because Projective Geometry cannot define this straight line. The equation $z^2 = 0$ represents the straight line that connects the limit points T and T' (Figure 14.4). Thus, when ϵ tends to zero, the conic section tends to coincide with the straight line TT'. But how does one obtain the Euclidean metric in this case? This point requires quite a delicate explanation. The geometry that results, when we take the expression $\lim(c \cdot \log(ABMN))$ as the measure of a line segment AB, where M and N are the points of intersection of the straight line AB with the straight line TT' (considered as a double straight line) and not with the straight line at infinity, is obviously not classical geometry. Actually this is isomorphic to classical geometry, under the assumption that we call **parallel straight lines** the straight lines that intersect on the straight line TT'. This statement means the following: If a person who studies classical geometry reads a theorem of this new geometry, he/she will not understand that this theorem was proved (belongs) in the new geometry; furthermore, two geometers who communicate through the radio or through the telephone (with no television), will perfectly understand each other, without any obstacles and without perceiving that one is referring to classical geometry and the other to the new geometry. The precondition (for the geometer of the new geometry) to call **parallel straight lines** the straight lines, that intersect on the straight line TT' is only a matter of "vocabulary" and does not affect any of the properties of these straight lines. Thus, for example, for both geometers, two parallel straight lines, that intersect two other parallel straight lines define on the latter equal line segments, i.e., segments whose measures are equal and where each geometer considers his/her own notion of length. Hence the point of intersection of two parallel straight lines lies, for both geometers, at an infinite distance from any other point of these straight lines.

Consequently, from a formal point of view, this new geometry does not differ from classical geometry and hence we have the right to call it Euclidean geometry.

Remark. The equation $\epsilon(x^2 + y^2) + z^2 = 0$ used above is the point-equation of the circle. If we want to find its tangential equation with the classical method, we obtain the following equation

$$\epsilon(u^2 + v^2 + \epsilon w^2) = 0 \quad \text{or if } \epsilon \neq 0, \text{ then } u^2 + v^2 + \epsilon w^2 = 0.$$

For $\epsilon = 0$ this equation takes the form $0 = 0$, while when ϵ tends to zero, the tangential coordinates u, v, w tend to satisfy the equation $u^2 + v^2 = 0$ or $(u+iv)(u-iv) = 0$. This last equation is equivalent to the pair of equations $1 \cdot u + i \cdot v + 0 \cdot w = 0$ and $1 \cdot u - i \cdot v + 0 \cdot w = 0$, which represent two bundles of rays whose vertices have coordinates 1, i, 0 and 1, $-i$, 0 respectively; these coordinates are the circular points, which are known from analytic geometry.

Consequently, when ϵ tends to zero, the circle (i.e., the set consisting of the conic-locus and the conic-envelope) tends to degenerate into a double range which lies on the straight line at infinity, and into two bundles of rays whose centers are the circular points.

It is not correct to say that the circle degenerates into the double straight line at infinity and the circular points, because these elements are the carriers of the degenerated circle and not the degenerated circle itself.

2. The absolute of the plane

Let us summarize the results obtained regarding the introduction of measure in 2-dimensional geometry, and let us underline simultaneously the observed phenomenon of duality.

We define the measure with the help of a conic called **the absolute conic** or **the absolute of the plane** for reasons that we cannot explain here.

The absolute conic consists of a set of points (the absolute conic-locus) and of a set of straight lines (the absolute conic-envelope), which have respectively the equations (see the Remark in 15.1)

$$\epsilon(x^2 + y^2) + z^2 = 0 \text{ and } u^2 + v^2 + \epsilon w^2 = 0.$$

A **segment** AB on a straight line s is (in the range whose carrier is the straight line s (see the Remark in 15.1)) the set of points which lie between point A and point B. We call measure of this segment the expression $c \cdot \log(ABMN)$, where M and N are the common points of intersection of the range whose part is AB with the absolute conic-locus. The value of the constant c depends on the measurement unit that we select.

For the measure of an angle ab we always use the expression $\frac{i}{2}\log(abmn)$, where m and n are the tangents of the absolute conic that are drawn from the vertex of ab, in the case that this conic is not degenerated (the corresponding geometries being the hyperbolic and the elliptic), while in the case that this conic is degenerated (the corresponding geometry being Euclidean geometry), m and n are the straight lines drawn from the vertex of ab that pass through the circular points. In both cases, m and n represent the straight lines that belong to the absolute conic-envelope $u^2 + v^2 + \epsilon w^2 = 0$ and to the vertex of ab (see the Remark in 15.1). So we can define the measure of angles in the following way, which follows by applying the law of duality to the definition of measure for line segments.

An **angle** ab with vertex S is (in the bundle of rays whose carrier is S) the set of straight lines that lie between ray a and ray b. We call **measure** of ab the expression $\frac{i}{2}\log(abmn)$, where m and n are the straight lines that belong both to the bundle of rays whose part is ab and to the absolute conic-envelope. The constant $\frac{i}{2}$ in this expression is not selected arbitrarily as in the case of the measure of line segments. It is due to the fact that angles have a natural unit, i.e., π, while nothing like that exists for line segments.

What we said about the measure of angles and line segments applies equally well to the cases that:

- The absolute conic is real and non-degenerated ($\epsilon < 0$, hyperbolic geometry).

- The absolute conic is imaginary and non-degenerated ($\epsilon > 0$, elliptic geometry).
- The absolute conic is degenerated ($\epsilon = 0$, Euclidean geometry).

In the last case we make a transition to the limit and the absolute conic is also a limiting case. Depending on the case considered, the elements M, N and m, n are real or imaginary.

CHAPTER 16

THE GEOMETRY OF EXPERIENCE

As mentioned before, the birth of non-Euclidean geometry is not due to the odd imagination of certain mathematicians. It was created in order to be used as a tool for investigating the foundations of geometry and then it was developed to become an autonomous field, which apparently was able to dethrone Euclidean geometry, at least concerning astronomy. Indeed, it appears that the results of certain astronomical calculations agree better with the observed phenomena, when we make use of elliptic geometry.

We wonder: Is it possible to determine unquestionably the nature of the geometry of the natural world, by observing various phenomena? Are there any observational criteria which permit us to give a final answer to the question whether the geometry of the surrounding world is Euclidean or non-Euclidean? Is there a way to verify through observations whether there exist really parallel straight lines?

For flat (2-dimensional) beings that live on the surface of a sphere, the great circles of the sphere have the fundamental properties of the straight line in the plane, because these lines constitute the shortest distance between two points. If the dimensions of these beings are very small compared to the ones of the sphere (which by the way can be considered to be the globe for reasons of clarity), if these beings live in a limited area of the sphere (for example, in one square kilometer) and if we assume that these beings do not have instruments with which they can measure distances with great accuracy, then they will have the impression that the points of two great circles are equidistant, i.e., that two great circles are parallel.

The geometry of these beings is not different from our plane geometry, because due to the lack of more accurate instruments, they cannot perceive the curvature of the surface on which they live. These beings would be surprised if they acquired the ability to make observations on a vast area of the sphere. Their geometry would become unfit for use and they would be forced to build a new geometry. But, one wonders, would they reject their old geometry in order to use only the new geometry after they made this discovery? Certainly not. The old geometry, with which they were familiar and which is deep-rooted in them, would still serve them in their daily needs because it is simpler and because the results of calculations based on the old geometry differ very little from the results obtained using the new geometry, provided that these calculations are restricted to a small area of the sphere.

Besides, we act in a similar manner when we construct a building or when we make calculations for the construction of a bridge where we consider all vertical lines as parallels. Also when we calculate the area of a lake we consider negligible the curvature of the surface of the water.

The flat inhabitants of the sphere would use their new geometry only for making calculations on given observations regarding very large areas.

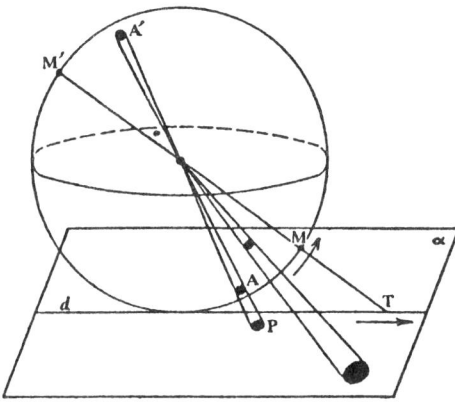

FIGURE 16.1

The same happens with us: for current use, Euclidean geometry satisfies us completely. But regarding astronomy, Euclidean geometry does not enjoy our confidence.

We saw that in elliptic geometry and in hyperbolic geometry the sum of the angles of a triangle is greater than π and smaller than π respectively. When we calculate with greater accuracy the angles of a triangle, we always find a sum which differs very little from π, and this difference is less important than the unavoidable errors which are due to the imperfection of our instruments of measurement. The observer mentioned above who lives on a sphere, would come to an analogous result, and would be wrong to conclude that the sum of the angles of a triangle is equal to π in his/her world.

Before examining closer what we must expect from the "geometry of the natural world", in order to make these theories clearer we will give an intuitive picture of the elliptic geometry of the plane.

The elliptic plane

Let us consider a crystal sphere lying on a plane α. In the center of this sphere we locate a source of light (Figure 16.1). Every opaque spot A lying on the lower hemisphere is projected on the plane α as a shadow P. A spot A' lying on the upper hemisphere does not project any shadow, of course, but if we extend the rays passing through the points of the spot A' until they intersect the plane α, then these rays determine on the plane α a portion which we will call the shadow of the spot A'. In other words, we call **shadow of the spot** A' the shadow of the spot A, which lies opposite the spot A' with respect to the center of the sphere.

Now let us consider an insect that circulates on the sphere: the shadow of the insect ranges over the plane α. But while the dimensions of the insect remain unchanged, its shadow sometimes gets bigger and sometimes gets smaller.

Let us assume that this shadow is an observer who wants to study the variations of its size and the variations of the length of the distances that it covers.

To this end, the observer is supplied with a measure, i.e., with the shadow of a measure with which the insect is supplied. But since the shadow of the insect and the shadow of the measure change in an analogous manner as the insect ranges over the surface of the sphere, the shadow of the insect will have the impression that its dimensions remain unchanged and that the steps that it makes on the plane α are all equal to each other.

When the insect ranges over a semi-meridian, from a point M to its antipodal point M', its shadow ranges the shadow of this semi-meridian, i.e., an entire straight line d, from the point T (the shadow of the point M) up to the point at infinity towards the right in order to come back from the point at infinity from the left to the point T. If the insect covered the semi-meridian by making a thousand equal steps, the shadow will have covered the entire straight line d in a thousand steps, which will appear to the shadow as equal to each other.

The geometry that the shadow perceives is exactly the elliptic geometry. When two arcs of a figure drawn on the sphere are equal with respect to the classical (i.e., Euclidean) geometry, then their shadows are equal if they are measured according to the elliptic geometry. The same thing happens with angles also. Moreover, not only two equal angles on the sphere have equal shadows (with respect to the elliptic measure), but the Euclidean measure of an angle formed by two arcs of the sphere is equal to the elliptic measure of its shadow on the plane.

We can obtain the same result for lengths: If we select a unit of measurement on the sphere such that the Euclidean measure of its radius is equal to $2c_e$, then the number that expresses the Euclidean measure of an arc is equal to the number that expresses the elliptic measure of its shadow.

Let us return for the moment to the angles. Since the sum of the angles of a spherical triangle is greater than π, the sum of the angles of a triangle drawn on the elliptic plane and measured with respect to the elliptic geometry will also be greater than π.

We saw that the shadow of the insect by making a finite number of steps ranges over an entire straight line of the elliptic plane. Furthermore, these steps appear to this shadow as equal to each other. This implies that in the elliptic plane every straight line has finite length. That is, if we express the length of an entire straight line using the elliptic metric, we find a finite number. For a walker whose height remains unchanged with respect to the elliptic metric, the plane does not extend to infinity. After walking for some time and always in the same direction, the walker reaches again the point of his/her departure. So for the walker the plane is finite, and since there is no obstacle restricting the area of his/her movements, this means that for the "elliptic" observer the plane is not only finite but also boundless.

So here is a plane which has infinite area for an observer of Euclidean geometry and finite area for an observer of elliptic geometry. What should we conclude? We conclude that the property of having finite or infinite area is not a property of this plane, but this property depends on the way in which the objects and, in particular, the observers move on this plane.

Remark. We note that if plane elliptic geometry permitted the proof of two propositions which are inconsistent with each other, then it would be possible, by going back to the sphere to prove two propositions which contradict each other with respect to Euclidean geometry. This remark is consistent with what was mentioned on page 104 regarding the equivalence between the logical values of non-Euclidean geometries and the logical value of Euclidean geometry.

The translations of solid bodies

One could have the objection that although elliptic geometry is consistent, it does not correspond to natural reality since it is obvious that the dimensions of the shadow of the insect do not remain constant during its movements.

But is this objection justifiable? Let us consider on the plane α an immobile "measure" (an object) M and two other measures which are placed at first in such a way that they coincide with M. Then they both start moving simultaneously in such a way that the Euclidean length of the first, say M', remains constant and the elliptic length of the second, say M'', also remains constant. At first, both M' and M'' coincide with M and consequently they coincide with each other. But after a translation, one of the measures covers, in general, only a portion of the other. From this we may conclude that one of these measures has changed with respect to the other.

It is of course natural for one to wonder which of the two measures has changed with respect to the immobile measure M. There exists only one way to answer this question: First we take the Euclidean measure M' and put it over M. We see that M' covers M exactly. Then we do the same thing for the elliptic measure M'' and we see that it also covers M exactly. Indeed, if A, B are the extremities of M and P, Q are the points at which the straight line AB intersects the imaginary absolute conic of the elliptic metric, and if we move the measure M'' in such a way that one of its extremities coincides with A and its other extremity B'' lies on the straight line AB, then B'' will necessarily fall on B. Indeed, since the elliptic length of M'' remains unchanged, we have the equality

$$c_e \cdot i \cdot \log(AB''PQ) = c_e \cdot i \cdot \log(ABPQ)$$

which holds only if the points B'' and B coincide. For the measure M' and its Euclidean metric we proceed in a similar manner. Now the following question arises: Does the above discussion imply (contrary to the most elementary common sense) that the measures M' and M'' remained unchanged, even though they don't cover exactly each other? Certainly not! Because our experience gives only the lengths of these measures at the moment that they were in contact with the measure M, and thus gives no information about these measures when they move away from M. Besides, how can we give meaning to the question whether two lengths are equal or not when they have no contact? Actually, the question that has a meaning, is whether the length of the measure changes or remains constant during a translation. But in this case we must clarify what kind of measure we use. If we take as a unit the measure M in both the Euclidean and the elliptic metric, then M' in the Euclidean sense will always remain equal to 1 (because we assumed that this measure moves in such a way that its Euclidean length remains constant), and the same will happen with M'' if it moves according to elliptic metric (i.e., it will be always equal to 1). Consequently, even though the lengths of the two measures remain unchanged, they do not represent the same meaning.

For simplicity, we discuss the elliptic geometry of the plane. In 3-dimensional space the situation is analogous. The length of a line segment AB is also expressed by $c_e i \log(ABMN)$, but in this case M and N are the points at which the straight line AB intersects an imaginary and non-degenerated second-order surface.

Now let us consider a material body, what we usually call a **solid body**, and let us allow it to move. In general, we say that solid bodies remain unchanged during a translation. We saw above that it is difficult to give a certain meaning to this phrase. On the contrary, the meaning of the phrase "**the measures** of solid bodies remain unchanged during a translation" is clear. But then the question arises whether the measures that remain unchanged are Euclidean or elliptic. However, since we talk about natural bodies, only our experience can answer this question.

Let us summarize: Space is neither Euclidean nor elliptic. Certain bodies, which we call solid, move in space. If these solids move in such a way that their sizes expressed through the Euclidean formula remain unchanged, then we say that they move in a Euclidean manner and that the natural world is Euclidean. On the contrary, if their sizes expressed through the elliptic formula remain unchanged, then we say that these bodies move in an elliptic manner and that the natural world is elliptic.

The difficulties of a direct experiment

But is it possible to determine experimentally which of the above two situations actually happens in reality, i.e., whether the natural world is Euclidean or elliptic?

From the above discussion it is obvious that a solid material object does not move on the plane α as the shadow of the insect does. From this it appears that we could conclude with no further discussion, that the natural world is not elliptic. But this reasoning only looks plausible. Let us make it clearer. The arguments we made about the insect and its shadow were based on purely theoretical arguments. The fact that when a figure moves on the sphere in such a way that its size remains unchanged with respect to Euclidean geometry, then the shadow of this figure preserves its size constant with respect to elliptic geometry, is proved through analytic geometry. In order to make our considerations more descriptive and more accessible intuitively, we assumed that the figure that moves on the sphere is an insect. But this is precisely the point where the flaw of the above "plausible" reasoning occurs. An insect is a natural solid body, of course. But nothing assures us that it moves in a Euclidean manner. By identifying the insect with a Euclidean figure of invariable size, we made a logical mistake which made the required result a demonstrative argument.

Besides, an experiment of this kind can teach us nothing. In order to see this, let us present another situation: Simultaneously with the insect (a natural solid body) that moves on the sphere, let us consider another insect (also a natural solid body), which is the twin brother of the first and which moves in such a way that it remains inside the shadow of the first. Our experience teaches us that sometimes the shadow barely covers the twin brother and sometimes it covers more than the twin brother. But it is easily recognized that this phenomenon takes place in both the elliptic as well as in the Euclidean space.

Indeed, if the world is Euclidean, i.e., if the solids (and hence the insects too) move by preserving their Euclidean size constant, then the insect on the sphere will preserve its Euclidean size unchanged and hence its shadow will preserve its elliptic size constant (as it follows from analytic geometry). The twin brother, which is also a solid, moves in this world by preserving its Euclidean size unchanged, and hence the shadow and the twin brother will not move in the same way, something that agrees with the experiment.

On the contrary, if the world is elliptic, i.e., if the solids move by preserving their elliptic size unchanged, then the insect on the sphere will preserve its elliptic dimensions constant, while its Euclidean size will vary. Thus, its shadow will not move in an elliptic way, because if this was not the case, then the insect would have been Euclidean. In addition, the twin brother, which moves in this case by preserving its elliptic size unchanged, as all solids do, will move in a different way from the shadow (because the shadow is not elliptic) and this result agrees again with our experience.

Thus, by observing how the insects move we learn nothing!

So let us forget these "insects" and let us assume that we have drawn on a piece of paper (with some medium) a Euclidean circle, such as the meridian MM' of Figure 16.1 (see page 112), and that we have also drawn a straight line d. Then let us place two arcs with equal Euclidean lengths on the lower quadrant of this meridian (i.e., on the quadrant that contains the point M), in such a way that one lies exactly after the other with only an extremity in common. Furthermore, let us project their shadow on d. Now if we consider a small and thin bar, which covers exactly the shadow of one of these arcs at first, and then slides along d. It is "obvious" from the figure that as this bar slides on d, it covers only a portion of the shadow of one of the two arcs. Can we conclude that this material bar does not move in an elliptic way?

Let us examine possible ways of transfering on the meridian arcs which are equal to each other, i.e., arcs which have equal Euclidean measures. One way is to put the legs of a compass on the extremities of an arc and then using this span of the compass to transfer the arcs in such a way that they are adjacent.

Another way, perhaps more geometric, is to divide the quadrant of the meridian into two equal parts using the construction taught to us by elementary geometry, then to divide each one of these two parts into two equal parts, and so on for a number of times. Of course, as in the first way, this method presupposes the use of a Euclidean compass, i.e., of an instrument that permits us to transfer a size of constant Euclidean length either on a straight line or on every other direction about a given point. This happens in both ways, because it presupposes that we draw Euclidean circles, i.e., curves whose points have the same Euclidean distance from some center.

If, in order to carry out this construction, we use a common compass, i.e., a natural solid, then again we make the logical mistake of taking the required result as a demonstrative argument (see page 115). (Because if our natural world was elliptic, then we would have to carry out the constructions of elementary geometry (i.e., of Euclidean geometry) by replacing the ordinary compass (which would no longer draw Euclidean circles) with an instrument consisting of thin bars which would not form a rigid solid.) Thus, the argument that solids do not translate in an elliptic manner, something that the figure drawn seems to indicate, is once more vacuous.

The above experiment would be more persuasive, if we could find a way to transfer on the meridian arcs which are equal in the Euclidean sense. But even then, how would we know that if the radius of the sphere was very large, the bar that slides on the straight line d would not cover exactly all successive shadows?

The sum of the angles of a triangle

Let us try to answer in a different way the question posed on page 115. We saw that even though the sum of the angles of a triangle is equal to two right angles according to the Euclidean metric, these same angles give a sum greater than two right angles when they are measured by the elliptic formula.

Let us emphasize again that the triangle and the angles are the same in both cases. What changes is the use of different metrics.

The main question is whether the solid moves in such a way that the angle formed by two edges (or two joints or two line segments) of this solid, preserves a constant Euclidean measure or a constant elliptic measure.

To this end, let us cut a triangle ABC from a sheet of laminated wood and then let us cut from it the three angles in order to place them in such a way that one is adjacent to the other with a point S as their common vertex. The sum of the Euclidean measures of these three material angles, which was equal to two right angles when they were still on the triangle, will still remain equal to two right angles provided they are moved in such a way that their Euclidean measures remain constant. This means that the three angles transferred to S will form together two right angles, i.e., the extreme sides will be on the extension of each other. On the other hand, if the elliptic measures remained constant, then the sum of the translated angles would be greater than two right angles, which means that the extreme sides would not be on the extension of each other.

Remark. In order to form a picture of the situation in the elliptic case, it is enough to perform the same construction on a spherical triangle of the sphere and to observe its shadow on the plane α. On the sphere, the sum of the angles transferred to S will not have its external sides on the same great circle, because this sum is greater than two right angles. Consequently, the sum of the shadows of the angles will not have its extreme sides on the same straight line.

In order to make the reader appreciate intuitively the fact that the elliptic measure of an angle is not equal to its Euclidean measure, it is enough to consider two great circles on the sphere that intersect each other at a right angle. Since these great circles form equal supplementary angles, the shadows of these supplementary angles are angles which are equal in the elliptic sense (i.e., they are right angles in the elliptic sense). But it is obvious that they are not right angles in the Euclidean sense.

Instead of cutting a triangle and its angles, we can transfer successively on the three angles a goniometer or a scaled circle of theodolite. This allows us to carry out this experiment even on triangles whose sides have lengths of tens of kilometers. The conclusions will be exactly the same, because the goniometer and the scaled circle of a theodolite are also solids (as are the pieces of the triangle), and hence it is indifferent whether we move the former or the latter parts.

Note. The way in which the instrument mentioned is scaled plays no role. The marked lines on the instrument are used only in order to place on it three angles, one after the other, equal to the ones of the triangle, and in order to check whether the extreme sides are on the extension of each other.

When we carry out the experiment in question, we find in general a sum which differs from two right angles by a quantity which does not exceed the expected errors which are due to the imperfection of the instruments and the imperfection of our senses. So what should we conclude?

Since we examine the natural properties of a body, let us discuss the scale of sizes occurring in nature. Based on the comparison between the observational data regarding certain phenomena, and the results obtained by calculations assuming that the world is either Euclidean or elliptic, the astronomers were led to believe that the universe is elliptic and that the total length of a straight line is 30 billion light years. This means that light, which propagates with the speed of 300,000 kilometers per second, covers an entire straight line in 30 billion years.

Under this assumption, what is the geometry of the universe that is immediately perceived through our senses?

The diameter of our galaxy has length around 300,000 light years. Thus, our galaxy would cover 1/100000 of an entire straight line. Our galaxy contains all the stars which are within eyeshot.

Now let us come back to the figure on page 112 and let us divide the quadrant of the meridian that contains the point M into 50,000 equal parts in the Euclidean sense. Each one of these parts will project on the straight line d, shadow which will cover 1/100000 of this straight line.

Now let us mentally place ourselves on one of these 100,000 parts of the straight line d and let us assume that all our observations are restricted in an area whose dimensions do not exceed 1/100000 of d in any direction. We can assume that this entire figure is drawn on a scale which is billion times larger. The area of our location and the area of the sphere whose shadow is our area will always correspond to the same, extremely small ($\pi/100000$), central angle of the sphere. A triangle T with the dimensions of our galaxy would be contained in this area and it would be seen from the center of the sphere under this extremely small angle. Also the triangle T' on the sphere that would correspond to T would have an extremely small area compared to the total area of the sphere.

Before proceeding further, we remind the reader that we call "angular or spherical excess" the difference between the sum of the angles of a triangle and two right angles. It is known that the ratio of the areas of two spherical triangles is equal to the ratio of their corresponding angular excesses. The angular excess of a spherical triangle with three right angles (which covers 1/8 of the surface of the sphere) is equal to $\pi/2$. Consequently, the angular excess of T' would be equal to an extremely small fraction of $\pi/2$. But since the elliptic measure of the angles of T is equal to the Euclidean measure of the angles of T', the angular excess of T would also be extremely small. The angular excess of an earthly triangle is thus so small that we are surprised by the fact that such angular excesses cannot be perceived through observation.

Before concluding our disucssion about the application of geometry to the natural world, we make the following remark: The geometric figures, on which mathematicians apply their logical arguments, are ideal figures. The material figures, on which we tried to apply the geometric theories, are not real geometric figures: material straight lines have a certain breadth and material points have dimensions. When we want to apply mathematics to natural phenomena, we must replace the material figures with ideal geometric figures trying to cover as accurately as possible the material figures. The thinner the natural lines of material figures and the smaller the dimensions of natural points, the better the approximation of ideal geometric figures. Our theories regarding the geometry of the natural world refer to these ideal figures and not to the natural figures.

The dimensions of the elliptic universe

We saw that it is difficult to use direct observation in order to decide if the natural world is Euclidean or elliptic. However, the consequences implied by the assumption that the natural world is either Euclidean or elliptic, give results which deviate sufficiently for their difference to be perceived via the observation of natural phenomena. Besides, the same thing happens with many theories in physics: if we accept the interpretation of natural phenomena, which accepts the existence of atoms, ions and electrons, it does not mean that we saw these particles through direct observation; it only means that the consequences implied by admitting their existence agree satisfactorily with the observed phenomena.

As mentioned above, the astronomers had in the past and have now even more reasons to accept the assumption that the natural world is elliptic. They even managed to calculate the total length of a straight line of the elliptic universe and as stated earlier, this length is equal to 30 billion light years.

Let us try to get an idea of such a length. If we represent the diameter of our galaxy (300,000 light years) by a line segment of length equal to one centimeter, then in the same scale the total length of the above straight line would be one kilometer. So this is the (linear) proportion between the universe within eyeshot and the entire universe. Now if we use one of our instruments, for example, a telescope, then we can observe nebulas at a distance of 500 million light years. In our figure, this distance would be represented by a line segment of length approximately equal to 15 meters. (The distance of the sun to the nearest star to us, called Proxima Centaurus, would be represented approximately by one thousandth of a millimeter.)

The curvature of space

Let us come back to the sphere that was considered on page 112. We saw that the angles drawn on the sphere and their shadows on the plane α have the same measure, provided that we apply the Euclidean metric to the first and the elliptic metric to the second. If the unit of length selected for the sphere is such that the length of its radius is equal to $2c_e$, then the Euclidean length of an arc of a great circle on the sphere is equal to the elliptic measure of its shadow, which is given by the expression

$$c_e \cdot i \cdot \log(ABMN).$$

This analogy led mathematicians to accept a terminology which however is not very successful. In geometry, we call curvature of a spherical surface of radius R the expression $\frac{1}{R^2}$. Mathematicians accept that the elliptic plane has **curvature** equal to $\frac{1}{4c_e^2}$. But of course it is wrong to think that as soon as we accept on a plane an elliptic metric, then the plane becomes curved.

Since, not only in the plane α but also in 3-dimensional space the elliptic metric is given by the expression $c_e i \log(ABMN)$, mathematicians also accept that the elliptic space is curved and its curvature is equal to $\frac{1}{4c_e^2}$.

In reality the elliptic metric of the universe is not as simple as presented so far. According to the theory of relativity, the metric is indeed given by the expression $c_e i \log(ABMN)$, but the constant c_e is not the same on the entire space. The value of c_e depends on the matter that exists in the neighborhood of every point. In the neighborhood of a star, this constant takes a smaller value, which depends not only on the matter of this star but also on the speed of this star. Since we call curvature the expression $\frac{1}{4c_e^2}$, this last property can be formulated by saying that the curvature increases in the neighborhood of matter.

If for the moment we assume that the total matter which is located in the part of the universe that is accessible through observation is uniformly distributed in the corresponding space, and that the mean density of matter is constant, and furthermore, we assume that matter is uniformly distributed in the part of the universe which is not accessible by the telescope, then the constant c_e will have everywhere the same value. This value can be considered as a mean value for the entire universe. The theory of relativity has the ability to calculate this mean value. The numbers given earlier refer to an elliptic space which corresponds to this mean value.

The fact that the calculations based on the above mean value may actually have a natural meaning which becomes clear with the following example: Let us consider a material sphere whose surface is rippled, as for example, the peel of an orange. Even though the curvature of this surface is different at every point (here we do not insist on the definition of the notion of curvature of a nonspherical surface at one of its points), we can still talk about a mean radius R of this surface, which is approximately a spherical surface, and consequently we can also talk about the mean curvature of this rippled surface being equal to $\frac{1}{R^2}$ (we can also say that the mean length of a great circle equals $2\pi R$).

The expansion of the universe

In order to give an explanation to the general phenomenon of spiral nebulas, which are moving away from us in all directions with speeds which become larger as they are moving away from us, the Belgian clergyman and astrophysicist **G. Lemaitre** (1894-1966) invented the theory of the Expansion of the Universe. According to this theory, the dimensions of the elliptic universe get larger as time elapses. At the beginning, the total matter of the universe was concentrated at a space whose dimensions tend asymptotically to zero (as we run time backwards). Then, during the Big Bang, matter started a movement of expansion by creating space and by drifting space simultaneously in this movement.

Even though the attraction of matter (Newton's law) acts as a hindrance to this expansion, the theory of relativity implies that this expansion is supported by a force which acts in a direction opposite to the one of the attraction of matter and which increases as the distance increases, but is negligible with respect to the dimensions of our planetary system.

Here the following legitimate question arises: Is it possible for the forces of attraction to prevail and force the universe to fold back before the opposite forces that favor the expansion become sufficiently large? The answer to this question is negative. Calculations show that this danger took place a long time ago. The critical moment at which these opposite forces balanced, took place at a time during which the relevant straight line was more than ten times shorter than it is today. Since then, the forces of expansion prevail over the forces of attraction and the expansion continues at an increasing rate. However, it must be stressed that during the critical time that the forces were at equilibrium, this equilibrium was very unstable and local perturbations could have destroyed it in favor of the forces of attraction in certain areas of the universe, so that the general expansion is followed by local contractions. Such phenomena could be the basis of the formation of nebulas. Regarding the general expansion, its speed today is such that the total length of a straight line increases by more than 5 million kilometers per second. This number is quite reasonable, since it represents only a lengthening per kilometer equal to half a thousandth of the centimeter per year.

The picture of a finite and boundless universe is indeed very fascinating. It is true that it is difficult to imagine such a universe. But is it easier to imagine an infinite universe?

Is the assumption of an infinite universe acceptable to us due to the fact that as long as our imagination makes us advance in space, we postpone the solution of this unsolved problem posed by our own imagination?

The same thing does not happen with the finite and boundless universe. Our thought cannot extend the journey in space ad infinitum. We must sometime imagine the return to the point of departure, when the direction of the travel has not changed.

It seems that the point of view tends to prevail that the universe is elliptic. Will we accept at some point this point of view? But let us not worry about such matters and let us answer this question using the words of a poet: "Maybe yes, maybe no".

The author apologizes to the reader for discussing matters which could be considered to be beyond the scope of the present book, even though such matters are favorite subjects to many people.

CHAPTER 17

ARISTARCHUS, ARCHIMEDES, APOLLONIUS, AND ERATOSTHENES

The mathematical school that Euclid founded in Alexandria produced, during the 3rd century B.C. mathematicians of exceptionally high caliber, including

- Aristarchus of Samos, 300-250 B.C.
- Archimedes of Syracuse, 287-212 B.C.
- Apollonius of Perga, 260-190 B.C.
- Eratosthenes of Cyrene, 275-195 B.C.

Aristarchus (300-250 B.C.). He was the first great astronomer of Alexandria, who also occupied himself with geometry, music and other branches of science. He gave a very interesting application of mathematics to astronomy. He knew that moonlight comes from the reflection of sunlight. Let $H\Gamma\Sigma$ be the triangle whose vertices are the sun (H), the earth (Γ), and the moon (Σ) (Figure 17.1). Aristarchus observed that when the moon is at its first quarter, the angle $H\hat{\Sigma}\Gamma$ is a right angle. For this reason we see exactly half of the surface of the moon facing the earth. When the moon is at its first quarter, one can see in the sky both the moon and the sun simultaneously. Thus, Aristarchus managed to measure the angle $H\hat{\Gamma}\Sigma$ and found that it equals 29/30 of the right angle. (A more exact value is 0.9981 of the right angle.) Then he constructed (using a ruler and a compass only) a right-angled triangle with acute angle equal to 29/30 of the right angle and he found the ratio of the smallest side over the hypotenuse approximately equal to 1/19. From this he concluded that the distance of the earth to the sun is approximately 19 times larger than the distance of the earth to the moon. If he could have made an exact measurement of the angle $H\hat{\Gamma}\Sigma$, he would have found that $H\Gamma$ is approximately 346 times larger than $\Gamma\Sigma$.

The above calculations would have been easier, of course, if one made use of trigonometry, which was actually discovered a year later. If $H\hat{\Gamma}\Sigma = 87°$, then $\Gamma\hat{H}\Sigma = 3° = \frac{\pi}{60}$ radians. Thus $\frac{\Gamma\Sigma}{\Gamma H} = \sin\frac{\pi}{60} \approx \frac{\pi}{60} \approx \frac{1}{19}$, using the fact that the sine

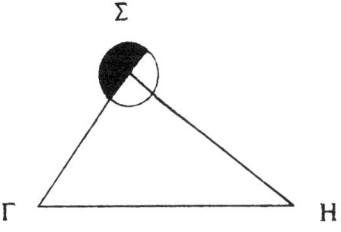

FIGURE 17.1

of a small angle is approximately equal to this angle, when the latter is expressed in radians.

Aristarchus, who knew these relative distances, calculated the relative sizes of the sun and the moon by measuring the sizes of the discs of the sun and of the moon, as they appear from the earth. He concluded that the sun is 7000 times larger than the moon. This estimate is far from the correct value, which is 64,000,000. He also estimated that the ratio of the diameter of the sun over the diameter of the earth is a number between 19/3 and 43/6, while the correct value of this ratio is approximately 109.

Aristarchus is also famous for being the first who suggested the heliocentric hypothesis, i.e., that the earth and the planets rotate on circular orbits about the sun which remains immobile, that the stars remain immobile and their apparent movement is due to the rotation of the earth about its axis, and that the moon rotates about the earth.

Copernicus withheld the fact that he knew the work of Aristarchus.

Archimedes (287-212 B.C.). During the Hellenistic era, even though Alexandria remained the center of mathematical activity, the greatest leading mathematical figure of that period as well of the entire antiquity, was not born in that city. Archimedes studied for a while in Alexandria and had as his teachers, students of Euclid. He also kept contact with the mathematicians in Alexandria, but he lived and died at Syracuse. The following phrase of the famous French writer **Voltaire** (1694-1778) is characteristic: "Archimedes' imagination was greater than Homer's".

There exist many stories that refer to Archimedes' life. One of them reports that, during the time that he was having a bath, he suddenly discovered a simple way to determine the proportion of gold over silver in an alloy of gold and silver. Full of joy for his discovery, he jumped out of the bathtub and rushed (naked!) into the streets of Syracuse crying "$εὕρηκα$[1]". The fact that Archimedes made this discovery while he was having a bath was not accidental. Indeed, let us assume that an object made of gold and silver weighs m grams, and we want to calculate the number x of grams of gold that the goldsmith put in this object. If g is the density of gold and s is the density of silver, then the volume of the object is $\frac{x}{g} + \frac{m-x}{s}$. Archimedes realized that by sinking the object in a rectangular bathtub and by observing the rise of the level of the water, he could determine the volume u of the object. Then by solving for x the equation

$$u = \frac{x}{g} + \frac{m-x}{s}$$

he could find the mass of gold that the object contains. Thus, using mathematics Archimedes was able to inform his friend King Hiero of Syracuse that the goldsmith had deceived him, by using only part of the gold given to him for the construction of the king's crown and by replacing the remaining gold with silver.

Another incident, which is not so well known, is the following: Hiero had built a ship, which was so big that it was impossible to launch it into the sea with the methods known then. Archimedes solved this problem by constructing a machine consisting of a cogwheel and a propulsive propeller. Hiero, who was present during the entire process, cried out a phrase which is translated into the contemporary language as follows: "I will always believe nothing unless Archimedes

---

[1]**Translator's note:** The meaning of this word in English is: *I have found it.*

accepts it". While Archimedes replied with the phrase[2]: "Δός μοι πᾶ στῶ καὶ τὰν γᾶν κινάσω".

During the siege of Syracuse from the Romans, Archimedes constructed various machines for the defense of the city.

The story that Archimedes burned the Roman ships using a system of concave mirrors was first reported centuries after his death and is considered to be inaccurate. However, many people believe that it might be true. According to tradition, every mirror was hexagonal and had around it 148 polygons, each consisting of 24 sides. In 1777 in Paris, someone named **Buffon** managed to burn a piece of wood at a distance of approximately 46 meters as well as to melt a piece of lead at a distance of approximately 43 meters, using a single mirror of the type mentioned above. This experiment was carried out during the April of 1777, when the sun in Paris is not very bright, which means that during summer in Sicily and with the use of many mirrors the operation became perhaps possible, provided that the ships were anchored near the city. It has been reported ([26]) that successful experiments similar to the ones of Buffon were carried out by the Greek mechanic **I. Sakkas**, and in particular the fourth experiment was carried out in Salamina in the 6th of November of 1973. Finally, it is interesting to mention that according to certain writers a similar system of mirrors was used for the defense of Constantinople in 512 A.D. Independently of whether the above story regarding the mirrors is accurate or not, there is no doubt that Archimedes invented the catapults and the other machines which delayed the landing operation of the Romans for approximately three years until the city was conquered in 212 B.C.

The Roman general Marcellus ordered his soldiers to bring to him Archimedes without hurting him. But the order was not executed. It is said that some soldier entered the area where the wise man was studying a geometric figure drawn on the sand and Archimedes told him not to erase it. But the soldier felt insulted by Archimedes' command and killed him without knowing the identity of his victim. The Romans built a magnificent grave for Archimedes, on which, following Archimedes' will, they inscribed the figure of a sphere inscribed in a cylinder, in remembrance of the proof that Archimedes had given that the volume of a sphere is equal to 2/3 of the volume of its circumscribed cylinder and that the area of the surface of a sphere is four times the area of one of its great circles.

The most well-known historical source for the above is Marcellus' biography written by Plutarch. Who would really know today anything about Marcellus if Archimedes did not exist?

In his book *Tusculam Disputations* (v. 23), **Cicero** describes vividly his successful efforts to rediscover Archimedes' grave in 75 B.C.

Archimedes' work is very broad and extremely important: circle–parabola–helices–sphere–cylinder–arithmetic–mechanics–statics– hydrostatics.

One of the most important of his writings, *The Method*, was discovered only in 1906 by **J. L. Heiberg** in Constantinople (see the book *A History of Math.*, 2nd edition, Carl. B. Boyer-Uta and C. Merzbach, J. Wiley and sons, 1989).

It is difficult to analyze in the present book, the entire work, and the discoveries of this unique and ingenious man, mainly because he occupied himself with almost all the subjects of mathematics in his time. Also because his works are contained

---

[2]**Translator's note:** The meaning of this phrase in English is: *Give me a place to stand on, and I will move the earth.*

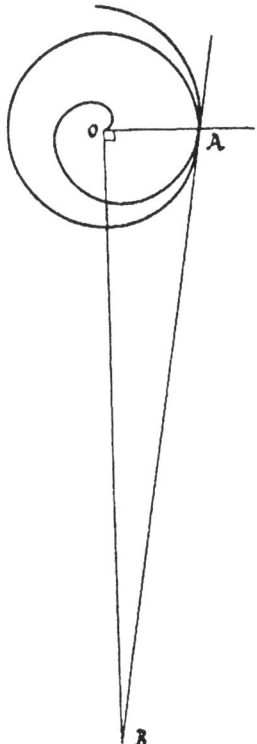

FIGURE 17.2

in a series of self-contained monographs. Hence we will restrict ourselves to only some of his works.

By considering a regular 96-gon inscribed in a circle, Archimedes proved that $\pi < 3\frac{1}{7}$. In addition, by considering a regular 96-gon circumscribed about a circle, he proved that $3\frac{10}{71} < \pi$. He knew that it is possible to approximate π up to any decimal digit by increasing the number of sides of the regular polygon ad infinitum.

He proved that the area of a circle of radius r equals πr^2 and that the volume of a sphere of radius r equals $\frac{4}{3}\pi r^3$.

In his book *On helices*, Archimedes studies the curve that today we call "Archimedean helix". If a half line with origin the point O rotates uniformly about O (counterclockwise) and simultaneously if a point P moves uniformly on this half line starting from O, then P traces a helix whose equation in polar coordinates is

$$r = \alpha \cdot \theta, \quad \theta \geq 0.$$

Among the most important properties of this helix is the following: Consider the curve from point O to point A in Figure 17.2 that results after the first full rotation of the half line (i.e., the rotation that corresponds to $0 \leq \theta \leq 2\pi$). It can be shown that the area contained between this curve and the line segment OA is equal to $1/3$ of the area of a circle of radius OA.

(We remind the reader that if $r = f(\theta)$ is the equation of a curve in polar coordinates and if f is differentiable, then the slope of the tangent at an arbitrary

point of the curve is equal to
$$\frac{f'(\theta)\sin\theta + f(\theta)\cos\theta}{f'(\theta)\cos\theta - f(\theta)\sin\theta}.$$

This result implies that Archimedes managed to square the circle without, of course, restricting himself to the use of only a ruler and a compass.

In his book *The quadrature of the parabola*, Archimedes proves (in three different ways) that the area of a part of a parabola (namely of the part contained between an arc of the parabola and a straight line that intersects the parabola and is not parallel to its axis) is equal to 4/3 of the area of the largest triangle inscribed in this part.

Archimedes calculated the volume of an ellipsoid of revolution, the volume of a part of a sphere, the center of gravity of a hemisphere and the volume of the common part of two equal circular cylinders which intersect at right angles.

Two of Archimedes' treatises have an arithmetical content. The first refers to the principles of arithmetic and is addressed to Zeuxippus, but unfortunately it has not been preserved. In the second treatise, which is entitled *Psammites* (from the word "ψάμμος" which means sand) and had a popularising character, Archimedes refutes an objection that was made against his first treatise. In particular, the object of his first treatise was the presentation of an appropriate number system, in which it is possible to write a number of any size. It appears that certain philosophers from Syracuse had expressed their doubts about whether such a system could ever exist.

"Many", Archimedes wrote, "talk about the sand that exists on the coasts of Sicily, as if it is something which is impossible to be calculated, although it is possible". With his method, he alleged that it is possible to find a number larger than the number of grains of sand with which one could fill up the universe (that is a sphere with center the earth and diameter equal to the distance of the earth from the sun) and that this number was 10^{63}.

But let's see in more detail how Archimedes proved that the area of a circle of radius α is equal to $\pi\alpha^2$. He started with the following assumptions and theorems:

1. Circles and parts of circles have areas.
2. The area of a set consisting of pairwise disjoint triangles and circular parts equals the sum of the areas of the triangles and of the circular parts which comprise it. Therefore if we partition a circle into triangles and circular parts, then the area of the circle equals the sum of the areas of the triangles and of the circular parts which comprise it. Furthermore, the area of a circle is larger than the sum of the areas of every proper subset of the considered set of triangles and parts of circle which comprise it.
3. Given a circle, there exists a line segment whose length is larger than the length of the perimeter of any polygon which is inscribed in the circle and less than the length of the perimeter of any polygon which is circumscribed about the circle. The length of this line segment is the length of the "circumference" of the circle.

Note. William Jones was the first who gave, in 1706, the name π to the length of the circumference of a circle of diameter equal to 1.

4. If e and f are two areas, then there exists a positive integer m such that $me > f$.

Note. Assumption 4 above is found at the beginning of Archimedes' book *On the sphere and the cylinder* and for this reason it is called Archimedes' axiom. But the same assumption is also found at the beginning of Book V of Euclid's *Elements*, as well as in Aristotle's *Physics* (266b).

  5. A regular 2^n-gon inscribed in a circle occupies more than $1 - (1/2^{n-1})$ of the area of the circle. A regular 2^n-gon circumscribed about a circle has area less than $1 + (1/2^{n-2})$ of the area of the circle.
  6. The area of a circle is proportional to the square of its diameter. (*Elements*, XII, 2).

Using the above assumptions and theorems, Archimedes derived the formula that gives the area of the circle, by proving first of all that each of the following assumptions (A) and (B) leads to a contradiction.

(A) The area of the circle is larger than the area of a right-angled triangle T whose perpendicular sides are equal to the radius and to the circumference of the circle respectively.

From (4) and (5) it follows that we can find a positive integer n such that:

$$\text{area of the circle } - \text{ area of the inscribed regular } 2^n\text{-gon}$$
$$< \text{ area of the circle } - \text{ area of } T.$$

Hence
$$\text{area of } T \ < \ \text{area of the } 2^n\text{-gon}.$$

Let AB be one of the sides of the inscribed regular 2^n-gon and let ON be the normal to AB drawn through the center O of the circle (where N is the midpoint of AB). Then ON is smaller than the radius of the circle. From (3) we find

$$\begin{aligned}\text{area of the } 2^n\text{-gon} &= 2^n \cdot \left(\tfrac{1}{2} \cdot AB \cdot ON\right) \\ &= \tfrac{1}{2} \cdot (2^n \cdot AB) \cdot ON \\ &< \tfrac{1}{2} \cdot \text{circumference} \times \text{radius},\end{aligned}$$

a contradiction. Hence (A) is not true.

(B) The area of the circle is smaller than the area of T.

From (4) and (5) it follows that there exists a positive integer n such that the area of T is larger than the area of a regular 2^n-gon that is circumscribed about the circle. But if AB is one of the sides of this 2^n-gon, then from (3) we have

$$\begin{aligned}\text{area of the } 2^n\text{-gon} &= 2^n \cdot \left(\tfrac{1}{2} \cdot AB \times \text{radius of the circle}\right) \\ &> \tfrac{1}{2} \cdot \text{circumference} \times \text{radius} \\ &= \text{area of } T,\end{aligned}$$

a contradiction. Hence (B) is not true.

Since (A) and (B) are not true, Aristotle's "Law of the Excluded Middle" implies that the following proposition is true:

(C) The area of the circle is equal to the area of the right-angled triangle whose perpendicular sides are equal to the circumference and to the radius of the circle respectively. That is

$$\text{area of the circle} = \frac{1}{2} \cdot \text{circumference} \times \text{radius}.$$

Obviously the expression on the right side of this equality is equal to πa^2 (see (6) above).

The proof given by Archimedes for the formula of the area of a circle constitutes the culmination of research regarding the circle which lasted approximately two centuries starting with Antiphon (425 B.C.).

Even though Archimedes was working in Syracuse, he was communicating his research results to the mathematicians of the University of Alexandria. In order to attract their attention, he posed to them the following problem (The Cattle Problem), whose "translation" into the modern algebraic notation is as follows:

God Sun had a herd consisting of w white bulls, g grey bulls, b brown bulls, s bulls with spots and w', g', b', s' cows of respectively similar colors as well. What is the total number of bulls and cows if

$$s = w - \frac{5}{6}g = g - \frac{9}{20}b = b - \frac{13}{42}w$$
$$w' = \frac{7}{12}(g + g') \ , \ g' = \frac{9}{20}(b + b')$$
$$b' = \frac{11}{30}(s + s') \ , \ s' = \frac{13}{42}(w + w')$$

and if the number $w+g$ is equal to the square of an integer, while the number $b+s$ is a triangular number?

Archimedes used the notation of the Egyptians and wrote the resulting fractions in their unitary form (see page 20). If we exclude the last two conditions, which refer to the numbers $w + g$ and $b + s$, we have here a system of seven equations with eight unknowns which cannot be solved with algebraic methods only. Since we seek solutions for the system which are positive integers, the problem is called "diophantine", because the Greek mathematician Diophantus of Alexandria (3rd century A.D.) was the first who introduced and occupied himself with such problems, but lived approximately 500 years after Archimedes.

We give an outline of the solution of the cattle problem and leave the details to the reader. The equations

$$s = w - \frac{5}{6}g = g - \frac{9}{20}b = b - \frac{13}{42}w$$

imply that for some positive integer m, we have

$$w = 2226m, \ g = 1602m, \ b = 1580m, \ s = 891m.$$

The remaining four equations imply that there exists a positive integer k such that

$$m = 4657k, \ w' = 7206360k, \ g' = 4893246k, \ b' = 3515820k, \ s' = 5439213k.$$

If $w + g$ is the square of an integer, then the number $4 \cdot 957 \cdot 4657 \cdot k$ is also the square of an integer. Since 4657 is a prime number and 957 is the product of two different prime numbers, it can be easily proved that $4 \cdot 957 \cdot 4657 \cdot k$ is the square of an integer only if k has the form $957 \cdot 4657 \cdot t^2$, where t is some integer.

If $b+s$ is a triangular number, then the number $2471m$ has the form $\frac{1}{2}n(n+1)$. Consequently, the number $8 \cdot 2471 \cdot 4657 \cdot 957 \cdot 4657 \cdot t^2$ has the form $4n^2 + 4n$. In other words, in order to find an integral solution of Archimedes' cattle problem, we must find an integral solution of the equation

$$(2n + 1)^2 - 8 \cdot 2471 \cdot 957 \cdot 4657^2 \cdot t^2 = 1.$$

The mathematicians of Alexandria did not manage to solve this problem, which actually remained unsolved until 1965, when **H. C. Williams, R. A. German** and **C. R. Zarnke** produced with the aid of a computer the solution, which is an integer with 206,545 digits. This solution was published in 1980-1981. The reader may find this number with all its 206,545 digits printed in **Harry L. Nelson**'s article entitled *A solution to Archimedes' Cattle Problem, Journal of Recreational Mathematics* 13, pp. 164-176.

Note. According to some old historians, the statement of the cattle problem is restricted only to the system of seven equations given above, i.e., it does not include the conditions referring to the sums $w + g$ and $b + s$. It is reported that the smallest positive integers which satisfy this system are

$$w = 10366482, \quad g = 7460514, \quad b = 7358060, \quad s = 4149387$$
$$w' = 7206360, \quad g' = 4893246, \quad b' = 3515820, \quad s' = 5439213$$

and that the solution given by Archimedes was the one formed by the numbers, each of which is the corresponding number above multiplied by 80.

The few remarks mentioned above regarding Archimedes' work are perhaps enough to appreciate the extent and the originality of his work. A comprehensive presentation of his work goes beyond the scope of the present book. If the reader looks through Archimedes' entire work, he will appreciate **Leibniz**'s statement:

"The people who know Archimedes' and Apollonius' work did not admire as much the discoveries of the greatest contemporary scientists."

Apollonius (260-190 B.C.). He was one of the great mathematicians of the 3rd century B.C. who was born in Pamphylia's Perga (Asia Minor). He is famous mainly because he wrote a systematic study, the *Conics*, which not only contains what was known until then about conic sections, but also greatly extends further the knowledge regarding these curves. His study was immediately accepted and became the classical textbook on the subject of conic sections, replacing the books of Menaechmus, Aristaeus and Euclid, which were used until then as textbooks on this subject. Apollonius studied mathematics in Alexandria and had as teachers, students of Euclid. He stayed a few years in Pampylia's Pergamus. This city had a university with a remarkable library, which was second in importance only to the University of Alexandria. The administration of this university belonged to Alexander the Great's general, Lysimachus, and to his successors later. Finally, Apollonius went back to Alexandria and remained there for the rest of his life.

We note that in Ancient Greece, there existed many distinguished scientists with the name Apollonius. The biographies of 129 of them are contained in Pauly-Wissowa, *Real-Enzyclopädie der Klassischen Altertumswissenschaft*. Our Apollonius is the one from Perga, who was called the "great geometer" due to his amazing mathematical skills. He was also famous as an astronomer.

His work *Conics* consists of 8 books and contains 487 propositions. Four of these books were preserved in their Greek original and three of them were preserved in their Arabic translation. The eighth book has not been preserved. Today, we no longer study conic sections from Apollonius' work, because it is easier to study conic sections using analytic geometry.

The contemporary definition of a conic section is: the set of points P in the plane which have the property that the ratio of the distance to a fixed point (called the "focus") over the distance to a fixed straight line (called the "directrix") remains constant. This constant ratio is called "eccentricity" and is used for the classification of conic sections. Ellipses have eccentricity less than 1 (in particular, a circle has eccentricity 0). Parabolas have eccentricity equal to 1 and hyperbolas have eccentricity larger than 1.

Another book of Apollonius, which has not been preserved, is entitled *Contacts*. As **Pappus** informs us, this book contained the problem that we call today "Apollonius' Problem", namely: "Given three objects, each of which can be either a point or a straight line or a circle, draw a circle which is tangent to everyone of these three objects (where a circle and a point are tangent when the point belongs to the circle)." This problem presents ten cases ranging from the easiest two cases (where the three objects are three points or three straight lines) to the most difficult case, where one seeks to draw a circle which is tangent to three given circles. The two easiest cases are contained in Euclid's *Elements* and concern the operation of inscribing a circle in a triangle and the operation of circumscribing a circle about a triangle. Six other cases are examined in Book I of *Contacts*, while the case of two straight lines and a circle and the case of three circles are both examined in Book II of *Contacts*. A comprehensive bibliography of Apollonius can be found in [26].

Eratosthenes (275-195 B.C.). He is one of the greatest Greek wise men of Antiquity, as well as the first geographer. He was born in Cyrene (North Africa), studied in Athens and lived in Alexandria where he spent most of his life. He was the administrator of Alexandria's Library. He occupied himself with philosophy, poetry, history, literature, astronomy and mathematics. We have already mentioned earlier the "Sieve of Eratosthenes", which involves the construction of a table of prime numbers.

Eratosthenes invented the Julian calendar, according to which a day is added every four years. He also calculated the size of the Earth. The reason that contemporaries called him "beta" was that, according to some people, he was considered as the second wise man of antiquity, or according to others, that he always taught at room number "two" of the University of Alexandria. It is said that near the end of his life he became blind and committed suicide.

One of Eratosthenes greatest achievements was the measurement of the circumference of the earth. He assumed correctly that, since the sun is very far from the earth, the rays that reach the earth can be considered to be parallel. Eratosthenes knew that the Syene River (today known as the Aswan) is located near the Tropic of Cancer, which means that the goniometer of a sundial makes no shadow at noon on the middle day of Summer (June 21) (this can be verified if one is at the bottom of a well). Eratosthenes observed that the sun in Alexandria was at an angular distance of $360°/50$ from the point that is vertically above the observer at the noon of the 21st of June. He alleged that this angle is equal to the angle whose vertex is the center of the earth and whose sides pass through Alexandria and Syene (which is located south of Alexandria) respectively. He knew from *Elements* (Theorem VI, 33) that the length of a circular arc is proportional to the size of the corresponding central angle. Consequently, all he had to do was to measure the distance between Alexandria and the Syene River, which he found to be equal to 5000 stadia (a stadium is the length of an Olympic field), i.e., 900 kilometers. From this, Eratosthenes concluded that the ratio of the circumference of the earth divided by 5000 equals the ratio divided by $360°$ over $360°/50$, i.e., that the circumference of the earth equals $5000 \times 50 = 250,000$ stadia (Figure 17.3), or 45000 kilometers. This calculation of the length of the circumference of the earth is impressive since the exact length is approximately 40000 kilometers.

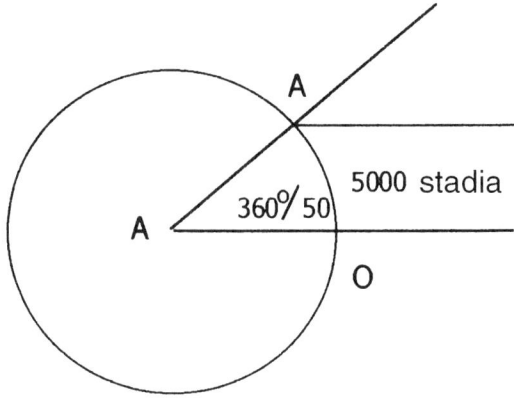

FIGURE 17.3

CHAPTER 18

THE PERIOD FROM 200 B.C. TO 500 A.D. IN ALEXANDRIA

The most important mathematicians of this period were the following:

- **Hipparchus**: He was born around 180 B.C. in Nicaea in Asia Minor, near Constantinople, where the 1st Ecumenical Synod of the Orthodox Church took place in 325 A.D.
- **Heron of Alexandria**: He was born around 60 A.D.
- **Menelaus of Alexandria**: He was born around 100 A.D.
- **Ptolemy** (Claudius) (108-168 A.D.): He was born in Ptolemais of Hermion on the banks of the Nile River. He lived and worked in Alexandria. He must not be confused with anyone of the Ptolemies, kings of Egypt.
- **Diophantus of Alexandria**: He was born around 250 A.D.
- **Pappus of Alexandria**: He was born around 320 A.D.

Less important as mathematicians but important from the point of view of history were the following:

- **Nicomachus** of Gerasa (Jordan): 50-120 A.D.
- **Hypatia**: She died in 415 A.D.
- **Proclus of Lycia** (the Diadochus): 410-485 A.D.
- **Boethius** (or **Boetius**): 475-524 A.D.

As mentioned before, Aristarchus of Samos was one of the first Greek astronomers who applied mathematics to astronomy and formulated the heliocentric theory of the solar system. For this reason Aristarchus is often called "Copernicus of Antiquity".

The next great Greek astronomer is Hipparchus. Even though he was the first who observed the spring equinox in 146 B.C. in Alexandria, he made his most important astronomical observations at the famous observatory of Rhodes. He was a very careful and accurate observer. The reader may find a detailed reference to Hipparchus' work in [26]. Among his achievements, we mention the determination of the mean lunar month, the duration of which as given by Hipparchus differs only by one second from the duration that we accept today, the exact calculation of "the inclination of the orbit to the ecliptic" (i.e., the angle formed by the orbit of the earth and the equator) and the discovery and the calculation of the annual precession of the tropical and spring equinoctial points. He calculated the parallax of the moon (57'), he determined the perigee of the moon and he classified 850 fixed stars. He calculated the duration of a year with an approximation of 6 minutes, and he made use of the geographic length and breadth for the determination of a position on the surface of the earth.

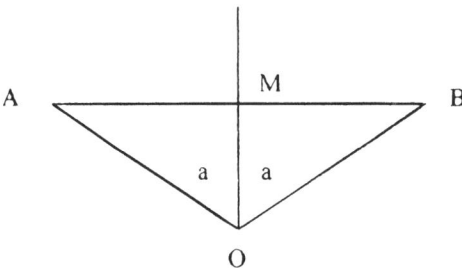

Figure 18.1

Regarding the motion of the moon, Hipparchus assumed that it moves uniformly on the circumference of a circle and that the earth is located near the center of this circle but not exactly on it. This is equivalent to saying that the orbit of the moon is a first-order epicycloid. He alleged also that something similar happens with the motion of the sun. Hipparchus believed that the use of higher order epicycloids was necessary for the description of the motion of the planets. He invented the instrument Astrolabe, which he used for his astronomical observations.

After the time of Hipparchus, no important progress was made in astronomy until the time of Copernicus, apart from the fact that Hipparchus' principles were analyzed and extended further by Ptolemy Claudius.

Theon of Alexandria (2nd half of the 4th century A.D.), the father of Hypatia, who was also a mathematician, an astronomer and a reviewer of books, reports a writing of **Hipparchus** that consists of 12 books, where he constructs a "table of chords". In this writing one can see the direct and very important contribution of Hipparchus to trigonometry. Although this original table has not been preserved, what is preserved is a table of chords given by Ptolemy Claudius, which constitutes essentially Hipparchus' table of chords.

Hipparchus' method is described by Ptolemy as follows:

The circumference of the circle is divided into 360° (degrees) and the diameter of the circle is divided into 120 equal parts. Every degree and every part of the diameter is subdivided into 60 equal parts (following the Babylonian number system with base 60), each of which is also divided into 60 equal parts and so on. Ptolemy's table gives the lengths of the chords of all central angles of a given circle starting from $(1/2)°$ and adding half a degree each time up to 180^0. Thus, if we denote by "chord α" the length of the chord of a central angle α, this table gives results as the following:

$$\text{chord } 36° = 37 \ 4' \ 5''$$

which means that the chord of a central angle of 36° is equal to 37/60 of the radius, plus 4/60 of a subdivision of the radius, plus 5/60 of a subdivision of the subdivision of the radius.

As it follows from Figure 18.1, this table of chords is equivalent to a table of "trigonometric sines" because

$$\sin \alpha = \frac{AM}{OA} = \frac{AB}{\text{diameter of the circle}} = \frac{\text{chord } 2\alpha}{120}.$$

Essentially the "table of chords" gives the sines of the central angles from 0° up to 90°. There is evidence and there are reports that Hipparchus made systematic use of the table of chords and knew formulas which are equivalent to the ones that we use today for the solution of right-angled spherical triangles. In other words, in order to construct the tables, Hipparchus made use of formulas which are equivalent to the following:

$$\sin(x \pm y) = \sin x \cos y \pm \cos x \sin y$$
$$2\sin^2(x/2) = 1 - \cos x.$$

Theon mentions a six volume writing of Menelaus regarding chords of a circle, which however has not been preserved. But there exists a three volume writing of Menelaus, the *Sphaerica*, which was preserved in Arabic language and which gives valuable information regarding the development of trigonometry by the Greeks. Menelaus is also known for the following theorem which can be found in *Sphaerica*.

Menelaus' Theorem

Let ABC be a given triangle. We assume that the points D, E, F are located on the straight lines defined by the pairs of points (B,C), (A,C), (A,B) respectively. We also assume that either none of the points D, E, F lies on some side of ABC or two of them lie on two sides of ABC. Then the points D, E, F are collinear if and only if the following relation holds

$$BD \cdot CE \cdot AF = CD \cdot AE \cdot BF.$$

Proof. Let us assume that the points D, E, F lie on the same straight line l. We can assume that l does not pass through any of the points A, B, C. Let A', B', C' be three points on l such that the straight lines AA', BB', CC' are perpendicular to l. Then

$$\frac{BD}{CD} = \frac{BB'}{CC'}, \quad \frac{CE}{AE} = \frac{CC'}{AA'}, \quad \frac{AF}{BF} = \frac{AA'}{BB'}.$$

The requested result follows by multiplying these three equalities by parts.

The converse of the theorem presents no difficulties and its proof is left to the reader as an exercise.

The loss of a great part of the initial work of the Greeks in astronomy is probably due to the fact that Ptolemy wrote a comprehensive work which overshadowed and made superfluous the previous works. This writing was presented by Ptolemy around 150 A.D. and constitutes a great and comprehensive Greek work in astronomy. It is called *Mathematical Syntaxis* and is distinguished for its completeness, its conciseness and its elegance, while at the same time it constitutes a crowning achievement in the history of mathematics. In order to distinguish this work from other inferior works, the subsequent reviewers gave to it the characterization "μεγίστη" (great (Mathematical Syntaxis)). Later, Arab translators added the prefix "Al" to the word "μεγίστη" and since then the work is known as *Almagest* and constitutes for astronomy what Euclid's *Elements* constitute for geometry.

Almagest consists of thirteen books. Book I contains, among other astronomical material, the "table of chords" which is accompanied by a brief explanation of how it was constructed based on a fruitful geometric proposition known as Ptolemy's Theorem: "In a quadrilateral, which is inscribed in a circle, the product of its diagonals is equal to the sum of the products of the two pairs of opposite sides".

Ptolemy denotes numbers by Greek letters. He curiously uses two letters of the Phoenician alphabet, the bigamma F (the capital letter) and s (the lower case letter), and the letter Koppa Φ, which correspond to the latin letters f and q, and

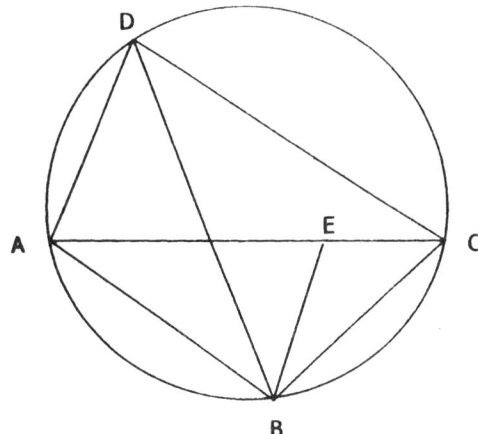

Figure 18.2

which were lost from the Greek alphabet. He also adds one more symbol at the end and hence he has a total of 27 symbols, which allow him to denote the numbers 1 through 9, 10 through 90, and 100 through 900. Ptolemy uses a small circle (o) to denote zero. In his tables he follows the Babylonian number system with base 60 in order to express not only angles (as we do today) but other quantities too, as Hipparchus did before him. Thus, he writes:

$$120° \ 0' \ 0'' = \rho\kappa/o/o$$
$$\text{chord } 1° = 1 \ 2' \ 50''$$

which means that he considers a circle of radius 1.

Below, we present a matrix which shows the arithmetic notation used by Ptolemy, and where we have replaced bigamma and koppa (which correspond to the numbers 6 and 90 respectively) with the latin letters f and q respectively, and where we have omitted the symbol (sanpi) for the number 900.

Ptolemy's Notation

1	α	10	ι	100	ρ
2	β	20	κ	200	σ
3	γ	30	λ	300	τ
4	δ	40	μ	400	υ
5	ϵ	50	ν	500	ϕ
6	f	60	ξ	600	χ
7	ζ	70	o	700	ψ
8	η	80	π	800	ω
9	θ	90	q	900	

Now we will prove the theorem of Ptolemy whose statement we gave above.

Let $ABCD$ (Figure 18.2) be a simple quadrilateral inscribed in a circle, and let E be a point on the diagonal AC such that $A\hat{B}E = D\hat{B}C$. From the similar triangles ABE and DBC, we have $\frac{AB}{AE} = \frac{DB}{DC}$, i.e., $(AB)(DC) = (DB)(AE)$. In addition, from the similar triangles ABD and EBC, we have $\frac{AD}{DB} = \frac{EC}{CB}$, i.e., $(AD)(CB) = (DB)(EC)$. Thus $(AB)(DC) + (AD)(CB) = DB \cdot (AE + EC) = (DB)(AC)$.

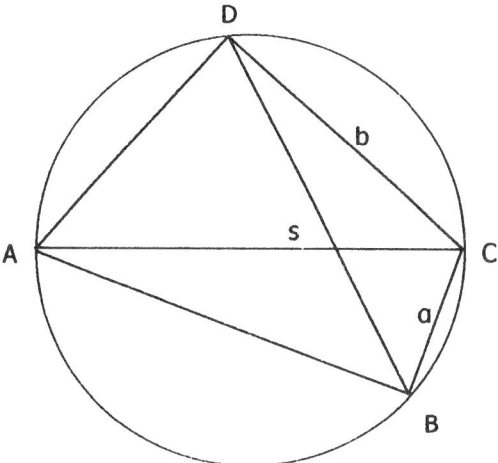

FIGURE 18.3

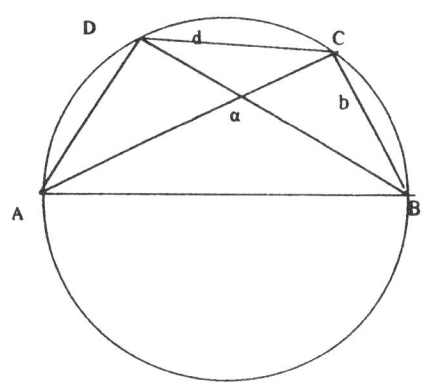

FIGURE 18.4

Conclusion 1. If a and b are the chords of two arcs of a circle of radius 1, then
$$s = (a/2) \cdot (4 - b^2)^{1/2} + (b/2) \cdot (4 - a^2)^{1/2},$$
where s is the chord that corresponds to the sum of the two arcs.

Proof. We apply Ptolemy's Theorem to the quadrilateral of Figure 18.3, where AC is a diameter, $BC = a$ and $CD = b$.

Conclusion 2. If a and b are the chords of two arcs of a circle of radius 1 with $a \geq b$, then
$$d = (a/2) \cdot (4 - b^2)^{1/2} - (b/2) \cdot (4 - a^2)^{1/2},$$
where d is the chord that corresponds to the difference of the two arcs.

Proof. We apply Ptolemy's Theorem to the quadrilateral of Figure 18.4, where AB is a diameter, $BD = a$ and $BC = b$.

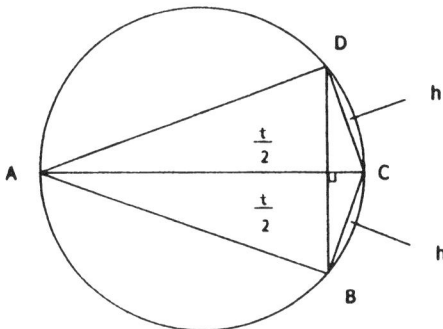

FIGURE 18.5

Conclusion 3. If t is the chord that corresponds to the smallest of the two arcs defined by two points on a circle of radius 1, then
$$h = [2 - (4 - t^2)^{1/2}]^{1/2},$$
where h is the chord of the half arc.

Proof. We apply Ptolemy's Theorem to the quadrilateral of Figure 18.5, where AC is a diameter, $BD = t$ and BD is perpendicular to AC.

We have
$$2t = 2h \cdot (4 - h^2)^{1/2}$$
from which we obtain by squaring that
$$h^4 - 4h^2 + t^2 = 0,$$
i.e.,
$$h^2 = 2 \pm (4 - t^2)^{1/2}.$$
Since h represents the chord of half the smallest arc of the chord BD, we keep the minus sign in the last equality and we have
$$h = [2 - (4 - t^2)^{1/2}]^{1/2}.$$

- We consider the isosceles triangle AOB of Figure 18.6, where $A\hat{O}B = 36°$. We draw the bisector AC of $B\hat{A}O$. From the similarity of the triangles AOB and BAC, we have $\frac{AC}{CB} = \frac{OB}{AB}$. We put $AB = x$ and we consider that $OB = 1$. We have
$$\frac{x}{1-x} = \frac{1}{x} \quad \text{or} \quad x^2 + x - 1 = 0$$
hence $x = \frac{\sqrt{5}-1}{2} \approx 0.6180$ (an approximation of four decimal digits).

  Therefore, in a circle of radius 1, we have chord $36° \approx 0.6180$. This value obviously gives the golden section of the radius.

- We know that in a circle of radius 1, we have chord $60° = 1$. Hence Conclusion 2 implies that
$$\text{chord } 24° = \text{chord } (60° - 36°) = 0.4158.$$

- From Conclusion 3, we can successively calculate the chords of $12°$, $6°$, $3°$, $90'$ and $45'$ of the unit circle. We have
$$\text{chord } 90' = 0.0262 \ , \ \text{chord } 45' = 0.0131.$$

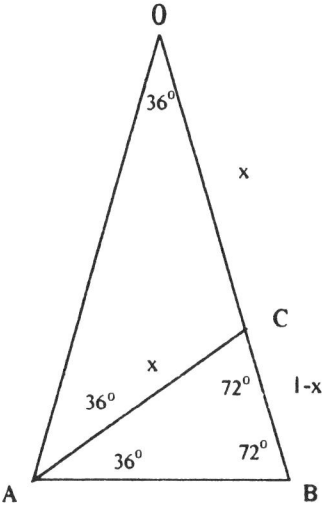

FIGURE 18.6

- From the relation $\frac{\sin a}{\sin b} < \frac{a}{b}$, where $b < a < 90°$ (an equivalent relation was also known to Aristarchus), we have

$$\frac{\text{chord } 60'}{\text{chord } 45'} < \frac{60}{45} = \frac{4}{3}$$

or

$$\text{chord } 1° < (4/3)(0.0131) = 0.01747.$$

Also

$$\frac{\text{chord } 90'}{\text{chord } 60'} < \frac{90}{60} = \frac{3}{2}$$

or

$$\text{chord } 1° > (2/3)(0.0262) = 0.01747.$$

Therefore, at an approximation of four decimal digits, we have

$$\text{chord } 1° \approx 0.0175.$$

- From Conclusion 3 we can calculate the chord of $(1/2)°$.

Now, using the above, we can construct a table of chords of the unit circle per intervals of $(1/2)°$.

Heron's personality is not only interesting from the point of view of history, but also expresses the characteristics of the Alexandrian period. Proclus calls Heron "engineer" and refers to him in relation to his teacher Ctesibius. An impressive element in Heron's work is that he combines rigorous mathematics with approximation methods and formulas. On the one hand, Heron reviewed Euclid's work, used exact results of Archimedes (to whom he often refers) and wrote an original work, where he proved a number of new theorems in Euclidean Geometry. On the other hand, he was interested in applied geometry and mechanics, and he gave many kinds of approximation results, but without supplying proofs.

In his works *Metrica* and *Geometrica*, Heron gives theorems and rules regarding the areas of plane figures, the areas of surfaces and the volumes of a large number of figures. The theorems contained in these books are not new. Regarding the figures with curvilinear boundaries, he uses the results of Archimedes. He also wrote the books *Geodaesia* and *Stereometrica* (calculation of volumes), which treat the same subject as the two books mentioned above. In all of these works, Heron's main interest is focused on arithmetical results.

In his work *Dioptra* (Theodolite), a writing regarding geodesy, Heron proves how one can calculate the distance between two points, only one of which is accessible, and how one can calculate the distance between two points which are visible but inaccessible. He also proves how one can draw from a given point a normal to an inaccessible straight line, and how one can calculate the area of a field without entering it.

The formula for calculating the area of a triangle whose sides a, b, c are given, i.e., the formula Area $= \sqrt{s(s-a)(s-b)(s-c)}$, where $s = \frac{a+b+c}{2}$, is attributed to Heron, even though other historians of mathematics believe that it is due to Archimedes. This formula appears in Heron's book *Geodesy*, while *Dioptra* and *Metrica* contain both this formula and its proof. In *Dioptra* he also proves how one can carry out the excavation of a tunnel inside a mountain, when the excavation starts simultaneously from both ends.

We present the way in which Heron finds the formula that gives the area of a triangle ABC with sides of lengths a, b, c respectively.

Let s be the semi-perimeter of the triangle ABC and let D, E, F be the points at which the circle that is inscribed in ABC is tangent to the sides of ABC. Let O be the center of the inscribed circle and let r be its radius (Figure 18.7). Heron extends BC and takes on its extension a segment $CH = AF$. It is $BH = s$. He draws the normal OK to BO and the normal CK to BC. The area of ABC is equal to the sum of the areas $OBC + OCA + OAB = \frac{1}{2}ar + \frac{1}{2}br + \frac{1}{2}cr = sr = BH \cdot OD$. Later he proves that $O\hat{A}F = C\hat{B}K$ and hence from the similarity of the triangles OAF and CBK he finds:

$$\frac{BC}{CK} = \frac{AF}{OF} = \frac{CH}{OD},$$
$$\frac{BC}{CH} = \frac{CK}{OD} = \frac{CL}{LD},$$
$$\frac{BH}{CH} = \frac{CD}{LD},$$
$$\frac{BH^2}{CH \cdot BH} = \frac{CD \cdot BD}{LD \cdot BD} = \frac{CD \cdot BD}{OD^2}.$$

Therefore

Area $ABC = BH \cdot OD = [CH \cdot BH \cdot CD \cdot BD]^{1/2} = [(s-a)s(s-c)(s-b)]^{1/2}$.

Even though many of the formulas given are accompanied with proofs, many other formulas are given which are only approximately valid, as is the formula $\sqrt{A} = \sqrt{a^2 \pm b} \approx a \pm \frac{b}{2a}$, which gives the square root of A and where a^2 is the square of the closest to A integer (larger or smaller than A) while b is the remainder.

Thus, Heron together with the exact formula mentioned above, gives an approximate formula for the area of a triangle. A reason for giving approximate formulas is that the exact formulas require the extraction of square or cubic roots, i.e., they require operations that the practical geometers could not perform. There was a distinction between "pure" geometry on the one hand and geodesy on the other. The calculation of areas and volumes belonged to geodesy and was taught

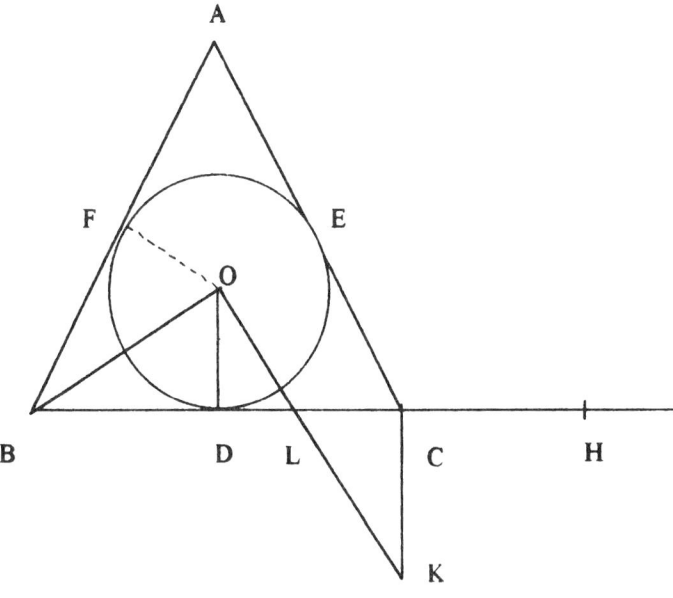

FIGURE 18.7

to practical geometers, builders, carpenters and other technicians. Heron's work regarding geodesy was used for many centuries.

Heron applied many of his theorems and rules to the designing of theaters, symposium areas and baths. His work regarding applications contains, among others, the *Mechanics*, the *Belopoiica* (construction of war machines), the *Pneumatica* (theory and application of "air pressure"), the *Automatica* and the *Water-clocks*.

Diophantus's work constituted the zenith of Alexandrian Greek algebra, but we do not know many things about his life. One of the problems in algebra found in some Greek collection gives the following information regarding his life: The duration of his childhood was 1/6 of the duration of his life. Then, 1/12 of his life elapsed before his beard started growing, and 1/7 of his life elapsed before he got married. His son was born 5 years after his wedding and lived only half the lifetime of his father. The father died 4 years after the death of his son. What is requested is at what age Diophantus died. It is easily calculated that he died at 84.

Even though Diophantus' work overshadowed the work of his contemporaries, it appeared too late to influence science greatly, because a destructive wave had already begun to cover the civilization of that time.

The most important among Diophantus works is *Arithmetica*, which consisted of 13 books. Up to 1973, only six of them were preserved. To these six books, three more were added, which were discovered in Arabic translation (see Jacques Sesiano, *Books IV to VII of Diophantus' Arithmetica*). Let it be noted that in Ancient Greece the word "$αριθμητική$" meant "number theory", and not calculations with arithmetic operations.

Arithmetica, as Rhind's Papyrus, is a collection of various problems with their solutions. One of Diophantus' great achievements is the beginning of the introduction of notation in algebra and this certainly constitutes a crowning moment for mathematics. Since we don't have at our disposal Diophantus' original manuscript but subsequent ones, we don't know exactly which symbols he introduced. It is believed that the symbol that he used for the unknown (i.e., the one that corresponds to x that we use today) was the Greek final σ, i.e., ς. This is only a conjecture, however, it is certain that the "square of the unknown" was denoted by Δ^Y, i.e., by the first two letters of the word $\Delta YNAMI\Sigma$[1]. Also the "cube of the unknown" was denoted by K^Y, i.e., by the first two letters of the word $KYBO\Sigma$[2]. Other powers of the unknown were denoted by $\Delta^Y\Delta$ (square-square) ($\delta\upsilon\nu\alpha\mu o\delta\acute{\upsilon}\nu\alpha\mu\iota\varsigma$), ΔK^Y (square-cube) and $K^Y K$ (cube-cube).

The presence of these symbols is, of course, remarkable, but the use of powers greater than three is indeed admirable, because the use of powers higher than three was not introduced until then, and this is because a product with more than three factors had no geometric meaning. For the symbol "minus" Diophantus used Λ, which is obtained from the letters Λ and I of the word $\Lambda EI\Psi I\Sigma$[3]. The addition of two terms was made by concatenating one after the other. The negative terms of an expression were grouped together and in front of them the symbol "minus" was placed. An arithmetical coefficient of an arbitrary power was denoted by the appropriate letter of the Greek alphabet and was placed after the symbol of the power. $\overset{\circ}{M}$ denoted a constant term and comes from the word $MONA\Delta E\Sigma$[4]. Examples of use of the above mathematical symbols are the following:

$$13 = \iota\gamma \ , \ 31 = \lambda\alpha \ , \ 742 = \psi\mu\beta.$$

For the notation of larger numbers they used "dashes" and "accents".

According to the notations that we mentioned above, the expressions $x^3 + 13x^2 + 8x$ and $x^3 - 8x^2 + 2x - 3$ are respectively written as

$$K^Y \alpha \Delta^Y \iota\gamma\varsigma\eta, \ \ K^Y \alpha\varsigma\beta \Lambda \Delta^Y \eta \overset{\circ}{M}\gamma$$

and are respectively read "unknown to the cube 1, unknown to the square 13, unknown 8" and "(unknown to the cube 1, unknown 2) minus (unknown to the square 8, units 3)".

The introduction of notation in algebra developed in three stages. The first stage corresponds to the so-called "rhetorical algebra" (where the problems and the various notions are formulated in general with words and without using any symbols), as is the algebra of the Babylonians. The second stage corresponds to the so-called "abbreviated algebra" (where abbreviation of words is made) to which Diophantus' notation belongs. And the third stage corresponds to "symbolic algebra". The development of symbolic algebra began mainly after the 16th century and finally it took its present form.

The first book of the work *Arithmetica* consists mainly of problems leading to first-order equations of determinate form, with one or more than one unknown, i.e., to equations which admit a single solution. The remaining five books treat mainly second-order equations which admit more than one solutions, i.e., equations of "indeterminate form" as we usually say.

In the case of equations of determinate form with more than one unknown, Diophantus makes use of certain facts in order to eliminate all the unknowns apart

---

[1] **Translator's note:** The meaning of this word in English is: *Power*.

[2] **Translator's note:** The meaning of this word in English is: *Cube*.

[3] **Translator's note:** The meaning of this word in English is: *Wanting*.

[4] **Translator's note:** The meaning of this word in English is: *Units*.

from one, and in many cases he obtains equations of the form $ax^2 = b$. For example, in Problem 27 of Book I: "Find two numbers whose sum is 20 and whose product is 96", Diophantus works as follows: Let $2x$ be the difference of the requested numbers. Thus, the numbers are $10 + x$ and $10 - x$. Consequently $100 - x^2 = 96$. Then $x = 2$ and the requested numbers are 12 and 8.

The most distinguishing feature of Diophantus' algebra is the solution that he gives to equations of indeterminate form. Such equations appeared before Diophantus, as, for example, in Pythagoras' work on the search for solutions of the equation $x^2 + y^2 = z^2$, or in Archimedes' "cattle problem" which leads to the solution of a system of seven equations with eight unknowns (plus two additional conditions), as well as in other works. Diophantus was largely and systematically occupied with equations of indeterminate form and is considered to be the founder of the branch of algebra that we call today "Diophantine Analysis".

We present some of the problems considered by Diophantus.

Book I, Problem 8. Express a given square number as the sum of two squares.

Here he considers 16 as the given number and he finds the numbers $256/25$ and $144/25$.

This problem was generalized by Fermat who formulated the proposition that the equation $x^m + y^m = z^m$ has no solution in positive integers if $m > 2$. (It is the famous "Last Theorem of Fermat", which was proved only recently.)

Book II, Problem 9. Express a given number, which is the sum of two squares, as the sum of two other squares. That is, if a and b are two rational numbers, find a (non-trivial) solution of the equation

$$x^2 + y^2 = a^2 + b^2.$$

Diophantus considers the special case in which $a = 2$ and $b = 3$, but the solution that he gives can be easily generalized. He writes:

"Take $(x+2)^2$ as the first square and $(mx-3)^2$ as the second (where m is an integer) and let it be $(2x-3)^2$. Consequently $(x^2 + 4x + 4) + (4x^2 + 9 - 12x) = 13$ or $5x^2 + 13 - 8x = 13$. Thus $x = 8/5$ and the requested squares are $324/25$ and $1/25$." We note that the general solution is obtained when m is an arbitrary rational number.

Book III, Problem 6. Find three numbers such that their sum and the sum of any two of them is a square number.

Diophantus gives the numbers 80, 320, 41.

Book IV, Problem 1. Express a given number as the sum of two cubes in such a way that the sum of the edges of these cubes (i.e., the cubic roots of these cubic numbers) is equal to a given number.

He considers the number 370 as the given number and the number 10 as the given sum, and he finds the numbers 343 and 27.

Book IV, Problem 29. Express a given number as the sum of four squares plus the sum of their sides (where sides are the square roots of these square numbers).

He considers the number 12 and he finds that the requested squares are $121/100$, $49/100$, $361/100$, $169/100$.

A very interesting question regarding the work of Diophantus' which occupied and continues to occupy the historians of mathematics, is the following: "Did

Diophantus know the algebraic rules on which the solutions of his problems are based?"

Diophantus remarks that: "the number 65 is expressed as the sum of two squares in two different ways since 65 is the product of 13 and 5, each of which is the sum of two squares". From this remark we may perhaps conclude that he knew that the product of two integers, each of which is the sum of two squares, is expressed as the sum of two squares in two different ways. Did Diophantus know the even more powerful proposition that

$$(a^2 + b^2)(c^2 + d^2) = (ac \mp bd)^2 + (ad \pm bc)^2?$$

T. L. Heath, based on the above remark of Diophantus (Book III, Problem 19), conjectures that Diophantus indeed knew the above identity (see page 105 of the translation of *Arithmetica* by Heath). However, **Abu Jafar al-Khazin** (950 A.D.) was the first to publish this identity, and **Fibonacci** reported it later in his work *Liber Quadratorum* (1225 A.D.).

According to other historians, Diophantus' work contains no general methods of solution of the problems mentioned in *Arithmetica*, and each one of the 189 problems is solved by a different method. There exist approximately 50 different types of problems, but Diophantus apparently made no effort to classify them into various types (as these historians allege).

Certainly, Diophantus' work contains results, which we could say are of a general character, as is the following: "There exists no prime number of the form $4n+3$ which can be written as a sum of two squares".

Euler believed that Diophantus followed in his work general methods of solution, but he could not present them as such because he made no use of coefficients of a general nature (i.e., letters of the alphabet and not concrete numerical values). In addition to Euler there exist other mathematicians who also recognize that Diophantus' work belongs to an abstract and fundamental science.

Anyway, despite the existing difference of opinions regarding Diophantus' work, it is accepted worldwide by the mathematical community that his work constitutes a magnificent landmark in Algebra.

In 320 A.D., the Roman empire acquires his first Christian emperor, Constantine the Great, after whom Constantinople, the capital of the Eastern Roman State, was named. During that time in Alexandria, the deacon and later Patriarch Athanasius, alleged that the origin of Jesus Christ is divine, thus overruling the viewpoints (or the heresy of Arius), who taught that Christ is "$o\mu o\iota o\acute{u}\sigma\iota o\varsigma$[5]" and not "$o\mu oo\acute{u}\sigma\iota o\varsigma$[6]" with the Father, as Athanasius alleged.

In mathematics the difference between these two words was preserved, since a distinction is made between the notions of "homeomorphism" and "homomorphism".

After the death of Apollonius, a recession in the progress of geometry is observed (with Menelaus work being the exception). Regarding the geometric works carried out during the beginnings of the Christian period, our knowledge comes from the works of Pappus, Theon of Alexandria (end of the 4th century A.D.) and Proclus. The geometers of that period probably occupied themselves with the study and clarification of the works of the great mathematicians that lived before them.

---

[5]**Translator's note:** The meaning of this word in English is: *of similar essence*.
[6]**Translator's note:** The meaning of this word in English is: *of one essence*.

Theon and Pappus mention the work of **Zenodorus** (between 200 B.C. and 100 A.D.), who studied figures which have equal perimeter and proved the following theorems.

(1) Among all polygons with n sides, which have the same perimeter, the regular n-gon has the largest area.
(2) Among regular polygons with equal perimeters, the one with the most sides has the largest area.
(3) The area of the circle is larger than the area of any regular polygon of the same perimeter.
(4) Among all solids with the same surface, the sphere has the largest volume.

The most important work of Pappus is *Mathematical Synagoge*, which is a massive work whose importance is summarized as follows:

First of all a historical retrospect of various subjects of mathematics of the Ancient Greeks is presented, which otherwise would have been unknown to the subsequent generations. The writing contains alternative proofs and complementary lemmas referring to propositions of Euclid, Archimedes, Apollonius and Ptolemy. But also, *Mathematical Synagoge* contains new discoveries and generalizations which are not found in previous works. The writing contains eight books, from which the first and the first part of the second book have not been preserved. The famous "Theorem of Pappus" constitutes proposition 139 of *Mathematical Synagoge* (Book VII). Essentially this theorem implies that multiplication is commutative and constitutes a fundamental proposition for Projective Geometry. Hilbert used it as a "key" theorem in his presentation of Euclidean Geometry. "Pappus' theorem" can be proved with the help of Menelaus' theorem as follows.

Pappus' theorem

Three points A, B, C are given on a straight line and three points A', B', C' are given on another straight line. Then the points of intersection $P = BC' \cap CB'$, $Q = AB' \cap BA'$, $R = CA' \cap AC'$ lie on the same straight line (Figure 18.8).

In the statement of the theorem we assume that the pairs of straight lines (BC', CB'), (AB', BA'), (CA', AC') are not parallel. We also assume that the straight lines ABC and $A'B'C'$ intersect at some point X and that no other pair of straight lines in the figure are parallel.

Proof (An outline). Let $A'B \cap B'C = U$, $AC' \cap A'B = V$, $B'C \cap AC' = W$.

We apply Menelaus' theorem five times to the triangle UVW. Since each one of the triples of points (A', C, R), (B, C', P), (A, B', Q), (A', B', C'), (A, B, C) lies on the same straight line, we have

$$VR \cdot WC \cdot UA' = RW \cdot CU \cdot A'V,$$
$$VC' \cdot WP \cdot UB = C'W \cdot PU \cdot BV,$$
$$VA \cdot WB' \cdot UQ = AW \cdot B'U \cdot QV,$$
$$VC' \cdot WB' \cdot UA' = C'W \cdot B'U \cdot A'V,$$
$$VA \cdot WC \cdot UB = AW \cdot CU \cdot BV.$$

If we multiply the sides of the first three equalities and we divide the resulting equation with the product of the last two, we obtain

$$VR \cdot WP \cdot UQ = RW \cdot PU \cdot QV.$$

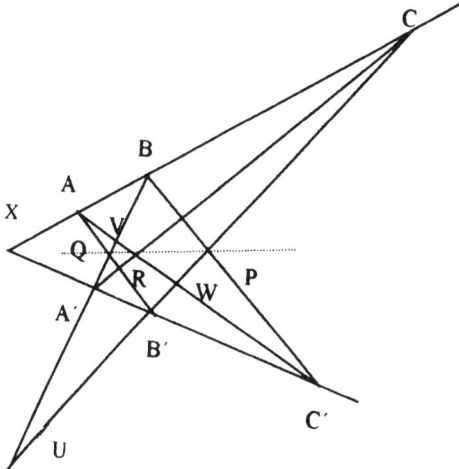

Figure 18.8

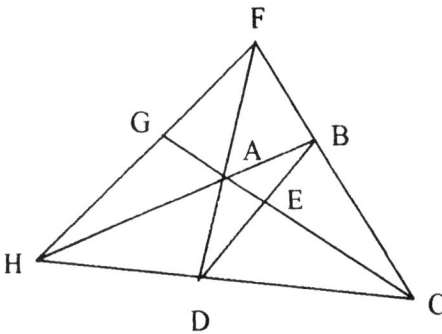

Figure 18.9

The required result follows by one more application of Menelaus' theorem. We note that the above calculations make use of the commutativity property of multiplication.

If the pairs of straight lines (BC', CB'), (AB', BA'), (CA', AC'), which were assumed to intersect, are parallel, then the proof of the theorem proceeds in the same way, but instead of using Menelaus' theorem we use the theory of similar triangles. The details of the proof are left to the reader.

Proposition 131 of Book VII can be stated as follows: "In every quadrilateral, one diagonal is intersected harmonically by the other diagonal and by the straight line connecting the points of intersection of the pairs of opposite sides".

Thus, if $ABCD$ is a quadrilateral (Figure 18.9) and CA is one diagonal, then CA is intersected at point E by the other diagonal BD and at point G by the straight line FH connecting the intersection of AD and BC with the intersection of AB and CD.

Then the conclusion of Proposition 131 is that the points C, E, A and G of the figure constitute a harmonic tetrad, which means that E divides internally AC into the same ratio that G divides externally AC.

Proposition 238 gives a fundamental property of conic sections: "The geometric locus of the points whose distances from a certain fixed point (the focus) and from a certain fixed straight line (the directrix) have a fixed ratio is a conic section". This property of the conic sections is not contained in Apollonius' writing *Conic Sections*, but it was probably known to Euclid.

Book VIII contains another important theorem, which also bears the name "Pappus' Theorem", but is sometimes called "**Paul Guldin**' s Theorem" since it was "rediscovered" by Guldin (1577-1643). This theorem asserts that the volume of a solid, which is generated by the revolution of a closed curve lying entirely on one side of the axis of revolution, is equal to the area surrounded by the curve times the length of the circumference of the circle traced by the center of gravity of the curve.

Nicomachus' work, which was preserved in a Latin translation in two volumes under the title *Introductio Arithmetica*, constitutes a movement to create "Arithmetic" independently of Geometry. The content of *Introductio* mainly refers to the arithmetical work of the first Pythagoreans. Nicomachus studied the even, the odd, the square, and the polygonal numbers. He also studied the composite numbers and the numbers of the form $n^2(n+1)$, and furthermore, he defined other kinds of numbers. The following proposition is due to Nicomachus: If we write in increasing order the sequence of the odd numbers

$$1, 3, 5, 7, 9, 11, 13, 15, 17, ...$$

then the first is the cube of 1, the sum of the next two is the cube of 2, the sum of the next three is the cube of 3, and so on. Nicomachus gives proportions of various kinds among which is the "musical proportion"

$$a : \frac{a+b}{2} = \frac{2ab}{a+b} : b.$$

The value of *Introductio* is that it constitutes a systematic, clear and easy to understand presentation of the arithmetic of the integers and of the ratios between integers, and this presentation is independent of geometry.

Hypatia, the daughter of Theon of Alexandria, took care of the publication of Euclid's *Elements* and also wrote a study where she comments on the work of Apollonius, the work of Diophantus and Ptolemy' s *Almagest*.

Proclus wrote a study where he comments on Book I of Euclid's *Elements* (Chapter 3, 2). This study is interesting because it gives us information about works which have not been preserved, such as the history of geometry of Eudemus of Rhodes (350-290 B.C.) and the book of Geminus of Rhodes (110-40 B.C.), which also refers to subjects of mathematics (and is entitled "The Doctrine" or "Theory of Mathematics").

Proclus studied in Alexandria and then went to Athens, where he was placed in charge of Plato's Academy. He was a neoplatonic philosopher and wrote books referring to the work and the philosophy of Plato in general. He occupied himself with poetry and mathematics. He believed, as Plato did, that mathematics serve philosophy, that mathematics constitute a preparatory course for philosophy, and that make lucid the eye of the soul by removing the obstacles placed by our senses in our effort to acquire knowledge.

Boethius, who was a Roman, studied in Athens and lived in Rome. He is mainly known for his writing *De Consolatione Philosophicae*, which he wrote while he was in prison, around 525 A.D. His Arithmetic and Geometry were used as school textbooks during the Middle Ages, although these books contain much less mathematics than Euclid's *Elements*.

Among many other commentators we will refer only a few, as is **Simplicius** who commented on Aristotle's work. He studied in Alexandria and in Plato's Academy, and he went to Persia when Justinian closed Plato's Academy in 529 A.D. Simplicius repeated a part of Eudemus' *History*, a great part of Antiphon's efforts to square the circle, as well as material that refers to the squaring of Hippocrates' menisci.

Isidorus of Miletus (6th century A.D.) was one of the architects of the Temple of Saint Sophia in Constantinople in 532-537 A.D., as was **Anthemius** also. Isidorus founded in Constantinople a school and wrote works, where he commented on and probably wrote part of the fifteenth book of Euclid's *Elements*. In addition, **Eutocius** (6th century A.D.), who was a student of Isidorus, wrote comments on Archimedes' work. It is believed that certain passages from Archimedes' work and the first four books of Apollonius were preserved until our days due to the Byzantine architects Anthemius and Isidorus and due to the commentator Eutocius.

The Middle Ages had already begun in Europe.

When the Arabs occupied Alexandria in 641 A.D., probably not many books were left at the Library. According to a repeated story that we find in Moslem sources of the 13th century, the Arabs destroyed this Library. **Amru**, who was in charge of the Arabs, wanted to keep the Library intact, but the Caliph **Omar I** had a different opinion and alleged that: "If the books of the Greeks reassert what the Koran writes, then they are superfluous and can be destroyed. While if they contradict the Koran, then they are dangerous." Consequently, either way they had to be destroyed and that is what happened. It is said that the books were used in order to heat the furnaces of the public baths of Alexandria for six months!

CHAPTER 19

A BRIEF REVIEW OF THE HISTORY OF GREEK MATHEMATICS

Let us see, first of all, in what condition was Mathematical Science before the Greeks.

In the civilizations of the Babylonians and the Egyptians, we find an arithmetic of the integer and rational numbers, the beginnings of algebra as well as certain empirical formulas of geometry. Notation is almost non-existent and abstract thought is also non-existent. One finds no formulation of any general method and not a trace of any "notion of proof" regarding the correctness of the empirical mathematical formula. There exists no concept of any kind of theoretical science. In these two civilizations, apart from certain isolated results of the Babylonians, mathematics didn't constitute a separate science and was not cultivated for its own sake. Mathematics was simply a tool consisting of a set of elementary rules which were used for the solution of certain practical problems of everyday life. Even though the mathematics of Babylonians was more advanced than the mathematics of the Egyptians, the best that one can say about both of them is that there was some activity and some diligent effort, but nothing brilliant.

Even though every evaluation presupposes the use of some criterion, it is natural, although perhaps unfair to compare these two civilizations with the Greek civilization that succeeded them. In some sense, the Egyptians and the Babylonians were "clumsy carpenters" while the Greeks were "magnificent architects".

Of course, there exist historians who give a more favorable judgement regarding the achievements of the Babylonians and the Egyptians. But these judgements were made by specialists who were perhaps unconsciously impressed more than necessary by the object of their interest.

The Greek civilization

In the history of civilization the Greeks were brilliant, while in the history of mathematics the Greek mathematics constitute the crowning event. They built a civilization and a system of education that greatly affected the development of contemporary Western Civilization as well as the foundation of mathematical science as perceived today.

One of the greatest problems in the history of civilization is how one can evaluate the greatness, the brilliance, and the creative spirit of the Ancient Greeks.

Even though our knowledge on this subject is constantly changing due to archeological research and findings, we have every reason to believe, based on Homer's *Iliad* and *Odyssey*, that the Greek civilization dates back at least to 2800 B.C. The

Greeks settled in Asia Minor which was the homeland, to Europe where contemporary Greece is now located, to South Italy, to Sicily, to Crete, to the Aegean, to Rhodes, to Delos, to North Africa, as well as to several other places.

Around 775 B.C. the Greeks formulated their alphabet which allowed them to record their history and their ideas.

Miletus, this city of Ionia in Asia Minor which was an important commercial center of the Mediterranean, is considered to be the birthplace of Greek Philosophy, of Mathematics, and of Exact Sciences in general.

The repression of the revolt of Ionia against Persia in 494 B.C., contributed to Ionia's decline. But Ionia became again Greek when the Greeks defeated the Persians in 479 B.C. In the meantime the cultural activity had already been shifted to the European Greece with Athens as its center. We will try to describe briefly but clearly the huge heritage that the Greeks left to mathematical science.

The Greeks elevated mathematics to an abstract science. This great contribution is of inestimable value and importance. For example, the fact that the same abstract triangle or the same "algebraic equation" can be applied to hundreds of situations of various kinds proved to be the secret of the great power of mathematics.

The Greeks insisted that the various propositions and theorems must have "proofs" based on deductive logic. This was a giant step forward. Aristotle had said referring to Thales that: "For Thales, the primitive question is not what we know but how we know it".

A small number of the civilizations that preceded the Greek civilization had developed a kind of primitive arithmetic and geometry. But no other civilization apart from the Greeks had the idea to demand and accept only those results whose proof followed by the correct use of deductive logic. The decision to accept only proofs of this kind was entirely in juxtaposition with the empirical methods used in every other field of activity. But the Greeks wanted to have "truths" and understood that the acquisition of truths could be achieved only through the application of unquestionable methods of deductive logic.

The Greeks realized further that in order to secure these "truths", they had to base them also on "truths" of unquestionable facts. Hence they used the so-called axioms which they formulated at the beginning of every theory in order to make the development of the theory both possible and easier.

The fact that they appreciated that the various notions must not contradict each other, and that it is impossible to build a consistent structure working just with non-existing figures (as, for example, with a regular "decahedron") constitutes an unprecedented insight.

The ability of the Greeks to produce theorems and proofs become evident by the fact that all 467 propositions contained in Euclid's *Elements* and all 487 propositions contained in Apollonius' *Conic Sections* follow from the ten axioms stated in *Elements*; and furthermore all these results are important, relevant and useful.

The Greeks greatly contributed to the following notions of mathematical science: plane geometry and stereometry, plane and spherical trigonometry, the principles of number theory, the extension of the arithmetic and of the algebra of the Babylonians and the Egyptians. One must also note that this huge progress is due only to a small number of scientists and was completed in a time interval of a few centuries. To these achievements we must add geometric algebra, which needed only the recognition of irrational numbers and their use into symbolic language in order to become the basis of elementary algebra.

The consideration of curvilinear figures and the use of the "method of exhaustion" for the calculation of their areas, constitute undoubtedly the first steps towards Infinitesimal Calculus.

An equally vital contribution and source of inspiration for the subsequent generations was the beliefs of the Greeks about nature.

The Greeks identified mathematics with the natural world, and saw in mathematics the final truth regarding the structure and the form of the universe. They founded the existing close relationship between mathematics and the objective study of nature, something that constituted the basis of contemporary science. They formulated also the belief that the universe is indeed formed in a mathematical way, i.e., it is subject to laws and may be understood by the human.

The aesthetic aspect of mathematics was not at all neglected. In Ancient Greece mathematics was considered to be also a work of art, since it was recognized that it has beauty, harmony, simplicity, clarity and order. Number theory, geometry and astronomy were considered to be the art of the spirit as well as music for the soul. Plato praised geometry. Aristotle did not distinguish mathematics from "aesthetics", because order and symmetry constituted for him important elements of beauty, and he ascertained that these elements existed in mathematics. Indeed, in Greek thought it is difficult to distinguish the elements that are related to rationalism, aesthetics and morality. One repeatedly reads in Greek texts that among all solid bodies, the sphere has the most beautiful shape and hence it is divine and good. Similar points of view regarding the aesthetic nature of the circle also prevailed. This was indeed the reason that it was considered obvious that the orbits of those celestial bodies which represented the unchangeable eternal order that prevails in the firmament had to be circular, while the notion that prevails on the imperfect planet Earth is linear. There is no doubt that the consideration of the aesthetic aspects of a subject led Greek mathematicians to investigate certain things further, beyond the point that was necessary in order to perceive the natural world.

Weaknesses of Greek Mathematics

Despite these marvellous worldwide accepted achievements of the Ancient Greeks, some historians of mathematics present, under the title "disadvantages" or "weaknesses" of Greek mathematics, comments and views which consist of remarks of what the Greeks could have also achieved and how mathematics would have looked in that case. We will mention some of these views.

The first "weakness" consists of the fact that the Greeks did not manage to understand completely the notion of an irrational number. This fact, in addition to being an obstacle for the development of number theory and algebra, was also the reason for giving greater emphasis to geometry because geometric thought avoids the direct involvement of irrational numbers. If the notion of an irrational number had been properly understood, then algebra and number theory would have been developed further, or at least the road for future generations for the further development of these topics would have been prepared. Even though Archimedes, Heron and Ptolemy began working with irrational numbers, they did not have sufficient influence over the thought of the mathematicians of their time, as a result of which number theory and geometry were considered as unrelated mathematical activities.

The Greeks not only restricted mathematical activity mainly to the study of geometry, but also they restricted geometry itself to the study of only those figures that result from straight lines and circles. Thus, the only acceptable surfaces were the ones that resulted from the revolution of straight lines or circles about an axis (such as the cylinder, the cone and the sphere). Exceptions are the plane, the prism and the pyramid. The conic sections were introduced as intersections of a cone with a plane. Curves, as is Hippias' "quadratrix", Nicomedes' conchoid, and Diocles' cissoid, were pushed out of the limelight of geometry and were called "mechanical" as opposed to "geometric" figures.

One of the reasons, for which the Greeks restricted geometry to straight lines, circles as well as figures that result from them, is perhaps that in this way they were able to prove the "existence" of the geometric figures they considered. In order to clarify this point, we remind the reader that Aristotle alleged that the notions introduced into a certain theory (in order to study a certain subject), must not contradict each other which means that first of all one must prove that these notions "exist". In order to ensure this last condition, the Greeks accepted only figures that they could construct. The straight line and the circle are figures, which were considered to be constructible, and hence they could be introduced into the axioms. Thus the remaining figures of the theory had to be constructed only by using straight lines and circles.

One more reason, for which geometry was restricted to straight lines, circles as well as figures that result from them, has its roots in Plato's viewpoints, who was teaching that ideas must be "clear" in order to become acceptable.

While the notion of an integer (even though it was not defined) was acceptable as an idea which is clear by itself, geometric figures as well had to be accurate and clear. The straight lines, the circles and the figures that result from them were considered to be clear, while the curves produced with the help of mechanical instruments (apart from the ruler and the compass) were not considered to be clear and hence these figures were not acceptable. The restriction to only "clear" figures produced a simple, well-ordered, harmonious and beautiful geometry.

The historians mentioned above believe that certain philosophical points of view of the Ancient Greeks retarded the development of mathematics. Such a viewpoint is that mathematics is not created by the human, because it exists a priori, and all a mathematician has to do is discover it and record it. In his work *Theaetetus*, Plato compares research for the acquisition of knowledge with bird-hunting in an aviary. Birds are already there and what is left to do is just to catch them. This viewpoint regarding the nature of mathematics, which does not seem to have prevailed in the past, is discussed again today and now has serious supporters [42].

Another weakness of Greek Mathematics is considered to be the fact that the Greeks did not understand or accept the following notions: "infinitely large", "infinitely small" and "procedure repeated ad infinitum". The Pythagoreans connect the notions of "good" and "evil" with the notions of "finite" and "infinite" respectively. Aristotle said that "infinity" is imperfect, incomplete, and hence incomprehensible, shapeless and vague. Only objects which have a clear outline and are distinguished, have a certain nature.

In order to avoid the notion of "infinite length", Euclid accepts that a line segment can be extended at will in both directions, i.e., it can be extended as much as necessary.

The notion "infinitely small" which comes up in the relation between a straight line and its points or in the relation between the "discrete" and the "continuum", as well as Zeno's paradoxes, seem to be the reasons that forced the Greeks to avoid the consideration of similar subjects. The relation between a straight line and a point occupied the Greeks, and led Aristotle to consider them as two separate things. Thus, Aristotle accepts that points lie on straight lines, but he alleges that a straight line does not consist of points and that the continuum cannot be constructed from the discrete (Chapter 3, 10). This distinction contributed also to the separation of "numbers" from geometry, because numbers were considered to be discrete, while geometry was dealing with continuous quantities.

Finally, the fact that the Greeks avoided the notion of a "procedure repeated ad infinitum" retarded the discovery of the notion of a limit.

During the process of approximation of a circle with a polygon, they were satisfied with the fact that the difference between the two figures could become smaller than any given quantity, despite the fact that the remainder was always some positive quantity. This procedure was intuitively clear, while the "procedure of taking a limit" implies the notion of being infinitely small.

The above constitute in brief the viewpoints of those historians who point out "weaknesses" in the mathematics of the Ancient Greeks. According to the writer's opinion, the above cannot and must not be considered as disadvantages or weaknesses of the mathematics of the Ancient Greeks. Indeed, only something that someone did, something that is included in his/her work, can be considered as a disadvantage of his/her work, and not something that one could have done, but did not do. The facts reported as disadvantages can be considered on the contrary as "advantages", in the sense that they paved the way, as we mentioned before, for future generations to continue the research and achieve the further development of this science. The work of the Ancient Greeks was unprecedented and "superhuman", and no voluntary or involuntary effort must be made to blemish it with arguments as the ones mentioned above.

CHAPTER 20

MATHEMATICS IN CHINA AND INDIA

Of particular interest is the fact that Egypt, Babylon and China developed some mathematical activity before 1000 B.C. However, this was followed by a period of inactivity and stagnation; the cause of this phenomenon is not known. One point of view is that it was due to the fact that every activity during that period was concentrated in Greece. In any case it is certain that between 1000 B.C. and 300 B.C. there was no significant mathematical activity in China. Actually, we do not know many things about the development of mathematics in China before China made some contact with the West.

An event of great importance in the history of mathematics in China was the burning of all books (213 B.C.), which was ordered by the emperor **Shi Huang-ti**, the founder of the Ts'in dynasty (221 B.C.). He wanted with this action to give the impression that he was the creator of a new era of knowledge. The people who did not comply with the emperor's order were pilloried and sentenced to five years hard labor on the Great Wall. It is reported that the emporer gave the order to bury alive 460 intellectuals who protested against this hideous law, in order to make an example of them.

The book *Arithmetic in Nine Volumes* (**Chiu Chang Suan Shu**) was written before 200 A.D. This book contains a set of problems and their solutions. In Chapter 8, the solution of a system of n linear equations with n unknowns is given and the method for the solution is essentially the same as the one we use today. One of the systems whose solution is given is the following:

$$3x + 2y + z = 39$$
$$2x + 3y + z = 34$$
$$x + 2y + 3z = 26$$

The interest of the Chinese for systems of linear equations is probably connected with their interest for "magic squares". The square

$$\begin{array}{ccc} 4 & 9 & 2 \\ 3 & 5 & 7 \\ 8 & 1 & 6 \end{array}$$

has the "magic" property that the sum of the numbers in every row, in every column, and in every diagonal is the same.

In addition, **Sun Tsu** (400 A.D.) gave a solution to the following problem:

> If you divide with 3, the remainder is 2.
> If you divide with 5, the remainder is 3.
> If you divide with 7, the remainder is 2.
> What is the number?

Even though Sun Tsu gave only one of the infinitely many solutions, his method allows one to find as many solutions as one wants.

In the book *Nine Areas of Mathematics* (1247 A.D.), **Ch'in Chiu Shao** calculated a root of the equation

$$x^4 - 763200 x^2 + 40642560000 = 0$$

using a method which was rediscovered by **W. Horner** in 1819.

In the book *Precious Mirror of the Four Elements* (1303 A.D.), **Chu Shih Chieh** gives "Pascal's triangle", which was also known in India and which was rediscovered by **Pascal** in 1653.

During the 16th and the 17th century, Christian missionaries from Europe went to China and introduced the Chinese to the mathematics of the West (for example, the theory of logarithms). Today, the Chinese mathematicians make important contributions in all branches of mathematics and, in particular, in number theory. For example, in 1985, **De Gang Ma** gave the first elementary solution of the Diophantine equation

$$x(x+1)(2x+1) = 6y^2.$$

Also, the Chinese mathematician **Chen Jing-Run** proved that every sufficiently large even number (say $> 10^{10^{10}}$) has the form $p + q$, where p is a prime number and q is either a prime number or the product of two prime numbers.

During the so-called "cultural revolution" in the sixties, this kind of mathematics was considered to be useles for applications in industry and agriculture, and Chen Jing-Run lost his position as an academic teacher. But later, when the circumstances changed, he was restored to his position and was also considered to be a "hero of the revolution"!

The most ancient Indian mathematical texts preserved are the so-called *Sulvasutras*, i.e., "rules for the chord". The word *sulva* (or *sulba*) refers to chords (ropes) that were used in measurements, while the word *sutra* means book with rules or aphorisms which refer to ceremonial or scientific matters. The writing of the above texts is dated between 500 B.C. and 200 A.D. We find in them rules for the construction of right angles with the use of "triples of chords", the lengths of which constitute Pythagorean triples, as are (3,4,5), (5,12,13), (8,15,17) and (12,35,37), which means that they knew the general method of finding Pythagorean triples. It is very likely that *Sulvasutras* were influenced by the mathematics of the Babylonians, who already knew Pythagorean triples.

Various parts of India presented some unexpected mathematical activity. About 800 B.C., **Pingala** wrote a book entitled *The science of metrical speech*. The syllables in the poems are separated into "light" ones and "heavy" ones, something similar to what we call in our language "short" syllables and "long" syllables. Number 1 corresponded to light syllables and number 0 corresponded to heavy syllables. Thus, a number written in the binary system corresponded to every verse of a poem. This number can be transformed into the decimal system and vice versa.

The following were four important Indian mathematicians:

- **Aryabhata** (the elder), who was born in 476 A.D.
- **Brahmagupta**, who flourished in 628 A.D.
- **Mahavita**, who lived around 850 A.D.
- **Bhaskara**, who lived from 1114 A.D. to 1185 A.D.

In their work we find a mixture of old and traditional Indian mathematics, with mathematics of the Alexandrian era. In contrast to the Ancient Greek mathematicians, the above mathematicians omitted the proofs of a large number of propositions and wrote their books in metrical poetic speech.

For example, **Bhaskara** gives the following problem:
The square root of half the number of a
swarm of bees
flew over a bush of jasmine.
Eight ninths of the swarm remained at their position
and a female bee flies around a male that
buzzes over the flower of a lotus.
During the night, captivated by the sweet smell of the flower,
he enters in it.
And now he is trapped.
Tell me, charming lady, what is the number of the bees?

The above formulation of the problem constitutes a poetic way of asking the solution of the equation $\sqrt{\frac{x}{2}} + \frac{8x}{9} + 2 = x$.

Aryabhata knew the formulas that give the sum $1^k + 2^k + ... + n^k$ for $k = 1, 2, 3$. He solved second-order equations as well as equations of indeterminate form (i.e., equations admitting more than one solution). He was one of the first to make use of the sine function and he defined $\sin A$ as $\frac{\text{chord } 2A}{2}$. He constructed a table of sines which is actually quite accurate. He wrote a short book entitled *Aryabhatiya* in metrical speech, where he occupies himself with subjects from astronomy and mathematics. In this book, the following proposition deserves particular attention:

"Add 4 to 100, multiply by 8 and add 62,000. The result is approximately the circumference of a circle of diameter 20,000."

This proposition indicates that π is assumed to take the value 3.1416, which is essentially the value also used by Ptolemy.

In the book *Aryabhatiya* we find a fact that influenced decisively the mathematics of the subsequent generations regarding the way of writing a number in the decimal system. We don't know exactly how Aryabhata performed the arithmetical operations, but his phrase "from position to position, every digit is ten times the preceding one" suggests that he knew the principle that the value of a digit changes according to its position in the number. Of course, this principle was already known to the Babylonians, by the Indians who applied it to the expression of a number in the decimal system.

Brahmagupta studied the Diophantine equation $x^2 - \rho y^2 = 1$, where ρ is a given positive integer which is not the square of an integer. This equation is incorrectly called "Pell's equation". A complete solution of this equation was given by **Lagrange** in 1767.

Brahmagupta calculated the area and the volume of pyramids and cones using $\sqrt{10}$ as an approximate value of π. Of particular importance is the fact that he was the first to give a systematic presentation of the rules concerning the use of zero and of the negative numbers. Actually it is likely that he was the one who introduced the negative numbers into the number system. He states that a positive number divided by a positive number and a negative number divided by a negative number both give a positive number, while a positive number divided by a negative number and a negative number divided by a negative number both give a negative number. Division with zero seems to have puzzled Brahmagupta.

Perhaps the most elegant result (without proof) in Brahmagupta's work is the expression
$$\sqrt{(s-a)(s-b)(s-c)(s-d)}$$
that gives the area of a quadrilateral with sides a, b, c and d, inscribed in a circle, where $s = \frac{a+b+c+d}{2}$. This formula generalizes Heron's formula for the triangle.

We recall that the area of a quadrilateral (which is not necessarily inscribed in a circle) is given by the expression
$$\sqrt{(s-a)(s-b)(s-c)(s-d) - abcd \cdot \cos^2\alpha},$$
where α is half the sum of two opposite angles of the quadrilateral.

Continuing Brahmagupta's work, Mahavita studies the properties of the number zero and writes
$$a + 0 = a, \ a - 0 = a, \ a \cdot 0 = 0.$$

He also writes that a number remains unchanged when it is divided by zero, a mistake that might also be due to a person who copied the work.

He studied negative numbers and concluded that they are not squares of other numbers. He also gave the rule for dividing a number by a fraction.

Indian mathematicians recognized that second-order equations have in general two roots, including negative and irrational ones.

Bhaskara was the most important Indian mathematician of the 12th century. His book studies the writing of numbers in the decimal system, the operations of arithmetic, linear and second-order equations, arithmetic and geometric progressions, triangles and quadrilaterals, the approximation of π, trigonometric formulas, volumes of solids, linear equations of indeterminate form and problems in combinatorics. The book contains the oldest preserved presentation of notation in the decimal system where use of the symbol 0 (zero) is made.

Bhaskara gives the following rules regarding the use of the symbol 0:
$$a \pm 0 = a, \ 0^2 = 0, \ \sqrt{0} = 0, \ \frac{a}{0} = \infty.$$

Bhaskara apparently dedicated his book to his daughter, Lilavati, in order to comfort her because she never married. According to the opinion of astrologers, there was only one appropriate moment during which Lilavati could get married. But unfortunately, something happened which prevented Lilavati from her only chance to have an astrologically successful wedding. Then Bhaskara decided to dedicate to Lilavati a mathematical writing by saying: "I will dedicate to you a book that will bear your name and that will remain for times to come. Because a good name constitutes a second life and the foundation of eternal existence."

The arithmetical operations of the Indians were similar to ours today. As we mentioned above, the Indians progressively introduced the negative numbers into their arithmetical operations. They also began to perform operations with irrational numbers following correct methods that allowed them to draw useful conclusions, despite the fact that in general they did not give proofs of these methods.

The sum and the difference of two irrationals were calculated as if they were integers. For example, for the sum of the irrationals $\sqrt{3}$ and $\sqrt{12}$ Bhaskara writes
$$\sqrt{3} + \sqrt{12} = \sqrt{(3 + 12) + 2\sqrt{3 \cdot 12}} = \sqrt{27} = 3\sqrt{3}.$$

Here the following general principle is used:
$$\sqrt{a} + \sqrt{b} = \sqrt{(a+b) + 2\sqrt{ab}}.$$
We remark that in calculations the Indians treated the irrationals as if they had the same properties with the integers. Thus, if c and d are integers, then of course we write
$$c + d = \sqrt{(c+d)^2} = \sqrt{c^2 + d^2 + 2cd}.$$
Now if in this relation we set $c = \sqrt{a}$ and $d = \sqrt{b}$, then we obtain the relation displayed immediately above.

Bhaskara also gives another rule for adding two irrationals which can be understood from the example below:
$$\sqrt{3} + \sqrt{12} = \sqrt{\left(\sqrt{\frac{12}{3}} + 1\right)^2 \cdot 3}.$$

Indian mathematicians were less profound than Greek mathematicians, because they could not appreciate the logical difficulties that come up in the notion of the irrational number. The fact that they concentrated mainly on performing operations, didn't allow them to observe that there existed certain distinctions of philosophical nature or at least distinctions based on principles that the Greeks considered to be fundamental. However, we can say that this "casual" way of treating the irrational numbers was very beneficial for the progress of mathematics.

Bhaskara died near the end of the 12th century and for many centuries there were very few mathematicians in India who could be compared with Bhaskara, until the ingenious **Srinivasa Ramanujan** (1887-1920). He was the most important Indian mathematician of the 20th century and he had the same mysterious ability as Bhaskara to handle easily problems in number theory and algebra. When the English mathematician **G. H. Hardy** visited Ramanujan at Putney's hospital, he mentioned to his friend that the taxi that brought him there had the uninteresting number 1729. Ramanujan, without hesitating at all, replied to him that this number is interesting because it is the smallest integer that can be written in two different ways as the sum of two cubes:
$$1^3 + 12^3 = 1729 = 9^3 + 10^3.$$

Ramanujan was self-educated and perhaps this is the reason for which his work, which contains many beautiful and ingenious mathematical formulas, does not constitute an organized set, but at the same time it shows an enormous power of "intuitive logic". Just like his ancestors, Ramanujan treats no geometric subjects in his work.

CHAPTER 21

MATHEMATICS OF THE ARABS

The fall of the empire of Saba in Arabia occurred about the time that Brahmagupta flourished in India. As a result of this fall a deep crisis began in the Arabic peninsula. Arabia was then inhabited by nomads of the desert, the Bedouins, who did not know how to read or write. Among them was the subsequent Prophet Mohammed, who was born in Mecca around 570. Mohammed traveled a lot and in his travels met many Christians and Judaeans. The "amalgam" of the various religious emotions created in him as a result of these contacts, led him to consider himself as God's emissary whose destiny was to become the leader of his people.

For approximately ten years Mohammed was preaching God's word in Mecca and in 622 when a conspiracy against his life was revealed, he replied affirmatively to an invitation and went to Medina. This "transition", known as "Hegira", which constitutes the beginning date for the Moslems (16th of July of 622 A.D.), also exercised a powerful influence to the development of mathematics. Mohammed became then the military and religious leader. Ten years later he founded a Moslem state with Mecca as its capital, where Christians and Judaeans, since they were also monotheists, were welcome, protected and free to work. Mohammed died in Medina in 632, exactly when he was preparing to attack Byzantium. But his sudden death was no obstacle to the expansion of Islam; actually the people who continued his work conquered neighboring territories with an amazing speed. In a few years the Arabs conquered Damascus, Jerusalem and a large part of Mesopotamia, while in 641 they conquered Alexandria which was, as mentioned earlier, for many years the mathematical center of the world. In previous chapters we mentioned the fate of the books of the famous Library of Alexandria. In approximately one century after Mohammed's death, the Arabs conquered territories that spread from India to Spain, including territories of North Africa and South Italy.

In 755 the Arabic state was divided into two independent kingdoms, the Eastern with Bagdad as its capital and the Western with Cordova in Spain as its capital.

After completing their conquests, the exnomads intensified their efforts to create civilization and to acquire education. Very quickly, the Arabs showed a vivid interest for art and science. Both capitals mentioned above attracted many scientists and supported their work. However, Bagdad proved to be more important, because an academy, a library and an observatory were founded in it and the city became the new center of mathematical interest.

The Arabs had substantial cultural sources at their disposal: They invited Indian scientists to settle in the Arab state; when Justinian closed Plato's Academy in 529, many Greek scientists went to Persia and the Greek education that flourished there became part of the Arabic education a century later. The Arabs also had many contacts with the Greeks of the Byzantine Empire and many caliphs bought Greek manuscripts from the Byzantines. Egypt, the center of Greek Education during the Alexandrian era, was conquered by the Arabs and the existing Greek education contributed to the strengthening of the scientific activity of the Arabs. In summary the Arabs had at their disposal both the intellectuals and the civilization of the Byzantine Empire of Egypt, Syria, Persia and other eastern territories including India.

We emphasize that when we talk about the mathematics of the Arabs, we mean mainly the mathematics in which use of the Arabic language was made. The majority of the scientists were Greeks, Christians, Persians and Judaeans. What the Arabs possessed was essentially "Greek knowledge" which they acquired from Greek texts which were translated into Syrian and Hebrew. Thus, they had at their disposal the most important writings. In 800 A.D. they acquired a copy of Euclid's *Elements* from the Byzantines and they translated it into Arabic. Ptolemy's *Mathematical Syntaxis*, the so-called *Almagest*, was translated into Arabic in 827 A.D. and became for the Arabs a writing of utmost importance and an almost divine origin. The Arabs also translated Ptolemy's *Tetrabiblos*, a writing on astrology which became very popular.

As time went by, the Arabs became acquainted with the works of Aristotle, Apollonius, Archimedes, Heron, Diophantus as well as with the works of certain Indian writers. The also wrote comments and made improvements on these translated works. These translations, some of which were preserved, came later to the possession of the Europeans which was very fortunate since the ancient Greek writings were lost. The Arabic civilization remained in a dynamical state until 1300 and was greatly disseminated.

The Arabs used and improved the arithmetical symbols of the Indians and, in particular, the idea that "the value of a digit depends on its position in the number" (units, tens, hundreds, etc.). Just like the Indians, the Arabs used the irrational numbers casually. Both **Omar Khayyam** (1048-1122) and **Nasir-Eddin** (1201-1274) state clearly that the ratio of two arbitrary quantities can be called a "number" no matter whether they are commensurable or incommensurable. Later, in 1707, **Newton** was obliged to accept this statement in his writing *Arithmetica Universalis*.

But in Arithmetic, the Arabs made a step backwards since they neither accepted the negative numbers nor the operations on negative numbers, which as we show were introduced by the Indians.

The name "Algebra" comes from a book of the astronomer **Mohammed ibn Musa al-Khowârismi** written in 830 and entitled *Al-jabr w'al muqâba-la*. The word *al-jabr* means "restore" and in this case it means that one restores the equilibrium in an equation by putting a term to one of its sides provided that it is subtracted from its other side; for example, if -7 is subtracted from $x^2 - 7 = 3$, then the equilibrium will be restored if we write $x^2 = 7 + 3$. *Al' muqâbala* means "simplification", i.e., what we do when we write $7x$ instead of $3x + 4x$ or when we subtract the same term from both sides of an equation.

Al-jabr also means the one who pieces together a broken bone. The same etymology explains the word "algebrista" which appears in ancient Spanish, which also means the one who pieces together a broken bone and which was written in the past in entrances of barbershops in Spain.

In a Latin translation (1875) of Al-Khowârismi's book, the author's name is written Algorithmi, from which originated the word "algorithm".

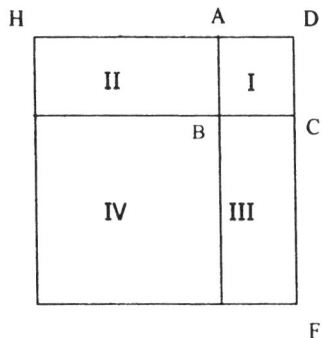

FIGURE 21.1

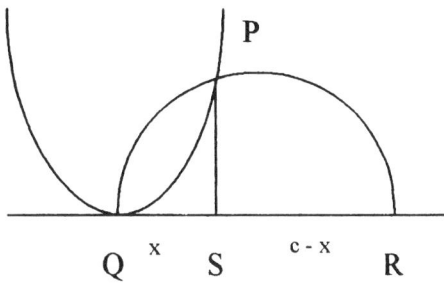

FIGURE 21.2

Despite its great fame, Al-Khowârismi's "Algebra" cannot be considered to be profound from a mathematical vantage point, because it contains nothing more than what was known by the Babylonians and the Ancient Greeks.

Even though the solutions to second-order equations given by the Arab mathematicians were algebraic, they also explained them geometrically. Since they were obviously influenced by the Greeks, they believed that algebraic proofs had to be accompanied by geometric proofs. Thus, Al-Khowârismi gives the following geometric solution of the equation $x^2 + 10x = 39$. Let AB (Figure 21.1) stand for the value of the unknown x. We construct the square $ABCD$. We extend DA, DC and we consider on them the points H, F, respectively such that $AH = CF = 5 =$ half the coefficient of x. We form the square with sides DH and DF. The areas I, II and III are equal to x^2, $5x$ and $5x$ respectively. Their sum equals the left side of the equation. We add to both sides of the equation the area IV, which is equal to 25. Thus, the entire square has area $39 + 25 = 64$ and its side is equal to 8. Hence $AB = 8 - 5 = 3$ and consequently $x = 3$. We remark that the above geometric proof is based on Proposition 4 of Book II of *Elements*.

The Arabs also gave solutions to certain third-order equations and these solutions were accompanied by geometric explanations similar to the ones mentioned above about second-order equations. The following solution of the equation $x^3 + Bx = C$, where B, C are positive numbers, given by Omar Khayyam (1079) constitutes an example.

Khayyam writes the above equation in the form $x^3 + b^2x = b^2c$, where $b^2 = B$ and $b^2c = C$. Then he considers the parabola $x^2 = by$ (Figure 21.2). Even though he cannot construct it with the help of a ruler and a compass, he can construct as many points of it as he wants. Next he constructs the semicircle with diameter $QR = c$. Then the intersection P of the parabola with the semicircle determines the normal PS and the solution of the equation is QS.

The proof given by Khayyam is synthetic. It follows from the equation $x^2 = by$ that
$$x^2 = b \cdot PS \tag{1}$$
or
$$\frac{b}{x} = \frac{x}{PS}. \tag{2}$$
From the right-angled triangle QPR we have
$$\frac{x}{PS} = \frac{PS}{c-x}. \tag{3}$$
It follows from (2) and (3) that
$$\frac{b}{x} = \frac{PS}{c-x}. \tag{4}$$
From (2) we have
$$PS = \frac{x^2}{b}. \tag{5}$$
If we substitute the value of PS in (4), we observe that x satisfies the equation $x^3 + b^2 x = b^2 c$.

Although the Greeks, and in particular Euclid, Archimedes and Heron, obviously influenced the Arabs in geometry, the solution given by the Arabs to third-order equations with the use of intersecting conic sections constitutes an important step forward in Algebra.

Finally, the Arabs made very small progress in trigonometry.

In summary, the most important contribution of the Arabs in mathematics was mainly the fact that they "absorbed" the mathematics of the Greeks and the Indians, they preserved it, and finally they transmitted it to Europe. The Arabs' activity reached its highest point during 1000 A.D.

The crusades that took place between 1100 and 1300 reduced the power of the eastern Arabic state, which was later conquered by the Mongols. After 1258, the caliph of Bagdad ceased to exist. The complete catastrophe came with the raid of the Tatars let by Tamerlane.

In Spain the Arabs were constantly under attack by the Christians. In 1492 the Christians conquered Spain and thus the mathematical and the general scientific activity of the Arabs in that area came to an end.

Summarizing, we can say that mathematics presented the following picture during the year 1300: There exist two basic traditional trends or directions. On the one hand, we have the Greek Creation, i.e., the set of acquired knowledge which resulted from proofs using deductive logic. This trend was aimed at helping the human understand and interpret nature. On the other hand, we have the empirical and practical trend in mathematics, which was founded by the Babylonians and the Egyptians. Its creation was aided by certain Alexandrian Greek mathematicians and it was extended by the Indians and the Arabs. The first direction favored geometry while the second favored arithmetic and algebra.

CHAPTER 22

EUROPE DURING THE MIDDLE AGES

Mathematics in central and western Europe began to develop at the same time the Arab civilization began to decline.

In order for the reader to appreciate the general situation in Europe during the Middle Ages, and how and in what directions European civilization began to develop, we must briefly refer to the beginnings of this civilization.

During the time of the Babylonians, the Egyptians, the Greeks and the Romans, only a primitive civilization existed in the area that we call today Europe (excluding Italy and Greece). The German tribes that lived in Europe were completely illiterate. The Roman historian **Tacitus** (1st century A.D.) describes these tribes during the time of Christ, as follows: They were honest and hospitable, they enjoyed drinking, they hated peace and they were proud of the honesty of their spouses. Their main occupation was cattle-breeding and the cultivation of grains.

The church had imposed Latin as the official language in Europe. Thus, Latin became the language of mathematics and of the exact sciences until approximately the middle of the 8th century A.D. Europeans were obliged to use Latin, i.e., Roman books, but since the mathematics of the Romans was insignificant, Europeans acquired from the Romans very little knowledge, which involved a primitive number system and certain matters of arithmetic.

The main translator of writings, whose books were in use until the 12th century, was **Amicius Manlius Severinus Boethius** (480-524). Boethius used Greek sources and collected in Latin various texts from elementary writings in arithmetic, geometry, and astronomy. From Euclid's *Elements* he translated between two to five books and these translations formed part of his writing *Geometry*. Boethius also published the book *Institutis arithmetica*, which was a translation of **Nicomachus'** writing *Introductio arithmetica*, from which, however, he omitted certain subjects. He also translated certain works of Aristotle and published a book on astronomy based on Ptolemy's work. Finally, Boethius published a book on music based on works of Euclid, Ptolemy and Nicomachus. The viewpoint exists that Boethius did not always understand everything that he translated!

The most known work of Boethius, which is read even today, is *Consolations of Philosophy*, which he wrote during the time that he was in prison, accused of treason. For this accusation he was finally decapitated.

All existing books which were written by mathematicians of the Middle Ages, contain only the four known operations on the integers. The use of fractions is rare. Nowhere did they use irrational numbers. But they made wide use of various kinds of abaci. The main reasons that mathematics in Europe did not show any essential progress during the time interval 400-1100, are: the lack of interest for the study of the natural world, religious fanaticism and the influence of various philosophers, such as the Stoicals, the Epicureans and the neo-Platonics, who urged people to stand above flesh and matter and prepare their soul for life after death.

Roman civilization did not favor the development of mathematics because the main interest was to obtain direct and practical results.

European civilization in the Middle Ages also did not favor the development of mathematics but for exactly the opposite reasons: There was no interest for the study of the natural world. Mundane affairs and problems were considered to be of no importance. Emphasis was given to life after death and to the preparation for this life.

Mathematics cannot thrive in civilizations such as the Roman and European in the Middle Ages. It only thrives in an atmosphere of complete intellectual freedom, where one is interested equally in problems that come from the natural world and in abstract ideas that these problems suggest, and where consideration of these problems does not necessarily promise a direct and useful practical result. Ideas must be studied for their own sake. Thus, oddly enough, a new picture of nature appears, which is richer and wider, and this results in the formation of a more powerful viewpoint about nature, which in turn gives birth to deeper mathematical activities.

Around the year 1100, European civilization was somewhat stabilized. Despite the domination of feudal societies, there existed trends for independent activities in commerce, over the creation of large or small industries, in arts, etc. Around that time the intellectual atmosphere in Europe was subject to new impacts. Travel and commerce brought the Europeans in contact with the Arabs in the areas of Mesopotamia and the Middle East, as well as with the Byzantines of the Eastern Roman State. The Crusades (1100-1300) brought the Europeans to Arabic territories. As a result of all these movements, Europeans began to study the works of the Ancient Greeks via the Arabs and the Greeks of Byzantium, and this fact marked the beginning of a great intellectual activity in Europe.

Europe became acquainted with the works of Euclid and Ptolemy, with the works of Al-Khowârismi in Arithmetic and Algebra, with Theodosius' *Sphaerica*, with many works of Aristotle and Heron, and with some works of Archimedes regarding mainly the measurement of the circle. The remaining works of Archimedes were translated into Latin in 1544 by **Hervagius of Basle**. None of the works of Apollonius or Diophantus were translated during the twelfth and the thirteenth century. But works on philosophy, medicine, exact sciences, theology and astrology were translated. Thus, since the Arabs had at their disposal almost all Greek works, the Europeans acquired through the Arabs a huge number of Greek writings.

Gerbert von Aurillac (940-1003) studied in Spain. He wrote excellent books on arithmetic and geometry (his contemporaries believed that he had made a "deal" with the devil!). Gerbert became Pope under the name Sylvester II and remained at his dignity from 999 until 1003.

A contemporary of Gerbert was another mathematician, **Hrotsvitha** of Saxony (932-1002), a clergyman who was interested in perfect numbers.

The Englishman **Adelard of Bath** (1090-1160) studied in Spain's Cordova, where he acquired a copy of Euclid's *Elements* in Arabic.

22. EUROPE DURING THE MIDDLE AGES

Abraham Ben Ezra (1095-1167) resided in Spain but made many trips to Egypt and England. He wrote a book *Sefer ha-Mispar* which contained the arithmetic of the Indians, where he used Hebrew letters for numbers and he also made use of the zero.

Jordanus Nemorarius (beginnings of the 13th century) wrote a book on Geometry, Arabic numbers, prime numbers, perfect numbers, polygonal numbers, ratios, powers, and progressions. He used letters to denote the unknowns in an equation. In his book *Tractatus de numeris datis*, he solved equations of the type: $x + y = 10$ and $x^2 + y^2 = 58$". Problems of this kind had greatly occupied the people of Mesopotamia.

Nicole Oresme (1320-1382) was a French bishop and was the first to prove that the harmonic series diverges.

We note that in a time interval of one thousand years, i.e., from 400 until 1400, there existed only one distinguished European mathematician, namely **Leonardo of Pisa** (1180-1250), known as **Fibonacci**. He studied mathematics in Algeria, and in 1202 he published the book *Liber abaci*, where he presents the Indian number system and introduces the famous "**Fibonacci sequence**":

$$1, 1, 2, 3, 5, 8, 13, 21, 34, 55, \ldots$$

every term of which, apart from the first two, is the sum of his two immediately preceding ones.

Another book of Fibonacci is *Liber quadratorum*. This original work in the area of indeterminate analysis gives the first proof of the identity

$$(a^2 + b^2)(c^2 + d^2) = (ac - bd)^2 + (ad + bc)^2.$$

We note that Diophantus (Problem 19, Book III, *Arithmetica*) gave a numerical example of this Fibonacci identity.

In 1225 the emperor Frederic II postponed his departure for a crusade in order to find the time to organize a mathematical conference. Fibonacci, attended the conference and solved successfully all the suggested problems. Two of these problems were the following:

(1) Find a rational number a/b such that the expressions $(a/b)^2 \pm 5$ are squares of rational numbers.
(2) Solve the equation $x^3 + 2x^2 + 10x = 20$.

Problem 1 can be solved using the property mentioned by Diophantus that in every right-angled triangle the following relation holds:

$$\left(\frac{1}{2} \cdot \text{hypotenuse}\right)^2 \pm \text{area} = \text{perfect square}.$$

Example. $a/b = 41/12$.

Fibonacci at some point in his work understood the importance of the negative numbers which he interpreted as "losses".

We note that the seeds of the plant helianthus are ordered in such a way that they form two kinds of arcs starting from the center. The number of seeds located on each of these arcs is 21 and 13 **which are successive Fibonacci numbers.**

Below we present the formula that gives the nth term u_n of the sequence of Fibonacci numbers.

Let $u_0 = 0$. We have $u_1 = 1$ and $u_{n+2} = u_n + u_{n+1}$ for $n \geq 0$. We set
$$U(x) = u_0 + u_1 x + u_2 x^2 + \ldots = \sum_{n=0}^{\infty} u_n x^n.$$
This formal power series (formal because we don't know whether it converges) is the "function" that gives the term u_n, even though we don't know whether it is indeed a function. It is easily recognized that
$$U(x)(1 - x - x^2) = u_0 + (u_1 - u_0)x$$
since $(u_{n+2} - u_{n+1} - u_n)x^{n+2} = 0$ for $n \geq 0$. Consequently
$$\frac{x}{1 - x - x^2} = \frac{A}{1 - \alpha x} + \frac{B}{1 - \beta x} = A \sum_{n=0}^{\infty} \alpha^n x^n + B \sum_{n=0}^{\infty} \beta^n x^n$$
where
$$(1 - \alpha x)(1 - \beta x) = 1 - x - x^2$$
$$A(1 - \beta x) + B(1 - \alpha x) = x.$$
The first identity implies
$$\alpha + \beta = 1, \quad \alpha\beta = -1,$$
thus
$$\alpha = \frac{1 + \sqrt{5}}{2}, \quad \beta = \frac{1 - \sqrt{5}}{2}.$$
The second identity implies
$$A = \frac{1}{\alpha - \beta} = \frac{1}{\sqrt{5}}, \quad B = \frac{1}{\beta - \alpha} = \frac{-1}{\sqrt{5}}.$$
By comparing the above two expressions of $U(x)$ we find that
$$u_n = A\alpha^n + B\beta^n = \frac{1}{\sqrt{5}} \left(\frac{1 + \sqrt{5}}{2} \right)^n - \frac{1}{\sqrt{5}} \left(\frac{1 - \sqrt{5}}{2} \right)^n.$$

Number α is the "golden section" and played an important role in Euclid's construction of the side of a regular pentagon.

Fibonacci numbers came up in relation to the following problem that can be found in the book *Liber abaci* mentioned above.

"Assume that a rabbit's pregnancy lasts one month and that every female rabbit becomes pregnant at the beginning of every month starting from the moment that it is one month old. Assume also that female rabbits always give birth to two rabbits, one male and one female. How many pairs of rabbits will exist on January 2, 1203 if we start with a newborn pair on January 1, 1202?"

The number of pairs of rabbits increases as follows:
$$1, 1, 2, 3, 5, 8, 13, 21, 34, 55, \ldots$$
hence if u_n is the nth Fibonacci number, then we have $u_1 = 1$, $u_2 = 1$ and $u_{n+1} = u_n + u_{n-1}$.

By summarizing, we could say that during the Middle Ages, activities were in general restricted to the consideration, the study and the review of existing works. The dominating spirit during that time forces the scientists to keep to the beaten track, i.e., to follow known and secure paths. Research for the formulation of a philosophy that would cover all the phenomena, which are related to the human, to nature, and to God, is a characteristic feature of the second half of the Middle Ages.

But this effort was unsuccessful, the ideas were vague and there existed mysticism and dogmatism.

As time went by it became gradually understood that it was necessary to revise the way of learning as well as certain existing beliefs. Before the time that Galileo demonstrated the value of experimentation and the value of experience, before the time that Descartes taught the human how to observe his/her internal "ego", and before the time that Pascal formulated the idea of progress, we find only a few independent and isolated thinkers. These few non-conformist intellectuals, tried to trace new roads, to question the existing crystallized views and to place more emphasis on the observation of nature (as the Ancient Greeks had done in the past).

Although one can observe some new trends, no essential progress was achieved in mathematics and in the exact sciences. Free research was forbidden. The few existing universities in 1400 were controlled by the Church and the professors were not free to teach what they considered to be right. The atrocities of Inquisition were unprecedented (the beginning of Inquisition dates back to the 13th century and is due to Pope **Innocent III**).

There existed also additional reasons, which delayed changes in Europe. Only few intellectuals had the time to consider again and to study the Greek Knowledge of older times. Manuscripts were expensive and hence inaccessible to the wide public.

During the period 1100-1500 Europe was divided into independent duchies, principalities and small states which depended on the Pope. The wars between these political units were absorbing the main activity of their people.

The crusades that began in 1100 caused the death of a huge number of persons. The epidemic of plague during the second half of the 14th century caused the death of approximately 1/3 of the population of Europe, thus delaying the progress of civilization. Fortunately, forces of revolutionary power began to influence the European intellectuals and politicians, as well as the general public.

CHAPTER 23

RENAISSANCE (1400-1600)

During the thousand years that followed Diophantus, apart from the discovery of the decimal system by the Indians with the symbol zero for the empty place, and apart from the work of few talented mathematicians like Pappus and Fibonacci, no other important progress was made in mathematics.

The main events, which led to the drastic revival not only of mathematical knowledge but of knowledge in arts and sciences in general, were the conquest of Constantinople (1453) by the Turks and the discovery of typography.

After the fall of Constantinople, many Greek intellectuals took refuge in the West and in particular in Italy, carrying with them manuscripts of the Ancient Greeks. Some historians allege that if the king of the Serbs, **Stephen Dusan** (1308-1356), had not died so prematurely or had a good heir, Constantinople would have been dominated by the West and not by the Turks. It would be interesting to study (if it hasn't been already) what the effects of such an event would have on a civilization in general and on the development of mathematical science in particular.

Before we review the development of mathematics during the Renaissance, we first make some general remarks.

Renaissance did not produce sensational new results in the science of mathematics. Progress in mathematics was relatively small as opposed to the great achievements in literature, in painting and in architecture, where we find masterpieces which form part of our civilization today. For mathematics this period was mainly a period of "absorption" of the Greek works, an "afresh" discovery of the ancient Greek civilization.

However, the Renaissance was very important for the "health" and preparation for further development of mathematics, because during that period the close relationship that exists between the exact sciences and technology was restored, and because people perceived that mathematics constitutes and will constitute the powerful force that is needed for the creation of new important achievements.

We will refer now in more detail to the development of mathematics during the Renaissance, and in particular to the following areas:

(1) Mathematical Notation
(2) The theory of equations
(3) The invention of logarithms
(4) Mechanics and Astronomy

(1) Mathematical Notation

Johannes Regiomontanus (1436-1476) was born in Königsberg which at the time belonged to Germany. He wrote a systematic study regarding plane and spherical trigonometry, in which he used "sines" and "cosines". In algebra, he denoted by "res" the unknown x and he denoted x^2 by "census".

Christopher Columbus took with him on his fourth trip a writing of Regiomontanus entitled *Ephemerides* and used the prediction written in it that "on February 29 of 1504 an eclipse of the moon will take place", to intimidate a hostile group of Red Indians in Jamaica.

Johannes Widman (1462-1500) came from Eger, which belongs today to the Czech Republic. He published a book in which the contemporary symbols + and − appear for the first time.

Luca Pacioli of Italy (1445-1517) was a Franciscan monk. He used the terms "res" and "census" of Regiomontanus. In 1509 he published the book *Divina Proportione*, where he studied the regular polygons, the regular polyhedra and the "golden section". The superb figures of the book are attributed to **Leonardo da Vinci** (1452-1519). We note that Leonardo is often reported as a mathematician. However, he was not in close contact with the main mathematical trends of his time.

A famous painting of **Jacobo de Barbari** exists in Napoli's Museum, which presents Pacioli and his friend Guidebaldo standing next to a regular dodecahedron.

One of the problems solved by Pacioli was the following:

"The radius of a circle inscribed in a triangle is 4, and the segments in which one of the sides of the triangle is divided by the point of contact with the circle are 6 and 8. Find the other two sides of the triangle."

Robert Recorde (1510-1558), who was a professor at Cambridge, was the most important English mathematician of his century. Recorde wrote the first English writing on algebra entitled *The Whetstone of Witte* (1557) and is the first who introduced the symbol = of equality. He had said that he had never known two things that were as similar to one another as two parallel straight lines and this is why two parallel straight lines had to denote "equality".

In 1525 the German mathematician **Christoph Rudolff** used $\sqrt{\ }$ to denote the extraction of square roots.

Michael Stifel (1487-1567) was one of the first German monks who followed Luther. He used for the first time negative integers as exponents and introduced the notation $1A$, $1AA$, $1AAA$ to denote the powers A, A^2, A^3 respectively. He used mathematics for the interpretation of the Bible. In this way Stifel concluded that Pope Leon X was the best among those mentioned in the Book of Apocalypse. He also made the prophecy that the end of the world would come on October 18, 1533. The inhabitants of his village believed his prophecy and spent all of their money! But when the prophecy was not fulfilled, Stifel was sent to prison.

The Englishman **Thomas Harriot** (1560-1621) wrote a, aa, aaa for a, a^2, a^3 respectively and introduced the symbols > and < to express "strict" inequality.

In 1603 Harriot calculated the area of a spherical triangle: The part ABC of a spherical surface, which is enclosed by three arcs of great circles, is called a **spherical triangle**. The points A, B, C are called **vertices**. The three arcs a, b, c are called **sides**, while the angles formed by the intersecting tangents to the

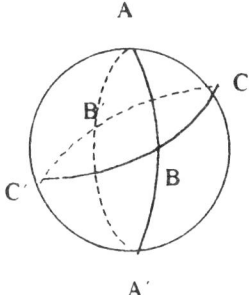

FIGURE 23.1

sides at the vertices are called **angles** of the spherical triangle. If we also denote these angles by A, B, C, then we have the relation $A + B + C - \pi = E > 0$. The quantity E is called **spherical excess**. Spherical triangles have properties which are analogous to the ones of plane triangles:

$$\frac{\sin a}{\sin A} = \frac{\sin b}{\sin B} = \frac{\sin c}{\sin C},$$
$$\cos a = \cos b \cdot \cos c + \sin b \cdot \sin c \cdot \cos A.$$

The study of spherical figures via trigonometric functions is called **spherical trigonometry** and is widely used in astronomy, in geodesy and in navigation.

The calculation of the area of a spherical triangle ABC is carried out as follows: We subtract 180° from the sum of the three angles of the triangle and we form the fraction whose numerator is $A + B + C - 180°$ and whose denominator is 360. This fraction expresses how much of the area of the hemisphere is occupied by the triangle. The correctness of the above calculation is proved as follows: If the radius of the sphere is equal to 1, then its area is equal to 4π or 720°. Consequently, the area that is contained between two great circles and lies on one hemisphere is $720°/360 = 2°$. From this it follows that the area contained between two great circles forming an angle A is equal to $2A$.

The spherical triangle ABC is surrounded by three great circles, as shown in Figure 23.1, where A', B', C' are the antipodal points of A, B, C respectively.

We have

$$\triangle ABC + \underline{\triangle A'BC} = 2A,$$
$$\triangle A'B'C' + \underline{\triangle A'BC'} = 2B,$$
$$\underline{\triangle ABC} + \underline{\triangle ABC'} = 2C.$$

The four underlined triangles represent the visible hemisphere (Figure 23.1), i.e., their areas sum up to 360°. Consequently, by adding the sides of the above equalities, we find

$$\triangle ABC + \triangle A'B'C' + 360° = 2A + 2B + 2C.$$

But $\triangle A'B'C' = \triangle ABC$, because the triangles ABC and $A'B'C'$ are antipodal to one another. Therefore,

$$\triangle ABC + 180° = A + B + C.$$

Consequently, the area of a spherical triangle ABC expressed in degrees is equal to its spherical excess, i.e., it is equal to $A + B + C - 180°$.

(2) The theory of equations

The solution of the second-order equation was known from the time of the ancient Babylonians.

Omar Khayyam (1100 A.D.) had developed a method with which he could draw a line segment whose length is equal to the positive real root of a given polynomial of degree three. In 1225 **Leonardo of Pisa** gave a numerical solution to the equation
$$x^3 + 2x^2 + 10x = 20.$$

Apparently **Scipione Ferro** of Bologna (1465-1526) was the first to obtain a comprehensive method of solution of a third-order equation, giving all its roots, no matter whether they are complex or real, positive real or negative real. Ferro was apparently able to solve any equation of the form $x^3 + bx = c$ and to give its solutions at an approximation of any number of digits. Ferro kept his method secret in order to have an advantage over the other mathematicians at mathematical competitions. Ferro entrusted his solution, just before he died, to **Antonio Fiore**.

In 1530 **Zuanne da Coi** sent to **Niccolo Tartaglia** (1500-1557) the following problems:
$$x^3 + 3x^2 = 5,$$
$$x^3 + 6x^2 + 8x = 1000.$$

Tartaglia declared that he could solve these equations and hence Fiore challenged him to a competition in solving problems. Every participant gave a certain amount of money to a notary and gave a number of problems to his opponent to solve. The participant that would solve in 30 days the most problems would take all the deposited money. Tartaglia, who suspected that Fiore would present to him equations of the form $x^3 + bx = c$, found in a short period of time a general method for solving such equations . Indeed, the problems that Fiore suggested to him were of this nature and Tartaglia was able to solve them all. Tartaglia himself had suggested to Fiore equations of the form $x^3 + ax^2 = c$, which proved to be very difficult for Fiore, while Tartaglia knew how to solve them.

Tartaglia was born in Italy's Brescia. This city was occupied by the French in 1512 and many of the inhabitants were slaughtered. At this battle, a soldier with a stroke of his sword injured Tartaglia's jaw, and this is how Tartaglia acquired his name meaning stutterer.

Tartaglia taught in Verona and in Venice, and after his victory over Fiore he became famous. In 1537 he published a book on "ballistics". In this book he mentions, but without proving it, that the maximum range of a shell is obtained when the shell is launched at an angle of $45°$. In 1500 Tartaglia wrote a book on number theory.

Near the end of his life Tartaglia quarreled with **Girolamo Cardano** (1501-1576), a famous Italian doctor who lived in Milano. Cardano occupied himself with mathematics, which he applied to mechanics, to games of chance, and to astrology. He is considered by many as the founder of probability theory. Cardano's elder son was executed for poisoning his wife, while Cardano himself was imprisoned in 1570 as a heretic accused of publishing a horoscope about Jesus Christ. Cardano was released from prison after he was made to recant his allegations.

Cardano convinced Tartaglia to entrust him with the secret method for solving a third-order equation, under the condition that Cardano would not reveal it to anyone. But after Cardano learned of Ferro's work he decided to publish it in his writing *Ars Magna* (1545). Cardano's indiscretion angered Tartaglia, who challenged Cardano to a public competition in solving mathematical problems. Cardano sent his student **Ferrari** (1522-1565) in his place, and he proved to be better than Tartaglia.

Cardano's book *Ars Magna* was the best of its time. Cardano uses in it a geometric method in order to prove the algebraic identity

$$(a-b)^3 = a^3 - b^3 - 3ab(a-b).$$

He avoids the use of negative numbers, something that forces him to consider as different cases the equations

$$x^3 + px = q, \ x^3 = px + q, \ x^3 + px + q = 0, \ x^3 + q = px.$$

Despite all that, the book *Ars Magna* contained a comprehensive study of the third-order equation and also, what is very important, he treats complex numbers (which will be discussed later).

Ferrari, who came from Italy's Bologna, extended the work of Tartaglia and Cardano by solving the general fourth-order equation. The solution that he gave is also published in the book *Ars Magna*. According to the historian **W. W. Rouse Ball**, Ferrari was murdered by the lover of his sister.

Rafael Bombelli (1526-1572) came from Bologna. He published a book on algebra, where he studied the history of algebra starting from the time of Diophantus. He remarked that the problem of trisecting an arbitrary angle using a ruler and a compass only, is reduced to the solution of a third-order equation. He also denoted the power x^n by $\overset{n}{\bigcup}$.

François Viète (or Vieta) (1540-1603) was a French lawyer and a member of Parliament. He wrote the book *In artem analyticem isagoge* (1591), where he applied algebra to geometry, as opposed to what mathematicians used to do until then. When King Henry IV asked him to solve a 45th-order equation of a special form, Viète solved it in only a few minutes, because he observed that this equation was satisfied by the length of the chord of an angle of $360°/45$. Viète also constructed the circles which are tangent to three given circles, using only Euclidean geometry. The solutions that Viète gave to third-order and fourth-order equations are similar to the ones that we give today. We will present these solutions immediately below.

The equations of order three and four

Cardano was the first to use complex numbers in his printed work. We will present below, using also complex numbers, the solution that Cardano gave to the third-order equation. We will present also the solution that Ferrari gave to the polynomial equation of order four.

We recall that $i = \sqrt{-1}$.

We set $\omega = \frac{-1+i\sqrt{3}}{2}$, hence it follows that $\omega^2 = \frac{-1-i\sqrt{3}}{2}$ and $\omega^3 = 1$.

The following propositions are known:

(α) Every complex number $x+iy$ can be written in the form $r(\cos A + i \sin A)$, where r is a non-negative real number and A is a real number.
(β) $[r(\cos A + i \sin A)]^3 = r^3(\cos 3A + i \sin 3A)$
(γ) The equation $z^3 = 1$ has exactly three solutions: $1, \omega, \omega^2$.
(δ) If b is a given non-zero number, then the equation $z^3 = b$ has exactly three solutions. If z_1 is one of them, then $z_1 \omega$ and $z_1 \omega^2$ are the other two.
(ϵ) If y and p are complex numbers, then there exist complex numbers u and v such that $u + v = y$ and $uv = p$.

The general equation of order three (with real coefficients) has the form
$$x^3 + ax^2 + bx + c = 0 \quad (a, b, c \text{ are real numbers}).$$
We put $x = y + k$, hence the equation takes the form
$$y^3 + (3k + a)y^2 + (\ldots)y + (\ldots) = 0.$$
We take $k = -\frac{a}{3}$, hence the coefficient of y^2 vanishes, and the equation that results has the form
$$y^3 - 3py - 2q = 0,$$
where p and q are real numbers, while the numbers -3 and -2 were introduced in order to make calculations easier.

If we put $y = u + v$, in the above equation, we find
$$u^3 + v^3 + 3(uv - p)(u + v) - 2q = 0,$$
where we put $v = \frac{p}{u}$ (see Proposition (ϵ)) and we obtain
$$u^3 + v^3 = 2q.$$
Since $u^3 v^3 = p^3$, it follows that the numbers u^3 and v^3 are the solutions of the equation
$$t^2 - 2qt + p^3 = 0.$$
Consequently
$$u^3 = q + \sqrt{q^2 - p^3},$$
$$v^3 = q - \sqrt{q^2 - p^3}.$$
Therefore $y = u + v = u + \frac{p}{u}$, where u is a cubic root of $q + \sqrt{q^2 - p^3}$.

If one of these cubic roots is u_1, then the other two are $u_1 \omega$ and $u_1 \omega^2$. So let $v_1 = \frac{p}{u_1}$. Then
$$u_1 \omega + \frac{p}{u_1 \omega} = u_1 \omega + v_1 \omega^2,$$
$$u_1 \omega^2 + \frac{p}{u_1 \omega^2} = u_1 \omega^2 + v_1 \omega.$$
Consequently, the equation $y^3 - 3py - 2q = 0$ has only the following three solutions:
$$u_1 + v_1, \quad u_1 \omega + v_1 \omega^2, \quad u_1 \omega^2 + v_1 \omega.$$

In order to investigate these solutions we consider the following three cases:

Case 1. $q^2 - p^3 > 0$.

Let $q^2 - p^3 = r^2$, where r is a real number. Let u_1 be the real cubic root of $q + r$, hence v_1 is the real cubic root of $q - r$ (the product $u_1 v_1$ must be a real number). Then the solutions are the real number $u_1 + v_1$ and the conjugate complex numbers $u_1 \omega + v_1 \omega^2$, $u_1 \omega^2 + v_1 \omega$ (we noted that ω^2 is the conjugate of ω).

Case 2. $q^2 - p^3 = 0$.

We obtain as u_1 the real cubic root of q and as v_1 the same root. Then the equation has only two distinct roots, $2u_1$ and $-u_1$ (because $\omega + \omega^2 = -1$), and we say that $-u_1$ is a double root.

Case 3. $q^2 - p^3 < 0$.

Let $q^2 - p^3 = -r^2$, where r is a real number. Then $u^3 = q + ir$ and $v^3 = q - ir$. We put $u_1 = \alpha + i\beta$, where α and β are real numbers. Then $(\alpha^2 + \beta^2)^3 = q^2 + r^2 = p^3$, hence $\alpha^2 + \beta^2 = p$ and consequently $v_1 = \frac{p}{u_1} = \alpha - i\beta$. After calculations we find

$$u_1\omega + v_1\omega^2 = \alpha(\omega + \omega^2) + i\beta(\omega - \omega^2) = -\alpha - \beta\sqrt{3}$$

and similarly

$$u_1\omega^2 + v_1\omega = -\alpha + \beta\sqrt{3}.$$

It can be easily proved that

$$\alpha = \sqrt{p} \cdot \cos \frac{\arctan \frac{r}{q}}{3},$$

$$\beta = \sqrt{p} \cdot \sin \frac{\arctan \frac{r}{q}}{3}$$

(we note that in Case 3, $p > 0$).

If we set $z = y^3 - 3py - 2q$, then the three cases above are depicted by the graphs of Figure 23.2.

The general equation of order four has the form

$$x^4 + ax^3 + bx^2 + cx + d = 0$$

and can be written as

$$x^4 + ax^3 = -bx^2 - cx - d.$$

If we add $\frac{a^2 x^2}{4}$ to both sides of the last equation, we obtain

$$\left(x^2 + \frac{1}{2}ax\right)^2 = \left(\frac{1}{4}a^2 - b\right)x^2 - cx - d.$$

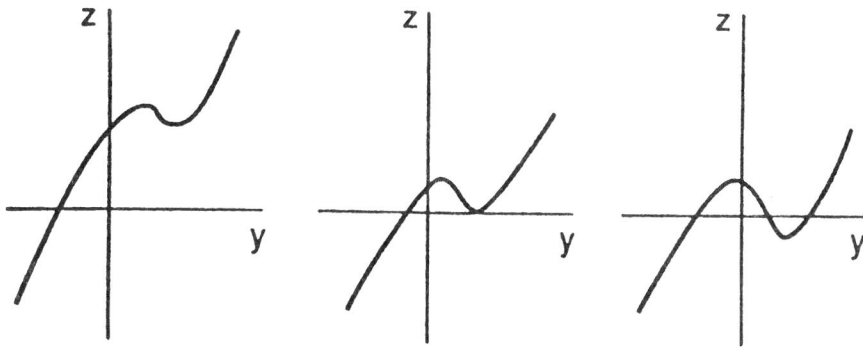

FIGURE 23.2

We will try to make perfect squares in both sides of the last equation by adding in both sides of the equation the expression

$$t\left(x^2 + \frac{1}{2}ax\right) + \frac{t^2}{4}.$$

Hence we have

$$\left(x^2 + \frac{1}{2}ax + \frac{t}{2}\right)^2 = \left(\frac{1}{4}a^2 - b + t\right)x^2 + \left(-c + \frac{at}{2}\right)x - d + \frac{t^2}{4}. \quad (\star)$$

We know that a trinomial $Ax^2 + Bx + C$ is a perfect square if and only if $B^2 - 4AC = 0$.

Consequently, in order to make the right-hand side of $(\star)$ a perfect square, it is enough to give to t such a value that it satisfies the equation

$$-d + \frac{t^2}{4} = \frac{\left(-c + \frac{1}{2}at\right)^2}{4\left(\frac{1}{2}a^2 - b + t\right)},$$

i.e., the equation

$$t^3 - bt^2 + (ac - 4d)t + 4bd - a^2d - c^2 = 0.$$

This is a third-order equation! Since we only need a single solution to this equation, in practice we can try to find the solution by speculation.

In summary, if x is a value that satisfies the equation $x^4 + ax^3 + bx^2 + cx + d = 0$, then there exists a number t which is a root of a third-order equation such that

$$\left(x^2 + \frac{1}{2}ax + \frac{1}{2}t\right)^2 = \left(x\sqrt{A} + \frac{B}{2\sqrt{A}}\right)^2,$$

where

$$A = \frac{1}{4}a^2 - b + t, \; B = -c + \frac{1}{2}at, \; C = -d + \frac{1}{4}t^2,$$

and hence we obtain the equation

$$x^2 + \frac{1}{2}ax + \frac{1}{2}t = \pm\left(x\sqrt{A} + \frac{B}{2\sqrt{A}}\right),$$

which is a second-order equation with respect to x.

(3) The invention of logarithms

During the 16th and the 17th century, the greatest progress in arithmetic was the invention of logarithms by **John Napier** who was not a professional mathematician. He was a Scottish landowner, a baron of Murchiston. In a comment in the Book of Apocalypse, he alleged that the Pope of Rome was the Antichrist. From a mathematical point of view, what is most important is that Napier was the inventor of logarithms.

As we know from our school years, logarithms constitute a very powerful tool for calculations, because by using them multiplication and division are reduced to the simpler operations of addition and subtraction respectively. It appears that the forerunner of this reduction was the trigonometric formula

$$\sin A \cdot \sin B = \frac{1}{2}[\cos(A - B) - \cos(A + B)],$$

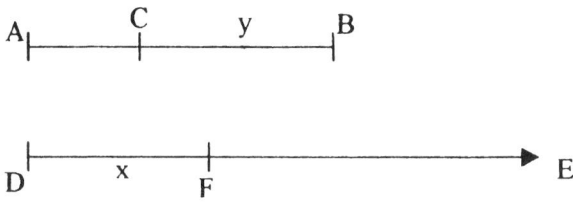

FIGURE 23.3

which first motivated Napier, because at first he restricted logarithms to logarithms of sines of angles. Napier worked for more than twenty years in order to develop his theory and came to the following definition of a logarithm.

We consider a line segment AB and an infinite half line DE (Figure 23.3). Points C and F start simultaneously with the same initial speed from points A and D respectively and move on these straight lines towards the same direction. Let us assume that C moves with a speed numerically equal to the distance CB and that F moves uniformly with constant speed.

Napier defines the length DF as the logarithm of CB. Thus, if we set $DF = x$ and $CB = y$, then we have
$$x = Naplog(y).$$

In order to avoid fractional numbers, Napier considered the length of AB to be equal to 10^7, because the best table of sines that he had at his disposal had numbers with up to seven decimal digits. From Napier's definition of the logarithm it follows that
$$Naplog(y) = 10^7 \log_{1/e}(y/10^7).$$

From this it is obvious that the statement found in various books that the Napierian logarithms are the natural logarithms (i.e., the logarithms with base e) is incorrect. We remark that the Napierian logarithm of a number decreases as the number increases, while the opposite happens with natural logarithms. We obtain the last relation above using infinitesimal calculus as follows:

We have $AC = 10^7 - y$, thus the speed of point C is equal to $-\frac{dy}{dt} = y$ and hence $\frac{dy}{y} = -dt$. By integrating we obtain that $\ln y = -t + k$, where k is some constant. Putting $t = 0$, we find $k = \ln 10^7$, hence
$$\ln y = -t + \ln 10^7.$$
We also have the *speed of* $F = \frac{dx}{dt} = 10^7$, i.e., $x = 10^7 t$. Thus
$$Naplog(y) = x = 10^7 t = 10^7 (\ln 10^7 - \ln y)$$
$$= 10^7 \ln(10^7/y) = 10^7 \log_{1/e}(y/10^7).$$

Napier published his invention of logarithms in 1614 in some tract entitled *Mirifici logarithmorum Canonis Descriptio* (Description of the Wonderful Law of Logarithms). The work includes a table of Napierian logarithms of sines of angles which differ successively by one minute.

In 1615 Napier and the mathematician **Henry Briggs** (1561-1631), who was a professor of geometry at Gresham College in London (and who had visited Napier in 1614 in order to obtain information on the invention of logarithms), decided that the logarithmic tables would be more useful if they were changed in such a way that the logarithm of 1 is 0 and the logarithm of 10 is an appropriate power of 10. Thus, the so-called **common logarithms** were born, i.e., the logarithms that we are taught at school. These logarithms have base 10 and are useful due to the fact that our number system has also base 10.

For some other number system with some other base b, it would be more advantageous to use logarithms with base b.

From the theoretical and the analytical point of view, the best base for a system of logarithms is the one that gives the simplest form to the derivative of the logarithmic function. In books on infinitesimal calculus it is proved that if

$$y = \log_b x,$$

then

$$\frac{dy}{dx} = \frac{\log_b e}{x}.$$

If we take $b = e$, then this derivative takes its simplest form, i.e., $\frac{dy}{dx} = \frac{1}{x}$.

The logarithms with base e are called **natural logarithms** and the notation that is generally used is $\ln x$ instead of $\log_e x$.

The word "logarithm" comes from the Greek words[1] "λόγος − αριθμός".

The "logarithmic ruler" is an instrument of fast calculation of logarithms that fell out of use after the appearance of personal computers. As opposed to that, the logarithmic function is and will always be useful for the simple reason that the logarithmic and the exponential variations constitute vital elements of nature and of mathematical analysis. Thus, the study of the mathematical properties of the logarithmic function and of its inverse, i.e., of the exponential function, will always constitute an important part of mathematical education.

(4) Mechanics and Astronomy

Nicolaus Copernicus was one of the most important astronomers of his time. He was born in Poland's Thorn in 1473. He studied mathematics and physics at the University of Cracow. At the age of 23 he went to Bologna where he became acquainted with Pythagorean and Greek thought in general, as well as with various astronomical theories. He also studied medicine and ecclesiastical law. In 1512 he returned to Poland where he was given an administrative position at the Cathedral temple of Frauenberg and he remained there until his death in 1543.

Without neglecting his administrative duties, Copernicus devoted himself to an intensive study of astronomy and to astronomical observations which led to the formulation of his truly revolutionary theory.

It is difficult to determine what led Copernicus to modify many points of Ptolemy's theory that existed for fourteen centuries. A classic work of Copernicus is the writing entitled *De Revolutionibus Orbium Coelestium* (On the Revolution of the Celestial Spheres, 1543).

Copernicus preserved certain principles of Ptolemy's astronomy. He used the circle as the basic curve through which he interpreted and constructed the way in which celestial bodies move. In addition, just like Ptolemy, he used the fact that the motion of the planets must be constructed as the resultant of motions having

---

[1]**Translator's note:** The meaning of these words in English is: *relation* or *ratio* and *number* respectively.

constant speed. The reason for which Copernicus alleged that the motions must have constant speed was that he believed that speed can change only if the driving force changes. But since the driving force was God, who is constant, the speeds of motion had to be constant too! He reconsidered the Greek viewpoint that the motions take place on epicyclic orbits.

Copernicus adopted Aristarchus' viewpoint that the earth moves around the sun, which permitted him to replace the existing until then diagrams of motion with much simpler ones.

The value of Copernicus' system is mainly due to the fact that by assuming that the earth moves around the sun, he interpreted the main singularities of the motion of planets without using a large number of epicyclic motions. He applied his theory to all the planets, as opposed to Ptolemy who made a distinction between the interior planets (Mercury, Venus) and the exterior planets (Mars, Jupiter, Saturn). In Copernicus' system the calculations for the determination of the position of celestial bodies were simpler than the ones that were used until then.

Simon Stevin (1548-1620) lived in Holland. He was one of the first who studied the theory of decimal fractions. In particular, he is known for his book *Statistics and Hydrostatics*. In this book he treats the "parallelogram of forces", the analysis of forces, stable and unstable equilibrium, and pressure. He also wrote works on algebra and geometry.

Galileo Galilei (1564-1642) is often called the father of contemporary physics. He was born in Italy's Pisa on the day that Michelangelo died. He entered the University of Pisa in order to study medicine. At that time the level of university courses was still at the level that it was during the Middle Ages. Galileo was taught mathematics from some practical mechanic, and at the age of 17 he switched from medicine to mathematics.

It is said that once, while Galileo was attending the divine service in a church, instead of paying attention to the divine service he was using his pulse to time the swings of a lamp that was hanging from the ceiling of the temple. Thus, he discovered that the period of swinging does not depend on its width. This discovery led to the construction of pendulums.

After publishing a book on hydrostatics and on the centers of gravity, Galileo was appointed professor. Galileo discovered that a body that is dropped from a certain height acquires a uniformly accelerated motion which is independent of its weight. He verified his conclusions by studying the motion of bodies on an inclined plane (apparently, he was also dropping objects from the Tower of Pisa).

Galileo was the first to use the already discovered telescope for studying the celestial sphere and he was awarded a prize for the discovery of Neptune's satellites This discovery reasserted Copernicus' viewpoints which were supported by Galileo.

In 1633, Inquisition made Galileo to recant Copernicus' viewpoint that the earth moves around the sun.

Of course, today we know that motion is relative (Einstein). One can select any celestial body as a point of reference in order to study the motion of the other celestial bodies. We can say that the earth moves around the sun or that the sun moves around the earth. From the mathematical point of view, it is more advantageous to consider the center of gravity of the solar system as the point of reference, because this choice makes the study of the solar system simpler. So from Einstein's vantage point, the dispute between Galileo and Inquisition as to whether the sun moves around the earth or the earth moves around the sun today seems to be somewhat irrelevant!

Johannes Kepler was born in 1571 in Weilderstadt near Stuttgart in Germany and began his studies at the University of Tübingen in order to become a Lutheran priest. As in Galileo's case, he didn't like his first choice, because he was interested in the exact sciences and particularly in astronomy. In 1594 he accepted a teaching position at the University of Graz in Austria. Five years later he became the assistant of the famous (Danish-Swedish) astronomer **Tycho Brahe**, who had settled in Prague as an astronomer in the court of Kaiser Rudolph II. In 1601, after Brahe's death, Kepler was assigned to his teacher's position, and he inherited Brahe's great collection which contained highly accurate results of measurements of the positions of the planets as they move in the firmament. With great fussiness and care, and with enormous persistence, Kepler tried to discover the way in which planets move in the firmament by examining this huge number of data that devolved to him.

We often hear people saying that "almost every problem can be solved if we occupy constantly our thought with it and we work on it for a long period of time". **Thomas Edison** believed that 1% of a discovery is due to "inspiration" and 99% due to persistent work. This viewpoint of Edison is confirmed completely in Kepler's case, who solved the problem of the motion of the planets around the sun. The solution was given under the form of the following three laws. The first two were discovered in 1609 and the third in 1619.

    I. Planets move around the sun on elliptical orbits and the sun lies on one of the foci.

  II. The radius connecting a planet with the sun scans (passes over) equal areas in equal time intervals.

 III. The square of the time interval needed for a planet to make a full revolution on its orbit around the sun is proportional to the cube of half the great axis of its orbit.

The empirical discovery of these laws constitutes one of the most remarkable events in the area of exact sciences. In his effort to obtain mathematically Kepler's laws, **Isaac Newton** was led to create Celestial Mechanics.

An event of great interest is the fact that it took 1800 years for the discovery of the theory of "conic sections" by the Ancient Greeks to find an important application. We never know when a discovery in theoretical mathematics will find its unexpected practical application!

In all of his scientific efforts and achievements, Kepler was driven by his zealous faith in Pythagorean theory, that there exist no phenomena that do not have a mathematical structure.

In his private life Kepler was very unfortunate. He died in 1630.

CHAPTER 24

THE SEVENTEENTH CENTURY

During the 17th century, a huge mathematical activity is observed. Some of the most important mathematicians of that period were the following:

- Martin Mersenne (1588-1648) of France,
- Gerard Desargues (1591-1661) of France,
- René Descartes (1596-1650) of France,
- Pierre de Fermat (1601-1665) of France,
- Blaise Pascal (1623-1662) of France,
- Bonaventura Cavalieri (1598-1647) of Italy,
- John Wallis (1616-1703) of England,
- Nicolaus Mercator (1620-1687) of Germany, who lived in England,
- Christian Huygens (1629-1695) of Holland,
- Isaac Barrow (1630-1677) of England,
- James Gregory (1638-1675) of Scotland,
- Isaac Newton (1642-1727) of England,
- Gottfried Wilhelm Leibniz (1646-1716) of Germany.

• **Mersenne** was a monk. The prime numbers of the form $2^n - 1$, are called "Mersenne-primes" in his memory. Mersenne's main contribution is that he corresponded with all other French mathematicians and informed each one of them about the ideas of the others.

• **Desargues** was one of the founders of Projective Geometry.

• **Descartes** was the first great contemporary philosopher, the founder of contemporary biology, and an important physicist and mathematician.

He was born in La Haye of Touraine in March 31, 1596. When he was 8 years old, his father sent him to the school La Flèche in Anjou. There, every morning he was allowed to remain and study in bed, because his health was delicate. But also during the rest of his life, Descartes carried out the greatest part of his intellectual work in bed. At the age of 16 he left La Flèche and at the age of 20 he graduated as a lawyer from the University of Poitiers and went to Paris where he became acquainted with Mydorge and Mersenne. He remained close to them for a year studying mathematics. Since he was the offspring of a "good" family, he had to make a career either in the Church or in the army. He chose the army. He served first in the army of Maurice of Orange, and then he served in the army of the Duke of Bavaria, after the outbreak of the Thirty Years' War (a series of conflicts during 1618-1648 between most of the countries of Western Europe, whose initial cause were disputes of a religious nature). In 1621 he resigned from the army and traveled for five years while he studied mathematics. Finally, he settled in Holland, where for thirty years he devoted himself entirely to the study of subjects regarding the intellect.

After insistent invitations by Queen Christine, he went to Sweden in 1649. There, the young queen insisted that Descartes' gave her lessons in mathematics at five o'clock in the morning. This schedule exhausted the invalid Descartes, and as a result he became ill and died of pneumonia within two months.

Descartes' first work, entitled *Regulae ad Directionem Ingenii* (Rules for guiding the spirit), was written in 1628, but it was published after his death in 1692.

Descartes' first book, entitled *Le Monde*, extends and improves Copernicus' model of the solar system. This work was completed exactly during the time that Inquisition condemned Copernicus' viewpoints. Because of this, Descartes did not publish his book, under fear of persecution by the Church.

In 1637 he published his famous writing *Discours de la Méthode pour bien conduire sa Raison et chercher la Vérité dans les sciences*. In Part II of this writing the following phrase exists:

"It is impossible for one to imagine a viewpoint, no matter how absurd and unbelievable it is, which is not supported by some philosopher."

In his writing *Discours ...* Descartes develops his program, according to which first a process of "systematic doubt" is followed, which then leads to the acceptance of knowledge. Later, a very careful accumulation of true knowledge takes place. This accumulation is based on certain clear principles, such as: (a) cogito ergo sum (I think therefore I exist), (b) every phenomenon must have a cause, (c) a result cannot be greater than its cause, (d) the ideas of perfection, space, time and motion are inherent in our spirit. The book includes three very important supplements. In Supplement I, **La Dioptrique**, studied Optics. In Supplement II, **Les Météores**, studied atmospheric phenomena. In particular, this book contains an interpretation of the shape of the rainbow. From the mathematical vantage point, Supplement III, **La Géométrie**, is the most important. It is divided into three books where the principles are given on which Analytic Geometry is based. Descartes is considered to be the creator of Analytic Geometry.

The revolutionary idea that lies on the basis of analytic geometry is the following: The points of the Euclidean plane can be represented by pairs of real numbers and consequently the straight lines and the conic sections can be described by sets of pairs (x, y) that satisfy equations of the form

$$Ax + By + C = 0$$

and

$$Ax^2 + Bxy + Cy^2 + Dx + Ey + F = 0$$

respectively. In this way one achieves a connection between geometry and algebra. Thus, it became possible to solve problems, which had remained open for centuries such as the problem of the duplication of the cube (even though the "solution" that we know today was given two centuries later (Chapter 10)). In particular, in the supplement *La Géométrie*, Descartes tries to solve problems of geometric constructions by using algebra. He remarks first of all that geometric constructions require the operations of addition, subtraction, multiplication and division between lines as well as the extraction of square roots (lengths), and that since these operations exist in algebra, they can be expressed in algebraic terms.

In order to solve a certain given problem, says Descartes, we must assume that we know its solution and then we must denote with letters all known and unknown lines that we judge to be necessary for the requested construction. Now, without making any distinction between known and unknown elements, we must "locate" the existing difficulty by pointing out in which way these lines are related to one another and by trying to express the same quantity in two different ways. This procedure leads to an equation. We must find as many equations such as the unknown lines, and finally we must obtain a single equation that gives the unknown line in terms of the known ones. Later, Descartes proves how the unknown line is constructed using the algebraic equation that is satisfied by it.

Example. Assume that it is requested to construct a line segment of length x, where $x^2 = ax + b^2$, while a and b are known lengths.

We know from algebra that:

$$x = \frac{a}{2} + \sqrt{\frac{a^2}{4} + b^2}.$$

Descartes (omits the negative root and) gives the following construction of x:

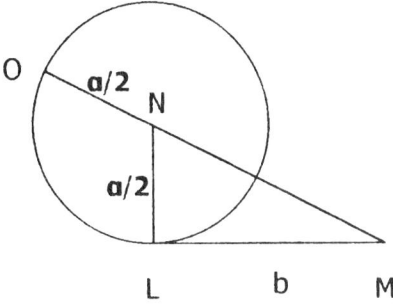

FIGURE 24.1

He constructs the right-angled triangle NLM (Figure 24.1), where $LM = b$ and $NL = \frac{a}{2}$, and he extends MN until he reaches a point O such that $NO = NL = \frac{a}{2}$. Then the solution x is the length OM. The proof that OM is the requested line segment is not given by Descartes, but follows easily from the relation

$$OM = ON + MN = \frac{a}{2} + \sqrt{\frac{a^2}{4} + b^2}.$$

Descartes' "rule of signs" states that: the number of positive real roots of a polynomial equation $f(x) = 0$ with real coefficients is equal to $v - 2k$, where v is the number of "sign changes" and k is a certain natural number. In order to calculate v we write $f(x)$ in decreasing powers of x omitting the terms having coefficient zero. Then v is the number of times that the sign changes as we proceed from term to term, from left to right. For example, the equation

$$x^6 - x^5 - x^4 + x - 4 = 0 \tag{1}$$

has three sign changes, which means that it has either one or three positive real roots. In some cases we can be certain about the number of real roots. Indeed, if we put $x = -y$ in (1), we find

$$y^6 + y^5 - y^4 - y - 4 = 0 \tag{2}$$

which presents a single sign change. This means that (1) has exactly one negative root.

Descartes also wrote the books *Principia Philosophicae* (1644), where he gives a theory for the interpretation of the motion of planets. He also wrote *Méditations*, which contains a "geometric proof" of the existence of God! In particular, in this last book one can find the following viewpoint:

"The notion of the existence of God cannot be separated from the essence of God, just like the idea of the notion "mountain" cannot be separated from the idea of a "valley", or the fact that "the sum of the angles of a triangle is equal to two right angles" cannot be separated from the essence of the triangle (*Méditations* V)."

In other words, Descartes' viewpoint is that God exists because "existence" is one of the defining properties of God. We note that since in hyperbolic geometry there exist triangles, the sum of the angles of which is not equal to two right angles, one must add in the above viewpoint of Descartes the word "Euclidean" in front of the word "triangle".

Descartes' scientific ideas were the dominating ideas of the seventeenth century. His teaching and his work were known even to non-scientists, because the presentation of his ideas was lucid and attractive. Only Church rejected him. However, Descartes was a devout Christian and was very satisfied, because he believed that he had proved the existence of God. He was teaching that the Bible does not constitute a source of scientific knowledge, that the intellect itself could prove the existence of God, and that the human must accept only what he/she can understand. The Church reacted against this teaching and included Descartes' books in the list of forbidden books, after his death. Moreover, on the day of his burial in Paris, the Church prohibited any funeral oration.

Descartes approaches the science of mathematics from three different directions, namely as a philosopher, as a student of nature, and as a man who is interested in the applications of science. It would be difficult but also unnatural for one to attempt to separate these three directions. Descartes lived in a period during which the dispute between Catholics and Protestants was at its zenith, while progress in science, which had already begun to discover laws of nature, constituted a provocation for the main religious dogmas. Thus, Descartes began to doubt the knowledge he had acquired in his school years and wondered whether he could find somewhere a "safe" set of knowledge on which one could rely. He soon came to the conclusion that "logic" itself was sterile and incapable of providing the fundamental truths, which therefore had to be sought elsewhere. He rejected the philosophy of his time, because he believed that it provided only means for discussion, thus giving the illusion that it provided the truth in all matters. Also, he believed that Theology indicates to the human the way to happiness, something that he and every other person would like to achieve, of course, but he was not certain that the suggested way was the correct one.

Finally, Descartes came to the conclusion that the mathematical method was the most advisable one for revealing and proving truths in all subjects. He believed that mathematics is the most powerful tool of knowledge compared to any other tool at the human's disposal, and furthermore it constitutes the basis of all other tools. The long chain of simple and easy arguments, with which geometers come to conclusions leading to the most difficult proofs, convinced me, says Descartes, that the human can acquire also every other knowledge in a similar way.

- Even though during the Renaissance almost all algebraists of that time made conjectures and remarks on various subjects, **Fermat** was the first whose contributions in Number Theory were extensive and impressive.

Fermat was a legal adviser at the parliament of Toulouse so he had little time for mathematics. He published almost nothing during his lifetime. His contributions in mathematics are contained in his correspondence (for example, with Mersenne) and in his notes that were found after his death.

Fermat's viewpoints regarding analytic geometry can be found in his work *Ad locos Planos et Solidos Isagoge*.

The historian Carl B. Boyer writes in his book *History of Analytic Geometry* (*New York: Scripta Mathematica*, 1956) the following:

"Analytic geometry was the invention of two men (who worked independently of each other), none of whom was a professional mathematician. Pierre de Fermat (1601-1665) was a lawyer with a vivid interest for the geometric works of classical antiquity. René Descartes (1596-1650) was a philosopher who found in mathematics a basis for logical reasoning."

Fermat also collaborated with Pascal and contributed to probability theory. In addition, he studied subjects regarding the tangents of curves, maxima and minima, and the areas of domains surrounded by curves. With these studies he came close to the discovery of Infinitesimal Calculus. The methods that Fermat used were influenced by the methods of his contemporaries, such as Cavalieri and Wallis, who we will discuss later.

As mentioned before, Fermat is mainly known for his great and valuable contributions in number theory. Diophantus mentions that a prime number of the form $4n-1$ cannot be expressed as the sum of two squares. Fermat (who studied Diophantus' works deeply) proved that: "every prime number of the form $4n+1$ can be uniquely expressed as the sum of two squares". He also remarked that: "every odd prime number p can be uniquely expressed as the difference of two squares". This last proposition is proved as follows:

$$p = \left(\frac{1}{2}(p+1)\right)^2 - \left(\frac{1}{2}(p-1)\right)^2,$$

where $\frac{1}{2}(p+1)$ and $\frac{1}{2}(p-1)$ are integers since p is odd. Assume that $p = x^2 - y^2$, then $p = (x+y)(x-y)$, hence $p = x+y$ and $x-y = 1$, since p is divided only by 1 and by p, because p is a prime. Consequently

$$x = \frac{1}{2}(p+1), \quad y = \frac{1}{2}(p-1).$$

In order to prove that every prime number of the form $4n+1$ can be uniquely written as the sum of two squares, Fermat used his method "of continuous descent", which he describes in his letter to Mersenne dated December 25, 1640, as follows:

"If there exists a prime number of the form $4n+1$ which does not have the above property, then there exists a smaller prime number of the form $4n+1$ which also does not have the above property. But then, since n is arbitrary, there exists an even smaller number, and so on. "Descending" in this way the positive integral values of n, we reach the value $n = 1$, i.e., we reach the prime number $4 \cdot 1 + 1 = 5$, which means that 5 cannot have the above property. This is a contradiction, because 5 can be uniquely expressed as the sum of two squares. Thus, every prime number of the form $4n+1$ can be uniquely expressed as the sum of two squares."

Fermat's (little) Theorem

If p is a prime number and a is an arbitrary integer which is not a multiple of p, then p divides $a^{p-1} - 1$.

One of the many existing proofs of this theorem is the following.

Proof. The binomial theorem implies
$$(x+1)^p = x^p + \binom{p}{1} x^{p-1} + \binom{p}{2} x^{p-2} + \ldots + \binom{p}{p-1} x + 1,$$
where the number
$$\binom{p}{k} = \frac{p(p-1)\ldots(p-k+1)}{k(k-1)\ldots 1}$$
is an integer. If $0 < k < p$, then p is not a factor of the expression $k(k-1)\ldots 1$. But since p is a factor of the expression
$$\binom{p}{k} k(k-1)\ldots 1 = p(p-1)\ldots(p-k+1),$$
it follows that p is a factor of $\binom{p}{k}$ if $0 < k < p$. Thus
$$(x+1)^p = x^p + 1 + mp,$$
where m is some integer.

If $x = 1$, then $2^p = 2 + mp$, which means that p divides $2^{p-1} - 1$. If $x = 2$, then $3^p = 2^p + 1 + m'p = 2 + mp + 1 + m'p = 3 + (m+m')p$, which means that p divides $3^{p-1} - 1$. If we continue in the same way (i.e., if we apply the principle of induction), we find that for all a, we have $a^p = a + np$, where n is some integer.

Thus, p divides $a^p - a = a(a^{p-1} - 1)$, which proves the theorem because p does not divide a.

The most famous theorem of Fermat is the so-called "Fermat's Last Theorem", which remained unproved for more than 350 years. This theorem was finally proved in 1994 by the English mathematician Dr. Andrew Wiles, a professor at Princeton University in the USA, who was then only 40 years old. The first announcement that the theorem was proved was made by Wiles on June 23, 1994. But this proof was incomplete and a complete proof was given by Wiles 18 months later with the help of Richard Taylor, a professor at Cambridge University in England, who was a former student of Wiles. But what does "Fermat's Last Theorem" say?

As he was studying Bachet's translation (from Greek into Latin) of Diophantus' writing *Arithmetica*, Fermat found the equation $x^2 + y^2 = z^2$ (Book II, Problem 8) and a solution to it (for example, $x = 2$, $y = 3$, $z = 5$). Fermat wrote on the margin of the book that he was reading the following phrase:

Cubum autem in duos cubos, aut quadratoquadratum in duos quadratoquadratos, et generaliter nullam in infinitum ultra quadratum potestatem in duos ejusdem nominis fas est dividere: cujus rei demonstrationem mirabilem sane detexi. Hanc marginis exiguitas non caperet.

(An approximate translation of this phrase is the following:)

[It is impossible to divide a cube into two cubes, or a fourth power into two fourth powers, and in general no power greater than two into two powers equal to it. I discovered a truly wonderful proof of this proposition, which I cannot write in the limited space of this margin.]

This conjecture of Fermat can be stated as follows:

Fermat's Last Theorem
If $n > 2$ is an integer, then there exist no positive integers x, y, z such that $x^n + y^n = z^n$.

The above theorem was called Fermat's "Great Theorem" or "Last Theorem", because it was Fermat's last conjecture, which mathematicians were not able to prove or disprove for a long time.

It is very likely that Fermat proved the theorem for $n = 3$. Certainly a proof of his exists for $n = 4$. But the dominating viewpoint is that Fermat did not have a proof for an arbitrary $n > 4$.

In 1832 Legendre proved this theorem for $n = 14$. In 1849, Kummer had a proof for all $2 < n < 100$ apart from the values 37, 59 and 67. Before 1994, with the help of computers, it was known that the theorem is true for all values $2 < n < 10^8$. Finally, Wiles gave a complete proof for all $n > 2$.

Regarding Fermat's theorem we note that Faltings "came very close" to a proof of Fermat's last theorem, because he proved that for any $n > 2$, the equation $x^n + y^n = 1$ has only finitely many rational roots.

Finally, let it be noted that Wiles proved something stronger and more important than Fermat's theorem. He proved the long-standing "Taniyama-Shimura conjecture" regarding the so-called elliptic and modular curves. Fermat's last theorem is a simple consequence of this conjecture. In particular, in June 23, 1994, what Wiles announced essentially during his talk at the Newton Institute in Cambridge in England was that: "many" elliptic curves are modular and in particular there exist so many that their existence implies the validity of Fermat's last theorem. Further discussion and development of Wiles' work goes beyond the scope of the present book.

• **Pascal** was born in Clermont in France. He was an invalid and his father wanted to keep him away from mathematics until the age of fifteen, when he would be sufficiently mature. But at the age of twelve, Blaise had already rediscovered many theorems of Euclid, and hence his father backed down and gave his son a copy of Euclid's *Elements*.

Pascal's family moved to Paris when he was eight years old. Even as a child, he was attending with his father the weekly meetings of "Académie Mersenne", which was later renamed "Académie Libre" and in 1666 it was renamed again "Académie des Sciences". Among its members were Father Mersenne, Descartes, Roverbal (a professor of mathematics at the Collège de France), Claude Mydorge (1585-1647) and Fermat.

At the age of 16, Pascal wrote an essay on conic sections and at the age of 18 he constructed the first calculating machine. Today the computer language PASCAL bears his name in his honor.

At 27 Pascal stopped doing mathematics in order to devote himself to the philosophy of religion and to the worship of God. Actually, Pascal was an excellent handler of prose. His works *Pensées* and *Lettres provinciales*, which were written during that time, are considered to be masterpieces of French literature.

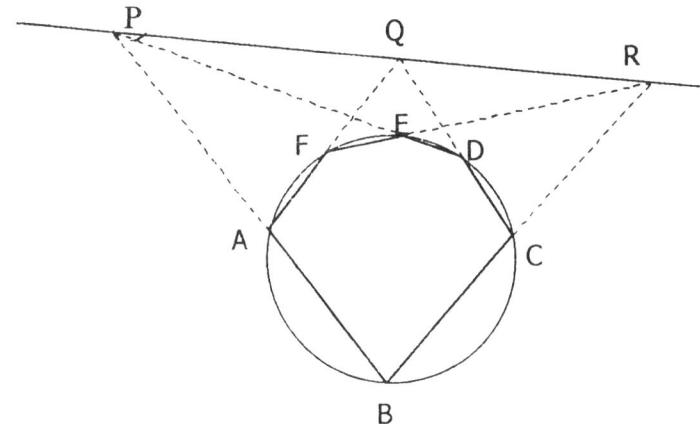

Figure 24.2

Later Pascal returned to the study of mathematics. Pascal died at 39. His last achievement was the creation of a system of public transportation, the profits of which were used for the benefit of the poor.

We now present some of Pascal's achievements in mathematics.

In 1639, using Desargues' ideas, he proved the following "Theorem of Pascal" (hexagramma mysticum) in projective geometry:

Theorem.
If a hexagon $ABCDEF$ **is inscribed in a conic section, then the three points, at which the pairs of opposite sides of the hexagon intersect, lie on the same straight line (Figure 24.2). Consequently, the points** P, Q, R **lie on the same straight line.**

There exist only indications about how Pascal proved the above theorem. He says that since the theorem is true in the case that the conic section is a circle, it must be true in the case of the remaining conic sections, since they are obtained from the circle via projection.

We note that Pappus' Theorem, where three points are considered on each one of two straight lines, constitutes a special case of the above theorem. When the conic section considered by Pascal degenerates into two straight lines, as when the hyperbola degenerates into its asymptotes, then we obtain Pappus' theorem.

The converse of Pascal's theorem: "if in a hexagon the points of intersection of the three pairs of opposite sides are collinear, then the vertices of the hexagon lie on a conic section" is also true, but Pascal didn't study this case.

In 1653 Pascal rediscovered what we call "**Pascal's Triangle**" (Figure 24.3).

We remark that every number in this construction is equal to the sum of the two numbers that lie immediately above it ($6 = 3 + 3$, $4 = 3 + 1$, and so on).

Pascal's triangle was known to the Chinese and to the Arabs, but Pascal was the first who gave clear and complete proofs of the basic properties of this triangle using mathematical induction. He proved that the kth term in the nth row is $\binom{n-1}{k-1}$, i.e., the number of ways that we can choose $k-1$ objects from a set of

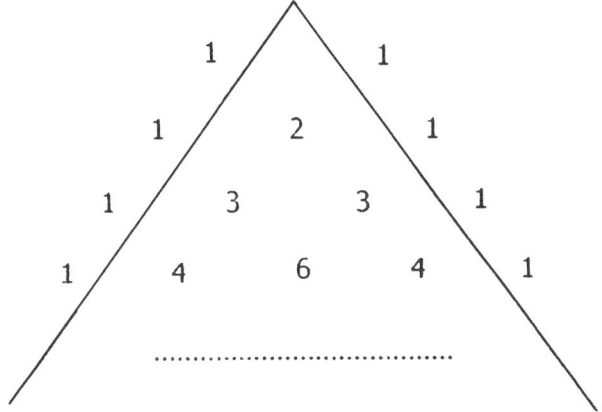

FIGURE 24.3

$n-1$ objects. Pascal proved that
$$\binom{n-1}{k-1} = \frac{(n-1)(n-2)...(n-k-1)}{(k-1)!}$$
and that
$$\binom{n}{k} = \binom{n-1}{k} + \binom{n-1}{k-1}.$$

This relation follows from the observation that in order to choose k objects from n objects, we can put aside one of the n objects and consider two cases: in the first case the object that is put aside is included in our choice, hence we have $\binom{n-1}{k-1}$ possibilities, while in the second case the object that is put aside is not included in our choice, hence we have $\binom{n-1}{k}$ possibilities.

Finally, Pascal proved that the number $\binom{n}{k}$ is the coefficient of the monomial $x^k y^{n-k}$ in the analytic expression of the binomial $(x+y)^n$.

In collaboration with Fermat, Pascal contributed to the development of probability theory as well as to the development of the notion of "mathematical expectation". In his work *Pensées*, Pascal using probability theory claimed that it is "wiser" to believe in the existence of God (Pascal, *Oeuvres Complètes*, p. 1212). His idea known as "Pascal's Wager" is as follows: "Even if the probability that "God exists" is very small (but not zero), if God rewards his believers with eternal happiness (whose value is considered to be infinitely large), then every rational human being must believe in this God".

• **Cavalieri** was a student of Galileo and a member of the religious order of the Jesuats which must not be confused with the religious order of the Jesuits. Cavalieri lived in Milano and in Rome until he became professor of mathematics in Bologna in 1629. He occupied himself with various subjects of theoretical and applied mathematics (geometry, trigonometry, astronomy, optics) and he was one of the first who appreciated the great value of the logarithms. Cavalieri is mainly known for his important writing *Geometria indivisibilibus continuorum* (1635). In this book Cavalieri develops the so-called "method of indivisibles", a method which has its roots in the works of the Ancient Greeks, as well as many other subjects of contemporary mathematics (ideas, methods, etc.). In particular, it has its roots in the works of Democritus (410 B.C.) and Archimedes (287-212 B.C.). It is also very likely that Cavalieri based his method on Kepler's ideas about "integration". In the above book of Cavalieri it is not so easy to understand what exactly he means by the word "indivisibilibus" (indivisibles). It seems that an "indivisible" of a plane figure is a chord of this figure, while the figure can be considered to consist of infinitely many parallel indivisibles. By complete analogy, an "indivisible" of a given solid is a planar section of this solid, while the solid can be considered to consist of infinitely many such indivisibles.

Cavalieri's notion of an "indivisible" was criticised by his contemporaries and Cavalieri tried to respond to their critique by arguments which were not so convincing. Despite all that, his method was widely applied by many mathematicians.

Cavalieri's method can be summarized into the following two **Principles of Cavalieri**.

(1) If each one of two plane figures is contained between two parallels, and if the ratio of the lengths of two line segments at which these figures intersect with two arbitrary straight lines which are parallel respectively to the straight lines enclosing these figures remains constant and equal to λ, then the ratio of the areas of the two figures is equal to λ.

(2) If each one of two solids is contained between two parallel planes, and if the ratio of the areas of two figures at which these solids intersect with two arbitrary planes which are parallel respectively to the planes enclosing these solids remains constant and equal to μ, then the ratio of the volumes of the two solids is equal to μ.

As an example of application of the above principles of Cavalieri we present the calculation of the area of the ellipse $y = \pm \frac{b}{a}\sqrt{a^2 - x^2}$. We consider first the circle $y = \pm\sqrt{a^2 - x^2}$ in the same orthogonal system of coordinates with the ellipse (Figure 24.4). We remark that the ratio of the corresponding ordinates of the ellipse and of the circle is equal to $\frac{b}{a}$. From this it follows that the ratio of the corresponding vertical chords of the ellipse and of the circle is equal to $\frac{b}{a}$. Consequently, it follows from Cavalieri's first principle that the ratio of the area of the ellipse over the area of the circle is equal to $\frac{b}{a}$. Thus

$$\begin{aligned}\text{area of the ellipse} &= \frac{b}{a} \cdot (\text{area of the circle}) \\ &= \frac{b}{a} \cdot \pi a^2 = \pi a b.\end{aligned}$$

The above approach is essentially the one used by Kepler in order to calculate the area of the ellipse with semiaxes a and b.

As a second example of Cavalieri's principles we give the calculation of the volume of a sphere of radius r.

In Figure 24.5 we have on the left a hemisphere of radius r and on the right a right circular cylinder of radius r and of height r, from which a cone has been subtracted whose base is the upper base of the cylinder and whose vertex coincides with the center of the lower base of the cylinder. The hemisphere and the incomplete cylinder have their bases on the same plane. Now we intersect both solids with a plane, which is parallel to the plane of the base and which lies at a distance h above the plane of the base. This plane intersects the solid on the left at a circle and the

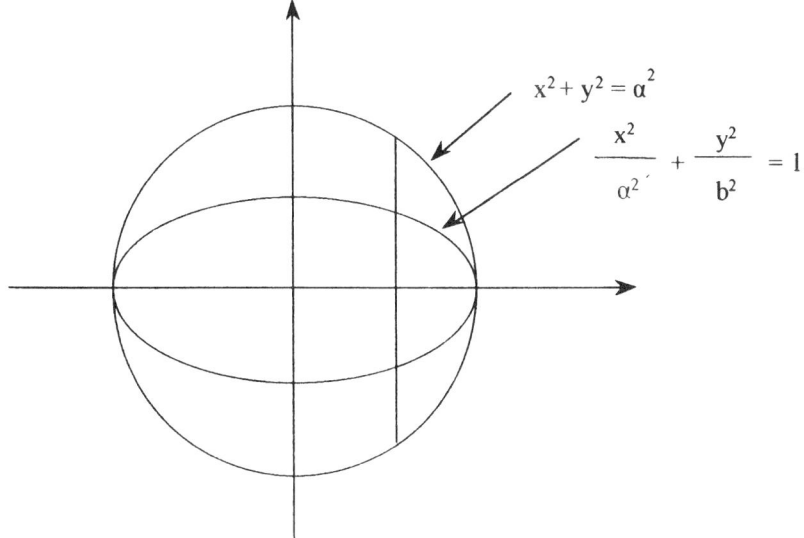

FIGURE 24.4

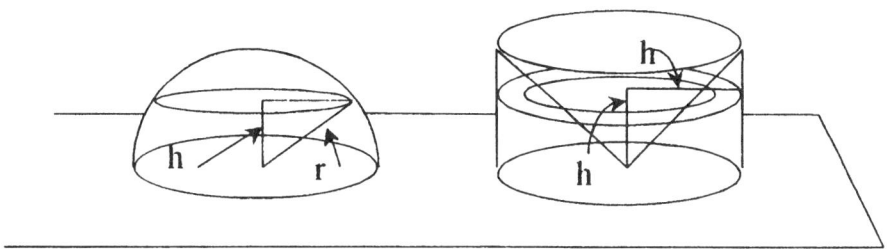

FIGURE 24.5

solid on the right at a ring. It is easily recognized that each one of these sections has area equal to $\pi(r^2 - h^2)$. Consequently, it follows from Cavalieri's second principle that the two solids have the same volume. Thus, the volume V of the sphere is given by the relation:

$$V = 2 \cdot (\text{volume of the cylinder } - \text{ volume of the cone})$$
$$= 2 \cdot \left(\pi r^3 - \frac{\pi r^3}{3}\right) = \tfrac{4}{3}\pi r^3$$

One of the most important theorems of Cavalieri was, perhaps, the theorem that is equivalent to the following formula, if we express it with the notation that we use today:

$$\int_0^a x^n dx = \frac{a^{n+1}}{n+1}.$$

Two plane figures, which can be placed in such a way that every straight line of a family of parallel straight lines intersects these figures in two equal segments, are called **equivalent figures in the sense of Cavalieri**.

Two solids, which can be placed in such a way that every plane of a family of parallel planes intersects these solids at two figures of the same area, are called **equivalent solids in the sense of Cavalieri**.

Among the strange phenomena regarding equivalences of Cavalieri's type are the following:

(1) Even though there exists no polygon which is equivalent in the sense of Cavalieri to a given circle, there exists a polyhedron (a tetrahedron) which is equivalent in the sense of Cavalieri to a given sphere.

(2) Even though there exist tetrahedra which have the same volume and which are not equivalent in the sense of Cavalieri, two triangles having the same area are always equivalent in the sense of Cavalieri.

• **Wallis** was a professor of geometry in Oxford. He was a royalist and was finally appointed priest of king Charles II. He had invented a method of teaching deaf-mute persons.

In algebra Wallis used negative and fractional numbers as exponents:
$$x^{-n} = \frac{1}{x^n}, \quad x^{\frac{p}{q}} = \sqrt[q]{x^p}.$$

Using Cavalieri's methods, he calculated the area of the domain that is enclosed by the curve
$$y = a_0 + a_1 x + \ldots + a_n x^n$$
and the x-axis.

Wallis managed to express the number π in the form of an infinite product, i.e.,
$$\pi = 2 \cdot \frac{2}{1} \cdot \frac{2}{3} \cdot \frac{4}{3} \cdot \frac{4}{5} \cdot \frac{6}{5} \cdot \frac{6}{7} \cdot \frac{8}{7} \cdot \frac{8}{9} \cdot \ldots,$$
but he did not give a rigorous proof of this formula.

We present a short proof of the formula
$$\pi = \lim_{n \to \infty} \left(2 \cdot \frac{2^2 \cdot 4^2 \cdot \ldots \cdot (2n)^2}{3^2 \cdot 5^2 \cdot \ldots \cdot (2n-1)^2 \cdot (2n+1)} \right).$$

We set
$$f(n) = \frac{2^2 \cdot 4^2 \cdot \ldots \cdot (2n)^2}{3^2 \cdot 5^2 \cdot \ldots \cdot (2n-1)^2 \cdot (2n+1)}.$$

The following two formulas are known:
$$I(2n) = \int_0^{\frac{\pi}{2}} \sin^{2n} x \, dx = \frac{1 \cdot 3 \cdot 5 \cdot \ldots \cdot (2n-1)}{2 \cdot 4 \cdot 6 \cdot \ldots \cdot (2n)} \cdot \frac{\pi}{2}, \quad n = 1, 2, \ldots$$

$$I(2n+1) = \frac{2 \cdot 4 \cdot 6 \cdot \ldots \cdot (2n)}{1 \cdot 3 \cdot 5 \cdot \ldots \cdot (2n+1)}, \quad n = 1, 2, \ldots$$

We have
$$I(2n) = ((2n+1)f(n))^{-\frac{1}{2}} \cdot \frac{\pi}{2}$$

$$I(2n+1) = \left(\frac{f(n)}{2n+1} \right)^{\frac{1}{2}}.$$

Since
$$\sin^{2n+1} x \leq \sin^{2n} x \leq \sin^{2n-1} x,$$

if we integrate these three functions of sine, we find

$$\left(\frac{f(n)}{2n+1}\right)^{\frac{1}{2}} \leq ((2n+1)f(n))^{-\frac{1}{2}} \cdot \frac{\pi}{2} \leq \frac{(f(n)(2n+1))^{\frac{1}{2}}}{2n},$$

i.e., we find

$$2f(n) \leq \pi \leq \frac{2f(n)(2n+1)}{2n},$$

and the requested result follows from this last double inequality.

• **Nicolaus Mercator** (1619-1687) was born in Holstein, but he spent most of his life in England. In 1683 he went to France where he designed and supervised the construction of the fountains of Versailles. But after the completion of the work, Louis XIV refused to give him the agreed payment, unless Mercator adopted Catholicism. Mercator died embittered and poor in Paris in 1687.

He wrote a book on logarithms entitled *Logarithmotechnica* (1668), where he calculates the area of the domain that is enclosed by the curve

$$y = \frac{1}{1+x} = 1 - x + x^2 - x^3 + ...$$

and the x-axis, and he obtains the formula

$$\log_e(1+x) = x - \frac{x^2}{2} + \frac{x^3}{3} - \frac{x^4}{4} + ...$$

(the series being convergent for $-1 < x < 1$).

Note. Gregory of St. Vincent (1584-1667) was the first who proved that the integral of the function $\frac{1}{x}$ is equal to the natural logarithm of x. Nicolaus Mercator must not be confused with Gerardus Mercator (1512-1594), to whom an important method of constructing maps is due, where he introduced the so-called "Mercator projection".

• **Huygens** was born in Hague in Holland and died in the same city. He was writing his name as Hugens. Huygens was an internationally famous scientist. He is known for "**Huygens' Principle**", which bears his name and concerns the theory of propagation of light, for the observation of Saturn's rings, and for his essential contribution to the invention of the pendulum. His research for the improvement of the science that concerns the measurement of time led him to a very important mathematical discovery. He knew that the oscillations of a simple pendulum are not absolutely isochronous and they depend on the width of the oscillation. In particular, if an object is left to slide on the interior smooth surface of a concave pot that has the shape of a hemisphere, then the time that is needed for the object to reach the bottom of the pot is almost (but not completely) independent of the height from which the object is left to slide. Huygens discovered that if the pot is such that every vertical section of it has the shape of an inverted cycloid, then always the same time is needed for the object to reach the bottom, independently of the height from which it is left to slide.

We remind the reader that a "cycloid" is the curve traced by a particular point of the circumference of a circle, when the circle rolls on a smooth plane surface.

But how is it possible to construct a pendulum that swings on a cycloid and not on a part of the circumference of a circle? Huygens answered this question in the following ingenious way: If we hang from a point P (Figure 24.6), which is the common extremity of two inverted halves of an arc of a cycloid, a pendulum of

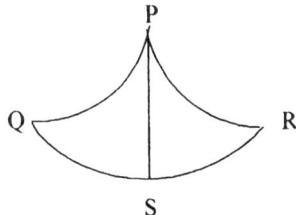

Figure 24.6

length equal to the length of one of these halves, then the bullet of the pendulum will move on the arc of a cycloid QSR, which has the same shape and length with the cycloid whose halves are the arcs PQ and PR. In other words, if the pendulum swings between two jaws having the shape of equal arcs of a cycloid, then the oscillations are absolutely isochronous.

Huygens also proved the interesting proposition that the geometric locus of the centers of curvature of a cycloid is also a cycloid.

Huygens' theory regarding the propagation of light constituted an important contribution in Physics. By assuming that light propagates in waves through "ether", he was able to interpret the laws of reflection and refraction of light. From a certain point of view his theory was better than Newton's theory, which states that light propagates in particles, but which cannot interpret the phenomena of "interference" of light. Even though Newton's theory superseded Huygens' theory for many years, the contemporary theory of Quantum Mechanics accepts both viewpoints regarding the propagation of light.

• **Barrow** was a professor of mathematics at Cambridge University. As an admirer of the Ancient Greeks and since he was well versed in Greek language, he translated some of the works of Euclid and improved other translations of works of Euclid, Apollonius and Theodosius. His main writing is *Lectiones Geometricae* (1670), where he uses geometric methods, and it constitutes an important contribution to infinitesimal calculus. Barrow gives methods of finding the tangents of a curve. In the writing of his Lecture 10 he gives an example of finding the tangent to the curve $C : y^2 = px$ at its point $P(x, y)$. To this end he replaces x with $x + \epsilon$ and y with $y + \alpha$. Then

$$y^2 + 2\alpha y + \alpha^2 = px + p\epsilon.$$

He subtracts (side from side) the equality $y^2 = px$ from the above equality and he obtains

$$2\alpha y + \alpha^2 = p\epsilon.$$

Next, he omits the higher powers of the quantities α and ϵ (wherever they appear), which means that he replaces PRP' in Figure 24.7 with PRP' in Figure 24.8, and he concludes that

$$\frac{\alpha}{\epsilon} = \frac{p}{2y}.$$

He remarks that $\frac{\alpha}{\epsilon} = \frac{PM}{NM}$, hence

$$\frac{PM}{NM} = \frac{p}{2y}.$$

But $PM = y$, hence by calculating NM he finds the position of point N.

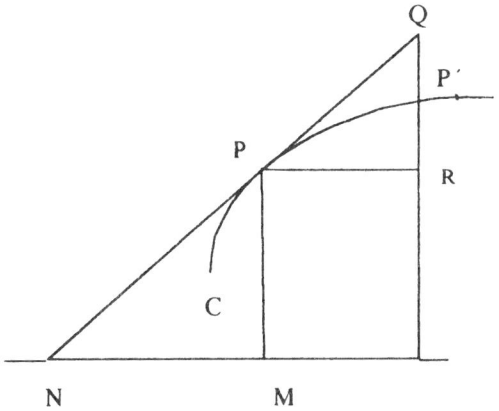

FIGURE 24.7

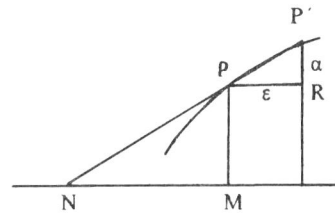

FIGURE 24.8

• **James Gregory** was born in Drumoak, near Aberdeen, and died in Edinburgh. Gregory used Wallis' method to obtain the series

$$x = \tan x - \frac{1}{3}\tan^3 x + \frac{1}{5}\tan^5 x - ...$$

Letting $x = \frac{\pi}{4}$, we obtain

$$\frac{\pi}{4} = 1 - \frac{1}{3} + \frac{1}{5} - \frac{1}{7} + ...$$

which gives the number π.

The series with terms powers of $\tan x$ was known to the Indian mathematicians and is found in the book *Tantrasangraha-uyakhya* (1530).

• **Isaac Newton**, one of the greatest intellects of mankind, was born (prematurely) in Woolsthrope, near Grantham, in the parish of Colsterworth in Lincolnshire, on December 25, 1642, about two months after the death of his father, who owned a small farm. Even though his family wanted him to become a farmer and to occupy himself with the farm, he spent most of his time solving problems, making experiments or inventing various mechanical devices. His mother, who perceived this situation, decided with the mediation of his uncle, to sent him to study at Cambridge, where he studied works of Euclid, Descartes, Kepler and Wallis, and he attended Isaac Barrow's lectures.

Newton studied at Cambridge's Trinity College, where he lived from 1661 to 1697. During that period he produced almost his entire research work. He obtained a B.A. degree (Bachelor of Arts) in 1664. In 1667 he was appointed Fellow of Trinity

College. He obtained a M.A. degree (Master of Arts) in 1667 and in 1669 he became professor of Trinity College, where he taught until 1701. Around 1699 he accepted an appointment to a very important governmental position with an annual salary of £1500, and hence he moved to London, where he remained until his death. In 1703 he was elected president of the British Royal Society.

Newton's published works in chronological order are:

(1) *Philosophiae Naturalis Principia Mathematica*, 1687.
(2) *Optics* (with supplements whose titles are: *Cubic curves*, *Quadrature of curves*, *Method of fluxions*), 1704.
(3) *Universal Arithmetic*, 1707.
(4) *Lectiones opticoe*, 1729.
(5) *Methodus differentialis*, 1736.
(6) *Analytical geometry*, 1736.

Before even obtaining his first degree (Bachelor of Arts), he had already discovered the derivatives of functions, which he called "**fluxions**", and he had already proved the "Binomial Theorem" and the "Generalized Binomial Theorem", where the exponent of the binomial can be any real number:

"If x is a real number between -1 and 1, and if m is an arbitrary real number, then
$$(1+x)^m = 1 + a_1 x + a_2 x^2 + a_3 x^3 + ...,$$
where
$$a_k = \frac{m(m-1)(m-2)...(m-(k-1))}{k!}, \qquad (k = 0, 1, 2, ...).$$

If m is a positive integer, then the terms vanish after some value of k, and the above formula gives the classical theorem of the binomial, which was known to the Chinese and to the Arabs. When m is not a positive integer, then we have a completely new result."

Note. Let $f(x) = 1 + a_1 x + a_2 x^2 + ...$, where the a_k are given by the above equality. Prove that $(1+x)f'(x) = mf(x)$, from which it follows that
$$\frac{d}{dx}\left(\frac{f(x)}{(1+x)^m}\right) = 0,$$
which means that for some constant c, we have $f(x) = c(1+x)^m$. For $x = 0$, we have $c = 1$. Therefore $f(x) = (1+x)^m$.

Newton used the binomial theorem in order to express functions $f(x)$ in power series. Thus, in order to calculate the area under the curve $y = f(x)$ he could apply Wallis' method and replace x^n with $\frac{x^{n+1}}{n+1}$. During that time mathematicians did not know many things about the conditions under which a series with infinite terms converges.

During the plague (1665-1666) Newton went to his family's farm, where he studied the theory of gravity. When he returned to Cambridge, he helped professor Barrow to write notes for his course and then, after Barrow's resignation, he was appointed to Barrow's chair in 1669.

Barrow's resignation in favor of the young Newton constituted (and still constitutes today) a lesson for certain professors, who stop working productively and offer nothing to science from the moment that they occupy a permanent academic position. About that period of time Newton discovered the "analysis of sunlight" with the use of a prism. From 1673 to 1683 he taught algebra and the theory of equations.

Newton's greatest contribution to human knowledge was the **Theory of Gravitation**, which gave a complete explanation of the motion of celestial bodies.

Before Newton, Hooke, Huygens, Halley and Wren had realized that Kepler's laws imply that every "gravitational force" must satisfy a law, according to which the force is inversely proportional to the square of the distance. Newton's contribution in this subject is that he proved the converse of the last proposition, i.e., that: every law according to which the force is inversely proportional to the square of the distance implies Kepler's laws.

The "**law of gravitation**" asserts that if two bodies with masses M and m are separated by distance x, then the force F that each of the two bodies exerts to the other is given by the formula

$$F = kMmx^{-2},$$

where k is a universal constant (i.e., a constant which does not depend on the material).

If the mass M is very large compared to m, we may assume that M exercises the force F on m. Taking into consideration the fact that Newton defined force as the rate of change of the quantity of motion, in this case $F = -m\ddot{x}$, it follows that the acceleration $\ddot{x} = -kMx^{-2}$ of the smaller body does not depend on its mass m. If M is the mass of the earth and we assume that it is concentrated in the center of the earth, while m is the mass of say an apple which lies on or near the surface of the earth, then $\ddot{x}$ practically remains constant, which is consistent with Galileo's initial observation.

At first Newton studied the motion of the planets by assuming that the sun and the planets were points. But he was reluctant to publish his results, because this assumption did not satisfy him. After approximately twenty years he was finally able to prove that the gravitational force of a solid sphere is the same as the one that the sphere would exert if its entire mass was concentrated in its center. He assumed that the density of matter in an interior point of the sphere depends only on the distance of this point from the center and that this is true for all the planets of our solar system.

After overcoming this last obstacle, Newton published in 1686 his celebrated writing *Philosophiae Naturalis Principia Mathematica*, which is of great historic importance. Unfortunately, in his writing, Newton replaced the method of infinitesimal calculus which he had initially used, with the method of classical Euclidean geometry, because the geometric method was better understood by his contemporaries. Because of this fact *Principia* is read today with difficulty.

Newton's mechanics, as stated in the *Principia*, summed up and systematized the work of the scientists of the 17th century and formed the groundwork on which the sciences of astronomy and mechanics were developed for the following two centuries. The mechanistic view of the universe, implicit in Newtonian mechanics and particularly in an orderly system of heavenly bodies rotating in their orbits in accordance with a mathematical law, also greatly influenced the trend of philosophic thinking in the following centuries. A more comprehensive view of the physical universe, which did away with the concepts of absolute space, absolute time, and forces acting at a distance (which was necessary to Newtonian mechanics), was introduced with the theory of relativity developed by the 20th-century German-American physicist Albert Einstein. Although today we accept the relativistic view, Newtonian mechanics is still adequate for predicting physical phenomena to a high degree of accuracy, and for solving practical problems in physics and engineering.

Newton's contribution to the "philosophic" program initiated by Galileo, which consisted of freeing the Exact Sciences from metaphysical considerations and arguments, was immense. The contemporary mathematicians, who dislike every kind of metaphysical thinking in relation to Mathematics, are perhaps not aware how much they owe in this respect to Galileo and Newton. Thanks to the efforts of Galileo and Newton, mathematicians are free today to make advances in mathematics, leaving other people to worry about Metaphysics. These two giants of Science worked hard for the establishment of two basic principles of scientific methodology, which we consider today as granted: a) The mathematical conclusions of a mathematically formulated scientific law, have the same scientific validity as the law. b) This way is the best and the most advisable way, that theoretical sciences must follow.

The second volume of *Principia* studies hydrostatics and hydrodynamics.

In 1692 Newton published two works on "fluxions", i.e., on derivatives. He denotes by $\dot{x}$ and $\ddot{x}$, the first and the second derivative of the function x with respect to the variable t (time). He denotes dt by o and dx by $\dot{x}o$.

In 1704, in the supplement of his book on Optics, he studies "**fluents**", i.e., integrals. In this study he presents the existing relation between "fluents" and "definite integrals". He also studies subjects regarding maxima and minima, tangents of curves, and lengths of curves.

In 1696 Newton resigned from Cambridge in order to accept a governmental position in London. However, in his new position he continued to study mathematics. He suggested the following problem:

"Assume that the grass grows in some field with constant speed. For $i = 1, 2, 3$, assume also that x_i oxen eat the grass that covers a_i hectars in t_i weeks. Prove that

$$a_1 a_2 x_3 t_3 (t_2 - t_1) + a_2 a_3 x_1 t_1 (t_3 - t_2) + a_1 a_3 x_2 t_2 (t_1 - t_3) = 0."$$

Note. Assume that the grass starts to grow from height h and assume that it grows by g every week. Prove first that in one week, one ox eats

$$\frac{ha_1 + gt_1 a_1}{x_1 t_1} \text{ units of grass.}$$

Newton taught as a professor for approximately thirty years. His lectures were not always attended by many students. William Dunham reports in his book *Journey through Genius* (*New York*: Wiley, 1990) that: "The duration of a lecture of Newton was half an hour. But if nobody came to listen, Newton remained in the room for only 15 minutes".

Newton's words below show how modest this ingenious scientist was.

"I don't know what image people have formed about me. I see myself as a small child who plays in the beach and enjoys himself finding here and there some pebble which is smoother, or some seashell which is more beautiful than the others, while the great ocean of truth lies in front of me unexplored."

Newton died on March 20, 1727 and was buried in the Abbey of Westminster.

Further information about the life and the work of Newton can be found in the author's talk *The life and the work of Isaac Newton, Proceedings of the Academy of Athens*, Vol. 63 (1988), while for additional reading we refer the reader to the following:

(1) D. T. Whiteside (ed.): *Sir Isaac Newton, Mathematical works, Jhonson Reprint*, I, 1964; II, 1969.

(2) D. T. Whiteside (ed.): *Sir Isaac Newton, Mathematical papers, Cambridge Univ. Press*, I, 1967; II, 1968; III, 1969; IV, 1971; V, 1972; VI, 1974; VII, 1976; VIII, 1981.

(3) I. Newton: *Philosophiae Naturalis Principia Mathematica, London*, 1687; English translation, *Mathematical principles of natural philosophy*, trans. by A. Motte *in* 1729, *Univ. of California Press*, 1934.

(4) D. Brewster: *Memoirs of the life, Writings, and Discoveries of Sir Isaac Newton, Constable*, 1855.

• **Leibniz**, was born on July 1, 1646 in Leipzig. His father died when he was six years old. He entered the University of Leipzig at the age of 15. In 1666 he was ready to claim the title of doctor of philosophy in law, but the university turned down his request, because he was considered too young to acquire this title. During the same year Leibniz conceived the idea of "symbolic logic", i.e., of a universal language through which it would be possible to express every rational thought.

Leibniz entered the diplomatic service and as a diplomat, he went to Paris, where he met Huygens who introduced him to the subjects of geometry and physics. Huygens posed to Leibniz the problem of finding the sum of the series

$$\sum_{n=1}^{\infty} \frac{2}{n(n+1)}$$

and Leibniz gave the following "solution": We have

$$\frac{2}{n(n+1)} = 2\left(\frac{1}{n} - \frac{1}{n+1}\right).$$

Hence the sum of the series "equals"

$$2\left(1 - \frac{1}{2} + \frac{1}{2} - \frac{1}{3} + \frac{1}{3} - \frac{1}{4} + \frac{1}{4} - \ldots\right) = 2(1+0) = 2.$$

Of course, a contemporary mathematician does not accept this method of solution, since both series involving $1/n$ and $1/n+1$ diverge.

In 1673 Leibniz visited England and was named "fellow" of the Royal Society. At that time he met Newton. Two years later, Leibniz developed his theory of Infinitesimal Calculus, where he introduced the notation that we use today. In particular, he formulated the Leibniz rule, i.e., the rule for differentiating a product of functions.

Leibniz's work in Infinitesimal Calculus became known in 1684, before the publication of Newton's results. This gave rise to a contention between Newton and Leibniz, as well as between their contemporaries, about who was the first to discover Infinitesimal Calculus, and this contention was continued for many years by the successors of these two giants of mathematics. For approximately 100 years the English mathematicians used faithfully the notation that Newton introduced into Infinitesimal Calculus; this harmed mathematics in England, because Leibniz's notation was certainly better and easier to use than Newton's.

Today, historical research has come to the conclusion that Newton and Leibniz discovered calculus independently using different paths. Hence they are both considered to be the creators of Infinitesimal Calculus.

Leibniz considered dy and dx, in the ratio $\frac{dy}{dx}$, as "infinitesimals". Thus, dx was an infinitely small increment of x while, for a given function $f(x)$, dy was defined by the relation
$$dy = f(x + dx) - f(x).$$
For example, if $f(x) = x^2$, then
$$dy = (x + dx)^2 - x^2 = 2x(dx) + (dx)^2.$$
This expressed the change of the function f that corresponded to the increment dx of x. Consequently, the slope of the tangent was
$$\frac{dy}{dx} = 2x + dx$$
which means that the tangent of the curve at the point (x, y) has slope $2x$.

The notion of an "infinitesimal", which is also implicit in Newton's theory of fluxions, was criticised by many people, including the philosopher and bishop **George Berkeley** (1685-1753). Berkeley raised the following questions: How can we divide by dx if dx is equal to zero? In the above example, how can we say that the slope of the tangent is $2x$ and not $2x + dx$, if dx is not equal to zero? Either dx is equal to zero or dx is not equal to zero. In both cases we have a problem!

The mathematicians of the 19th century **Augustin Cauchy** and **Karl Weierstrass** also agreed that problems indeed existed, and constructed the foundation of Infinitesimal Calculus on its present secure bases. Today the symbol $\frac{dy}{dx}$ is not considered to be the ratio of two numbers, but the limit of a ratio, namely,
$$\frac{dy}{dx} = \lim_{h \to 0} \frac{f(x+h) - f(x)}{h}.$$
In other words, $\frac{dy}{dx}$ is the number $L(x)$ with the property that for any given $\epsilon > 0$, there exists a number $\delta > 0$ such that
$$\left| \frac{f(t) - f(x)}{t - x} - L(x) \right| < \epsilon \text{ if } |t - x| < \delta.$$
Today there exists a "rigorous" definition of the notion of an "infinitesimal". In 1966 **Abraham Robinson** introduced the so-called "non-standard" real numbers, in which some real numbers exist which are smaller than any positive rational number and greater than zero (see Chapter 6).

Leibniz's notion of an "infinitesimal" is simple and intuitively easy to understand, but it is not rigorous. On the contrary, Robinson's notion of an "infinitesimal" is rigorous, but it is neither simple nor intuitively easy to understand.

Leibniz occupied himself also with philosophy. He is known for his viewpoint that: "the universe in which we belong is the "best" of all possible universes". Voltaire satirized as utopian this viewpoint in his book *Candide* (1759). But it is doubtful whether Leibniz meant by the word "best" the most pleasant, which is the interpretation given by Voltaire.

Leibniz's most original contribution in philosophy is contained in his writing *Monadology*. In this work he claims the idea that the universe consists of simple substances, which he calls "monads" and which have the ability to "perceive". Every monad represents an individual microcosm and acts independently of all the remaining monads. Monads are "related" to one another only through a harmony that exists a priori and has been designed by God.

Leibniz was a theist, i.e., a supporter of the theory which accepts the existence of a single God, who created the world and who acts upon nature and upon the life of beings. In part XXIII of his writing *Discours de Métaphysique* he writes the following:

"An exquisite privilege of "divine nature" is the fact that just the possibility of existence of "divine nature" is sufficient to imply its existence."

In other words, either God cannot exist or God necessarily exists. Using various arguments, Leibniz alleged that God's existence is possible, and hence he concluded that God necessarily exists. At the end of his book *Discours de Métaphysique*, Leibniz claims that God desires to have a personal relationship with all human beings.

Additional Reading

(1) M. Cantor, *Vorlesungen über Geschichte der Mathematik*, Teubner, II, 1892; III, 1898.

(2) P. L. Boutroux, *L'idéal scientifique des mathématiciens dans l'antiquité et les temps moderns*, Presses Universitaires de France, new edition, 1955.

(3) D. T. Whiteside, *Patterns of mathematical thought in the later seventeenth century*, Arch. History Exact Sci. 1 (1961), 179-388.

(4) H. Freudenthal, *Zur Geschichte der vollständigen Induktion*, Arch. Internat. Histoire Sci., 6, no. 22 (1953), 17-37.

(5) K. Hara, *Pascal et l'induction mathématique*, Rev. Histoire Sci. Appl., 15 (1962), 287-302.

(6) N. Bourbaki, *Éléments d' histoire des mathématiques*, Hermann, 1960.

CHAPTER 25

THE EIGHTEENTH CENTURY

The greatest achievement of the 17th century was the discovery of infinitesimal calculus (differential and integral calculus), which is the origin of many new branches of mathematics including differential equations, series with infinitely many terms (which we will call simply "series"), differential geometry, calculus of variations, functions of complex variables, and others. Many of these branches started with the very work of Newton and Leibniz. The further development of some of these branches of mathematical analysis covers a great part of the activities of the 18th century. But for this development to take place, infinitesimal calculus had to be extended even further. Newton and Leibniz had invented certain basic methods, but many other developments were still needed. New functions of one variable and functions of two or more variables still had to be either expressed through some explicit formula or constructed otherwise. The techniques of differentiation and integration had to be extended to some of the existing functions as well as to others that had to be introduced. The rigorous foundation of infinitesimal calculus was not yet achieved.

In order for the reader to understand and appreciate the work and the arguments of the thinkers of the 18th century, we must emphasize that they made no distinction between algebra and analysis, because they did not perceive the necessity of the notion of a limit and they did not recognize the problems that were created by the use of series. They considered infinitesimal calculus as a simple extension of algebra.

We mention some of the most important mathematicians of the 18th century.

- Brook Taylor (1685-1731)
- Colin Maclaurin (1698-1746)
- Abraham de Moivre (1667-1754)
- Leonard Euler (1707-1783)
- Joseph Louis Lagrange (1736-1813)
- Pierre Simon Laplace (1749-1827)
- Adrien Marie Legendre (1752-1833)

• In his book *Methodus Incrementorum Directa et Inversa* (1715), **Taylor** tried to clarify the basic ideas of infinitesimal calculus, but he restricted himself only to algebraic functions and to algebraic differential equations. He is mainly known for the discovery of the so-called "Taylor series". Archimedes was the first who occupied himself with the sum of a series. In his writing *Quadrature of Parabola* he proved that the sum of the geometric series

$$1 + \frac{1}{4} + \frac{1}{4^2} + ... + \frac{1}{4^{n-1}} + ...$$

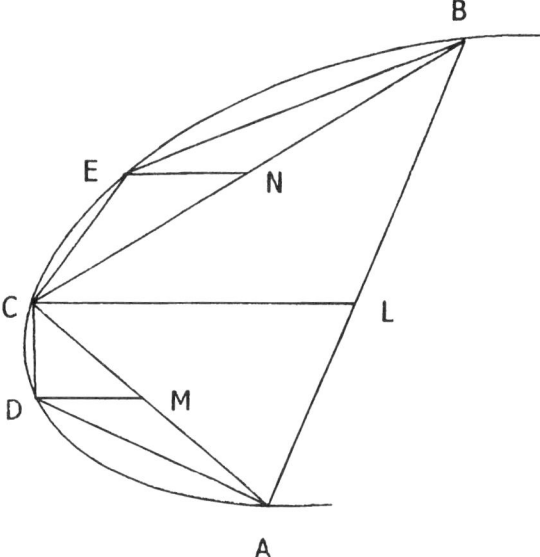

FIGURE 25.1

is equal to $\frac{4}{3}$. It is interesting to examine how this series came up in Archimedes' work: Let A and B be two points of an arc of a parabola (Figure 25.1). From the midpoint L of the segment AB we draw a line parallel to the axis of the parabola, which intersects the arc at the point C. From the midpoints M, N of AC, BC respectively we draw the parallels to LC. Thus, points D and E are defined. From the geometric properties of the parabola, Archimedes deduced that

$$\triangle CDA + \triangle CEB = \frac{\triangle ACB}{4}.$$

The repetition of this procedure ad infinitum shows that the area of this part of the parabola is equal to the sum

$$\triangle ABC + \frac{\triangle ABC}{4} + \frac{\triangle ABC}{4^2} + ... + \frac{\triangle ABC}{4^{n-1}} + ...$$
$$= \triangle ABC \cdot \left(1 + \frac{1}{4} + \frac{1}{4^2} + ... + \frac{1}{4^{n-1}} + ...\right).$$

A series of the form: $a_0 + a_1 x + a_2 x^2 + ... + a_{n-1} x^{n-1} + ...$ is called a **power series** with respect to x.

From elementary books on infinitesimal calculus we know that if a power series with respect to x converges for $x = b$, where b is a positive constant, then it will converge for $|x| < b$, while if it diverges for $x = b$, it will diverge for $|x| > b$. Consequently, the set of values for which the series converges consists of an interval of the form $-b < x < b$ and possibly of one or both of the values $-b$ and b. Obviously, a convergent power series with respect to x represents a function of x defined on the interval of convergence of the series. Consequently, we have

$$f(x) = a_0 + a_1 x + a_2 x^2 + ... + a_{n-1} x^{n-1} + ...$$

We know from infinitesimal calculus that if a function $f(x)$ has derivatives of any order and can be represented by a power series with respect to x, then, for values of x that belong to the interval of convergence of the series, we have

$$f(x) = f(0) + \frac{f'(0)}{1!}x + \frac{f''(0)}{2!}x^2 + \ldots + \frac{f^{(n-1)}(0)}{(n-1)!}x^{n-1} + \ldots .$$

In general, if a given function with derivatives of any order can be represented by a power series with respect to $x - a$, then, in the interval of convergence of the series, we have

$$f(x) = f(a) + \frac{f'(a)}{1!}(x-a) + \ldots + \frac{f^{(n-1)}(a)}{(n-1)!}(x-a)^{n-1} + \ldots$$

The expansion of a function $f(x)$ into a power series with respect to $x - a$ was first given by Taylor in the writing mentioned above, and since then it is known as the expansion of $f(x)$ into a "Taylor series" at the point $x = a$.

Taylor applied his series to the problem of finding roots of equations, as follows:

Let a be an approximate root of $f(x) = 0$. We put $f(a) = k$, $f'(a) = k'$, $f''(a) = k''$ and $x - a = h$. We expand $f(x)$ into a Taylor series at the point $x = a$ and we omit in it the terms where the powers of h are greater than 2. We substitute the values of k, k', k'' and we solve the resulting equation with respect to h. Then the number $a + h$ is a value that gives a better approximation of the requested root. By repeating this procedure we obtain values which every time give a better approximation of the requested root.

• A Taylor series with $a = 0$ is known as a Maclaurin series. **Maclaurin**, who succeeded James Gregory as a professor at Edinburgh, mentions this special case in his writing *Treatise of Fluxions* (1742) and points out that it is a special case of the Taylor series. Historically this special case is considered to have been discovered by Maclaurin as a separate theorem. Furthermore, Stirling mentions this special case for algebraic functions (1717) and for general functions in his work *Methodus Differentialis* (1730).

Maclaurin's method of proof of the above special case is the following:
Let
$$f(x) = A + Bx + Cx^2 + Dx^3 + \ldots$$
We have
$$f'(x) = B + 2Cx + 3Dx^2 + \ldots$$
$$f''(x) = 2C + 6Dx + \ldots$$
$$\ldots$$

For $x = 0$ the above equations give respectively A, B, C, ...

Maclaurin used the above result without being concerned about the convergence of the obtained series.

Maclaurin was one of the most competent mathematicians of the 18th century. His work in geometry and, in particular, his work on the plane curves of higher order is also important. In addition, he applied successfully classical geometry to problems in physics.

• **De Moivre** was a French Protestant. In 1688, three years after the invalidation of the decree *Edict of Nantes*, he left France and settled in England. We remind the reader that *Edict of Nantes* was a decree that gave restricted freedoms to Huguenots. Its declaration was made in 1598 by the king of France Henry IV and its invalidation was made in 1685 by the king of France Louis XIV.

De Moivre is known for the discovery of the formula

$$(\cos x + i \sin x)^n = \cos nx + i \sin nx$$

and for his work in probability theory. In 1718 he published the book *Doctrines of Chances*, which contained a series of solved problems, such as the following: "We assume that three out of 40,000 tickets of a lottery will win three prizes, one for each winning ticket. What is the probability that one wins at least one prize if one buys 8000 tickets?"

De Moivre was the first who discovered the formula that gives the nth Fibonacci number. Moreover, De Moivre, and not James Stirling, was the one who discovered "Stirling's Formula", i.e., the formula

$$n! \approx \left(\frac{n}{e}\right)^n \cdot \sqrt{2\pi n}.$$

For example, for $n = 10$, we have $10! = 3628800$, while the above formula gives 3598695.6.

Despite letters of recommendation from Newton and Leibniz, De Moivre did not find a dignified position as a mathematician. He had a low income that came from private lessons and from predictions that he gave to gamblers. According to a story, as he was nearing his end, every day he was sleeping fifteen minutes more than he had slept the day before, until he ended up sleeping continuously for 24 hours and then he died.

• **Leonard Euler** was Swiss. He was born and grew up in a town named Riehen, near Basel in Switzerland. He was urged by his father to attend the lessons of Johann I. Bernoulli, who immediately recognized Euler's great mathematical talent. Euler obtained his first degree from the university of Basel at the age of 15. At 19, at an announced competition, he was awarded a prize from the Acadèmie des Sciences in Paris for his work *Meditationes super problemate nautico...*

Encouraged by Nicholas and Daniel, sons of Johann Bernoulli, he went to the Academy of Saint Petersburg in Russia, a very important research center, which was suitable to host a mathematical genius such as Euler. There, he met his compatriots Jacob Hermann and Daniel Bernoulli, and soon he became a friend of the diplomat and amateur mathematician Christian Goldbach. During his stay in Saint Petersburg (1727-1741), Euler wrote more than 100 original research papers as well as his fundamental writing on Mechanics. In 1741, after an invitation by king Frederick the Great, he went to the Academy of Berlin. During the 25 years that he stayed at Berlin, his incredibly high productivity in mathematics continued. Among his works we distinguish the Calculus of Variations, his writing *Introductio in Analysin infinitorum* (1748), and Benjamin Robins' writing on "artillery", which he translated and rewrote.

The deterioration of Euler's relations with the king and his court, led Euler to accept in 1776 a much more favorable invitation by Catherine II to return to Saint Petersburg, where she extended a "princely" welcome to him.

Despite the fact that during the last 17 years of his life he was completely blind, he wrote with the help of his students, his famous writing *Algebra* as well as another 400 original research papers, while he left many other papers unpublished.

During the last decades, important material regarding Euler was discovered in the archives of the former USSR Academy of Sciences.

Euler recognized the very favorable conditions of work that he was offered by the Academy of Saint Petersburg and he indeed expressed his gratitude. In relation to that, Judith Kopelevič writes the following:

"Euler's grave which was built in the Academy, his bust which is located in the Presidium building of the Academy, and the fact that for two centuries the Academy took care and made efforts to collect and publish his intellectual heritage clearly show that Euler's collaboration with the Academy of Sciences of Saint Petersburg was beneficial to both sides."

Euler's entire scientific work consists of 75 bulky volumes.

For a detailed account of Euler's life, work, etc., we refer the reader to the following:

(a) *Mathematics Magazine*, Vol. 56, No 5, November 1983.
(b) *The American Mathematical Monthly*, Vol. 91, No 9, November 1984.

We present some of the research results of Euler.

(1) If a convex polyhedron has V vertices, F faces and E edges, then $V + F - E = 2$.

(2) $e^{i\pi} = -1$, where e is the "Euler's number":
$$e = \lim_{n \to \infty} \left(1 + \frac{1}{n}\right)^n.$$

(3)
$$\frac{1}{1^2} + \frac{1}{2^2} + \frac{1}{3^2} + \frac{1}{4^2} + \ldots = \frac{\pi^2}{6}.$$

(4) Every perfect number has the form $2^{n-1}(2^n - 1)$, where $2^n - 1$ is a prime number.

(5) Let n be a positive integer and let $\phi(n)$ be the number of natural numbers which are $\leq n$ and prime to n. Then n divides the number $a^{\phi(n)} - 1$, if a is positive integer which is prime to n.

We note that "Fermat's Little Theorem" is a corollary of the above proposition.

(6) In any triangle, the center of the circumscribed circle, the point of intersection of the heights, and the point of intersection of the medians lie on the same straight line, which is called **Euler's straight line**.

This proposition is proved as follows:

Let ABC be a triangle and let A' be the midpoint of BC. Assume that H is the point of intersection of the heights, O is the center of the circumscribed circle, and G is the point of intersection of the medians. Let H' be the point on the straight line OG which lies between O and H' and satisfies the condition $\frac{H'G}{GO} = 2$. From the known relation $\frac{GA}{GA'} = 2$ it follows that the triangles AGH' and $A'GO$ are similar. Consequently, AH' is perpendicular to BC. Similarly we have that CH' is perpendicular to AB. Therefore the points H' and H coincide.

(7) Fermat's conjecture that all natural numbers of the form $2^{2^n} + 1$ are primes is not true. Indeed, we have:
$$2^{2^5} + 1 = 4294967297 = 6700417 \times 641.$$

In one of his papers, Euler recognized the fact that the use of a series is allowed only when it converges. Despite that, in the same paper (Comm. Acad. Sci. Petrop., 11, 1739, 116-127, pub. 1750 = Opera, (1), 14, 350-363) he obtains the relation
$$... + \frac{1}{x^2} + \frac{1}{x} + 1 + x + x^2 + x^3 + ... = 0$$
and he justifies this result as follows: He writes
$$\frac{x}{1-x} = x + x^2 + x^3 + ...$$
$$\frac{x}{x-1} = \frac{1}{1-\frac{1}{x}} = 1 + \frac{1}{x} + \frac{1}{x^2} + \frac{1}{x^3} + ...$$
and he adds the sides of these two equalities.

The following proof of the formula $V + F - E = 2$ is due to H. S. M. Coxeter.

Let O be a point that lies in the interior of the convex polyhedron and assume that a source of light is placed on O. The rays of light, which are emitted by O, project the polyhedron on the sphere of radius 1 and map every plane polygon on a spherical polygon, whose sides are arcs of great circles (it is said that this idea is due to the Arab mathematician Abu'l Jafa). In the interior of every spherical polygon we choose a point and we connect it with the vertices of the spherical polygon with arcs of great circles. Thus, we divide every spherical polygon in as many spherical triangles as are the sides of the spherical polygon. Then

$$\begin{aligned}
720° &= \text{area of the surface of the sphere} \\
&= \text{sum of the areas of the spherical triangles} \\
&= \text{sum of the angles of the spherical triangles} \\
&\quad -180° \times \text{number of the triangles} \\
&= \text{sum of the angles at the interior points} \\
&\quad + \text{sum of the angles at the vertices } - 180° \times 2E \\
&= 360° \times F + 360° \times V - 360° \times E
\end{aligned}$$

If we divide by $360°$ both sides of this equality, we obtain Euler's formula.

• **Lagrange** was born in Italy. His parents were of French and Italian origin. Joseph-Louis was the youngest of the eleven children of the family and the only child who survived after childhood. At his youth he was appointed professor of mathematics at the Military Academy of Torino, and later he was supported by powerful people of the courts of Frederick the Great of Prussia and Louis XVI of France. The Condorcet family (1743-1794) had members who were serving in the cavalry and in the church, and hence Lagrange's education presented no problems.

At the schools of the Jesuits and later at the Collège de Navarre, Lagrange acquired an enviable fame as a mathematician. Finally, instead of becoming a captain in the cavalry, he lived the life of a scientist as did Voltaire, Diderot and D'Alembert. His decision to devote himself to mathematics was also influenced by the study of a work of Halley.

At 23, based on Newton's theory of gravitation, Lagrange was in position to explain why the moon always shows its same side to the earth. He used the equations that now bear his name. His success encouraged the Paris Academy of Sciences in 1766 to propose, as a problem, the theory of the motions of the satellites of Jupiter. The prize was awarded to Lagrange, and he won the same distinction in 1772, 1774, and 1778. In 1766, on the recommendation of Euler and the French mathematician Jean d'Alembert, Lagrange went to Berlin to fill a position at the academy vacated by Euler, at the invitation of Frederick the Great, who expressed the wish that "the greatest king in Europe" must have "the greatest mathematician in Europe" at his court.

Lagrange stayed in Berlin until 1787. His productivity in those years was prodigious: he published papers on the three-body problem, which concerns the evolution of three particles mutually attracted according to Sir Isaac Newton's law of gravity; differential equations; prime-number theory; the fundamentally important number-theoretic equation that has been identified (incorrectly by Euler) with John Pell's name; probability; mechanics; and the stability of the solar system. In his long paper *Réflexions sur la résolution algébrique des équations* (1770; *Reflections on the Algebraic Resolution of Equations*), he inaugurated a new period in algebra and inspired Évariste Galois to his group theory.

A kind and quiet man, living only for science, Lagrange had little to do with the factions and intrigues around the king. When Frederick died, Lagrange preferred to accept Louis XVI's invitation to Paris. He was given apartments in the Louvre, was continually honored, and was treated with respect throughout the French Revolution. From the Louvre he published his classic *Mécanique analytique*, a lucid synthesis of the hundred years of research in mechanics since Newton, based on his own calculus of variations, in which certain properties of a mechanistic system are inferred by considering the changes in a sum (or integral) that are due to conceptually possible (or virtual) displacements from the path that describes the actual history of the system. This led to independent coordinates that are necessary for the specifications of a system of a finite number of particles, or "generalized coordinates". It also led to the so-called Lagrangian equations for a classical mechanical system, in which the kinetic energy of the system is related to the generalized coordinates, the corresponding generalized forces, and the time. The book was typically analytic; he stated in his preface that "one cannot find any figures in this work".

The revolution that began in 1789 pressed Lagrange to work on a committee with the responsibility to reform the metric system, and then pressed him into teaching. In order to mock those who praised the number system with base 12, he supported ... the number system with base 11! When the great chemist Antoine-Laurent Lavoisier was guillotined, Lagrange commented: "It required only a moment to sever that head, and perhaps a century will not be sufficient to produce another like it". When the École Polytechnique was opened in 1795, he became, with Gaspard Monge, its leading professor of mathematics. His lectures were published as *Théorie des fonctions analytiques* (1797; *Theory of Analytic Functions*) and *Leçons sur le calcul des fonctions* (1804; *Lessons on the Calculus of Functions*) and were the first textbooks on real analytic functions. In them Lagrange, worried about the weak foundation of the calculus having to do with the ratios of small quantities and the limits of such ratios (i.e., the derivatives), tried to base his work on algebra and thus to eliminate the infinitesimal — a gallant but unsuccessful attempt. He also continued to work on his *Mécanique analytique*, but the new edition appeared only after his death.

Lagrange was a mathematical genius. His interests ranged from number theory to physics. From his various achievements we mention the following:

(1) He proved Wilson's theorem: "If p is a prime number, then p divides $(p-1)! + 1$, and vice versa".

(2) He gave the first complete solution of the Diophantine equation
$$Ax^2 + Bxy + Cy^2 + Dx + Ey + F = 0,$$
where A, B, C, D, E, F are given integers.

(3) In 1770 he proved the following theorem: "Every natural number can be written as the sum of four squares of natural numbers".

(4) He wrote a treatise on differential equations.

(5) He wrote his book *Mécanique* at 19, but published it only at 52. In it he characterizes the dynamics of a solid system by the equations
$$\frac{d}{dt}\left(\frac{\partial T}{\partial \dot\theta}\right) - \frac{\partial T}{\partial \theta} + \frac{\partial V}{\partial \theta} = 0,$$
where T is the total kinetic energy, V is the potential energy, t is time, θ is any coordinate, and $\dot\theta = \frac{d\theta}{dt}$. Lagrange remarked that these equations express the fact that the total **action** $\int_a^b (T-V)dt$ of the system is minimal. This led him to invent the "calculus of variations".

After the death of his wife (in 1783), Lagrange devoted himself to the preparation of *Mécanique Analytique* (1788). The death of his wife and the turmoil of the French Revolution contributed to him suffering depression. But in 1792 he married a twenty year old girl and he spent the remaining twenty years of his life happy and productive.

Proof of the four squares theorem

Proposition 1 (Euler) If
$$p = ae + bf + cg + dh$$
$$q = af - be + ch - dg$$
$$r = ag - bh - ce + df$$
$$s = ah + bg - cf + de,$$
then $(a^2 + b^2 + c^2 + d^2)(e^2 + f^2 + g^2 + h^2) = p^2 + q^2 + r^2 + s^2$.

Note. From this proposition of Euler it follows that if every prime number can be written as the sum of four squares, then every natural number can be written as the sum of four squares.

Proposition 2 (Euler) If p is a prime number greater than 2, then there exists an integer m such that $0 < m < p$ and mp can be written as the sum of four squares.

Proof. The numbers
$$0^2, \ 1^2, \ 2^2, \ ..., \ \left(\frac{p-1}{2}\right)^2$$
give different remainders when they are divided by p. Indeed, assume that this is not true and let $A^2 = ap + r$ and $B^2 = bp + r$, where $A > B$. Then p divides the number
$$A^2 - B^2 = (A-B)(A+B).$$
But since $A - B > 0$ and $A + B < p$, p divides neither $A - B$ nor $A + B$, which is a contradiction. Similarly, the numbers
$$-1 - 0^2, \ -1 - 1^2, \ -1 - 2^2, \ ..., \ -1 - \left(\frac{p-1}{2}\right)^2$$

give different remainders when they are divided by p. This last set and the set
$$0^2, \ 1^2, \ 2^2, \ ...., \ \left(\frac{p-1}{2}\right)^2$$
each contain $\frac{p+1}{2}$ numbers. Consequently, the two sets together contain $p+1$ integers. But since there exist only p possible remainders of division by p, it follows that some number x^2 from the first set and some number $-1-y^2$ from the second set give the same remainder when they are divided by p. This means that p divides the difference
$$x^2 - (-1-y^2) = x^2 + y^2 + 1.$$
Consequently, for some integer m, we have
$$mp = x^2 + y^2 + 1^2 + 0^2.$$
But since $0 \leq x, y \leq \frac{p-1}{2}$, it follows that $0 < m < p$.

An immediate consequence of Proposition 2 is that in order to prove that every natural number can be written as the sum of four squares it suffices to prove that $m = 1$.

Proposition 3. If p is a prime number greater than 2 and if m is the smallest integer for which $0 < m < p$ and $mp = a^2 + b^2 + c^2 + d^2$ for some natural numbers a, b, c, d, then m is odd.

Proof. Indeed, let us assume that m is even. Then, for the relation
$$a^2 + b^2 + c^2 + d^2 = mp$$
to hold, either none of a, b, c, d is odd or only two of them are odd or all four of them are odd. If we take the odd numbers in pairs, we have the relation
$$\left(\frac{a+b}{2}\right)^2 + \left(\frac{a-b}{2}\right)^2 + \left(\frac{c+d}{2}\right)^2 + \left(\frac{c-d}{2}\right)^2 = \frac{mp}{2},$$
which constitutes an expression of the integer $\frac{mp}{2}$ as the sum of four squares. But this is a contradiction, because $\frac{m}{2} < m$ and m is the smallest integer for which mp can be written as the sum of four squares. Therefore m is odd.

By using Propositions 1, 2, 3, Lagrange proved (in 1770) the following:

Theorem.
Every natural number can be written as the sum of four squares of natural numbers.

Proof. Since $2 = 1^2 + 1^2 + 0^2 + 0^2$, it follows from Proposition 1 that, in order to prove the above theorem, it is enough to prove it for every prime number $p > 2$. So let p be any odd prime number and let m be the smallest integer between 0 and p for which mp is the sum of four squares. We remind the reader that the existence of such an m follows from Proposition 2 and that m is odd (Proposition 3).

We will work by reductio ad absurdum, assuming that $m \geq 3$. Let $mp = a^2 + b^2 + c^2 + d^2$ and let x be the closest to $\frac{a}{m}$ integer. Then $\left|\frac{a}{m} - x\right| < \frac{1}{2}$ and the number $x' = a - mx$ lies between $-\frac{m}{2}$ and $\frac{m}{2}$. Let y, z and w be the closest to $\frac{b}{m}$, $\frac{c}{m}$ and $\frac{d}{m}$ integers respectively. Then

$$y' = b - my$$
$$z' = c - mz$$
$$w' = d - mw$$

and each number y', z', w' lies between $-\frac{m}{2}$ and $\frac{m}{2}$. We put $Z' = (x')^2 + (y')^2 + (z')^2 + (w')^2$. Then $Z' < 4\left(\frac{m}{2}\right)^2 = m^2$. We have $Z' \neq 0$, because if $Z' = 0$, then m divides a, b, c, d, which means that m^2 divides $a^2 + b^2 + c^2 + d^2 = mp$, but this is impossible, because p is a prime number and $1 < m < p$. We put $Z = x^2 + y^2 + z^2 + w^2$ and $T = xx' + yy' + zz' + ww'$. Then the relation $mp = a^2 + b^2 + c^2 + d^2$ implies that

$$mp = m^2 Z + 2mT + Z',$$

because $a = x' + mx$, and so on. We put

$$M = \frac{Z'}{m} = p - mZ - 2T = \text{integer}.$$

Since $Z' \neq 0$, it follows that $M \neq 0$. In addition, since $Z' < m^2$, it follows that $M < m$. Thus, since $M = \frac{Z'}{m}$, we have

$$\begin{aligned} Mp &= \left(\frac{M}{m}\right) mp \\ &= \left(\frac{M}{m}\right)(m^2 Z + 2mT + Z') \\ &= ZMm + 2MT + \frac{MZ'}{m} \\ &= ZZ' - T^2 + (T+M)^2. \end{aligned}$$

From Proposition 1 we have

$$ZZ' = T^2 + q^2 + r^2 + s^2$$

for some natural numbers q, r, s. Consequently

$$Mp = q^2 + r^2 + s^2 + (T+M)^2,$$

i.e., the number Mp can be written as the sum of four squares, which is a contradiction because $M < m$.

• **Laplace** was born in the town Beaumond of Normandy. He entered the University of Caen at 16 and studied there. During his studies he wrote his first paper on the "Calculus of finite differences". After the completion of his studies, equipped with recommendation letters, he went to Paris, where he presented himself to d'Alembert, but d'Alembert ignored him. After that, Laplace wrote to d'Alembert a letter, in which he presented his views regarding the general principles of mechanics. This time Laplace got d'Alembert's attention, and invited Laplace to visit him. With d'Alembert's recommendation, Laplace was appointed professor of mathematics at the École Militaire in Paris.

Even from his youth Laplace was very productive in his science. Shortly after his election at the Acadèmie des Sciences in Paris, in 1773, people expressed the opinion that until then nobody that young had presented so many research works on so many difficult subjects. In 1783, he replaced Bézout as an examiner at the "Artillery" and examined Napoleon Bonaparte, who was then a student. During the French Revolution he was appointed member of the Committee on Weights and Measures, but he was later relieved of this post, as it happened with Lavoisier and others, because he was considered a bad "democrat". When Laplace retired, he went to Melun, a small town near Paris, where he devoted himself to the writing of his famous book *Exposition du Système du Monde* (1796). After the end of the Revolution, Laplace was elected professor at the École Normale, where Lagrange was already teaching, and he served also as a member of various governmental committees. Later, he served successively as Minister of the Interior, as a member of the Senate, and later as the president of the Senate. Despite the fact that Napoleon honored him with the title of the count, Laplace voted against Napoleon in 1814 and sided with Louis XVIII, who awarded him the title of the marquis and the title of the "peer" of France. During the years that these political events took place, Laplace continued to work scientifically. Between the years 1799 and 1825 the five volumes of his writing *Mécanique Céleste* were published. In this work Laplace presents the solar system, using as few experimental data as possible. The writing *Mécanique Céleste* contains discoveries and results of Newton, Clairaut, d'Alembert, Euler, Lagrange, and of Laplace himself. This work was so comprehensive that the immediate successors of Laplace added to it only a few things. Perhaps the only drawback of this work was that Laplace neglected to recognize the sources of the results that he presented in it, thus giving the impression that they were all his.

In 1812 he published his writing *Théorie analytique des probabilités*. The introduction of the second edition (1814) of this work constitutes a famous essay entitled *Essai philosophique sur les probabilités*, which contains Laplace's celebrated viewpoint that "the future of the universe is completely determined by the past; if we possess the mathematical knowledge (the respective equations) of the situation of the universe at any given moment, then we can predict its future."

Laplace made many discoveries in the area of mathematical physics. Essentially, he was interested in everything that could help in the interpretation of the universe. He worked on subjects involving hydrodynamics, the propagation of sound, waves, and tides. In Chemistry, his work regarding the fluid state of matter constitutes a classic. Moreover, the papers that he wrote in collaboration with Lavoisier are also very interesting.

Laplace dedicated the greatest part of his life to Celestial Mechanics. He died in 1827. It is said that his last words were the following: "What we know is minimum. What we don't know is boundless." But de Morgan alleges that Laplace's last words were the following: "The human follows only the ghosts."

Laplace and Lagrange present no similarities, when we compare their character and their work. Laplace's vanity did not allow him to recognize the works of the people that he considered as his competitors. He used many of Lagrange's ideas as his own.

Lagrange, as a mathematician, was very "elegant" in his writings, very clear and very careful. Laplace invented many methods, which later developed into branches of mathematics. He showed interest for mathematics only when it was helping him in the study of nature. For this reason he occupied himself more with mathematical physics than with theoretical mathematics. He introduced the notion of the potential V and he proved that it satisfies the relation

$$\frac{\partial^2 V}{\partial x^2} + \frac{\partial^2 V}{\partial y^2} + \frac{\partial^2 V}{\partial z^2} = 0.$$

In mathematics, Laplace often made use of the expression: "It is easily recognized that ...". The American astronomer Nathaniel Bowdich (1773-1838), who translated into English a great part of *Mécanique Céleste*, writes the following:

"I have never met Laplace's expression "It is obvious that ..." without having the feeling that many hours of hard work await me in order to fill in the gap and prove why "it is obvious that ...""

• **Legendre** was born in Toulouse. He studied at the Collège Mazarin in Paris. He served as a professor of mathematics at the École Militaire and at the École Normale in Paris. He was elected member of the Académie des Sciences (1783) and Fellow of the Royal Society (1789).

Legendre was a great devotee and supporter of Euclid. His efforts to prove the Axiom of Parallelism can be seen in the successive editions of his writing *Éléments de Geometrie*, whose first publication in 1794 was followed by 12 editions until 1823. This writing constitutes an improvement of Euclid's *Elements* from the pedagogical point of view, because the propositions which are contained in Euclid's *Elements* are reordered and simplified in Legendre's *Éléments de Geometrie*. This writing became easily accepted in both continental Europe and the USA. In his work Legendre proves that the Axiom of Parallelism is a consequence of the assumption that the plane contains quadrilaterals having four equal sides and four angles equal to a right angle.

Legendre is known for his work in number theory. He was the first who proved that the Diophantine equation

$$x^5 + y^5 = z^5$$

has no non-trivial solutions.

The most original discovery of the 18th century in number theory that has many consequences is perhaps the so-called "**Law of quadratic reciprocity**". This law uses the following notion of "quadratic residues". Given two integers p and q, if there exists an integer x such that p divides $x^2 - q$, then q is said to be a quadratic residue of p. We express this fact by writing (following Gauss' notation) $x^2 \equiv q$ (mod p) (which is read: x^2 is congruent to q modulo p). Legendre rediscovered a beautiful theorem, which was given before by Euler and which is known as the "Law of quadratic reciprocity". This theorem is stated as follows:

If p and q are two different odd prime numbers, then either both congruences $x^2 \equiv q$ (mod p) and $x^2 \equiv p$ (mod q) have solutions or both of these congruences have no solutions, except if p and q have the form $4n + 3$, a case in which one of the above congruences has a solution, while the other has no solution.

For example, $x^2 \equiv 13$ (mod 17) has the solution $x \equiv 8$ (mod 17) and $x^2 \equiv 17$ (mod 13) has the solution $x \equiv 11$ (mod 13). Also, it is proved that $x^2 \equiv 5$ (mod 13) and $x^2 \equiv 13$ (mod 5) have no solutions. In addition, $x^2 \equiv 19$ (mod 11) has no solution, while $x^2 \equiv 11$ (mod 19) has the solution $x \equiv 7$ (mod 19).

The above theorem was stated using the contemporary terminology. If we follow the notation that was used by Legendre, then the conclusion of the theorem is written as follows:

$$\left(\frac{p}{q}\right)\left(\frac{q}{p}\right) = (-1)^{\frac{(p-1)(q-1)}{4}},$$

where the symbol $\left(\frac{p}{q}\right)$, which was introduced by Legendre, is equal to 1 or -1, depending on whether the congruence $x^2 \equiv p$ (mod q) has a solution or not.

Based on Legendre's notation, the above conclusion is interpreted as follows:

If the exponent of -1 is even, then either both congruences $x^2 \equiv q$ (mod p) and $x^2 \equiv p$ (mod q) have solutions or both of these congruences have no solutions.

If the exponent of -1 is odd, something that happens when p and q are of the form $4n+3$, then one of these congruences has a solution and the other has no solutions.

Note. Euler had stated (without proof) the "law of quadratic reciprocity" as a conjecture. Legendre gave proofs of it for some cases only. Gauss was the first who gave a complete proof of the theorem. Later, Gauss gave five more different proofs of the same theorem. Today, there exist more than 50 different proofs.

Now we will review and comment on the mathematical achievements of the 18th century.

The notion of a function, as well as the most simple algebraic and transcendental functions were already introduced during the 17th century. As a result of the research conducted regarding concrete problems, such as the motion of the pendulum, the shape of a heavy string which is suspended from two fixed points, etc., the mathematicians of that time, taking into account also works of a more general nature that were carried out in Infinitesimal Calculus, treated these subjects almost in the same way as we treat them today.

For example, even though the logarithmic function started initially from the existing relation between the terms of a geometric progression and the terms of an arithmetic progression and it was defined during the 17th century as the integral of the function $\frac{1}{1+x}$, it was introduced again during the 18th century on a new basis. The studies of the exponential function by Wallis, Newton, Leibniz and Johann Bernoulli proved that the logarithmic function is the inverse of the exponential function, whose properties are relatively simpler.

Euler defines these functions in his work *Introductio* as follows:

$$e^x = \lim_{n \to \infty} \left(1 + \frac{x}{n}\right)^n \qquad \log x = \lim_{n \to \infty} \left(n \cdot \left(x^{\frac{1}{n}} - 1\right)\right).$$

The mathematics regarding the trigonometric functions was more systematized. Newton and Leibniz had given expansions of these functions into series. The formulas regarding the functions $\sin(x+y)$, $\sin(x-y)$, etc., are due to various mathematicians among who are Johann Bernoulli and Thomas Fantet de Lagny (1660-1734). The periodicity of the trigonometric functions becomes clear in Euler's work, where he introduces the measurement of angles in radians.

The study of the hyperbolic functions began from the moment that people perceived that the area of the part of the plane that lies below the arc of a circumference of a circle is given by an integral of the form $\int \sqrt{a^2 - x^2} dx$, while the area of the part of the plane that lies below the arc of a hyperbola is given by an integral of the form $\int \sqrt{x^2 - a^2} dx$. Since the two values differ only by a sign and the area below the arc of a circle can be expressed by a trigonometric function (for example, $x = a \sin \theta$), while the area below the arc of a hyperbola is related to the logarithmic function, this observation gave rise to the idea that some relation must exist between the trigonometric functions and the logarithmic function, which involves complex numbers. The hyperbolic functions were studied systematically by Johann Heinrich Lambert (1728-1777).

The notion of a function was formulated by Johann Bernoulli. In the beginning of his writing *Introductio*, Euler defines as a function every analytic expression that can be formed, in any way, using a variable and constants. In this definition he includes polynomials, power series, logarithmic and trigonometric expressions. He also defines a function of many variables. Later, Euler introduces the notion of an algebraic function, where the operations on the independent variable are only algebraic, but he divides the algebraic functions into two classes: the "rational" functions, where the use of only the four usual operations is allowed, and the "irrational" functions, where the extraction of roots is also allowed. Then he introduces the transcendental functions, i.e., the trigonometric, the logarithmic, and the exponential functions, the powers of various variables with irrational exponents, as well as some integrals. The main difference between functions, as Euler writes, is the way in which the variables and the constants are combined together. Thus, the transcendental functions differ from the algebraic functions in the fact that the first can be given under the form of series with infinitely many terms. Euler and his contemporaries did not consider as necessary to check whether it is "legitimate" to use expressions that result from an interminable application of the four rational operations.

Euler makes a distinction between explicit and implicit functions.

In the 18th century, when Euler, Leibniz, and all their contemporaries talk also about a "continuous" function, they essentially mean an "analytic function", i.e., a function that can be expressed by an analytic formula (where some isolated discontinuity might exist, as it happens, for example, with the function $y = \frac{1}{x}$).

In general, the dominating notion of a function during the 18th century was that a function is given only by an analytic expression with finitely many or infinitely many terms.

The basic method of integrating a function during the 18th century was the technique introduced by Newton, i.e., writing the function in the form of a series and integrating every term of the series separately. As time went by, certain particular techniques were developed and perfected.

Following Newton, Johann I. Bernoulli considered the integral as the inverse operation of differentiation. Thus, if $dy = f'(x)dx$, then $y = f(x)$. In other words, Newton's "antiderivative" was considered to be the integral, while differentials were used instead of Newton's "derivatives".

Some examples of the "developed techniques" of integration are the following:

Calculate the integral
$$\int \frac{a^2 dx}{a^2 - x^2}.$$

Jacob Bernoulli puts $x = a\frac{b^2 - t^2}{b^2 + t^2}$ and hence the integral takes the form $\int \frac{dt}{2at}$, which is equal to the logarithmic function.

In 1702, in his writing *Mèmoires* (Académie des Sciences), Johann I. Bernoulli remarks that
$$\frac{a^2}{a^2 - x^2} = \frac{a}{2}\left(\frac{1}{a+x} + \frac{1}{a-x}\right),$$
hence the integration is directly feasible.

In this last way the integrals of the form
$$\int \frac{dx}{ax^2 + bx + c}$$
are calculated.

It should be noted that these integrals lead to logarithms of complex numbers when the first-order factors of the expression are complex numbers, because then one finds integrals of the form $\int \frac{dx}{cx+d}$, where at least d is a complex number. Despite the presence of complex numbers, both Leibniz and Johann I. Bernoulli performed integrations using the "logarithmic way", which made the presence of logarithms of complex numbers unavoidable. Despite the confusion that existed regarding the complex numbers, none of them hesitated to perform the integration in this way. Leibniz was saying that the presence of complex numbers hurts no one. The investigation of the nature of the logarithm of a negative number constituted the object of serious discussions and conflicts between Johann I. Bernoulli, Leibniz and other mathematicians of that time.

Later, the problem of the integration of irrational functions came up. The problems of calculating the arc length of the lemniscate and of the ellipse gave rise to the so-called **elliptic integrals**. The mathematicians of that time were not able to express these integrals in terms of the known algebraic and transcendental functions. A basic paper on the study of elliptic integrals is Legendre's *Traité des Fonctions Elliptiques* (1825-1826). Further studies of the elliptic integrals were carried out later by Abel and Jacobi, who introduced the so-called **elliptic functions**, through which the study of elliptic integrals becomes simpler.

• The indefinite elliptic integrals are new transcendental functions. In the course of time, other transcendental functions were also discovered, the most important of which is the **Gamma function**, which came up from works on two problems, the problem of "interpolation" and the problem of "antidifferentiation".

In the problem of interpolation one tries to give a meaning to $n!$ when n is not a natural number. Euler had observed that

$$n! = \left[\left(\frac{2}{1}\right)^n \frac{1}{n+1}\right] \cdot \left[\left(\frac{3}{2}\right)^n \frac{2}{n+2}\right] \cdot \left[\left(\frac{4}{3}\right)^n \frac{3}{n+3}\right] \cdots$$

$$= \prod_{k=1}^{\infty} \left(\frac{k+1}{k}\right)^n \frac{k}{k+n}.$$

This last relation has a meaning for all values of n apart from the negative integers.

Euler had also observed that for $n = \frac{1}{2}$ the right-hand side of this relation can be written as follows:

$$\frac{\pi}{2} = \left(\frac{2 \cdot 2}{1 \cdot 3}\right) \cdot \left(\frac{4 \cdot 4}{3 \cdot 5}\right) \cdot \left(\frac{6 \cdot 6}{5 \cdot 7}\right) \cdot \left(\frac{8 \cdot 8}{7 \cdot 9}\right) \cdots$$

This expression is Wallis' infinite product.

If we make use of the notation $\Gamma(n+1) = n!$, which was introduced later by Legendre, we can state the formula $\Gamma(n+1) = n\Gamma(n)$ that Euler proved, from which one obtains the values $\Gamma(3/2)$, $\Gamma(5/2)$, etc.

The observed connection between the above product and Wallis' result led Euler to study the same integral that was studied by Wallis, namely the integral

$$\int_0^1 x^e (1-x)^n dx,$$

where, for Euler, e and n are arbitrary numbers. Euler expanded $(1-x)^n$ according to the law of the binomial and then calculated the above integral. He obtained the following:

$$\int_0^1 x^e(1-x)^n dx = \frac{1}{e+1} - \frac{n}{1 \cdot (e+2)} + \frac{n(n-1)}{1 \cdot 2 \cdot (e+3)} - \frac{n(n-1)(n-2)}{1 \cdot 2 \cdot 3 \cdot (e+4)} + \cdots$$

For $n = 0, 1, 2, 3, 4, ...$ the sums on the right-hand side are respectively equal to
$$\frac{1}{e+1}, \quad \frac{1}{(e+1)(e+2)},$$
$$\frac{1 \cdot 2}{(e+1)(e+2)(e+3)}, \quad \frac{1 \cdot 2 \cdot 3}{(e+1)(e+2)(e+3)(e+4)}, \quad ...$$
Consequently, for any positive integer n, Euler discovered the formula
$$\int_0^1 x^e (1-x)^n dx = \frac{n!}{(e+1)(e+2)...(e+n+1)}.$$

Then, Euler tried to find an expression for $n!$ when n is an arbitrary real number. After applying a series of transformations, which are not entirely acceptable today, Euler obtained the relation
$$n! = \int_0^1 (1 - \log x)^n dx.$$

This integral, which has a meaning for almost all n, was later called "Gamma function" by Legendre and is denoted by $\Gamma(n+1)$. In 1781 Euler gave to this integral its contemporary form by putting $t = -\log x$, hence
$$\Gamma(n+1) = \int_0^\infty x^n e^{-x} dx.$$

The integral $\int_0^1 x^e (1-x)^n dx$ gives the so-called "Beta function", which is defined by the relation
$$B(m,n) = \int_0^1 x^{m-1} (1-x)^{n-1} dx.$$

In addition, Euler discovered the existing relation between the two integrals above, i.e., the relation
$$B(m,n) = \frac{\Gamma(m)\Gamma(n)}{\Gamma(m+n)}.$$

The development of infinitesimal calculus for functions of two and three variables takes place in the beginnings of the 18th century. The creation of the theory of partial derivatives is mainly due to Alexis Fontaine des Bertin (1705-1771), Euler, Clairaut and d'Alembert.

The following result is due to Clairaut:

The expression $pdx + qdy$ is an exact (or perfect) differential (which means that there exists a function $f(x,y)$ such that $p = \frac{\partial f}{\partial x}$ and $q = \frac{\partial f}{\partial y}$) if and only if $\frac{\partial p}{\partial y} = \frac{\partial q}{\partial x}$.

In one of his papers Euler proved in 1734 that if $z = f(x,y)$, then
$$\frac{\partial^2 z}{\partial x \partial y} = \frac{\partial^2 z}{\partial y \partial x}.$$

Multiple integrals can be found even in Newton's *Principia*, but the methods used there are geometric. During the 18th century, his work is examined and developed from the analytical point of view. We meet multiple integrals whose use is made in order to express the solutions of the equation
$$\frac{\partial^2 z}{\partial x \partial y} = f(x,y).$$

In 1770 Euler had a clear understanding of certain double integrals, where the integration is carried out over a domain defined by arcs of curves, and he gave methods of calculating them via repeated integrations.

Lagrange expressed the "attraction between ellipsoids of revolution" via triple integrals (1775). Then, he tried to find transformations of multiple integrals, while Laplace was the one who gave their transformation into spherical coordinates (1776).

During the 18th century, there existed many oppositions and controversies regarding the sound and rigorous foundation of infinitesimal calculus. The most powerful opposition came from the bishop George Berkeley (1685-1753), who was worried by the observed growing trend towards mechanistic and deterministic conceptions, because he believed that this trend was dangerous for religion!

Berkeley criticised many of Newton's arguments. For example, in Newton's writing *De Quadratura*, where he claims that he has avoided the notion of infinitely small, the following procedure is followed: the writer gives an increment to x, which he denotes by o, expands $(x+o)^n$, subtracts from it x^n, divides the difference by o, and finds the ratio of the change of x^n over the change of x. Later, he omits the terms containing o and obtains the fluxion (derivative) of x^n. Berkeley criticises this procedure by saying that Newton gives first to x an increment which he later makes equal to zero. This, continues Berkeley, is absurd and the derivative obtained in this way is in reality equal to $0/0$.

We will not discuss any further the subject of controversies regarding the rigorous foundation of infinitesimal calculus, which was achieved during the 19th century.

As we saw before, series with infinitely many terms were known for a long time. Aristotle recognized that these series have a "sum".

Dresne (1360) had proved that the harmonic series

$$1 + \frac{1}{2} + \frac{1}{3} + \frac{1}{4} + \frac{1}{5} + \dots$$

is divergent. The method that he had used was the same with one of the methods that we use today, i.e., he replaced this series with another series having smaller terms, namely, with the series

$$\frac{1}{2} + \frac{1}{2} + \left(\frac{1}{4} + \frac{1}{4}\right) + \left(\frac{1}{8} + \frac{1}{8} + \frac{1}{8} + \frac{1}{8}\right) + \dots$$

and observed that the latter series diverges.

Newton and Mercator discovered the series

$$\log(1+x) = x - \frac{1}{2}x^2 + \frac{1}{3}x^3 - \dots$$

People then commented that for $x = 2$, the right-hand side becomes infinite, while the left-hand side takes the value $\log 3$.

In 1702 Jacob Bernoulli (1657-1705) expressed as a series both $\sin x$ and $\cos x$. These series were used for the calculation of the numbers π and e, and for the calculation of trigonometric series.

In 1674 Leibniz discovered the celebrated result

$$\frac{\pi}{4} = 1 - \frac{1}{3} + \frac{1}{5} - \frac{1}{7} + \dots,$$

where the series is slowly convergent, as it happens with the series that gives $\log(1+x)$. Due to the slow convergence, the problem emerged of finding a corresponding to it rapidly convergent series.

James Gregory (*Exercitationes Geometricae*, 1688) discovered the series

$$\frac{1}{2} \log\left(\frac{1+z}{1-z}\right) = z + \frac{1}{3}z^3 + \frac{1}{5}z^5 + ...,$$

which was used for calculating logarithms.

One of the problems that was treated by mathematicians near the end of the 17th and the beginning of the 18th century was the construction of tables of values of functions using the so-called method of interpolation.

The progress in navigation, in astronomy, and in geography demanded greater accuracy in the values given by the trigonometric, the logarithmic, and the nautical tables, whose construction was made using the method of interpolation.

The method of interpolation that is usually followed is called linear, because according to this method it is assumed that the function is a linear function of the independent variable in the interval between two known values. In reality though, these functions are not linear, and hence a better method of interpolation was necessary.

The method that we will describe below is due to Briggs (1624).

Let $f(x)$ be a function whose values are known at the points a, $a + c$, $a + 2c$, $a + 3c$, ..., $a + nc$. We set

$$\Delta f(a) = f(a + c) - f(a)$$
$$\Delta f(a + c) = f(a + 2c) - f(a + c)$$
$$\Delta f(a + 2c) = f(a + 3c) - f(a + 2c)$$
$$...$$

We also set

$$\Delta^2 f(a) = \Delta f(a + c) - \Delta f(a)$$
$$\Delta^3 f(a) = \Delta^2 f(a + c) - \Delta^2 f(a)$$
$$...$$

Then the Gregory-Newton formula asserts that

$$f(a + h) = f(a) + \frac{h}{c}\Delta f(a) + \frac{\frac{h}{c} \cdot \left(\frac{h}{c} - 1\right)}{1 \cdot 2}\Delta^2 f(a) + ...$$

Newton gave a proof of this formula, something that Gregory did not do.

In order to calculate the value $f(x)$ for any x between two known values, we give to h the value $x - a$. The value obtained is not necessarily the exact value of the function. What this formula gives is the value of a polynomial with respect to h, which coincides with the value of the function at the points a, $a + c$, $a + 2c$, ...

The Gregory-Newton formula was used for the approximate integration of functions.

Let $g(x)$ be a function to be integrated in order to calculate, for example, the area of the domain that lies below the graph of $g(x)$. To this end, we use the values $g(a)$, $g(a + c)$, $g(a + 2c)$, ..., their differences, and their higher-order differences, which we substitute in the formula in question. Then this formula gives a polynomial approximation of $g(x)$, and if we integrate this approximation of $g(x)$, we obtain an approximate integral of $g(x)$.

In 1750, Euler proved in one of his papers the following very beautiful result:
$$\sum_{\nu=1}^{\infty} \frac{1}{\nu^{2n}} = (-1)^{n-1} \frac{(2\pi)^{2n}}{2(2n)!} B_{2n},$$
where B_{2n} are the Bernoulli numbers
$$\left(B_2 = \frac{1}{6},\ B_4 = -\frac{1}{30},\ B_6 = \frac{1}{42},\ B_8 = -\frac{1}{30},\ B_{10} = \frac{5}{66},\ \ldots \right).$$

Euler worked also on subjects regarding harmonic series, i.e., series, the inverses of the terms of which form arithmetic progressions. In particular, he proved (1740) how one can calculate the sum of finitely many terms of the usual harmonic series, using the logarithmic function. The proof is as follows: We have
$$\log\left(1 + \frac{1}{x}\right) = \frac{1}{x} - \frac{1}{2x^2} + \frac{1}{3x^3} - \frac{1}{4x^4} + \ldots$$
i.e.,
$$\frac{1}{x} = \log\left(\frac{x+1}{x}\right) + \frac{1}{2x^2} - \frac{1}{3x^3} + \frac{1}{4x^4} - \ldots$$

For $x = 1, 2, 3, \ldots, n$, we find
$$\frac{1}{1} = \log 2 + \frac{1}{2} - \frac{1}{3} + \frac{1}{4} - \frac{1}{5} + \ldots$$
$$\frac{1}{2} = \log \frac{3}{2} + \frac{1}{2 \cdot 4} - \frac{1}{3 \cdot 8} + \frac{1}{4 \cdot 16} - \frac{1}{5 \cdot 32} + \ldots$$
$$\frac{1}{3} = \log \frac{4}{3} + \frac{1}{2 \cdot 9} - \frac{1}{3 \cdot 27} + \frac{1}{4 \cdot 81} - \frac{1}{5 \cdot 243} + \ldots$$
$$\ldots$$
$$\frac{1}{n} = \log \frac{n+1}{n} + \frac{1}{2n^2} - \frac{1}{3n^3} + \frac{1}{4n^4} - \frac{1}{5n^5} + \ldots$$

If we add the sides of the above equalities, we find
$$\frac{1}{1} + \frac{1}{2} + \frac{1}{3} + \ldots + \frac{1}{n} = \log(n+1) + \frac{1}{2}\left(1 + \frac{1}{4} + \frac{1}{9} + \ldots + \frac{1}{n^2}\right)$$
$$- \frac{1}{3}\left(1 + \frac{1}{8} + \frac{1}{27} + \ldots + \frac{1}{n^3}\right)$$
$$+ \frac{1}{4}\left(1 + \frac{1}{16} + \frac{1}{81} + \ldots + \frac{1}{n^4}\right)$$
$$\ldots$$
or
$$1 + \frac{1}{2} + \frac{1}{3} + \ldots + \frac{1}{n} = \log(n+1) + c,$$
where c is the sum of the infinitely many terms of a set, each member of which is the sum of finitely many numbers. Euler calculated the value of c (which depends on n, but when n is very large, then the value of n does not affect the result) and found it approximately equal to 0.577218. The number c is now known as "Euler's constant" and is denoted by γ.

A more exact expression of γ is given today as follows: We subtract $\log n$ from both sides of the last relation. Since the difference
$$\log(n+1) - \log n = \log\left(1 + \frac{1}{n}\right)$$
tends to 0 as $n \to \infty$, we have
$$\gamma = \lim_{n \to \infty} \left(1 + \frac{1}{2} + \frac{1}{3} + \ldots + \frac{1}{n} - \log n\right).$$

Let it be noted that for the constant γ, no other simpler expression has been found, while for the numbers e and π, there exist various other expressions. Also, **up to now it is not known whether γ is a rational or irrational number**.

- The mathematicians of the 18th century worked extensively on trigonometric series. In particular, they persevered in works regarding astronomy, because trigonometric series represent periodic functions and the astronomical phenomena are in their majority periodic phenomena.

The representation of functions in terms of trigonometric series arose from the need to solve problems using the method of interpolation and, in particular, from the need to determine the positions of planets which are located between the positions of other known planets, whose positions are also known and already determined through observation. The trigonometric series were also introduced in the first works regarding partial differential equations. It is a wonder how these two directions of research remained independent of one another, even though there existed mathematicians who worked on both problems.

We call trigonometric series a series of the form

$$\frac{1}{2}a_0 + \sum_{n=1}^{\infty}(a_n \cos nx + b_n \sin nx),$$

where a_n and b_n are constants. If such a series represents a function $f(x)$, then

$$a_n = \frac{1}{\pi}\int_0^{2\pi} f(x)\cos nx\, dx, \quad b_n = \frac{1}{\pi}\int_0^{2\pi} f(x)\sin nx\, dx$$

for $n = 0, 1, 2, 3, \ldots$ The finding of these formulas, which give the coefficients a_n and b_n, constitutes one of the most important results in the theory of trigonometric series.

Even since 1729, Euler had occupied himself with the following interpolation problem: A function $f(x)$ is given, whose values for $x = n$ (n being a positive integer) are known, and one seeks to find the values of $f(x)$ for other values of x. In 1747, he applied his method to a function that came up in astronomy and managed to express this function as a trigonometric series. But only in 1753 he published the method that he had discovered in 1729.

In 1754, d'Alembert occupied himself with the problem of expanding the inverse of the distance between two planets into a series of cosines and sines of the angle formed by the half lines that connect the origin of the coordinate system with the two planets.

In 1757, Clairaut, who was working on some problem in astronomy, made the statement that any function f can be expressed in the form

$$f(x) = A_0 + 2\sum_{n=1}^{\infty} A_n \cos nx.$$

He considers this problem as a problem of interpolation and he uses the values of the function that correspond to the values $\frac{2\pi}{k}, \frac{4\pi}{k}, \frac{6\pi}{k}, \ldots$ of x. Hence, after some calculations, he obtains

$$A_0 = \frac{1}{k}\sum_{\mu} f\left(\frac{2\mu\pi}{k}\right),$$

$$A_n = \frac{1}{k}\sum_{\mu} f\left(\frac{2\mu\pi}{k}\right)\cos\frac{2\mu n\pi}{k}.$$

For $k \to \infty$, Clairaut obtains the formula
$$A_n = \frac{1}{2\pi} \int_0^{2\pi} f(x) \cos nx\, dx$$
which is the correct formula that gives A_n.

Even though many kinds of functions were expressed by trigonometric series during the time that the above works on trigonometric series were carried out (as well as other similar works), Euler, d'Alembert, and Lagrange never abandoned the viewpoint that "arbitrary" functions cannot be represented by trigonometric series. Since then, the question of whether an "arbitrary" function can be represented by a trigonometric series remained one of the central questions in the theory of trigonometric series.

• The use of continued fractions for the approximate calculation of irrational numbers dates back to the 16th century. In his writing *Algebra* (1572), **Bombelli** was the first who used continued fractions for the approximate calculation of square roots.

In order to approximate $\sqrt{2}$ he works as follows:
$$\sqrt{2} = 1 + \frac{1}{\psi} \tag{1}$$
from which it follows that
$$\psi = 1 + \sqrt{2} \tag{2}$$
We add 1 to both sides of (1) and, by taking into account (2), we find
$$\psi = 2 + \frac{1}{\psi}. \tag{3}$$
From (1) and (3) we have
$$\sqrt{2} = 1 + \frac{1}{2 + \frac{1}{\psi}},$$
while from (3) we have again
$$\sqrt{2} = 1 + \frac{1}{2 + \frac{1}{2 + \frac{1}{\psi}}}.$$
By substituting the value of ψ repeatedly, Bombelli deduces that
$$\sqrt{2} = 1 + \frac{1}{2 + \frac{1}{2 + \frac{1}{2 + \frac{1}{2 + \cdots}}}}$$
The right-hand side of the last equality is also written as
$$\sqrt{2} = 1 + \frac{1}{2+} \cdot \frac{1}{2+} \cdot \frac{1}{2+} \cdots$$

Euler too occupied himself with the subject of continued fractions. In his first paper on this subject entitled *De Fractionibus Continuis* he obtained many results, for example, that every rational number can be expressed as a continued fraction with finitely many terms.

He gave the expansion:
$$e - 1 = 1 + \frac{1}{1+} \cdot \frac{1}{2+} \cdot \frac{1}{1+} \cdot \frac{1}{1+} \cdot \frac{1}{4+} \cdot \frac{1}{1+} \cdot \frac{1}{1+} \cdot \frac{1}{6+} \cdots$$

which had already appeared in Cotes' writing *Philosophical Transactions* (1714). He also gave the expansion:

$$\frac{e+1}{e-1} = 2 + \frac{1}{6+} \cdot \frac{1}{10+} \cdot \frac{1}{14+} \ldots$$

Euler proved that the numbers e and e^2 are irrational.

In his writing *Introductio* Euler presented the foundation of a theory of continued fractions and proved that any given series can be expressed as a continued fraction and vice versa.

Johann Heinrich Lambert (1728-1777) used Euler's works on continued fractions in order to prove that if x is a non-zero rational number, then the numbers e^x and $\tan x$ are not rational. This means that, not only e^x is irrational when x is a non-zero integer, but also all positive rational numbers have irrational natural logarithms (i.e., logarithms with base e). Thus it follows from the result regarding $\tan x$ that since $\tan \frac{\pi}{4} = 1$, neither $\frac{\pi}{4}$ nor π can be a rational number.

Lagrange used continued fractions in order to approximate the irrational roots of equations, while, in some other work of his, he managed to calculate approximate solutions of differential equations in the form of continued fractions.

In 1768 Lagrange proved the converse of a theorem that Euler had proved in 1744. The converse of this theorem of Euler asserts that a real root of a second-order equation can be written as a periodic continued fraction.

The work on series that was carried out during the 18th century was extensive and very important. But people ignored the problem of the convergence of a series. They considered series as a kind of polynomials with infinitely many terms. But two ideas regarding divergent series proved to be useful for the future. The first is that divergent series can be used for the numerical approximation of functions. While the second is that even if a series is divergent, it can represent a function when we perform analytical operations.

The mathematicians of the 18th century used infinitesimal calculus for the solution of a large variety of problems. The simplest of these problems were solved by integrating elementary functions, while somewhat more difficult ones led to integrals whose calculations were impossible, as for example, it happened with the elliptic integrals. The solution though of even more complicated problems demanded the existence of more specialized techniques, which led to the birth of Differential Equations. A category of such complicated problems was the one that came up in elasticity theory. A body is called elastic, if, whenever it changes shape under the action of a force, it returns to its initial shape when the force stops acting on it.

The most practical problems concern the shape that (vertical or horizontal) bars take when weights act on them. These problems, which were treated only empirically in the past, are treated in a new way by scientists of the 18th century, who were now equipped with more mathematical knowledge. Their study begins with problems regarding the shape that a non-elastic but flexible string takes, when it is suspended from two fixed points, or when it is suspended from a single fixed point and has been set into an oscillatory motion, etc.

Mathematicians were still interested in the pendulum. The study of the differential equation of the circular pendulum, i.e., of the equation

$$\frac{d^2\theta}{dt^2} + \frac{g}{l \cdot \sin\theta} = 0,$$

presented difficulties even if one replaces in it $\sin\theta$ with θ, which makes things simpler.

As mentioned before, the period of the motion of a circular pendulum is not completely independent of the width of the motion. Huygens solved this problem geometrically by using a cycloid (Chapter 24), but the analytic solution of this problem was not given.

The problem of the pendulum was closely related to two other directions of research in the 18th century, namely the shape of the earth, as well as the verification of the law of gravitation which states that attraction is inversely proportional to the square of the distance.

The solution of ordinary differential equations had already begun since the recognition of the existing inverse relation between integration and differentiation. One of the achievements of the 18th century was the discovery of various types of differential equations, whose solution was possible through relatively simple methods.

The Bernoulli equations of the form $y' + P(x)y = Q(x)y^n$ constitute such a type and are solved by the substitution $z = y^{1-n}$.

Another type of equations was introduced by Alexis Claude Clairaut (1713-1765). These are equations of the form $y = xy' + f(y')$. Here, if we set $p = y'$ and differentiate the terms of the equation with respect to x, we obtain a first-order equation with respect to x, p and $\frac{dp}{dx}$, whose general solution is $y = cx + f(c)$.

One of the most interesting equations of the 18th century is the equation

$$y' = p(x)y^2 + q(x)y + r(x)$$

of Ricatti, which was studied by many mathematicians. Euler was the first who remarked that if a partial solution $v = f(x)$ of Ricatti's equation is known, then the substitution

$$y = v + \frac{1}{z}$$

transforms Ricatti's equation into a linear differential equation with respect to z, and hence the general solution can be found.

During the middle of the 18th century, the subject of ordinary differential equations constituted an independent branch of mathematical science.

At first, mathematicians were seeking solutions that are given by elementary functions. Later, they were also satisfied with solutions that are given as integrals whose calculation is not feasible. And some time after that, they also accepted solutions that are given in the form of series.

The efforts to find general methods of solution of ordinary differential equations seem to stop in 1775. During the next one hundred years no important new methods were found, until the introduction of methods via operators, via the Laplace transform, etc., at the end of the 19th century. General principles for the solution of ordinary differential equations still do not exist. But efforts are made to find various techniques, which can be applied to the types of problems that appear in practice.

• As it happened in the case of ordinary differential equations, mathematicians did not proceed consciously to the creation of the object of study that we call today partial differential equations. Instead, by continuing to study the same natural phenomena, they reached mathematical problems whose study falls into the area of partial differential equations. Thus, even though the displacement of a string had been studied until then separately as a function of time and separately as a function of the distance of a point of the vibrating string from one of its extremities, now the effort to study the displacement as a function of these two variables and understand all possible motions led to the consideration of a partial differential equation.

Until 1765, partial differential equations appear in the solution of problems from physics. We will not occupy ourselves with these equations here.

The first paper that treats partial differential equations from the point of view of purely theoretical mathematics is Euler's paper *Recherches sur l'intégration de l'équation*

$$\left(\frac{ddz}{dt^2}\right) = aa\left(\frac{ddz}{dx^2}\right) + \frac{b}{x}\left(\frac{dz}{dx}\right)$$

(*Misc. Taur.* 3, 1762/65, 60-91, *pub.* 1766=*Opera*, (1), 23, 47-73).

Later, Euler published a paper on this subject in the third volume of his work *Istitutiones Calculi Integralis* (1770=*Opera*, (1), 13).

Before d'Alembert's paper on the vibrating chord (1747), partial differential equations were known merely as equations that represented certain conditions, and the matter at hand was to find particular solutions to them. After d'Alembert's work, mathematicians realized the difference that exists between "particular" and "general" solutions. As soon as this difference became understood, mathematicians also perceived that the general solutions were the most important.

What became also understood was that the study of partial differential equations did not demand new techniques of calculation, and that their difference from ordinary differential equations is that arbitrary functions may appear in the solutions of partial differential equations.

The main achievement of the 18th century in the area of differential equations was the fact that their importance in problems of elasticity, hydrostatics and gravitation became understood. However, the entire theory in this area remained at its first stages.

• The investigation of natural problems led to the acquisition of more knowledge regarding curves and surfaces. The impressive achievements of the 18th century in the area of analytic geometry, in combination with the new direction of research that arose from the application of infinitesimal calculus to problems in geometry, led to the creation of **Differential Geometry**.

Plane analytic geometry was developed during the 18th century. Newton and Jacob Bernoulli introduced the system of polar coordinates, Jakob Hermann (1678-1733) applied it widely in the study of curves, and Euler extended its use further and made use of trigonometric notations.

In his writing (1748), Euler introduced the parametric representation of curves, i.e., he expressed the coordinates x and y as functions of a third variable, and made a systematic study of analytic geometry.

In a letter addressed to Leibniz (1715), Johann I. Bernoulli introduced the 3-dimensional system of coordinates, i.e., of the system that we use today.

Surfaces, such as the sphere, the cylinder, the paraboloid, the two-sheet hyperboloid and the ellipsoid, were, of course, known from the geometric vantage point before 1700. Some of them appear already in the work of Archimedes. In his book published in 1731, Clairaut gave the equations of some of these surfaces. He also proved that a homogeneous equation with respect to x, y, z (i.e., an equation that all its terms have the same degree with respect to these variables) represents a cone whose vertex is the origin of the coordinate system. Jakob Hermann added (1732) to this result that the equation $x^2 + y^2 = f(z)$ represents a surface of revolution about the z-axis.

In 1748 in *Introductio*, Euler developed systematically the 3-dimensional analytic geometry and studied the general second-order equation with respect to x, y, z using a transformation in order to bring it into the normal form.

Gaspard Monge's works include a large part of 3-dimensional analytic geometry. In particular, we mention the paper entitled *Application de l'Algèbre à la Géométrie* (1802), which he wrote in collaboration with his student Jean-Nicolas-Pierre Hachette (1769-1834).

The analytic geometry mentioned so far concerned curves and surfaces of first and second order. Hence, it was natural for the mathematicians of that time to also examine curves whose equations were of a higher order.

The curves that were studied during the 18th century were algebraic, which means that their equations $f(x,y) = 0$ were given by polynomials of the variables x and y, where the degree or the so-called order of the curve is given by the highest degree of the terms.

Newton made the first systematic and extensive study of higher-order curves and used axes with both positive and negative x, y, working in all four quadrants (1704). He showed that the general third-order equation can be reduced to four other forms. He studied important points of these curves, i.e., points of inflexion, cusps, isolated points, multiple points, etc. Analogous studies of these curves were also made by Leibniz and his associates.

In his book *Geometria Organica* (1720), which he wrote at the age of 19, Maclaurin proved that the maximum number of double points of an irreducible curve of order n is $\frac{(n-1)(n-2)}{2}$ and introduced the notion of the "genus of a curve" as the difference between the maximum possible number of double points of the curve and the actual number of double points.

Gabriel Cramer (1704-1752) studied the problem of the determination of the components of a higher-order curve (1750).

Subjects that also occupied the mathematicians of the 18th century were the intersections of a curve and a straight line or the intersections of two curves.

In his paper *Lineae* (1717), Stirling proved that an algebraic curve of order n (with respect to x and y) is determined by $\frac{n(n+3)}{2}$ points of it, because essentially this is the number of its coefficients. He also alleged that two parallel straight lines intersect a given curve at the same number of real or imaginary points and he proved that the number of components of a curve that extend to infinity, is even.

In his work *Geometria* mentioned above, Maclaurin founded the theory of intersections of higher-order plane curves. By generalizing results that hold in particular cases, he came (without proof) to the general conclusion that two curves of orders m and n respectively intersect at $m \cdot n$ points.

In 1748 Euler and Cramer tried (unsuccessfully) to prove the above conjecture of Maclaurin. Unsuccessful were also the efforts made by other mathematicians. The complete solution of the problem was given by Georges-Henri Halphen (1844-1889) in 1873 (*Bull. Soc. Math. De France* 1873, 1875).

In 1750 Cramer occupied himself with a "paradox" that was observed by Maclaurin regarding the number of common points of two curves:

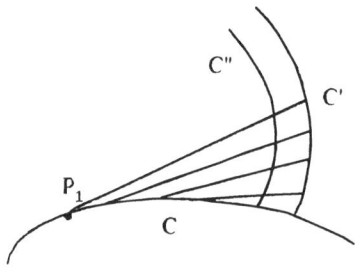

Figure 25.2

A curve of order n is determined when $\frac{n(n+3)}{2}$ of its points are given. Two curves of order n intersect at n^2 points. When $n=3$ the number of points that determine the curve is 9. Also two curves of order three intersect at $3 \cdot 3 = 9$ points. Since every third-order curve is determined by 9 points and the two third-order curves pass from the same 9 points, this means that these 9 points do not determine a single third-order curve. A similar paradox is observed when $n=4$. The explanation that Cramer gave to this paradox was that the n^2 equations that determine the n^2 points of intersection are not independent. All the third-order curves that pass from 8 fixed points of a given third-order curve must pass from the same ninth fixed point. In other words, the ninth point depends on the 8 first points. The same interpretation to this paradox was also given by Euler in 1748.

The second volume of Euler's *Introductio* (1748) and Cramer's *Lignes courbes algébriques* (1750) were the best two books of the 18th century regarding results on higher-order curves.

Differential geometry is the result of the application of infinitesimal calculus to subjects of analytic geometry and, in particular, to properties of various curves and surfaces that change from point to point, thus proving that the techniques of infinitesimal calculus are the appropriate techniques for studying these properties.

The term "differential geometry" was first used by Luigi Bianchi (1856-1928) in 1894.

The first applications of infinitesimal calculus regarding curves were made on plane curves. In 1673 Huygens examined the evolute, the involute and the centers of curvature of a plane curve. We give the definitions of the "evolute" and the "involute" of a plane curve C: Let us imagine that a thread is wound around the curve C starting from point P_1 and going to the right of P_1 (Figure 25.2). The end of the thread at the point P_1 is fixed and the other end is wound off by keeping the thread tight. Huygens proved that the thread at its free end is perpendicular to the curve C' that this end traces; C' is called an involute of C. In addition, every point of the thread traces an involute of C. Thus, C'' is also an involute of C and Huygens proved that the involutes of a curve do not intersect each other.

Since the thread is tangent to C exactly at the point from which it is left loose, it follows that every orthogonal trajectory of the family of the tangents of the curve is an involute of the curve.

Huygens examined also the "evolute" of a plane curve. Assume that a fixed normal to the curve at an arbitrary point of it, say P, is given. Then, when a normal to the curve that lies in the neighborhood of the fixed normal moves towards P, the point of intersection of the two normals takes a limiting position on the fixed normal which is called **center of curvature of the curve** at the point P. Huygens proved that the distance of the arbitrary point P of the curve from the corresponding center of curvature, as it is measured along the normal to the curve at this point, is equal (in contemporary notation) to

$$\left[1+\left(\frac{dy}{dx}\right)^2\right]^{\frac{3}{2}} \cdot \left(\frac{d^2y}{dx^2}\right)^{-1}.$$

This length is the **radius of curvature of the curve** at the point P. The geometric locus of the centers of curvature as P moves on the given curve is called the **evolute** of the curve. Thus, the curve C is the evolute of any of its involutes. Huygens proved also that the evolute of a cycloid is again a cycloid.

Clairaut (1713-1765) is one of the first researchers of space curves in the framework of 3-dimensional Differential Geometry. In 1731 he published his work *Recherche sur les courbes à double courbure*. He had completed the writing of this work in 1729 when he was 16, and for another work of his he was elected member of the Académie des Sciences in Paris at the age of 17!

Clairaut called the curves of 3-dimensional space "curves of double curvature", because he was considering projections of a space curve to two planes perpendicular to each other, and he was examining this space curve in relation to the curvatures of the two plane curves that he was obtaining via the projection of the space curve on these two perpendicular planes. From the geometric point of view, he considered a space curve as the intersection of two surfaces, where every surface was expressed by an equation with three variables. Thus, Clairaut perceived that a space curve can have infinitely many normals at any given point of it, which lie on a plane that is perpendicular to the tangent of the curve at the given point. The expression of the arc length of a space curve and the calculation of the area of a part of a surface are also due to Clairaut.

Euler was the one who made the next important step in differential geometry of space curves. The formulas that give in polar coordinates the components of the acceleration of a particle that moves on a plane curve are due to Euler. These are

$$a_r = \frac{d^2r}{dt^2} - r\left(\frac{d\theta}{dt}\right)^2 \quad \text{and} \quad a_\theta = r\frac{d^2\theta}{dt^2} + 2\frac{dr}{dt}\frac{d\theta}{dt},$$

where a_r is the component along the radius and a_θ is the component along the normal to the radius.

In 1760, in one of his works Euler laid the foundations of the theory of surfaces. In this work, Euler represents a surface by an equation $z = f(x,y)$ and introduces the notation $p = z'_x$, $q = z'_y$, $r = z''_{xx}$, $s = z''_{xy}$, $t = z''_{yy}$ that we still use today. Later, he calculates the radius of curvature of a curve that is the intersection of a surface with a plane.

The theory of surfaces was advanced even further by **Meusnier de la Place** (1754-1793) in 1776.

The so-called "developable" surfaces, that is, the surfaces that can be placed on a plane (i.e., that can be unfolded on a plane) without being deformed, attracted the interest of the cartographers. Given that the sphere cannot be cut and fitted on a plane, the problem arose of finding surfaces with shapes similar to the one of the sphere, which can be placed on a plane without being deformed. The developable surfaces were studied by Euler, who introduced the parametric equations of a surface $x = x(t,u)$, $y = y(t,u)$, $z = z(t,u)$ and tried to find sufficient conditions on these functions for the surface to be developable.

The problem of developable surfaces was also studied by Monge, who used both geometric and analytic methods. The revival of pure geometry after Desargues is due to Monge.

The work of **Gaspard Monge** (1746-1818) in the areas of descriptive geometry (which serves mainly architecture), analytic geometry, differential geometry, ordinary and partial differential equations excited Lagrange's admiration and "jealousy". At the end of a scientific talk of Monge Lagrange said "Dear colleague, you presented to us very elegant things. I wish I had discovered them myself."

In the differential geometry of 3-dimensional space, Monge obtained results which went even further than the results of Euler. His paper entitled *Mémoire sur les développées, les rayons de courbure et les différents genres d' inflexions des courbes à double courbure* (1771), which was published much later, was followed by the paper entitled *Feuilles d' analyse appliquée à la géometrie* (1795, second edition in 1801). Monge recognized that a family of surfaces which have a common property or which are produced in the same way, must satisfy a partial differential equation. Monge's work on developable and ruled surfaces was continued by **Charles Dupin** (1784-1873).

An important part of the differential geometry of the 18th century has its origin in problems of geodesy and cartography. We remind the reader that the construction of maps dates back to older times and is related to the stereographic projection and to other methods of the times of Ptolemy and Mercator (16th century).

As it was known that an arbitrary mapping of a sphere into a plane does not preserve all the geometric properties, mathematicians turned their attention to mappings which preserve angles. If two curves of a surface form a certain angle at a point of their intersection and if the corresponding angle, formed by their images via a mapping, remains the same in both size and direction, then this mapping is called **conformal**. Both the stereographic and Mercator's projection are conformal mappings.

J. H. Lambert inaugurated a new era in theoretical cartography. He was the first who studied the conformal mapping of the sphere onto the plane in its full generality and gave formulas for this mapping (1772).

Lagrange (1779) managed to give all the conformal transformations of a part of the surface of the earth onto a plane area, which transform circles of geographic length and breadth into circular regions.

The subject of the construction of maps was advanced further with the further development of differential geometry and of the theory of complex functions.

• As it happened with the case of series with infinitely many terms as well as with the case of differential equations, the first works on the "Calculus of Variations" were merely works that fall into the area of infinitesimal calculus.

A few years after Newton's death (1727), people understood that a new branch of mathematics was emerging, which had its own characteristic problems and its own methodology. It was the Calculus of Variations, which perhaps was as interesting for mathematical physics, as differential equations were for mathematics and physics.

Historically, the first important problem in the calculus of variations was posed and solved by Newton. In Book II of *Principia* Newton studies the motion of

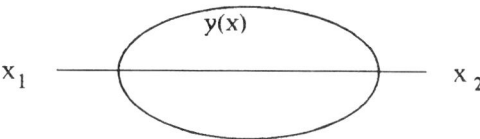

FIGURE 25.3

bodies in the water. Later, in the third edition of *Principia* he studies the shape that a ruled surface which moves with constant speed along the direction of its axis must have, in order that the resistance that it faces during its motion be the minimum possible. Newton assumed that the resistance of the fluid at every point of the surface of the moving body is proportional to the component of the speed that is perpendicular to the surface. In *Principia* Newton gives a geometric characterization of the requested shape, and later in a letter addressed to David Gregory (1694) he also gives the analytic solution.

In contemporary form, Newton's problem consists of finding the minimum value of the integral

$$J = \int_{x_1}^{x_2} \frac{y(x)[y'(x)]^3}{1 + [y'(x)]^2} \, dx.$$

The minimum value of J is obtained by an appropriate selection of the function $y(x)$ so that when the curve represented by $y(x)$ is rotated about the x-axis it gives the requested surface (Figure 25.3). The peculiar nature of this problem (and of the calculus of variations in general) is that it seeks the value of an integral whose value depends on an unknown function $y(x)$, which appears in the expression under the integral and which must be determined in such a way that the value of the integral becomes either maximum or minimum.

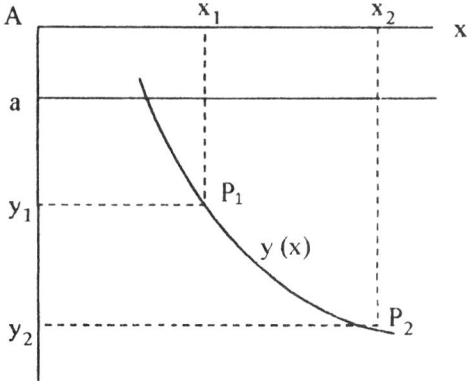

FIGURE 25.4

In *Acta Eruditorum* (1696), Johann I. Bernoulli had suggested to the mathematical community a problem, which is known today as the problem of the "brachistochrone" and consists of the following: Find the (downwards) trajectory that a particle must follow, when it slides (on a vertical plane) from a given point to another point, which does not lie exactly below the first, in the shortest time possible (Figure 25.4).

The initial speed v_1 at P_1 is given, while friction and the resistance of the air are assumed negligible. In contemporary form, the problem consists of choosing $y(x)$ in such a way that the value of the integral J representing the time of descent becomes minimum:

$$J = \frac{1}{\sqrt{2g}} \int_{x_1}^{x_2} \sqrt{\frac{1+[y'(x)]^2}{y(x)-a}} dx,$$

where g is the acceleration of gravity and $a = y_1 - \frac{v_1^2}{2g}$.

A solution to this problem was given by Newton, Leibniz, L'Hôpital, Johann I. Bernoulli and his elder brother Jakob. The answer is that the curve that the particle must follow is the cycloid that connects P_1 with P_2 and is convex upwards. The straight line l, on which the circle rolls, must be exactly at the height $y = a$ above the initially given point that slides. There exists a single cycloid that passes through the points P_1 and P_2.

Another important class of problems refers to geodesy, i.e., to the finding of the shortest distance between two points of a surface. If the surface is a plane, then the problem reduces to the minimization of the value of the integral

$$J = \int_{x_1}^{x_2} \sqrt{1+[y'(x)]^2} dx$$

and hence the answer is obviously a line segment.

The problems that we mentioned so far have the following analytic form

$$J = \int_{x_1}^{x_2} f(x, y(x), y'(x)) dx,$$

and one tries to find an appropriate function $y(x)$ for which J becomes either maximum or minimum.

Another category of problems in the calculus of variations are the so-called "isoperimetric", in which one tries to find the maximum area that is surrounded by closed plane curves having a given fixed length.

The existence of isoperimetric problems dates back to the time of the Ancient Greeks. Apart from Zenodorus' work (Chapter 18, page 145), until the end of the 17th century, there exists no other work regarding perimetric problems.

The basic isoperimetric problem is formulated analytically as follows:

The curves that one considers have the parametric representation

$$x = x(t), \quad y = y(t), \quad t_1 \leq t \leq t_2,$$

and since they are closed,

$$x(t_1) = x(t_2), \quad y(t_1) = y(t_2).$$

Moreover, each of the curves must not intersect itself. Then, what is requested is to find the functions $x(t)$ and $y(t)$ for which the length

$$L = \int_{t_1}^{t_2} \sqrt{[x'(t)]^2 + [y'(t)]^2} dt$$

is equal to a given constant and the area, which is given by the integral

$$J = \int_{t_1}^{t_2} [x(t)y'(t) - x'(t)y(t)] dt,$$

is maximum.

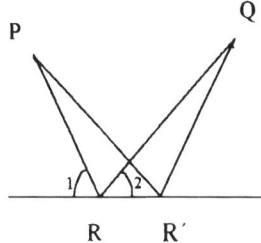

FIGURE 25.5

A new stimulus to the development of the calculus of variations came again from physics and is the **Principle of Least Action**. The origins of this principle are in the work of the Ancient Greeks. In his work *Catoptrica*, Euclid had proved that the path of a light ray, that starts from a point P, falls on a mirror, and then reaches point Q (Figure 25.5), is such that $\angle 1 = \angle 2$.

Later, Heron of Alexandria proved that the path PRQ that the light ray follows is the shortest among all other paths (such as, for example, the path $PR'Q$ that the ray could have followed). When the medium in which the light travels above the straight line RR' is homogeneous, the fact that the light follows the shortest path means that, since it moves with constant speed, implies that the path requires the shortest time. Heron applied this principle of the shortest path and of the shortest time in problems of reflection of light on concave and convex mirrors.

Based on the phenomenon of reflection of light, as well as on philosophical, theological and aesthetic principles, the philosophers and the scientists that followed the ancient Greeks promoted the doctrine that nature acts in the best possible way. As Olympiodorus (6th century A.D.) mentions in his work *Catoptrica* "nature makes no work which is superfluous or unnecessary". Leonardo da Vinci alleged that nature is economical and its economy is quantitative, while the philosopher Robert Grosseteste (1168-1253) believed that nature acts in the mathematically shortest and best way. During the Middle Ages, it was generally accepted that indeed nature behaves in this way.

The above problem of the mirror constitutes one of the first examples of the so-called principle of least action. Fermat formulated this principle as follows: light always follows the path that requires the minimum time (1657, 1662). Later, Pierre-Louis Moreau de Maupertius (1698-1759) supported the same principle (1744) and Euler formulated it as an exact theorem of dynamics. Euler went further than Maupertius and believed that all natural phenomena behave in such a way that they maximize or minimize a certain function, and hence the basic physical principles must be expressed in such a way that a certain function is maximized or minimized. In particular, this must be true in dynamics, which studies the motions of bodies which are propelled by forces acting on them. These viewpoints of Euler are not very far from the truth.

In 1755 Lagrange introduced a general systematic and uniform method for the solution of a large range of problems in the calculus of variations and worked on this method for several years. He published his results in his writing *Essai d' une nouvelle méthode pour déterminer les maxima et les minima des formules intégrales indéfinies* (1762). Lagrange's method regarding the basic problem of the calculus

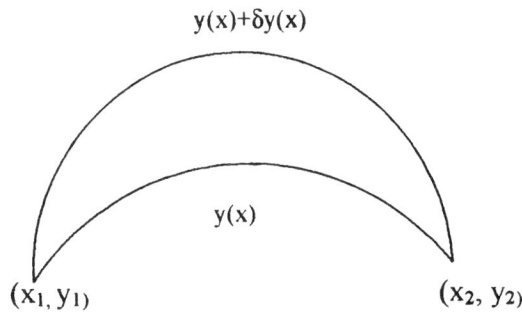

FIGURE 25.6

of variations, i.e., the maximization or minimization of the integral

$$J = \int_{x_1}^{x_2} f(x, y(x), y'(x))dx,$$

where one seeks to determine the function $y(x)$, is outlined as follows:

Lagrange introduces new curves which pass through the extreme points (x_1, y_1) and (x_2, y_2). He represents these curves (Figure 25.6) in the form $y(x) + \delta y(x)$, where δ is a special symbol introduced by Lagrange that shows a change of the entire curve $y(x)$. The introduction of a new curve in the expression under the integral in J changes the value of J. We denote by ΔJ this change, which is

$$\Delta J = \int_{x_1}^{x_2} [f(x, y(x) + \delta y(x), y'(x) + \delta y'(x)) - f(x, y(x), y'(x))]dx.$$

Then Lagrange considers f as a function of three independent variables (namely x, y, y') and since x remains unchanged, the quantity under the last integral can be expanded into a Taylor series as a function of two variables. By following a procedure that we will not present here, Lagrange obtains the following **necessary but not sufficient condition** that f must satisfy in order to maximize or minimize J,

$$f'_y - \frac{d}{dx}\left(f'_{y'}\right) = 0,$$

where, in evaluating f'_y and $f'_{y'}$, $f(x, y, y')$ is considered to be a function of the independent variables x, y and y', but the derivative $\frac{d}{dx}\left(f'_{y'}\right)$ means the derivative of $f'_{y'}$ with respect to x, where $f'_{y'}$ depends only on x through x, $y(x)$ and $y'(x)$.

During 1760-1761 Lagrange considered problems of minimization or maximization involving multiple integrals, and in 1770 he used simple and multiple integrals in which the functions under the integrals had derivatives of higher orders.

Lagrange applied the calculus of variations to dynamics and was the first to express the Principle of Least Action in a concrete form: For a particle, the integral of the product of its mass times its speed and times the distance between two fixed points is either maximum or minimum. In other words, $\int mvds$ must be either maximum or minimum at the path followed by the particle. The quantity mv^2 (today $\frac{1}{2}mv^2$) is called **kinetic energy**, while in the time of Lagrange it was called vital force. Lagrange also asserted that the above principle also holds for a set of particles.

• Even though during the 18th century people made no clear distinction between algebra and analysis because the importance of the notion of the limit was still obscure and vague, as we examine these matters from the contemporary vantage point, it is preferable that we make a distinction between these two areas. During the 17th century, the main interest was in algebra, while during the 18th century, algebra became of secondary importance, so that apart from number theory the main stimulus of algebraic works came from analysis.

Since the number system constitutes the basis of algebra, it is advisable to examine in what situation this system was.

At the beginning of the 18th century, integers, fractions, irrationals, negative numbers and complex numbers were known. But during this century, people never ceased to oppose the acceptance of new types of numbers. For example, negative numbers were not completely understood until the contemporary era. During the second half of the 18th century, Euler still believed that the negative numbers were "greater" than ∞. He alleged that $(-1) \cdot (-1) = +1$ because this product had to be equal to either $+1$ or -1 and since $1 \cdot (-1) = -1$, it follows that $(-1) \cdot (-1) = +1$. Carnot, who was a distinguished French geometer, believed that the use of negative numbers leads to erroneous conclusions.

During 1831, Augustus de Morgan (1806-1871), who was a professor of mathematics at London's University College and a famous researcher in the areas of mathematical logic and algebra, mentions in his writing *On the Study and Difficulties of Mathematics* the following: "The imaginary expression $\sqrt{-a}$ and the negative expression $-b$ are similar in the sense that whenever any of these expressions appears in the solution of a problem, it constitutes evidence of the existence of some inconsistency or of some contradiction. Both the expression $0 - a$ and the expression $\sqrt{-a}$ are incomprehensible." In order to support this viewpoint de Morgan considers the following problem:

"A father is 56 years old and his son is 29 years old. When will the father's age be twice that of his son?" To this end, de Morgan solves the equation $56 + x = 2 \cdot (29 + x)$, finds $x = -2$ and says that this result is absurd. He goes on to say that if we change x to $-x$ and solve the equation $56 - x = 2 \cdot (29 - x)$, then we obtain $x = 2$. From this he concludes that the formulation of the problem was incorrect and that this is the reason for which we obtained an unacceptable negative answer. De Morgan insisted that it is absurd to consider numbers which are smaller than zero.

Even though, during the 18th century, the notion of an irrational number was not clarified further, in 1737 Euler proved, as mentioned before, that the numbers e and e^2 are incommensurable, while Lambert proved that the number π is irrational. The effort to prove that π is incommensurable was stimulated by the desire to solve the problem of the squaring of the circle.

Legendre's conjecture that π might not be a root of an algebraic equation with rational coefficients led mathematicians to make a distinction between the various types of irrational numbers.

Every real or complex root of an algebraic (polynomial) equation with rational coefficients is said to be an **algebraic number**.

Thus, the roots of an equation

$$a_0 x^n + a_1 x^{n-1} + ... + a_{n-1} x + a_n = 0,$$

where the a_i are rational numbers, are called algebraic numbers. Consequently, every rational number and certain irrational numbers are algebraic numbers, because every rational number c is the root of $x - c = 0$, while, for example, the number $\sqrt{2}$ is also algebraic because it is a root of the equation $x^2 - 2 = 0$.

Every non-algebraic number is called **transcendental**, because, as Euler said, it "transcends" the power of the algebraic methods. But despite the existing distinction between algebraic and transcendental numbers, during the 18th century there existed no example of a transcendental number. The problem of the existence or not of transcendental numbers remained open during the 18th century.

The mathematicians of the 18th century considered the complex numbers as a kind of "poison".

For Euler, the complex numbers were useful only for the performance of operations, and he considered them as "impossible numbers". When Euler was saying that the complex numbers were useful, he meant that they may appear when we try to solve a problem, for which we don't know whether we can give an answer or not. For example, when we are asked to divide 12 into two parts so that their product is equal to 40, we find that these parts must be $6 + \sqrt{-4}$ and $6 - \sqrt{-4}$, and then we conclude that this problem has no solution.

In general, during the 18th century, the use of complex numbers in operations proved to be useful, but these numbers were not recognized because their nature was not understood. Here one observes the phenomenon that, despite the tremendous progress made in various areas, such as differential equations, the calculus of variations, etc., mathematicians of very high caliber had naive viewpoints for subjects such as, for example, the theory of complex numbers, which are considered today simple. This phenomenon is due perhaps to the fact that "these simple things", such as the nature of numbers, etc., are fundamental notions, which can only be understood after a long process.

In 1799 Carl Friedrich Gauss (1777-1855) gave the first proof of the fundamental theorem of algebra and, since this theorem depends necessarily on the recognition of complex numbers, he strengthened their position. We must mention though that even after the development and the further use of the theory of complex numbers, one observes that a certain antipathy exists towards $\sqrt{-1}$ and that people invent various "mechanisms" in order to avoid the presence of this symbol in calculations.

The mathematicians of the 18th century were not concerned with the logical foundation of the real and the complex numbers. They didn't take into account what Euclid does in Book V of *Elements* in order to determine the properties of incommensurable quantities. That is, they didn't try to do something similar to what Euclid did in order to found the real and the complex numbers.

- One of the researches that began in the 17th century and was continued throughout the 18th century was the research regarding the solution of polynomial equations. It is a fundamental mathematical subject and hence it was natural that people showed great interest in obtaining better methods for solving equations of any order, and methods for approximating of the roots of equations, as well as great interest in proving that every polynomial of degree n has exactly n roots.

Moreover, the use of the method of partial fractions for the calculation of integrals gave rise to the question whether an arbitrary polynomial with real coefficients can be expressed as a product of linear and second-order factors with real coefficients in order to avoid (in the case of the appearance of second-order factors) the use of complex numbers.

Leibniz and Nicolas Bernoulli (1687-1759) did not believe that every polynomial with real coefficients can be expressed as a product of linear and second-order factors. On the contrary, Euler believed that this was possible.

The heart of the problem of factorizing a polynomial in linear and second-order factors was to prove that every polynomial has at least one real or complex root, which means that the fundamental theorem of algebra had become the primary aim of the researchers. Incomplete proofs of this fundamental theorem were given by d'Alembert and Euler, while, as mentioned before, the first essentially complete proof was given by Gauss in his doctoral dissertation at Helmstädt. The method that Gauss followed in order to prove this theorem consisted of establishing the existence of the root and not in calculating the root explicitly. This method constituted the beginning of a new way of approaching the problem of existence from the mathematical vantage point. In order to become more clear, we remind the reader that when the properties of an object are considered in some problem, very wisely the ancient Greeks demanded that first of all the existence of this object must be proved. However, for them, the criterion of existence of an object was its constructibility. For example, the existence of the roots of a second-order equation is proved by presenting quantities which satisfy this equation. But in the case of equations of order greater than four, this method does not apply.

At the same time that the research for proving the proposition that every polynomial with real coefficients has at least one root was carried on and finally was successful, mathematicians never ceased in making great efforts to solve equations of order greater than four using algebraic methods. The reader may consult Chapter 9, where reference is made to the same subject.

The great research work that was carried out during the 18th century regarding the solution of equations by radicals can be found in a work (1771) of Alexandre-Théophile Vandermonde (1735-1796) and in Lagrange's voluminous work *Reflexions sur la résolution algébrique des équations* (*Nouv. Mém. de l'Acad. Berlin*, 1770, 1772, 1773).

We will present some of Lagrange's viewpoints:

Regarding the third-order equation

$$x^3 + nx + p = 0,$$

Lagrange remarks that if we use the transformation

$$x = y - \frac{n}{3y},$$

then we obtain the so-called "reduced" equation

$$y^6 + py^3 - \frac{n^3}{27} = 0,$$

which is a second-order equation with respect to r, where $r = y^3$, namely

$$r^2 + pr - \frac{n^3}{27} = 0.$$

Now we can calculate the roots r_1, r_2 of the last equation in terms of the coefficients of the initial one. But in order to express y in terms of r, we must introduce cubic roots, i.e., we must solve the equation $y^3 - r = 0$. Now if we present the cubic root of unity by

$$\omega = \frac{-1 + \sqrt{3}}{2},$$

then we have:

$$x_1 = \sqrt[3]{r_1} + \sqrt[3]{r_2}, \quad x_2 = \omega\sqrt[3]{r_1} + \omega^2\sqrt[3]{r_2}, \quad x_3 = \omega^2\sqrt[3]{r_1} + \omega\sqrt[3]{r_2}.$$

Thus, the roots of the initial equation are expressed in terms of the roots r_1, r_2 of the reduced equation.

Lagrange proved that the various ways in which the mathematicians before him tried to solve the third-order equation, are reduced to the method described above. He also went on to stress that we should not consider x as a function of y but y as a function of x, because the equation that finally leads us to the solution of the initial equation is the reduced equation, and the "secret" must be in the relation that expresses the solutions of the reduced equation in terms of the solutions of the initial equation.

Lagrange remarked that (since $1 + \omega + \omega^2 = 0$) each of the values of y can be written as follows:

$$y = \frac{1}{3}(x_1 + \omega x_2 + \omega^2 x_3),$$

where x_1, x_2, x_3 are taken in some order. By examining this last expression, we remark that the reduced equation of y has the following two properties. First, the roots x_1, x_2, x_3 that appear in the expression of y are not presented according to some concrete order. In other words, this expression of y is in some sense indeterminate. This means that x_1 can be any of the three values of x, hence x_2 can be any of the remaining two values of x, and what is left is x_3. There exist therefore 3! permutations of the values of x. Consequently, there exist 6 values for y, which means that y satisfies a sixth-order equation. Therefore, the order of the reduced equation is determined by the number of permutations of the roots of the initial equation.

The second property is that the relation

$$y = \frac{1}{3}(x_1 + \omega x_2 + \omega^2 x_3)$$

shows the reason that the sixth-order equation can be reduced to a second-order equation: the three (including the identity permutation) out of the six permutations come from interchanging all the x_i between themselves, while the remaining three permutations come from interchanging only two of the x_i and keeping the third constant. But then, if we take into account the value of ω, the resulting six values of y are connected as follows:

$$y_1 = \omega^2 y_2 = \omega y_3, \quad y_4 = \omega^2 y_5 = \omega y_6.$$

Hence, by taking the cubes of the sides of these equalities, we obtain

$$y_1^3 = y_2^3 = y_3^3, \quad y_4^3 = y_5^3 = y_6^3.$$

Another way of formulating the above result is that the expression

$$(x_1 + \omega x_2 + \omega^2 x_3)^3$$

can take only two values when we perform the six permutations of the x_1, x_2, x_3, and this is the reason why the equation that y satisfies is a second-order equation with respect to y^3. Moreover, the coefficients of the sixth-order equation satisfied by y are rational functions of the coefficients of the initial third-order equation.

In the case of the general form of the fourth-order equation, Lagrange considers the function
$$y = x_1 x_2 + x_3 x_4.$$
This function of the four roots of the equation takes only three distinct values when one performs all 4! = 24 permutations of the four roots. Consequently, a third-order equation must exist, which is satisfied by y and whose coefficients are rational functions of the coefficients of the given equation. Indeed, these conditions are satisfied by a fourth-order equation.

Then, following approximately the same method, Lagrange occupies himself with the general equation of order n, i.e.,
$$x^n + a_1 x^{n-1} + ... + a_{n-1} x + a_n = 0,$$
but now his efforts are unsuccessful. Despite this failure, his efforts shed some light and helps to understand the fact that for $n \leq 4$ the above equation can be solved, while for $n > 4$ it cannot be solved.

- The study of a system of linear equations, which we denote today by
$$x_i = \sum_{j=1}^{n} a_{ij} y_j, \qquad i = 1, 2, 3, ..., m,$$
where the x_i are known and the y_j are unknown, had began with Leibniz before 1678.

In 1693 Leibniz used a systematized set of indices for the coefficients of a system of three linear equations with two unknowns, x and y. He eliminated the two unknowns from the system of three equations and he obtained a determinant, the so-called **eliminant** of the system. If this determinant vanishes, this means that there exist x and y which satisfy the three equations.

The solution of a system of linear equations with two, three and four unknowns using the method of determinants is due to Maclaurin (1729) and was published after his death in the book *Treatise on Algebra* (1748).

Cramer gave the rule of determining the coefficients of the general conic section
$$A + By + Cx + Dy^2 + Exy + x^2 = 0$$
that passes though five given points. He calculated the resulting determinants in the same way that we calculate them today.

In 1764 Bézout proved that given n homogeneous equations with n unknowns, the vanishing of the determinant of their coefficients constitutes the necessary and sufficient condition for the existence of non-zero solutions.

The systematic and clear formulation of the theory of determinants is due to Vandermonde, who is considered to be the founder of the theory of determinants independently of their application to the solution of linear systems.

We saw above that the vanishing of the eliminant constitutes a necessary and sufficient condition for the existence of solution to a system of three linear equations with two unknowns. Obviously the same condition expresses also the result of the elimination of the two unknowns from the equations of the system. But the problem of elimination also extends to other directions. Given two polynomials
$$f = a_0 x^m + ... + a_m$$

and
$$g = b_0 x^n + ... + b_n$$
one could pose the question: under what conditions do $f = 0$ and $g = 0$ have a common root? If a value of x for which $f = 0$, satisfies $g = 0$, then, by substituting to $g = 0$ this value of x that we obtained from $f = 0$, we obtain a condition that refers to the a_i and b_i. This condition was first investigated by Newton. In his work *Arithmetica Universalis* he gives rules for eliminating x from two equations whose order is from two to four.

In his work *Introductio* Euler gives two methods of elimination. The second method is the forerunner of the so-called "multiplicative method" of Bézout, which was presented by Bézout in his book *Cours de mathématique* (1764-1769).

Bézout considers two equations of order n:
$$f(x) = a_n x^n + a_{n-1} x^{n-1} + ... + a_0 = 0,$$
$$g(x) = b_n x^n + b_{n-1} x^{n-1} + ... + b_0 = 0.$$

He multiplies f by b_n and g by a_n, and he subtracts one resulting equation from the other. Next, he multiplies f by $b_n x + b_{n-1}$ and g by $a_n x + a_{n-1}$, and again he subtracts one from the other resulting equation. Then he multiplies f by $b_n x^2 + b_{n-1} x + b_{n-2}$ and g by $a_n x^2 + a_{n-1} x + a_{n-2}$, and he subtracts again one from the other resulting equation, and so on. Each of the resulting equations is of order $n - 1$ with respect to x. Now we can consider this set of equations as a system of n linear homogeneous equations with respect to the unknowns x^{n-1}, x^{n-2}, ..., 1. The eliminant of this system, which is equal to the determinant of the coefficients of the unknowns, is the eliminant of the initial equations $f = 0$ and $g = 0$. Bézout gave also a method for finding the eliminant when the two equations are not of the same order.

The theory of elimination was applied also to two equations $f(x,y) = 0$, $g(x,y) = 0$ of order greater than 1. The stimulus for this work was the search for common solutions of the two equations, i.e., the finding of the number of points of intersection of the curves that correspond to these equations. In 1764 Bézout gave an important method of elimination of one of the unknowns from $f(x,y) = 0$, $g(x,y) = 0$ in a work of his. This method is included in his writing *Théorie générale des équations algébriques* (1779) and is outlined as follows:

By multiplying $f(x,y)$ and $g(x,y)$ by appropriate polynomials $F(x)$ and $G(x)$ respectively, he tried to form the expression
$$R(y) = F(x)f(x,y) + G(x)g(x,y).$$

He also tried to choose the F and G in such a way that the degree of $R(y)$ is the minimum possible.

The question of the degree of the eliminant was answered by Bézout in the above mentioned book *Théorie ...*, and it was also answered independently by Euler in 1764. They both gave the answer that the degree of the eliminant is equal to mn, i.e., to the product of the degrees of f and g.

Jacobi and Minding also gave the same method of elimination that was given by Bézout without mentioning him. It is likely that they didn't know Bézout's work.

- If one calls the 17th century as the century of ingenuity, the 18th century is the century of inventiveness and mathematical production. Without introducing original and fundamental principles, such as infinitesimal calculus, but by exercising only their skill in inventing and applying various techniques, the mathematicians of the 18th century exploited and advanced the power of infinitesimal calculus, and managed to produce main branches of contemporary mathematics, such as series with infinitely many terms, ordinary and partial differential equations, differential geometry, and calculus of variations. By extending infinitesimal calculus to the above areas, they managed to build the wide area of mathematics that we call "Mathematical Analysis".

The progress in analytic geometry and in algebra is a mere extension of what was already known during the 17th century. However, the main problem of algebra, namely the solution of equations of order n, was examined. This problem was useful in analysis, for example, in the calculation of integrals via the method of partial fractions.

During the 18th century, the geometric methods that were used until then were gradually replaced by analytic methods.

Near the end of the 18th century, Monge revitalized geometry and at the same time he used it in order to provide intuitive notions and directions in analysis. But the great importance that was given to mathematical analysis had also negative effects, such as the departure of "numbers" from geometry. It harmed indirectly the matter of the rigorous foundation of the number system, of algebra and even of mathematical analysis itself. The problem of the rigorous foundation of the number system constituted one of the most crucial problems that occupied research in the 19th century.

More than any time before, the inspiration for the mathematical work of the 18th century came from problems in physics. One could say that the exclusive goal of mathematical research was the solution of problems in physics. It was believed that the mathematical work of scientists, such as Descartes, Pascal and Newton, was exhausted and that the main interest should be turned towards Mechanics. But what happened in reality was the opposite. Physics was becoming all the more mathematical, something that laid the foundations of mathematical physics.

The notion of proof and the notion of "axiomatic" thought were not dominating elements of the 18th century. The discovery of mathematical propositions was based on intuition and self-evidence, while the verification of the correctness of a result was judged by its applications on physical sciences.

A widely applied method during the 18th century was the *pro rata* extension of certain results to other situations. For example, the rules of the operations on polynomials were applied also to series. In addition, whatever concerned the real roots of an equation was extended also to complex roots. In other cases, mathematicians ignored basic principles, such as the convergence or divergence of series, the change of the order of differentiation and integration, etc. As a result of their indifference for supplying proofs to propositions, the mathematicians of the 18th century criticised even Euclid, because, as they claimed, he gave proofs to "obvious" propositions!

• During the 18th century, research was not carried out in universities, but in the various academies that were established which supported the publication of mathematical journals.

Around 1795 the Académie des Sciences in Paris constituted one of the three subdivisions of the Institut de France.

Before 1800, in Germany no research was carried out in universities. The great mathematicians belonged to the Berlin Academy of Sciences. In 1810, Alexander von Humboldt (1769-1859) founded the University of Berlin and introduced the basic idea that professors must teach their research works and students must take courses that interest them.

Until the French Revolution (1789), the French universities of the 18th century were not better than the German ones. But the new government decided to upgrade them. Nicolas de Condorcet was in charge of the relative organizational work. In 1794 the École Polytechnique was founded, where Monge and Lagrange taught. In 1808 the French government founded the École Normale Supérieure, a forerunner of which was the École Normale, which was founded in 1794 and run for few months only.

During the 18th century, France was the leading country in the area of mathematics, with Switzerland in the second place. Germany remained relatively inactive. In England we distinguish Brook Taylor, Mathew Stewart (1717-1785) and Colin Maclaurin. The English mathematicians not only isolated themselves from their colleagues in Continental Europe (due to the conflict between Newton and Leibniz), but also suffered from their insistence on Newton's geometric methods.

The great production in mathematics during the 18th century and the unfounded extension of mathematics left many problems "open", which were inherited by the 19th century.

Additional Reading

(1) M. Cantor, *Vorlesungen über Geschichte der Mathematik* III, *Teubner*, 1898.

(2) D. J. Struik, *A concise history of mathematics*, Dover, 1948, *third edition*, 1967.

(3) N. Bourbaki, *Éléments d' histoire des mathématiques*, Hermann, 1960.

PART II

INTRODUCTION

The presentation in Part II of the *History of Mathematics*, which concerns the 19th and the 20th centuries, will be made in a way that differs from the one followed in the previous chapters. This is because, during the last two centuries, mathematics tends to become more specialized, and hence it requires a deeper and specialized knowledge of the reader, who, as we mentioned earlier, is mainly assumed to be near the end of his/her undergradate studies or at the beginning of graduate studies.

In Part II we *selectively* examine certain subjects of great importance, some of which concern contemporary discoveries of great interest. In addition, in the short biographies of certain mathematicians of the last two centuries that we will present, the reader will have the opportunity to follow the evolution of the course of research that was followed in certain basic areas of mathematics.

- The 19th century deserves to be called the Golden Century of Mathematics and if we exclude the ancient Greek era, it deserves to be considered as the most revolutionary century in the history of mathematics. From the technical point of view, the most original creation of this century was the **theory of functions of a complex variable**. This new branch of mathematics dominated the 19th century in the same way that infinitesimal calculus dominated the 18th century.

The introduction in the mathematician's "repertoire" of notions, such as the non-Euclidean geometries, the n-dimensional spaces, the non-commutative algebras, the "procedures repeated ad infinitum", and the non-quantitative structures, brought about a radical reformation that changed completely the view of the mathematical edifice.

- The most profound research activity of the 20th century was the research that refers to the axiomatic foundation of mathematics. Here, mathematicians study not only problems regarding the nature of mathematical science, but also the validity of the mathematical propositions obtained through the application of deductive logic. A characteristic feature of a great part of the mathematics of the 20th century is that it follows the general directions which began to appear near the end of the 19th century. These trends include the emphasis on the study and unification of mathematical structures that exist in various areas of mathematical science which until then were considered to be unrelated. In addition, these trends include the increasing communication between the research mathematicians living in various places on the Earth. Indeed, Despite the existing economical and political differences, during the greatest part of the 20th century mathematicians communicated with each other and informed themselves about the progress of the research works of their colleagues much more than they did during the previous century.

CHAPTER 1

SHORT BIOGRAPHIES OF MATHEMATICIANS OF THE 19th AND 20th CENTURIES

Carl Friedrich Gauss (1777-1855)
He was the son of a builder, who lived in the German town Braunschweig and seemed at first to be intended for manual labor. But the principal of the primary school, where Gauss studied, was impressed by the ingenuity of the young Carl and recommended him to the Duke Karl Wilhelm, who undertook Gauss' further education. The following incident is characteristic of the abilities of young Carl: One day, in order to keep his pupils busy, the teacher of the primary school that Gauss was attending asked them to calculate the sum of the whole numbers from 1 to 100, and then to place their papers on the desk. Almost immediately after the statement of the problem Carl turned in his paper. The teacher gave an incredulous look at his pupil, while the other pupils were still working with great attention. But when the teacher checked Gauss' paper, he observed stunned that it was the only one that gave the correct answer, i.e., 5050. The ten year old boy had calculated the requested sum via the formula $\frac{m(m+1)}{2}$, which he had obviously discovered at that moment.

Gauss studied at the University of Göttingen (1795-1798). At the age of 18 he invented the "**method of least squares**" and at the age of 19 he proved that the regular polygon with 17 sides can be constructed using a ruler and a compass only.

In 1798 he enrolled at the University of Helmstädt where he carried out his doctoral dissertation. Later he returned to Braunschweig, where he wrote his most important scientific works. In 1807 he was elected professor of astronomy and director of Göttingen's observatory, where he remained for the rest of his life.

Gauss' first important work was his doctoral dissertation where he gave the most rigorous (until then) proof of the fundamental theorem of algebra. In 1801 he published his classic work *Disquisitiones Arithmeticae*. His writing on differential geometry, entitled *Disquisitiones Generales circa Superficies Curvas* (General investigations of curved surfaces, 1827), constitutes a mathematical landmark. He also contributed to the development of algebra, of the theory of complex functions, and to potential theory. Among his non-published works, one finds original work on elliptic functions and non-Euclidean geometry.

His interest for subjects in physics was also great. When Giuseppe Piazzi (1746-1826) discovered the asteroid "Ceres" (1801), Gauss occupied himself with finding the orbit of this asteroid, and this effort constituted the beginning of his research work on astronomy that lasted for about twenty years. A writing of his work in this area is *Theoria Motus Corporum Coelestium* (Theory of Motion of Celestial Bodies, 1809). Gauss' works on the magnetism of the earth constitute a model of research works in physics and provide methods of measuring the magnetic field of the earth. His work in astronomy and in magnetism initiated a brilliant period of collaboration between mathematics and physics.

Gauss also occupied himself with problems in optics and his work (1838-1841) constituted a new basis for treating these problems.

Despite the fact that Gauss was considered to be the greatest mathematician after Newton and was named the "prince of mathematics", he was more a transitional figure from the 18th to the 19th century, than an innovator.

He published a relatively small number of his works, because he insisted on elaborating his texts, trying to give to them the greatest possible elegance and conciseness without affecting their mathematical rigor.

In the case of non-Euclidean geometry he published no final work. He was absolutely convinced that the efforts to prove Euclid's postulate of parallelism were in vain, and very early he conceived the idea of the existence of a geometry where this postulate does not hold. In 1813 Gauss began to develop a new geometry, which he first called anti-Euclidean geometry, later stellar geometry, and finally non-Euclidean geometry. We will not mention which were the non-Euclidean theorems of Gauss, because, as we said before, he wrote no complete text on this subject and the theorems that he proved look very much like the theorems in the works of Lobachevsky and Bolyai, who are considered to be the creators of non-Euclidean geometry (Part I, Chapter 15).

Augustin-Louis Cauchy (1789-1857)

He was born in Paris in 1789. In 1805 he was admitted at the École Polytechnique, where he studied to become a mechanic. Since his health was precarious, Lagrange and Laplace advised him to devote himself to mathematics. He served as a professor at the École Polytechnique, at Sorbonne, and at the Collège de France. Politics had unpredictable consequences on his scientific career. He was a zealous royalist and supporter of the Bourbons. When in 1830 some distant branch of the Bourbons took control of France, Cauchy refused to support this new monarchy. He resigned from his professorship at the École Polytechnique and went self-exiled to Torino, where he taught Latin and Italian for a few years. In 1833 he returned to Paris, where he taught as a professor at various ecclesiastic institutes, until the government that resulted from the revolution abrogated the so-called "oaths of faith". In 1848 Cauchy occupied the chair of mathematical astronomy at Sorbonne. Even though in 1852 Napoleon III restored the "oath of faith", Cauchy was exempted from it. Cauchy responded to this condescending behavior of the emperor by giving his salary to the poor inhabitants of the town Seaux, where he lived. Cauchy was an admirable teacher and a great mathematician. But he also had other interests; he knew the poetry of his time and he was the author of a work on Jewish prosody.

In mathematics he wrote more than 700 papers, thus being the second most prolific researcher after Euler. In a contemporary edition, his total work consists of twenty six volumes and refers to all branches of mathematics.

In mechanics he wrote important papers on elastic membranes, on the theory of waves (which was initiated by Fresnel), and on light diffusion and polarization. He greatly advanced the theory of determinants and contributed with basic theorems to the theory of ordinary and partial differential equations.

From the educational point of view, he suffered greatly for the rigor with which he pursued his work. His students did not like this "rigor", because it meant that they had to attend more lectures and study harder. At some point Cauchy began to teach the course of infinitesimal calculus to an audience of thirty students, who gave up the course one after the other, except one student who remained until the end! Even though he knew that he was not popular with his students, Cauchy remained faithful to his ideal, namely to giving complete and very careful proofs to all his theorems.

Cauchy was an idealist. But this virtue was not appreciated by his contemporaries, which made him either sometimes lose his position or lose various opportunities to obtain an important position. For example, in 1843, Guglielmo Libri (1803-1869) was chosen over Cauchy to become a professor at the Collège de France, not because Libri was considered to be a better mathematician than Cauchy, but because Libri had expressed his opposition against the Jesuits while Cauchy supported them. Later, Libri was forced to leave France when it was discovered that he was stealing books from various libraries!

Cauchy's contribution to the development of mathematical analysis was immense. First we will mention only a few of the subjects with which he occupied himself, and then we will comment briefly on these subjects.

(1) He gave the first rigorous definition of a continuous function.
(2) He gave the "Cauchy criterion" regarding the convergence of a sequence.
(3) He proved the generalized mean value theorem.
(4) He proved the theorem of residues in the theory of complex functions.

(**1'**) In the beginnings of the 1800s, mathematicians were seriously concerned about the existing lack of rigor in various mathematical notions and in various proofs of theorems in many branches of mathematical analysis. Even the notion of a "function" was not clear. The use of series, without taking into account whether they converge or diverge, had created many paradoxes; furthermore, the dispute about the possibility of representing a function by a trigonometric series made the situation even worse. In addition, inappropriate definitions were given to the fundamental notions of the derivative and the integral. In the area of mathematical analysis, rigor began with the works of Bolzano, Cauchy, Abel and Dirichlet, and it was later advanced by Weierstrass.

Cauchy's basic work regarding the foundation of mathematical analysis is found in his writings: *Cours d' analyse algébrique* (1821), *Résumé des leçons sur le calcul infinitésimal* (1823), and *Leçons sur le calcul différentiel* (1829).

The definition of a (single-valued) function that we often use even today was given by Peter Gustav Lejeune Dirichlet (1805-1859). Gradually, the matter of continuity or discontinuity of a function came up. The careful study of the properties of a function was initiated by Bernhard Bolzano (1781-1848), who came from Bohemia and was a priest, a philosopher and a mathematician.

Cauchy occupied himself with the notion of a limit and the notion of continuity. In the preface of his work of 1821 he says that, in order to talk about the notion of continuity of a function, one must first know the main properties of "infinitely small quantities". "We shall say (*Cours ...*, page 5) that a variable quantity becomes infinitely small, when its numerical value constantly decreases and converges to the limit zero". Cauchy called these variable quantities infinitesimals. In this way Cauchy clarifies the notion of an infinitesimal, which was introduced by Leibniz, and frees it from of any metaphysical element. Then, Cauchy adds the following: "A variable quantity becomes infinitely large, when its numerical value constantly increases and converges to the limit ∞". Here ∞ does not denote a fixed quantity but something that is constantly increasing. Then, Cauchy states the following: "Let $f(x)$ be a function of x such that for every value that lies between two extreme values, the function takes a single finite value. $f(x)$ is and remains continuous with respect to x, if between these values, an infinitely small increment of the variable always produces an infinitely small increment of the given function". He also states that "the function $f(x)$ is a continuous function of x in the neighborhood of a given value of x if the function is continuous between two extreme values that define this neighborhood, no matter how close to each other these values are". Then, he mentions that "$f(x)$ is discontinuous at some point x_0 if it is not continuous at any line segment that contains x_0 in its interior".

In his writing *Cours ...* (page 37) he claims that if a function of more than one variables is separately continuous with respect to each of the variables, then it is a continuous function of these variables. But, as it is known, this claim is incorrect.

(**2′**) The mathematicians of the 18th century used series without being concerned about their convergence or divergence. During the end of the century, some ambiguous and absurd results that were obtained in works where the use of series was made, motivated the mathematicians to study whether the operations that they applied on series were always permissible. Around 1810, Fourier, Gauss and Bolzano began to make correct use of series.

Gauss made the first important and rigorous study regarding the convergence of a series in his writing *Disquisitiones ...* (1812), where he studied the so-called hypergeometric series. In his work of 1817, Bolzano formulated the correct condition for the convergence of a series (a condition which is attributed today to Cauchy).

Cauchy's work regarding the convergence of a series was the first essential and extensive study on this subject. In *Cours d'analyse algébrique*, he writes the following:

"Let $s_n = u_0 + u_1 + ... + u_{n-1}$ be the sum of the first n terms of a series with infinitely many terms, where n is a positive integer. If s_n approaches a certain limit s, as n increases ad infinitum, then the series is called **convergent**, and the limit s is called the **sum** of this series. On the contrary, if s_n approaches no fixed limit, as n increases ad infinitum, then the series is called **divergent** and has no sum."

After defining the convergence and the divergence of series, Cauchy formulated (*Cours ...*, page 125) the so-called "**Cauchy criterion of convergence**": "A sequence $\langle s_n \rangle$ converges to a certain limit s if and only if $|s_{n+r} - s_n|$ can be made smaller than any given quantity for all r and for sufficiently large n." Cauchy proves that this condition is necessary and "simply" remarks that if this condition holds, then the sequence converges. The fact that he does not prove that this condition is "sufficient" is due to the fact that he ignores certain properties of the real numbers.

Later, Cauchy formulates and proves certain particular criteria for the convergence of series with positive terms. He remarks that the general term u_n must approach zero. Another criterion that he gives asks one to find the limit or the limits of the expression $\sqrt[n]{u_n}$ as n becomes infinitely large, and he denotes the larger of these limits by k; the series converges if $k < 1$ and diverges if $k > 1$. He also gives the so-called "**ratio test**", which makes use of the limit

$$\lim_{n \to \infty} \left(\frac{u_{n+1}}{u_n} \right).$$

If this limit is less than 1, then the series converges, but if it is larger than 1, the series diverges. He gives particular criteria for the case that this limit is equal to 1.

Then, the "comparison tests" and a "logarithmic test" follow. He proves that the sum $u_n + v_n$ of two convergent series converges to the sum of their separate sums and that an analogous result holds for the product of two series.

Cauchy proves that a series that also includes negative terms converges if the series of the absolute values of the terms converges. From this he deduces Leibniz's criterion regarding series with terms of alternating sign.

Cauchy also examines the sum of series of the form

$$\sum_{n=1}^{\infty} u_n(x) = u_1(x) + u_2(x) + \ldots + u_n(x) + \ldots,$$

where all the terms are single-valued, continuous, real functions. The theorems regarding the convergence of series with constant terms are used here for the determination of the interval of convergence of the above series. He also occupies himself with series whose terms are complex numbers.

Lagrange was the first who stated Taylor's theorem "with a remainder". But Cauchy makes the important remark, in his works of 1823 and 1829, that the infinite Taylor series converges to the corresponding function if the "remainder" has limit zero. He gives the example $e^{-x^2} + e^{-\frac{1}{x^2}}$ of a function whose Taylor series does not converge to this function. He gives (in 1823) the example $e^{-\frac{1}{x^2}}$ of a function, which has derivatives of any order at $x = 0$, but does not expand into a Taylor series at $x = 0$. With this last example, Cauchy disproves Lagranges' viewpoint that: "if f has derivatives of any order at some point x_0, then f can be expanded into a Taylor series at points x near x_0".

But Cauchy made mistakes too. In his writing *Cours d'analyse algébrique* (pages 131-132), he states that $F(x) = \sum_{n=1}^{\infty} u_n(x)$ is continuous, if the series $\sum_{n=1}^{\infty} u_n(x)$ converges and the functions $u_n(x)$ are continuous. In his work *Résumé des leçons sur le calcul infinitésimal*, he also states that if the $u_n(x)$ are continuous and the series $\sum_{n=1}^{\infty} u_n(x)$ converges, then "it is allowed" to integrate this series term by term, i.e.,

$$\int_a^b F(x)dx = \sum_{n=1}^{\infty} \int_a^b u_n(x)dx.$$

Here Cauchy neglects the fact that one must also assume that the series is uniformly convergent. He also claims that for continuous functions, the following formula holds

$$\frac{\partial}{\partial u} \left(\int_a^b f(x,u)dx \right) = \int_a^b \frac{\partial f}{\partial u}dx.$$

(**3'**) In Cauchy's infinitesimal calculus the notions of a function and of a limit of a function are considered to be fundamental. In order to define the derivative of

$y = f(x)$ with respect to x, he gives to the variable x an increment $\Delta x = \alpha$ and forms the fraction
$$\frac{\Delta y}{\Delta x} = \frac{f(x+\alpha) - f(x)}{\alpha}.$$
He defines the derivative $f'(x)$ of y with respect to x to be the limit of this expression (if it exists) as α tends to zero. If dx is a finite quantity, then he defines the **differential** dy of $y = f(x)$ to be the product $f'(x)dx$.

We saw above that Cauchy introduced the notion of a continuous function. During the 18th century, "integration" was considered to be the inverse operation of differentiation. It follows from the above definition of the derivative that the derivative does not exist at a point where the function is discontinuous.

However, the notion of the integral presents no difficulties. Even discontinuous curves can surround domains whose areas are clearly determined. Thus, Cauchy defined the "**definite integral**" to be the limit of sums in a way which does not differ much from the one that we use in contemporary textbooks, with the difference that he always considered the value of the function at the left endpoint of the interval.

If $s_n = (x_1 - x_0)f(x_0) + (x_2 - x_1)f(x_1) + ... + (X - x_{n-1})f(x_{n-1})$, then the limit s of the sum s_n, as the lengths $x_i - x_{i-1}$ of the intervals tend to zero, is the definite integral of the function $f(x)$ for the interval from $x = x_0$ to $x = X$.

Many useful generalizations of the integral followed from Cauchy's definition of the integral as a limit of a sum (independently of the notion of an antiderivative).

After defining the integral independently of the notion of an antiderivative, Cauchy had to prove the known existing relation between the integral and the antiderivative. He was able to give this proof using the mean value theorem. If $f(x)$ is continuous in the closed interval $[a, b]$ and differentiable in the open interval (a, b), then there exists $x_0 \in (a, b)$ such that $f(b) - f(a) = f'(x_0)(b-a)$. This proposition is a generalization of Rolle's theorem and it was known for approximately one hundred years. Even though the mean value theorem had played a useful role in mathematical analysis until the days of Cauchy, it did not catch the mathematicians' attention until then.

A theorem of a more general form, stating that
$$\frac{f(b) - f(a)}{g(b) - g(a)} = \frac{f'(x_0)}{g'(x_0)},$$
where the functions f and g satisfy certain conditions, is known as "**the mean value theorem of Cauchy**".

(**4′**) The foundation of the theory of complex functions of one complex variable is essentially due to Cauchy. Even though the contributions of Gauss, of Poisson, and of other mathematicians, were important, none of them published any basic work on this theory.

In 1814, Cauchy presented to the French Academy his first important work in the theory of complex functions, entitled *Mémoire sur la théorie des intégrales définies*, which he did not publish until 1827. In 1825, Cauchy wrote his paper *Mémoire sur les intégrales définies prises entre des limites imaginaires*, which was published in 1874. Although, for a long time Cauchy did not appreciate the value of his work, many people consider it as the most important work of Cauchy and one of the most beautiful works in the history of exact sciences. In this work Cauchy occupies himself with the integral

$$\int_{x_0+iy_0}^{X+iY} f(z)dz \tag{1}$$

where $z = x + iy$, and he gives the definition of this integral as the limit of the sum

$$\sum_{\nu=0}^{n-1} f(x_\nu + iy_\nu)[(x_{\nu+1} - x_\nu) + i(y_{\nu+1} - y_\nu)],$$

where $x_0, x_1, ..., x_n = X$ and $y_0, y_1, ..., y_n = Y$ are the points at which the curve that connects (x_0, y_0) with (X, Y) is subdivided. Here $x + iy$ is a point of the complex plane and the integral is taken along a curve that lies on the complex plane. Cauchy also proves that if $x = \phi(t)$ and $y = \psi(t)$, where t is a real number, then the value of the integral is independent of the choice of the functions ϕ and ψ, under the assumption, of course, that $f(z)$ has no point of discontinuity between the two different curves over which the integral is taken.

Cauchy formulates his theorem as follows:

"If $f(x + iy)$ is finite and continuous for $x_0 \leq x \leq X$ and $y_0 \leq y \leq Y$, then the value of the integral (1) is independent of the form of the functions $x = \phi(t)$ and $y = \psi(t)$." The proof that he gives to this theorem makes use of the calculus of variations and presents some imperfections.

In his work of 1825 Cauchy examines the case where $f(z)$ is discontinuous at some point of the plane that either lies in the interior of the domain surrounded by the two curves that connect the points (x_0, y_0) and (X, Y) or on one of these curves. Then the two values of the integral (1) may be different. If $f(z)$ becomes infinite at the point $z_1 = a + ib$, while the limit

$$F = \lim_{z \to z_1} ((z - z_1)f(z))$$

exists (in other words, if z_1 is, as we say today, a simple pole of f), then the difference of the two values of (1) is $\pm 2\pi F\sqrt{-1}$. Thus, for the function $f(z) = \frac{1}{1+z^2}$, which becomes infinite for $z = \sqrt{-1}$ (i.e., for $a = 0$ and $b = 1$), we have

$$F = \lim_{\substack{x \to 0 \\ y \to 1}} \frac{x + (y-1)\sqrt{-1}}{[x + (y+1)\sqrt{-1}][x + (y-1)\sqrt{-1}]} = \frac{-\sqrt{-1}}{2}.$$

In his writing *Exercises de mathématiques*, Cauchy called this quantity F **résidu intégral**. In Greek, we use the term "ολοκληρωτικό υπόλοιπο[1]".

In addition, when the function has more than one pole in the domain surrounded by the curves on which the integration is performed, Cauchy remarks that one must consider the sum of the residues in order to calculate the difference of the two integrals.

In the above mentioned work of his *Exercises ...*, Cauchy mentions that the residue of $f(z)$ at z_1 is equal to the coefficient of the term $(z - z_1)^{-1}$ in the expansion of $f(z)$ into a power series. In a subsequent work (1841) he gives a new expression for the residue that corresponds to a pole, namely

$$F(z_1) = \frac{1}{2\pi i} \int f(z)dz,$$

---

[1]**Translator's note:** The meaning of this phrase in English is: *integral residue*.

where the integral is taken over the circumference of an appropriately small circle that contains $z = z_1$.

The introduction of the integral residue (résidu intégral) and the development of the corresponding theory constitute a great contribution in the theory of complex functions.

Karl Weierstrass (1815-1897)

He was born in Ostenfelde (Westphalen) and he studied law for four years at the University of Bonn. In 1838 he changed the direction of his studies towards mathematics, thus leaving unfinished his work for obtaining a degree in law. Instead of obtaining a doctorate degree in law, he chose to obtain permission from the state so that he could teach in secondary education (gymnasium), where he taught "writing" and "gymnastics" (1841-1854). During the time that he was working in secondary education he made no contact with the mathematical community but he worked hard carrying out research in mathematics. The few research results that he did publish secured for him a teaching position at the Industrial Institute of Berlin in 1856, where he taught technical courses. The same year he was promoted to a lecturer at the University of Berlin and later in 1864 he was elected professor at the same university where he stayed until the end of his life. Among Weierstrass' students are the famous mathematicians, Georg Cantor, Sonya Kovalevsky, and David Hilbert.

Weierstrass was a methodical, hardworking and very careful person. He did not trust intuition, and he tried to place mathematical thought on a rigorous basis.

Cauchy's theory regarding the notion of the "limit", on which the notions of continuity, differentiation, and convergence depend, is based on a simple intuitive-geometric perception of the system of real numbers. Indeed, Cauchy considered the system of real numbers as given, whose entire properties are known. (Besides, the same happens today with almost all elementary textbooks on infinitesimal calculus.) But it became later understood that the notions of limit, continuity, differentiation, and convergence are based on much more obscure properties of the real numbers than it was thought until then. The need for a deeper understanding of the foundations of mathematical analysis became apparent in 1874, when Weierstrass gave an example of a continuous function, which is nowhere differentiable, that is an example of a continuous curve which does not have a tangent at any of its points, something that is in conflict with our intuition. This example was a hard blow to the practice of using geometric intuition in the study of subjects of mathematical analysis. Similar examples made obvious the fact that Cauchy had not been entirely successful in creating a sound foundation of mathematical analysis. Still there existed fundamental properties of the system of real numbers, which needed to be understood. For these reasons Weierstrass suggested a research program where, first of all, mathematicians had to achieve the logical construction of the theory of the system of real numbers, and then based on this construction, they had to develop the notions of the limit, continuity, differentiation, convergence and divergence, as well as the notion of integration. This very important program is known as "**arithmetization of analysis**" (a name given to it by Felix Klein in 1895) and, despite the fact that it was very complicated, Weierstrass together with other mathematicians were able to carry out this program successfully near the end of the 19th century.

The success of Weierstrass' program had far-reaching consequences. First of all, since the entire mathematical analysis can be "constructed" from the system of real numbers, it follows that mathematical analysis is consistent (i.e., it contains no contradiction) if the system of real numbers is consistent. Moreover, since Euclidean geometry is interpreted via analytic geometry, which is also based on the system of real numbers, it follows also that Euclidean geometry is consistent if the system of real numbers is consistent. In addition, since the system of real numbers, or part of it, can be used to interpret many branches of algebra, the consistency of a great part of algebra depends also on the system of real numbers. Consequently, a very large part of the science of mathematics is consistent if the system of real numbers is consistent. The above discussion leads to the conclusion that the system of real numbers plays a fundamental role in the foundation of mathematics, and hence it is not at all surprising that a group of great mathematicians worked hard in order to carry out the program of Weierstrass.

The second part of Weierstrass' program was realized with the introduction of detailed methods, the so-called "ϵ - δ" methods, which we use today. For example, the following clear and exact definition was given to the notion of a limit:
"If given any $\epsilon > 0$, there exists $\delta > 0$ such that $|f(x) - L| < \epsilon$ when $0 < |x - c| < \delta$, then the number L is the limit of $f(x)$ as $x \to c$."

This definition eliminates the vague phrases that were used until then, such as the phrases "successive values", "taking values which are as small as one desires", etc. What finally remains in the above notation, and in the language that is used in the definition of a limit, are the real numbers, the operation of addition (and its inverse, i.e., subtraction), and the relation "less than" (and its inverse, i.e., "greater than").

In a way similar to the above, all basic notions of mathematical analysis were formulated very carefully using the real numbers and the fundamental operations and relations between them.

There were two directions to the realization of the first part of Weierstrass' program. The first was to formulate a set of axioms that characterize the system of real numbers. Indeed, it was proved that the system of real numbers can be uniquely determined as a complete ordered field. We note that in the language of contemporary abstract algebra, when we say that a system is uniquely determined, we mean that we make no distinction between "isomorphic" systems.

The second direction to the realization of Weierstrass' program involved starting from a very simple system of axioms that characterizes the system of natural numbers, and then gradually proceeding to the construction of the system of real numbers. This second direction was followed successfully by Weierstrass, Richard Dedekind (1831-1916), Georg Cantor (1845-1918), Giuseppe Peano (1858-1932), and others. This achievement made mathematicians feel certain that mathematics is consistent. This feeling of certainty was due to the fact that the system of natural numbers was distinguished for its intuitive simplicity (something that lacks in the majority of the remaining mathematical systems), as well as for its wide use for a long time without ever leading to any contradictions.

After the construction of the theory of the system of real numbers (1841), Weierstrass directed his research towards the theory of analytic functions, which he based on the theory of power series and the method of "analytic continuation". He carried out this work during 1840, but he published it later. He also contributed to the development of many subjects of the theory of functions, and he also studied the ν-body problem in astronomy, and the theory of propagation of light.

It is difficult to determine the exact date of Weierstrass' works, because they were not published immediately after their completion. Many of his research results, were already known to the mathematical community before his publication, because he had taught them in his courses at the University of Berlin. During the decade of 1890, when he published his work *Werke*, he was not concerned about the priority of the publication of his results; actually many of them were already published by others. What interested him in particular was the presentation of his method regarding the development of the theory of functions.

The representation of complex functions by power series was already known. But Weierstrass was the one who solved the following problem: given a power series representing a function in some concrete domain of the plane, we need to find other power series, which define the same function in other domains of the plane.

A power series with respect to $z-a$, which converges in some disc C with center a and radius r, represents a function $f(z)$ which is analytic at every interior point of C. Now if we consider a point b in the interior of C, and if we use the values of the function $f(z)$ and of its derivatives as they are given by the initial series, then one obtains a new power series with respect to $z-b$, whose disc of convergence C', in general, intersects the first disc C and contains points which lie inside C and points which lie possibly outside C. At the common points of the two circles the two series give the same value of the function $f(z)$. But at points of C' which lie outside C, the values of the second series constitute an "analytic continuation" of the function defined by the first series. If we continue in the same way, as far as possible, we go from C' to other discs and in this way we obtain the entire analytic extension of $f(z)$. The complete function $f(z)$ consists of the set of values it takes at all points z which lie at the union of all the constructed discs. Everyone of the above series is called an "element" of the complete function.

It should be noted that during the process of extending the domain of definition of an analytic function by adding more discs of convergence, it might happen that one of the new discs covers part of a disc which is not the one immediately preceding in the chain of discs; if this happens and if the values of the function at this common part between the new disc and some disc preceding it do not coincide, the function is called "**multi-valued**".

The singular points (poles or branch points) that may appear during the above process necessarily lie on the boundary of the discs of convergence of the power series and (as Weierstrass points out), if they are of finite order, then they belong to the function; indeed, it is possible at such a point to have an expansion into powers $(z-z_0)^n$ having only a finite number of terms with negative exponents.

In order to obtain the expansion of a function at $z=\infty$, Weierstrass uses power series with respect to $\frac{1}{z}$. If the element of the function converges in the entire plane, Weierstrass calls this function entire, and if it is not a polynomial, then we say that it has an "essential singularity" at $z=\infty$ (as it happens, for example, with the function $\sin z$).

Weierstrass gave an example of a power series for which the boundary of the disc of convergence is a natural boundary. This means that all the points of the boundary of the disc of convergence are singular points of the function. In addition, he gave an example of an analytic expression, which may represent different analytic functions in different parts of the plane.

Niels Henrik Abel (1802-1829) - **Carl Gustav Jacob Jacobi** (1804-1851)

During the first half of the 19th century, the development of the subject of elliptic functions and later of the subject of abelian functions takes place in parallel with the development of the basic theorems of the theory of functions of a complex variable.

Undoubtedly Gauss had already obtained some basic results in the theory of elliptic functions, because after his death many of them were found in his notes, which were never published. However, today Abel and Jacobi are considered to be the founders of the theory of elliptic functions.

Abel was a son of a poor priest. As a student in Norway's Christiania (Oslo), he was fortunate to have Bernd Michael Holmboe (1795-1850) as his teacher. Holmboe recognized Abel's ingenuity and, when Abel was seventeen years old, Holmboe made the prediction that he would become the greatest mathematician of the world. After his studies in Christiania and in Copenhagen, Abel obtained a scholarship that allowed him to travel to other countries. In Paris, he presented himself to Legendre, Laplace, Cauchy, and Lacroix, but all ignored him. Later, he went to Berlin, where he remained from 1825 to 1827 and he worked with Crelle. He returned to Christiania physically exhausted, where he gave private lessons in order to improve his finances. After the publication of some of his works, he became more known, which allowed Crelle to secure for Abel a professorship at the University of Berlin. But it was already too late because Abel was affected by tuberculosis. He died in 1829 at the age of 27.

Abel knew the work of Euler, Lagrange, and Legendre regarding elliptic functions, and it is also likely that he took some ideas from Gauss' work *Disquisitiones Arithmeticae*. Abel started writing papers in 1825. He submitted for publication the most important of them at the Académie des Sciences in Paris on October 30, 1826. This paper entitled *Mémoire sur une propriété générale d' une classe très étendue de fonctions transcendantes* includes the basic and important theorem of Abel in the theory of elliptic functions. After reading the introduction of this paper, Fourier, who was the secretary of the French Academy at that time, forwarded it for evaluation to Legendre and, in particular, to Cauchy, who was more qualified to evaluate it. The paper was lengthy and difficult, because it included new ideas. Cauchy neglected it and occupied himself with his own works, while Legendre forgot to look at it. After Abel's death, when his fame had already spread, the Academy searched and found this paper, and published it in 1841.

The other founder of the theory of elliptic functions was Jacobi. As opposed to what happened with Abel, Jacobi lived a peaceful life. He was born in Potsdam, he studied at the University of Berlin and in 1827 he was elected professor at Königsberg. In 1842 he resigned for reasons of health. The government of Prussia gave him a pension with which he went to Berlin, where he died in 1851. Jacobi's fame was already great when he was still alive and his students spread his ideas to many research centers.

Jacobi taught the subject of elliptic functions for many years. He did research on functional determinants (known as Jacobians), on ordinary and partial differential equations, on dynamics, on celestial mechanics, on fluid mechanics, and on hyperelliptic integrals and hyperelliptic functions.

When Abel was working on subjects regarding elliptic functions, Jacobi, who had also studied the works of Legendre on elliptic integrals, began, in 1827, to work on the corresponding elliptic functions. He submitted for publication to the journal *Astronomische Nachrichten* a paper, in which, however, he didn't include the proofs of the various propositions that he was presenting. Almost simultaneously Abel published independently the paper *Recherches sur les fonctions elliptiques*.

Abel and Jacobi came to the basic conclusion that it is preferable to work with the inverse functions of those defined by elliptic integrals.

Now, we will only make some general remarks regarding the wide subject of elliptic integrals and functions.

The general elliptic integral is given by the relation

$$u = \int R\left(x, \sqrt{P(x)}\right) dx,$$

where $P(x)$ is a polynomial of degree three or four with distinct roots and $R(x,y)$ is a rational function of the variables x and y. The efforts to obtain some general results for the function u of x didn't bring satisfactory results, because the deeper meaning of the integral was not completely understood by Euler and Legendre. The coefficients of $P(x)$ were taken to be real numbers, the values of x were taken to be real and $P(x) \neq 0$. A better knowledge of the theory of complex functions would have allowed perhaps more progress, but such a knowledge still did not exist.

Abel and Jacobi had the following original idea, which proved to be very fruitful in the study of elliptic integrals and functions. Legendre had proved that the general elliptic integral can be reduced to three types. One of these types is

$$u = \int_0^x \frac{dy}{\sqrt{(1-y^2)(1-k^2y^2)}} = \int_0^\phi \frac{d\theta}{\sqrt{1-k^2\sin^2\theta}},$$

where $x = \sin\phi$ and $0 < k < 1$. Abel remarked that in studying the above integral one encounters the same difficulties as in studying the integral

$$u = \int_0^x \frac{dy}{\sqrt{1-y^2}} = \arcsin x.$$

Elegant relations arise when one views x as a function of u and not u as a function of x. For this reason Abel suggested, in the case of elliptic integrals, to study x as a function of the independent variable u. Since $x = \sin\phi$, ϕ can also be studied as a function of u. But a further presentation of the subject goes beyond the scope of the present book.

Georg Friedrich Bernhard Riemann (1826-1866)

Riemann was a student of Gauss and Wilhelm Weber. In 1846 he went to Göttingen to study theology, but he quickly turned over to mathematics. His doctoral dissertation (1851), under the guidance of Gauss, entitled *Grundlagen für eine allgemeine Theorie der Funktionen einer veränderlichen Komplexen Grösse* constitutes a basic work in the theory of complex functions. Three years later he was nominated Privatdozent at the University of Göttingen, which means that he had the right to teach and collect payment from the students. In order to obtain this title he wrote the so-called **Habilitations - Schrift**, *Über die Darstellbarkeit einer*

Funktion durch eine trigonometrische Reihe and gave an established talk (Habilitationsvortrag) entitled *Über die Hypothesen welche der Geometrie zu Grunde liegen* (On the existing assumptions on the foundations of Geometry).

These works were followed by a large number of Riemann's important works. Riemann succeeded Dirichlet as professor of mathematics at Göttingen in 1859. He died of tuberculosis in 1866.

A period of research in the theory of complex functions ended in the middle of the 19th century. Riemann's works were the beginning of a new period of discoveries in the theory of algebraic functions and their integrals, as well as in the theory of inverse functions. Riemann developed a broader theory in which he treated multi-valued functions, which were earlier treated only by Cauchy and Puiseux, thus opening the road of progress in many other directions.

Riemann's contributions to theoretical mathematics were indeed immense. He was also greatly interested in physics and in the relations that exist between mathematics and the physical world. He wrote papers concerning heat, light, the theory of gases, magnetism, the dynamics of fluids, and acoustics. He tried to unify the theories of gravity and light, and he investigated the mechanism of the human ear.

• As a mathematician, Riemann made extensive use of geometric intuition as well as of arguments from physics. Felix Klein's claims that Riemann's ideas regarding complex functions came up from his study of the problem of the flow of electric currents on the plane. The equation for the potential constitutes a central point of this problem and for this reason it played a major role in Riemann's development of the theory of complex functions.

Riemann's basic idea regarding multi-valued functions is the notion of a "**Riemann surface**". The function $w^2 = z$ constitutes an example of a multi-valued function of the independent variable z, because at every value of z there correspond two values of the function w, namely $\sqrt{z}$ and $-\sqrt{z}$. Riemann wanted to keep these two sets of values of w separate, i.e., he wanted to separate the two branches of w. To this end, he assigned a z-plane to every branch. In addition, (depending on the case considered) he defined on such a z-plane a point which corresponds to $z = \infty$. These two planes are placed one above the other and they are glued together first of all at those values of z where the branches give the same w-values. Thus, for the function $w^2 = z$, the two planes or "sheets", as they are also called, are glued at the points $z = 0$ and $z = \infty$. The branch $w_1 = \sqrt{z}$ is represented by the z-values of the upper sheet, while the branch $w_2 = -\sqrt{z}$ is represented by the z-values of the lower sheet. When z moves on the upper sheet, on the circumference of a circle centered at the origin (Figure 1.1), in such a way that θ takes values from 0 to 2π in the expression $z = \rho(\cos\theta + i\sin\theta)$, then the image of z ranges over a semicircle of the complex plane on which the w_1-values are mapped. Now let as assume that z moves on the lower sheet and let z intersect the positive semi-axis of x. When z moves on the lower sheet, w takes the values $w_2 = -\sqrt{z}$. As z, staying on the lower sheet, rotates one more time about the origin so that θ takes values from 2π to 4π, then, for this trajectory of z, we obtain the set of values $w_2 = -\sqrt{z}$, while the polar angle of the corresponding w-values changes from π to 2π. When z intersects again the positive semi-axis of x, we view it as moving on the upper sheet. In this way, with the two rotations (one at every sheet) of the z-values about the origin, we obtain the set of the w-values of the function $w^2 = z$. Here, the most important

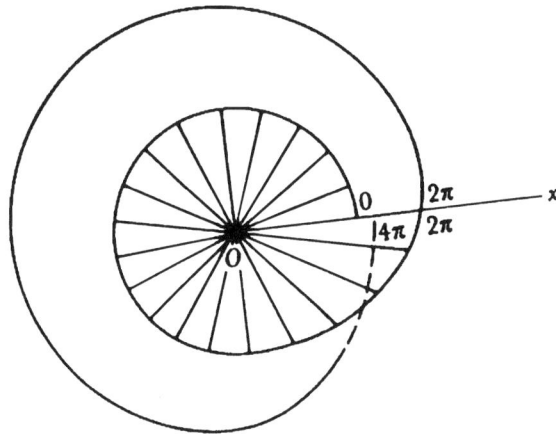

Figure 1.1

fact is that the function w becomes a single-valued function of the variable z, when z moves on the Riemann surface consisting of the two sheets mentioned above.

In order to distinguish the paths traced on one sheet from the paths traced on the other (in the case of the function $w^2 = z$), we consider the positive semi-axis of x as a **branch cut**. The positive semi-axis of x connects the points $z = 0$ and $z = \infty$. This means that every time the path traced by z intersects the positive semi-axis of x, z goes from one sheet to the other. The branch cut need not be the positive semi-axis of x, but in the example considered it must connect the points 0 and ∞. The points 0 and ∞ are called **branch points**, because the branches of $w^2 = z$ are interchanged when z traces a closed curve around any of these points.

The function $w^2 = z$ and the Riemann surface that corresponds to it are very simple. Let us consider the function $w^2 = z^3 - z$. This function has also two branches whose values for $z = 0$, $z = 1$, $z = -1$, $z = \infty$ coincide. If we argue in the way that we did in the previous example, we note that these four points are branch points, because when z traces a curve around any of these points, the function w changes the sheet on which it takes a value. As branch cuts, one can consider the line segments connecting 0 with 1, 0 with -1, 1 with ∞, and -1 with ∞. When the curve traced by z intersects any of the branch cuts, the value of w changes from the one that it had on one sheet to the one that it takes on the second sheet.

The Riemann surface becomes more complicated when the corresponding function becomes more complicated. A Riemann surface with ν sheets corresponds to a function, where to each z there correspond ν values. Many branch points may exist, and hence many branch cuts must be introduced, which connect the branch points pairwise. It should be noted that the sheets which meet at some branch point are not necessarily the same with the ones which meet at some other branch point. If k sheets meet at some branch point, then we say that the **order** of this branch point is $k - 1$. In addition, even though two sheets of a Riemann surface may have a common point, the branches of the corresponding function may not change when z makes a rotation about this common point. In this case, this common point is not a branch point.

We point out that it is not possible to construct an exact representation of a Riemann surface in 3-dimensional space. For example, when the two sheets of $w^2 = z$ are represented in 3-dimensional space, they must intersect along the positive semi-axis of x so that one can go smoothly from the upper sheet to the lower sheet and then, after a rotation about $z = 0$, one returns again to the upper sheet, according to the description that we made above.

The Riemann surface not only constitutes a way of representing a multi-valued function, but it also makes these functions single-valued which allows us to extend to multi-valued functions the theorems that hold for single-valued functions. For example, Riemann extended to the case of multi-valued functions Cauchy's theorem, which asserts that the integral of a single-valued function is equal to zero when the integration is performed over a curve which encloses some domain of the complex plane at which the function is analytic. The domain where the function is analytic must be simply connected, which means that it is possible to contract this domain continuously to a single point [7].

The study of a function on a Riemann surface consisting of ν planes which are connected to each other, presents serious difficulties. For this reason, mathematicians have invented other equivalent models of these surfaces, on which the study becomes simpler. But, we will not extend the discussion in that direction.

Riemann gave the following definition of a single-valued analytic function $f(z) = u + iv$:

"The function f is analytic at some point and in some neighborhood of this point, if f is continuous and differentiable, and it satisfies the so-called Cauchy-Riemann equations:

$$\frac{\partial u}{\partial x} = \frac{\partial v}{\partial y} \quad \text{and} \quad \frac{\partial u}{\partial y} = -\frac{\partial v}{\partial x}."$$

These equations first appeared in the works of d'Alembert, Euler and Cauchy.

Riemann was the first to state the condition that the existence of the derivative $\frac{dw}{dz}$ must mean that the limit of the expression $\frac{\Delta w}{\Delta z}$ must be the same no matter in which way $z + \Delta z$ tends to z. (We note that this last condition distinguishes complex functions from real functions of two real variables, because in the case of a real function $u(x, y)$, the existence of all directional derivatives of u at some point (x_0, y_0) does not imply that it is analytic at this point.) Later, Riemann tried to find as few conditions as possible so that a function of the variable $x + iy$ that satisfies them is completely determined. It follows from the Cauchy-Riemann equations that u and v satisfy the 2-dimensional equation for the potential

$$\frac{\partial^2 w}{\partial x^2} + \frac{\partial^2 w}{\partial y^2} = 0.$$

Starting from this observation, Riemann thought that a complex function could be determined at its entire domain of definition simply by the fact that u satisfies the above equation for the potential. In particular, Riemann assumed that a function w whose Riemann surface is given, can be completely determined to within an additive constant, from the real function $u(x, y)$ if u satisfies certain conditions. We will not mention which are the conditions stated by Riemann, because we will not extend the discussion of this subject any further. We will only add that if a function u satisfies the conditions stated by Riemann, then the Cauchy-Riemann

equations imply that
$$v = \int \left(-\frac{\partial u}{\partial y}dx + \frac{\partial u}{\partial x}dy\right),$$
which means that v is determined too and hence $w = u + iv$ is also determined.

• A very large part of Gauss' work is in geodesy and cartography (1816). The complete work of Gauss, entitled *Disquisitiones Generales Circa Superficies Curvas* (1827), constitutes a great contribution in differential geometry. In addition to this work in the theory of surfaces in 3-dimensional space, Gauss contributed to the development of the following completely new notion: "a surface is (i.e., can be considered as) also a space". Riemann generalized this notion, thus opening new horizons in non-Euclidean geometry.

The study of the works of Gauss, Lobachevsky and Bolyai raises certain doubts about what we can believe about the geometry of natural space. These doubts provided the stimulus for the creation of one of the most important theories of the 19th century, namely of the Riemannian Geometry.

The details of the work of Lobachevsky and Bolyai were, of course, not known to Riemann, but they were known to Gauss. Moreover, Riemann knew Gauss' doubts regarding the truth and the applicability of Euclidean geometry. We remark that in the area of geometry, Riemann followed Gauss, while in the area of the theory of functions, he followed Cauchy and Abel.

Riemann's research in geometry was also influenced by the teachings of the psychologist Johann Friedrich Herbart (1776-1841).

The geometry of space which was presented by Riemann, did not constitute simply an extension of Gauss' differential geometry. Riemann examined from the beginning the way in which we must approach the study of space by posing the following question: "About what can we be certain regarding natural space, that is, which are the conditions which are assumed to hold in space (before even we determine the concrete axioms which are true in space)?" One of Riemann's objectives was to prove that the axioms of Euclidean geometry were a result of experience and not self-evident truths, as it was believed. Riemann followed the analytic method in order to avoid the danger that exists in geometric proofs which are sometimes led by figures to erroneous conclusions.

Riemann developed for every space a kind of "intrinsic" geometry. He introduced the notion of a "n-dimensional manifold". A point of a n-dimensional manifold is represented by giving concrete values to n variable parameters x_1, x_2, ..., x_n. The set of all these "possible" points that one obtains constitutes the manifold itself, just like the set of points on a surface constitutes the surface itself. The n variable parameters are called coordinates on the manifold. When the x_i change continuously, the points range over the manifold.

Since Riemann believed that we know a space only locally, he began his theory with the definition of the distance between two neighboring points whose coordinates differ by infinitesimal quantities. He assumes that the square of the distance is given by the relation
$$ds^2 = \sum_{i=1}^{n}\sum_{j=1}^{n} g_{ij}dx_i dx_j,$$
where the g_{ij} are functions of the coordinates x_1, x_2, ..., x_n and $g_{ij} = g_{ji}$, while the right-hand side of the equality that gives ds^2 is always positive for all possible values of the dx_i. The expression that gives ds^2 constitutes a generalization of the

formula that gives the Euclidean distance
$$ds^2 = dx_1^2 + dx_2^2 + ... + dx_n^2.$$

By allowing the g_{ij} to be functions of the coordinates, Riemann provides the possibility that the nature of space can change from point to point.

A curve on a Riemannian manifold is given by the following set of n functions
$$x_1 = x_1(t), \quad x_2 = x_2(t), \quad ..., \quad x_n = x_n(t),$$
while its length between the points that correspond to the values $t = \alpha$ and $t = \beta$ is defined to be:
$$l = \int_\alpha^\beta ds = \int_\alpha^\beta \frac{ds}{dt} dt = \int_\alpha^\beta \sqrt{\sum_{i,j=1}^n g_{ij} \frac{dx_i}{dt} \frac{dx_j}{dt}}\, dt.$$

The shortest curve (geodesic) between two given points that correspond to the values $t = \alpha$ and $t = \beta$ can be defined using the method of the calculus of variations.

If θ is the angle formed by two curves that intersect at some point $(x_1, x_2, ..., x_n)$, where one curve is determined by the directions $\frac{dx_i}{ds}$, $i = 1, 2, ..., n$, and the other by the directions $\frac{dx'_i}{ds'}$, $i = 1, 2, ..., n$ (where the primes indicate that these values belong to the second direction), then θ is defined by the formula
$$\cos \theta = \sum_{i,j=1}^n g_{ij} \frac{dx_i}{ds} \frac{dx'_j}{ds'}.$$

By following the methods that Gauss had introduced for the study of surfaces, using the above definitions, we can develop a metric n-dimensional geometry; all metric properties are determined via g_{ij}.

The next basic notion in Riemannian geometry is the notion of the **curvature of a manifold**. Through this notion Riemann was able to characterize the Euclidean space and in general the spaces in which figures can move without changing their shape and size. The notion of curvature for an n-dimensional manifold constitutes a generalization of the notion of total curvature of surfaces introduced by Gauss, and is defined with the use of quantities which can be determined on the manifold itself, i.e., it is not necessary that the manifold is considered to be a part of another manifold of more dimensions.

Joseph Fourier (1768-1830)

The theory of partial differential equations occupied and still occupies a central place in the science of mathematics. One of the reasons for this prominence is its importance for natural sciences.

From the pure mathematical vantage point, the efforts to solve partial differential equations created the need to develop the theory of functions, the calculus of variations, the expansions into series, the ordinary differential equations, algebra and differential geometry. The first great step during the 19th century was made by Fourier.

Fourier was born in the town Auxerre and died in Paris. He became an orphan at eight and studied at the local military school which was ran by the Bénédictins. Due to his origin (he was the son of a tailor), he was given no rank in the army, but because he excelled in his studies, he was given a position of lecturer of mathematics at the military school.

After the year 1789, when many important events in France started, Fourier joined the popular front and supported with enthusiasm the French Revolution. The state rewarded him by electing him professor at the École Polytechnique. He resigned from this position in order to follow Napoleon at his expedition to Egypt together with the mathematician Gaspar Monge. In 1798 he was appointed the Governor of Lower Egypt. In 1801 he returned to France, where he was appointed Prefect of Grenoble. In 1816 he returned to Paris.

As many other scientists of his time, Fourier occupied himself with the problem of propagation of heat, i.e., with the problem of finding the temperature at the various points in the interior of the earth and finding the variations of temperature as time varies; he also studied other matters of a similar nature. In 1807 he submitted to the Paris Academy of Sciences a basic paper regarding the propagation of heat. His paper was examined by Lagrange, Laplace and Legendre, who rejected it. But the Academy wanted to encourage Fourier in his work and announced a great prize, which was scheduled to be awarded in 1812 on the study of propagation of heat. Fourier resubmitted an improved version of the same paper, which was again forwarded to the above mentioned Academicians for examination. The prize was awarded to Fourier, but his paper was judged to be non-publishable in the Academy's *Mémoires* because of its lack of mathematical rigor.

Fourier did not stop working on the same subject until 1822, when he published one of the classic works in the Science of Mathematics, entitled *Théorie analytique de la chaleur*. This work constitutes the main source of Fourier's ideas on this subject. After two years, Fourier was elected secretary of the Academy and managed to publish in the *Mémoires* the paper of 1811, whose publication in its original form was initially not accepted. What follows is a brief presentation of the subject of propagation of heat.

Temperature is distributed in the interior of a body, which gets either warmer or cooler, and changes at every point of it as time elapses. This means that the temperature T is a function of both space and time. The exact form of this function depends on the shape of the body, on its density, on the specific heat of the material, on the initial distribution of T (i.e., on the distribution of temperature at time $t = 0$), and on the conditions on the surface of the body. The first problem that occupied Fourier in his book mentioned above was the determination of the temperature in a homogeneous and isotropic body as a function of x, y, z and t. Using the principles of physics, he proved that T must satisfy the following partial differential equation, which is called the heat equation in 3-dimensional space:

$$\frac{\partial^2 T}{\partial x^2} + \frac{\partial^2 T}{\partial y^2} + \frac{\partial^2 T}{\partial z^2} = k^2 \frac{\partial T}{\partial t},$$

where k^2 is a constant which depends on the material from which the body is constructed.

Then, Fourier gave a solution to certain particular problems of propagation of heat.

We will treat the solution of the above equation in the case where the body is a cylindrical bar of length l, whose extremities are preserved at temperature $0°$ and whose cylindrical surface is insulated so that we have no loss of heat. Due to the one-dimensional shape of the bar, the above equation takes the form

$$\frac{\partial^2 T}{\partial x^2} = k^2 \frac{\partial T}{\partial t} \qquad (1)$$

and is subject to the boundary conditions
$$T(0,t) = 0 \quad \text{and} \quad T(l,t) = 0 \quad \text{for} \quad t > 0$$
and to the initial condition
$$T(x,0) = f(x) \quad \text{for} \quad 0 < x < l.$$

In order to solve this problem Fourier used the method of separation of variables. He assumed that
$$T(x,t) = \phi(x)\psi(t).$$
By substituting in the differential equation, we have
$$\frac{\phi''(x)}{k^2\phi(x)} = \frac{\psi'(t)}{\psi(t)}.$$
Then, he assumed that each one of these ratios has constant value, say $-\lambda$, hence
$$\phi''(x) + \lambda k^2 \phi(x) = 0 \quad \text{and} \quad \psi'(t) + \lambda \psi(t) = 0.$$
The boundary conditions that we mentioned above imply that
$$\phi(0) = 0 \quad \text{and} \quad \phi(l) = 0.$$
The general solution of $\phi''(x) + \lambda k^2 \phi(x) = 0$ is
$$\phi(x) = b \sin\left(xk\sqrt{\lambda} + c\right).$$
The condition $\phi(0) = 0$ implies that $c = 0$. From the condition $\phi(l) = 0$ it follows that $\sqrt{\lambda}$ must be an integral multiple of $\frac{\pi}{kl}$. Consequently, there exist infinitely many acceptable values λ_ν of λ, namely
$$\lambda_\nu = \left(\frac{\nu\pi}{kl}\right)^2, \text{ where } \nu \text{ is an integer.}$$
Today we call the values λ_ν **eigenvalues** or **characteristic values**.

We know that the general solution of $\psi'(t) + \lambda\psi(t) = 0$ is the exponential function. But since the values of λ are restricted to the λ_ν, while $T(x,t) = \phi(x)\psi(t)$, it follows that
$$T_\nu(x,t) = b_\nu e^{-\left(\frac{\nu^2 \pi^2}{k^2 l^2}\right)t} \sin\frac{\nu\pi x}{l},$$
where b_ν replaces the constant b and $\nu = 1, 2, 3, ...$

But since (1) is a linear equation, it follows that a sum of particular solutions is also a solution. Therefore, we have
$$T(x,t) = \sum_{\nu=1}^{\infty} b_\nu e^{-\left(\frac{\nu^2 \pi^2}{k^2 l^2}\right)t} \sin\frac{\nu\pi x}{l}.$$

In order that the initial condition $T(x,0) = f(x)$ is satisfied for $0 < x < l$, for $t = 0$ we must have
$$f(x) = \sum_{\nu=1}^{\infty} b_\nu \sin\frac{\nu\pi x}{l}.$$

Then Fourier posed the following questions: Can $f(x)$ be represented by a trigonometric series? Is it possible to determine the constants b_ν? Fourier tried to answer these questions. Even though he started to realize that the methods he was using lacked "mathematical rigor", he went on and applied "typically" these methods, thus following the spirit that dominated the 18th century (when the mathematical rigor of the methods applied was not the main concern of the researchers). In order to simplify the calculations, Fourier assumed that $l = \pi$, thus

$$f(x) = \sum_{\nu=1}^{\infty} b_\nu \sin \nu x, \quad \text{where} \quad 0 < x < \pi,$$

and by following bold steps and methods, which were often mathematically not rigorous, he obtained the formula

$$b_\nu = \frac{2}{\pi} \int_0^\pi f(s) \sin \nu s \, ds.$$

In a sense, this result was not new. Clairaut and Euler had expanded certain functions into trigonometric series and had obtained the formulas

$$a_n = \frac{1}{\pi} \int_{-\pi}^{\pi} f(x) \cos nx \, dx, \qquad b_n = \frac{1}{\pi} \int_{-\pi}^{\pi} f(x) \sin nx \, dx. \tag{2}$$

However, Fourier made certain remarkable observations. He observed that every b_ν can be interpreted as representing the area of the part of the plane that lies below the curve $y = \frac{2}{\pi} f(x) \sin \nu x$ and for x between the values 0 and π. But such an area has a meaning even for arbitrary functions, i.e., functions which are not necessarily continuous. This observation led him to the conclusion that "every" function $f(x)$ can be represented by a trigonometric series, namely

$$f(x) = \sum_{\nu=1}^{\infty} b_\nu \sin \nu x \tag{3}$$

for $0 < x < \pi$. This conclusion was rejected by all the mathematical authorities of the 18th century, except Daniel Bernoulli.

When Fourier "came" to the above simple conclusion, he understood that every b_ν can be calculated if we multiply both sides of (3) by $\sin \nu x$ and we integrate the series term by term from 0 to π. In addition, Fourier remarked that this process can be applied also to the expression

$$f(x) = \frac{a_0}{2} + \sum_{\nu=1}^{\infty} a_\nu \cos \nu x.$$

Later, Fourier occupied himself with the problem of representing an arbitrary function by a trigonometric series in the interval $(-\pi, \pi)$. The series $f(x) = \sum_{\nu=1}^{\infty} b_\nu \sin \nu x$ represents an odd function (because $f(-x) = -f(x)$), while the series $f(x) = \frac{a_0}{2} + \sum_{\nu=1}^{\infty} a_\nu \cos \nu x$ represents an even function (because $f(-x) = f(x)$). But every function can be represented as the sum of an odd function $f_o(x)$ and an even function $f_e(x)$, where

$$f_o(x) = \frac{f(x) - f(-x)}{2} \quad \text{and} \quad f_e(x) = \frac{f(x) + f(-x)}{2}.$$

Consequently, an arbitrary function $f(x)$ can be represented in the interval $(-\pi, \pi)$ by the series

$$f(x) = \frac{a_0}{2} + \sum_{\nu=1}^{\infty} (a_\nu \cos \nu x + b_\nu \sin \nu x),$$

where the coefficients can be calculated by multiplying both sides of this equality by $\cos \nu x$ or by $\sin \nu x$ and by integrating from $-\pi$ to π.

The trigonometric series

$$\frac{a_0}{2} + \sum_{n=1}^{\infty} (a_n \cos nx + b_n \sin nx),$$

where the coefficients are given by (2), is called the **Fourier series** of $f(x)$.

At this point the following question naturally arises:

"For which functions $f(x)$ does the Fourier series of $f(x)$ converge to $f(x)$ in the interval $(-\pi, \pi)$?" As mentioned before, Fourier alleged that the Fourier series of $f(x)$ converges to $f(x)$ for any function $f(x)$, but his arguments were not convincing. Even though the Fourier series of $f(x)$ represents $f(x)$ for a very wide class of functions $f(x)$, Fourier's claim that this happens for an arbitrary function is an exaggeration.

Later, in 1829, the German mathematician Gustav Lejeune Dirichlet (1805-1859) proved the following important theorem:

Dirichlet's Theorem. If in the closed interval $[-\pi, \pi]$ the function $f(x)$ is single-valued and bounded, and it has only finitely many discontinuities and finitely many maxima and minima, then the Fourier series of $f(x)$ converges to $f(x)$ at all points where $f(x)$ is continuous, while at every point where $f(x)$ is discontinuous, the series converges to half the sum of the limit on the right and the limit on the left of $f(x)$ at this point of discontinuity.

Even though the conditions stated in this theorem are sufficient and not necessary, they are satisfied by almost all periodic functions that one meets in physics and in mechanics. Fourier series, which were and still are the object of study of many mathematicians, proved to be very useful in the study of many areas of the exact sciences, including acoustics, optics, and electrodynamics. They also play a major role in harmonic analysis and in the solution of differential equations, and help us examine deeper and understand better the notion of a function.

In many applications of Fourier series in problems in mechanics, it is requested to expand a function into a series in an interval of length different from 2π, say equal to $2L$. The interval $(-L, L)$ can be considered as a dilation (or contraction) of the interval $(-\pi, \pi)$ if one multiplies the length of $(-\pi, \pi)$ by $\frac{L}{\pi}$. Thus, if we call z the variable that ranges over the interval $(-\pi, \pi)$, we have $\frac{x}{z} = \frac{L}{\pi}$ or $x = \frac{Lz}{\pi}$. Let us assume that we represent the function $f(x) = f\left(\frac{Lz}{\pi}\right)$ as a function of z by a Fourier series for $-\pi < z < \pi$. By putting $z = \frac{\pi x}{L}$ in this series, we obtain a representation of $f(x)$ for $-L < x < L$. Fourier came up with the idea to assume that L tends to infinity in this last series and hence he obtained one of his most remarkable creations, the **Fourier integral**, which is the limit of the Fourier series when the length of the period of the function tends to infinity.

Lord Kelvin (William Thomson, 1824-1907) said that his entire career in mathematical physics was influenced by Fourier's work on heat, while Clerk Maxwell (1831-1879) called Fourier's writing a "great mathematical poem".

Constantin S. Carathéodory (September 13, 1873–February 2, 1950)[2]

Carathéodory, the son of Stephen Carathéodory and Despina Petrokokkinou, was born in Berlin on September 13, 1873. His father was then a member of the diplomatic delegation in the conference of Berlin for the determination of the frontiers of the Ottoman Empire, a delegate of Turkey.

Carathéodory was the offspring of a noble and historic Greek family from Andrianoupolis. His grandfather was a nephew of the Patriarch Cyril the 6th who as it is known, was martyrly put to death by the Turks in 1821. His father and his grandfather had high administrative and diplomatic positions in the Turkish state.

Carathéodory's mother died when he was about six years old. Thus, his grandmother Euthalia A. Petrokokkinou undertook the upbringing of Carathéodory and his sister Julia until her death. As Carathéodory wrote in his autobiography (which is included in his *Collected Works* which were published by the Bavarian Academy), he learned simultaneously both the Greek and the French language. Moreover, before even going to school, he started learning the German language.

His elementary and secondary education was carried out mainly in Belgian schools, because during that time his father served as the Turkish ambassador in Brussels. Then, as Carathéodory himself writes in his autobiography, "my great love for mathematics began to come into sight in the course of Geometry".

On October of 1891 he entered the Military School of Brussels, from which he graduated in the fall of 1895. Starting in 1898, he worked as a mechanic in Egypt, where the construction of the great works in Aswan and Asyut was carried out. But his great love for mathematics brought him back in Europe, where he enrolled in the spring of 1900 in the University of Berlin. After two years he continued his studies at the University of Göttingen, whose department of mathematics was considered to be one of the most important research centers of the world at that time. During the period that Carathéodory was studying at Göttingen, Felix Klein, David Hilbert and Hermann Minkowski were giving lustre to the department of mathematics of that university. Carathéodory's great career began in Göttingen. I quote certain characteristic dates.

1908. He becomes a salaried reader at the University of Bonn.
1909. He becomes a regular professor at the University of Hannover.
1910. He becomes a regular professor at the Technical University of Breslau.
1913. He succeeds Felix Klein at the University of Göttingen.
1918. He is called back to the University of Berlin.
1919. He becomes a regular member of the Prussian Academy of Berlin.
1920. He becomes a regular professor at the University of Athens. (He did not assume duties.)
1922. He becomes a regular professor at the University of Athens and at the National Technical University of Athens.
1924. He resigns from his positions in Greece and accepts a professorship at the University of Munich, where he succeeds Lindemann.
1925. He becomes a member of the Academy of Munich.
1926. He becomes a member of the Academy of Bologna.

---

[2]The author would like to inform the reader that because Carathéodory is of Greek origin, the author felt obliged to present a more extensive biography of this important Greek Mathematician, and to include events from the contemporary history of Greece.

1926. Immediately after its foundation, the Academy of Athens elects him regular member during the meeting of the 26th of November of 1926, even though he resided permanently abroad and occupied a public position at a foreign country.
1927. He was awarded the German title Geheimrat (Privy Councillor).
1929. He becomes a member of the Society dei Lincei in Roma.
1930. He becomes emeritus professor at the University of Athens and at the National Technical University of Athens.
1934. He becomes a member of the Papal Academy of the Vatican.

The year 1920 was very important for Carathéodory's career. His research activity was extraordinary. His works were internationally recognized and he was considered to be one of the greatest mathematicians of the world. During that time Eleutherios Venizelos invited Carathéodory in Greece and assigned to him the organization of the University of Smyrna, which was about to be founded. (Venizelos' aim was also the foundation of a third university in Thessaloniki, as he had announced later in his talk at the Parliament on December of 1929.) Carathéodory went to Smyrna and began the groundwork for the foundation of the university, which would have occupied a place of honor in Greek science, if the destruction of the Greeks in Asia Minor had not followed.

Later, in 1930, when Carathéodory was invited again by the Greek Government to serve as a governmental trustee of the Universities of Athens and Thessaloniki, he submitted accordingly a 39-page long document regarding the organization of the University of Athens. We quote a characteristic extract from this document:

"In Greece, the viewpoint prevails that a legislative intervention, that is a simple modification of the organization of the University, can improve the functioning of the institution. However, in contrast to this, the experience during the last twenty years of almost all foreign Universities and of the University of Athens, teaches that these measures cannot be effective and will probably bring about the opposite of the requested result. The University cannot be compared with a system of telephones, which is easily replaced by another more advanced one. On the contrary, it is a very delicate living organism which reacts to every external intervention sometimes in an unusual way."

A quite detailed treatise regarding Carathéodory's life, work and actions in general was published in 1973 by the Greek Mathematical Society (GMS) and written by Michael Kasiouras (who was a gymnasium principal at that time) on the occasion of the international symposium organized by the GMS for the 100 years from Carathéodory's birth. This book contains 21 pages of Carathéodory's autobiographical notes which were taken from the 5th volume of his *Collected Works*.

There exist, of course, many other biographical sources about Carathéodory, among which are his *Collected Works* consisting of 5 bulky volumes published by the Bavarian Academy, each of which is prefaced by Heinrich Tietze, who was a member of the Academy of Munich (1st, 1954 - 2nd, 1955 - 3rd, 1955 - 4th, 1956 - 5th, 1956).

In what follows we will restrict ourselves to a short account of Carathéodory's scientific work. The areas in which he worked are the following: Calculus of variations - Conformal mappings - Functions of one complex variable - Functions of real variables (measure and integration theory) - Functions of more than one complex variables - Sequences and convergence of functions of complex variables - Differential geometry - Mathematical physics - Optics.

In the calculus of variations he published fundamental theorems regarding discontinuous solutions. In his doctoral dissertation (1904) he treats the theory of discontinuous solutions in the calculus of variations. In a supplement of his dissertation he generalizes an idea of Bernoulli and uses it in the solution of the problem of the brachistochrone.

Immediately after the publication of his doctoral dissertation he began research on the theory of functions of complex variables. He published classical theorems in the theory of conformal mapping, in which he gave the general solution of the problem of mapping a simply connected domain onto a disc. By combining methods from analysis and topology, he founded a new chapter on the geometry of analytic functions.

In 1905 he submitted to the Paris Academy an important announcement regarding Picard's theorem, where he opens new horizons for further generalizations. Similar generalizations were published by him from time to time during his entire scientific career.

In 1909 he published in *Math. Annalen* a paper entitled *Investigation of the bases of Thermodynamics*, which impressed among others the distinguished physicists Max Planck and Sommerfeld.

During the academic year 1912-1913 he began research in the area of conformal mapping. This research became more known in English speaking countries in 1928, when he was invited to teach at Harvard University. The results of this research were published in 1932 by Cambridge University Press under the title *Conformal Representation*.

In 1918 he published his classic 700 pages long book entitled *On real functions*, which contains also his university lectures. The work was first released in Germany, and after the end of the First World War it was distributed all around the world, while in 1927 a second edition of it followed.

In 1922 he published in *Ber. Pr. Akad. Wiss.* a paper in physics entitled *Über die Bestimmung der Energie und der abs. Temperatur mit Hilfe von reversiblen Prozessen*, which supplements the earlier paper of his regarding the foundation of thermodynamics. In 1924 he published another paper in physics, which was of an axiomatic nature.

In 1925 he published the treatise *Variationsrechnung*, which was included in the team-work that was published by Frank and Von Mises under the title *The differential and integral equations of mechanics and physics*.

In 1926 he published a very important study entitled *Schwarz's lemma in analytic functions of two complex variables*, and the same year he published two more papers in physics (Optics).

In 1935, a decade after the publication of his book on the calculus of variations, he published another book entitled *Calculus of variations and partial differential equations of the first order* (400 pages long), where he treats subjects related to problems in mechanics, geometric optics, and differential geometry, and where he uses partial differential equations of the first order.

In 1937 he published a paper on geometric optics.

In 1940 he published two papers, from which the first is an investigation of the parabolic mirror telescope published in *Naturf. Ges. Zürich*, while the second concerns Schmidt's telescope and was published in *Hamb. Math. Einzelschr.*

The third edition of his monumental writing *On real functions* was published updated, containing all the achievements in that area of mathematics until then in three volumes. The first volume entitled *Real functions, Volume 1. Numbers, Pointsets, Functions* was circulated in 1939. The second volume was printed and ready for circulation in 1943, but during that year the allies bombed Leipsig and all copies of it were burned. Carathéodory elaborated the second volume from the beginning and had almost finished it, when death cut the thread of his life on February 2, 1950 in Munich, where he was also buried.

During the end of the 19th century, it was obvious that in addition to the content of mathematics, many other things had changed since 1800. The number of mathematical journals as well as the number of higher educational institutions had both increased. The interchange of ideas between the mathematicians of various countries was aided by the foundation of national mathematical societies and by the organization of international mathematical congresses.

The London Mathematical Society (1865) and the Société Mathématique de France (1872) were the first mathematical societies, to be founded. In 1880, the foundation of the Edimburgh Mathematical Society in Scotland, the foundation of the Circolo Mathematico Palermo in Italy, and the foundation of the New York Mathematical Society followed. The latter was soon renamed American Mathematical Society, while the Deutsche Mathematiker Vereinigung was founded in 1890.

The Greek Mathematical Society (GMS) was founded on the 18th of March in 1918 by 93 mathematicians, among them were the late professors D. Aeginitis, B. Aeginitis, I. Hatzidakis, N. Hatzidakis, G. Remoundos, K. Maltezos, P. Zervos, N. Sakellariou, S. Sarantopoulos, T. Varopoulos, A. Economou, S. Plakidis, and others. The first board of directors was the following:

President: Nikolaos Hatzidakis (then professor at the University of Athens)
Vice-presidents: George Remoundos and Panagiotis Zervos (then professors at the University of Athens)
General secretary: Nilos Sakellariou (then professor at the University of Athens)
Special secretary: Athanasios Arvanitis (then professor in secondary education)
Treasurer: George Antonopoulos (then professor in secondary education)
Members: John Vainopoulos (then educational adviser), Konstantinos Lambiris (then director of a commercial school), Konstantinos Maltezos (then professor at the University of Athens), Dionisios Rigopoulos (then director of a commercial school), John Romanos (then professor in secondary education), and Stephen Stamatis (then professor in secondary education).

For more details regarding the history and the activities of the GMS the reader is referred to the volume of 1988 published by the GMS on the occasion of its 70th anniversary.

The first International Congress of Mathematicians took place in Chicago in 1893. In 1897 followed the second international congress of mathematicians that took place in Zurich and which was the first of a series of congresses that take place every four years and were only interrupted by the two world wars. The congresses that followed the second world war are the following: 1950 Cambridge

(Mass., USA), 1954 Amsterdam, 1958 Edinburgh, 1962 Stockholm, 1966 Moscow, 1970 Nice, 1974 Vancouver, 1978 Helsinki, 1982(1983) Warsaw, 1986 Berkeley, 1990 Kyoto, 1994 Zurich, 1998 Berlin, 2002 Beijing.

After the death of Gauss in 1855, it was thought that there could not exist another mathematician, who would be equally familiar and productive in all branches of mathematics, no matter whether they are theoretical or applied. Poincaré was the mathematician whose scientific activity contradicted the above viewpoint.

Henri Poincaré (1854-1912)

Poincaré was born in Nancy and must not be confused with his cousin Raymond Poincaré, who was the President of the French Republic during The First World War.

Poincaré was a professor at Sorbonne. His publications are almost as many as those of Euler and Cauchy, and cover a wide spectrum of mathematics and mathematical physics. His research papers in physics (with which we will not occupy ourselves here) refer to "capillary" attraction, elasticity, optics, electromagnetic theory, theory of relativity, and, in particular, celestial mechanics.

Poincaré was gifted with a penetrating intuition that enabled him to isolate the essential character of every problem that he studied. He examined in detail every problem and insisted on its qualitative study.

His doctoral dissertation on differential equations (existence theorems) led him to the discovery of the theory of "**automorphic functions**".

A function $f(z)$ of the complex variable z, which is analytic in some domain D (the only singular points of $f(z)$ might be poles), is called **automorphic** if it remains invariant by the action of a discontinuous subgroup of the group of transformations of the form

$$z' = \frac{az+b}{cz+d},$$

where a, b, c, d are real or complex numbers such that $ad-bc = 1$. A subgroup of this group of transformations is called **discrete** (the terminology is due to Poincaré), when all the images of an arbitrary point of the plane via transformations that belong to this subgroup which lie in a closed and bounded domain of the plane, are finitely many.

Automorphic functions constitute generalizations of cyclic, hyperbolic, elliptic and other functions of elementary mathematical analysis. Recall that the function $\sin z$ remains invariant if instead of z we put $z+2m\pi$, where m is an arbitrary integer. We can also say that $\sin z$ remains invariant if we apply on z any transformation from the group of transformations $z' = z + 2m\pi$. The hyperbolic function $\sinh z$ remains invariant if we apply on z any transformation from the group of transformations $z' = z + 2m\pi i$. An elliptic function $E(z)$ remains invariant if we apply on z any transformation from the group of transformations $z' = z + m\omega + m'\omega'$, where ω and ω' are the two periods of $E(z)$, i.e., $E(z+\omega) = E(z)$ and $E(z+\omega') = E(z)$. Poincaré remarked that "periodicity" is a particular case of a more general property, namely: the value of certain functions remains invariant if the variable z is replaced by any other variable z', where $z' = \frac{az+b}{cz+d}$. When z is replaced by z', then for any countably infinite set of values of a, b, c, d, there exist functions $F(z)$ such that

$$F\left(\frac{az+b}{cz+d}\right) = F(z).$$

If a_1, b_1, c_1, d_1 and a_2, b_2, c_2, d_2 are two quadruples of values of a, b, c, d and first we replace z by $\frac{a_1 z + b_1}{c_1 z + d_1}$ and, then we replace z in this last expression by $\frac{a_2 z + b_2}{c_2 z + d_2}$, then we obtain an expression of the form $\frac{Az+B}{Cz+D}$ and we have

$$F\left(\frac{a_1 z + b_1}{c_1 z + d_1}\right) = F(z), \quad F\left(\frac{a_2 z + b_2}{c_2 z + d_2}\right) = F(z) \quad \text{and} \quad F\left(\frac{Az+B}{Cz+D}\right) = F(z).$$

Besides, the set of transformations $z \mapsto \frac{az+b}{cz+d}$ (where the symbol $\mapsto$ must be read "maps to"), which leave the value of $F(z)$ invariant constitute (i.e., they have the structure of) a group. The result of the application of two successive transformations from this set, say

$$z \mapsto \frac{a_1 z + b_1}{c_1 z + d_1} \quad \text{and} \quad z \mapsto \frac{a_2 z + b_2}{c_2 z + d_2},$$

also belongs to this set. Moreover, this set contains an identity transformation, namely $z \mapsto z$ (where $a = 1, b = 0, c = 0, d = 1$), and every transformation in this set has a unique "inverse", which means that to every transformation that belongs to this set, there corresponds a unique transformation that also belongs to this set such that if we compose the two transformations we obtain the identity transformation. In other words, $F(z)$ is a function whose value remains invariant under the action of a countably infinite group, each of the elements of which is a fractional linear transformation. Poincaré called these functions **automorphic**.

We make two remarks in order to perceive the value of this marvellous discovery of Poincaré. First that elliptic functions constitute a particular case of his theory, and second that he proved the following two propositions: Two automorphic functions, which remain invariant under the action of the same group of linear fractional transformations, are related to one another through an algebraic equation. Conversely, the coordinates of an arbitrary point of any algebraic curve can be expressed via automorphic functions of a single variable.

We remind the reader that a curve is called algebraic when its equation is of the form $P(x, y) = 0$, where P is a polynomial in x and y. Such a curve is the circle with center the origin and radius a, i.e., $x^2 + y^2 = a^2$. It follows from the above proposition of Poincaré that in this equation of the circle it is possible to express x and y as automorphic functions of a single variable (say t). Indeed, if $x = a \cos t$ and $y = a \sin t$, then by eliminating t we obtain $x^2 + y^2 = a^2$, where the functions $\sin t$ and $\cos t$ are particular cases of elliptic functions, which are particular cases of automorphic functions.

Next we will examine another research activity of Poincaré regarding mathematical astronomy. His first great success (1889) in this direction concerns the so-called "ν-body problem". For $\nu = 2$ the problem was completely solved by Newton. We will refer to the famous 3-body problem ($\nu = 3$) a little later. Certain conditions regarding the case $\nu = 3$ are also applied in the case that ν is larger than 3.

Newton's law of gravitation asserts that two particles with masses m and M respectively, which lie at a distance D from one another, are mutually attracted by a force which is proportional to the quantity $\frac{m \times M}{D^2}$. Let us consider a set of material particles which are arbitrarily distributed in space. It is assumed that the masses, the initial velocities and the distances of the particles from one another are known at some moment. If these masses are mutually attracted according to the above law of Newton, then "which will be the relative positions and the velocities of these particles after some time interval?" In the problems that occupy astronomers, the stars of a nebula or a galaxy or a nebula of galaxies can be considered as material particles which are mutually attracted according to Newton's law.

What is requested in the "ν-body problem" is to find what will be the picture of the firmament after a year or after one billion years from today. Of course, in order to answer this question, we must have adequate observational data that give us the picture of the firmament as it is today. It is obvious that the problem becomes much more complicated if we take into account the existing radiation, because the masses of the stars do not remain constant as millions of years go by. But a complete solution of the ν-body problem, in its Newtonian form, which can be calculated, would give satisfactory results for the needs of mankind, because the human species will probably become extinct before the above mentioned radiation causes significant inaccuracies to the results obtained.

The above problem was more or less the problem that was announced for a prize by the king of Sweden, Oscar II, in 1887.

Poincaré did not solve the problem. But in 1889, the examining committee, consisting of Weierstrass, Hermite and Mittag-Leffler, awarded him the prize, because the paper that he submitted contained a general consideration and study of the differential equations and the dynamics of the problem, as well as a study of the 3-body problem which is usually considered to be the most important case of the ν-body problem, because the sun, the earth, and the moon is an obvious instance of the 3-body problem.

In his recommendation, which was addressed to Mittag-Leffler, Weierstrass wrote the following: "You may say to the candidate that, even though his study cannot be considered as a complete solution of the suggested problem, it is so important that its publication will mark the beginning of a new era in the history of celestial mechanics."

In order not to appear inferior to the king of Sweden, the French government named Poincaré "Knight of the Legion of Honor". Of course, this recognition of this young mathematical genius was much less expensive than the prize of 2500 kronas and the gold medal that the king of Sweden gave to Poincaré.

We will not discuss any further Poincaré's biography and work.

The most original work of Poincaré in mathematics is contained in his three-volume writing *Les méthodes nouvelles de la méchanique céleste* (1892, 1893, 1899). This writing was followed by another three-volume writing (1905-1910) of a more direct and practical interest entitled *Leçons de mécanique céleste*, while a little later he published his works *Sur les figures d' équilibre d' une masse fluide* and *Sur les hypothéses cosmogoniques*.

Andrei Nikolaevich Kolmogorov (1903-1987)

Kolmogorov was born in Tambov in Russia on April 25, 1903. In 1920 he enrolled as a first-year student at the University of Moscow. He presented his first scientific paper entitled *An example of a Fourier-Lebesgue series which diverges almost everywhere* at the Conference of the Moscow Mathematical Society in 1922. This paper revealed that Kolmogorov was an exceptional mathematician.

He obtained his degree in 1925 and in 1928 he was elected professor at the University of Moscow, which was the center of his scientific activities until his death. He was a member of the Academy of Sciences of the Soviet Union.

In 1924 Kolmogorov turned his interest towards Probability Theory, where his contributions and his influence were immense. He founded Probability Theory on axioms based on Measure Theory and the Theory of Functions of a real variable. His classic monograph of 1933, entitled *Foundations of the theory of probability*, constitutes the beginning of a new era in the development of Probability Theory as a branch of mathematics, and laid the foundations for the creation of the theory of random processes.

The contemporary theory that bears the name "Markov random processes" has its origins in Kolmogorov's paper of 1931, entitled *Analytic methods in probability theory*. This new branch of Mathematics that promises many applications is already used by Physicists, Biologists, Chemists, and Mechanics, and has become one of the most powerful existing research tools in mathematics.

During the decade of 1930, Kolmogorov occupied himself with various subjects from Projective Geometry, Mathematical Statistics, Approximation Theory, Mathematical Biology. He was also interested in the Philosophy and the History of Mathematics. During that period Kolmogorov and the American topologist Alexander formulated simultaneously and independently the notion of "cohomology" and founded the theory of "cohomology operations".

Around the end of the decade of 1930, Kolmogorov turned his interest towards the mechanics of "turbulence". The word turbulence refers to the irregular flux of a fluid and corresponds to a concrete mathematical notion.

The works of Kolmogorov and his students constitute the first attempt to make an exact and mathematical study of the theory of turbulence by applying measure theory in function spaces. In his main work of 1941 regarding the mechanics of turbulence one can see Kolmogorov's penetrating physical intuition that led him to the discovery of fundamental qualitative relations in this subject.

After the Second World War Kolmogorov continued his research in the Theory of Functions of a real variable, and in Mathematical Logic as well as in the Foundations of Mathematics. In particular, in the theory of real functions, Kolmogorov and his student Arnol'd solved the so-called Hilbert's 13th Problem. I will discuss this subject in somewhat more detail. In the International Congress of Mathematicians that took place in Paris in 1900, the German mathematician David Hilbert, who was one of the greatest mathematicians of the first half of the 20th century, suggested 23 problems, which he characterized as the aims of mathematical science in the 20th century. Some of these problems, among which is the famous Riemann Hypothesis regarding the roots of the ζ-function, still remain unsolved at the end of the 20th century.

Hilbert's 13th problem is the following: Show the impossibility of the solution of the general algebraic equation of the seventh degree by compositions of continuous functions of two variables.

This problem was solved negatively by Kolmogorov and Arnol'd! During my studies in France, I was very fortunate to attend a talk of Kolmogorov at Sorbonne, where he presented the solution of Hilbert's 13th problem at an audience consisting of French and foreign mathematicians.

During the period that Kolmogorov and Arnol'd solved Hilbert's 13th Problem, Kolmogorov broadened the scope of his research by carrying out research also in Classical Mechanics, Theory of Functions, Information Theory, and Theory of Algorithms.

During the decade of 1950, Kolmogorov worked in the area of Dynamical Systems and formulated a general theory of Hamiltonian Systems. The method that he used was later improved by his student Arnol'd and by Moser. Today it is known as the Kolmogorov-Arnol'd-Moser theory, or KAM-theory, and has many applications.

In particular, the theory of dynamical systems began with the study of the motion of planets and dates back to antiquity. Later, with the application of methods from mathematical analysis, we have the foundation of the so-called "Analytical Dynamics". Then the famous 3-body problem came up and occupied many mathematicians from the end of the 18th century until the end of the 19th century, when H. Bruns and H. Poincaré proved that the general solutions of the 3-body problem cannot be obtained through "integration", i.e., through the performance of finitely many changes of variables and integrations on elementary functions. This discovery caused a crisis in the theory of Analytical Dynamics, which was resolved by Poincaré himself via methods that we will not discuss here. Then going back to the applications of the KAM-theory, we emphasize that it gave remarkable and very interesting results regarding the existence of quasi-periodic solutions of the ν-body problem, which resolved the long-standing problem of the "stability of the solar system".

Kolmogorov's ideas on Information Theory were applied to Dynamical Systems and paved the way for the further development of the Ergodic Theory of Dynamical Systems.

During the decade of 1950, Kolmogorov provided new algorithmic bases for Information Theory, thus creating the branch called "Algorithmic Information Theory".

In parallel with his wide interests in research, Kolmogorov had a feeling of responsibility about the future of scientific research, and hence he occupied himself widely with activities of an educational nature. Many of his students distinguished themselves in various branches of science, as are Probability Theory, Logic, the Theory of Functions, Mechanics, the Physics of the Atmosphere, Oceanology, and Cybernetics.

During the last 25 years of his life, Kolmogorov dedicated a great part of his efforts to the Mathematics of Secondary Education. During the decade of 1960, he served as the president of the committee of the Academy of Sciences of the Soviet Union on the Education of Mathematics. From that position he played a central role in the reform of the teaching of mathematics that took place in his country. He wrote school textbooks and gave many lectures on Mathematics, Logic, Arts, and Literature. He served as the director of the Institute of Mathematical Research (1933-1939) and of the Institute of Mathematics and Mechanics (1951-1953) of the University of Moscow. He held chairs in Probability, Mathematical Statistics and Mathematical Logic, and he was the director of the Laboratory of Statistical Methods (1966-1976) of the University of Moscow.

In conclusion, Kolmogorov was a great mathematical figure of the 20th century.

Additional Reading

(1) F. Klein, *Vorlesungen über die Entwicklung der Mathematik im 19. Jahrhundert*, Springer, I, 1926; II, 1927 (*Chelsea*, 1956).

(2) N. Bourbaki, *Les éléments d' histoire des mathématiques*, Hermann, second edition, 1969.

(3) *Enzyklopädie der mathematischen Wissenschaften mit Einschluss ihrer Anwendungen*, Teubner, 1898-1934.

CHAPTER 2

REVIVAL OF SYNTHETIC GEOMETRY

After the discovery of analytic geometry by Descartes and Fermat, the methods that dominated geometry for more than 100 years were analytic and algebraic, as opposed to the use of pure geometric methods, the so-called **synthetic** methods.

During the beginning of the 19th century, certain famous mathematicians of that time thought that it was unfair and unreasonable to neglect synthetic geometry, and tried to revive and extend it.

For example, although Lagrange studied exclusively analysis, he stated the following: "Even though analysis has in general the advantage over the old geometric methods, which we incorrectly call "synthetic", there exist problems where synthetic methods have the advantage, either because they are naturally clearer or because they provide more elegant and easier solutions. Moreover, there exist problems whose algebraic analysis is insufficient and hence the use of synthetic methods is necessary." Lagrange gave as an example the problem of computing the attraction that an ellipsoid of revolution exerts on a unit mass that lies either in the interior or on the surface of the ellipsoid. This problem was solved by Maclaurin using synthetic methods.

Other opponents of the analytic methods were M. Chasles (1793-1880) and Carnot who wished that "geometry will be freed from the hieroglyphic symbols of analysis"

The opposition against the application of analytic methods in geometry was not only based on personal preferences and tastes. Indeed, the question was raised in what sense analytic geometry was a geometry, since the essence and the effectiveness of this method was based on algebra, which hides the underlying geometric meanings. Furthermore, Chasles complained that the steps made by analysis prevent us from realizing what has been achieved; the relation between the point of departure and the final result is not clear. Chasles posed the following question: "In a philosophic and basic scientific study, is it enough to know that something is true if we don't know why it is true, and what is the position that this truth occupies in the series of truths where it belongs?" In contrast to that, pure geometric methods allow us to construct intuitively obvious proofs and conclusions.

The supporters of analytic methods refuted the above arguments by saying that synthetic methods lack elegance.

As a result of the above dispute, the pure geometers reaffirmed their role in mathematics. An example of the intense rivalry between the supporters of analytic geometry and the supporters of synthetic geometry is provided by the geometer Steiner who threatened to stop publishing his works in Crelle's journal, entitled *Journal für Mathematik*, if Crelle continued to publish Plücker's works, which were based on analytic geometry.

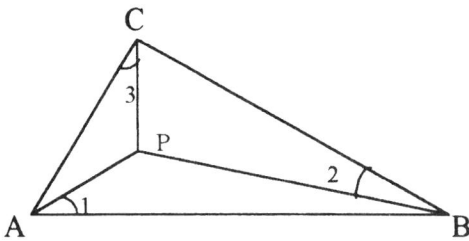

Figure 2.1

G. Monge played an important role in the revival of synthetic Geometry. In his well-known lectures at the École Polytechnique during 1795–1809 he tried to bring synthetic geometry into the limelight of mathematics as a method of interpretation of analytic results. Furthermore, he insisted that the two ways of thinking must coexist. His enthusiasm about analytic and differential geometry inspired his students Charles Dupin, François-Joseph Servois, Charles-Julien Brianchon, Jean-Baptiste Biot (1774-1862), Lazare-Nicolas-Marguerite Carnot, and Jean-Victor Poncelet, to make many contributions to the effort to revive pure geometry.

Monge's contributions to pure geometry are included in his writing *Traité de géométrie descriptive* (1799). Descriptive geometry studies how one can project orthogonally a 3-dimensional object on two planes (a horizontal one and a vertical one) in such a way that the mathematical properties of the object can be deduced from this representation. This method, which is useful in architecture and in the designing of fortifications, is the first method of studying a 3-dimensional figure by analysing two of its 2-dimensional projections. However, the methods of descriptive geometry did not lead to a further development of geometry. Most of the geometers, who responded to the stimulus created by Monge's work, occupied themselves with Projective Geometry.

Next, we will mention certain new results regarding synthetic Euclidean geometry. There exist hundreds of new theorems from which we will mention only a few as examples.

In every triangle ABC we distinguish the following nine points: the midpoints of the sides, points of intersection of the perpendicular from a vertex to the opposite side, and the midpoints of the line segments that connect the vertices of the triangle with the latter three points. It can be shown that these nine points lie on the circumference of a circle which is called the "nine-point circle". This theorem is due to Gergonne and Poncelet, although it is usually reported that it is due to Karl Wilhelm Feuerbach (1800-1834) who gave the proof of this theorem in his book *Properties of certain particular points of a linear triangle*, 1822. In this book, Feuerbach also gave another result regarding the nine-point circle: We call "escribed circle" a circle which is tangent to one side of the triangle and tangent to the extensions of the other two sides of the triangle. The nine-point circle is tangent to the inscribed circle and (tangent to) the three escribed circles of the triangle.

In the book *On certain properties of plane linear triangles* (1816), Crelle proved that there exists a point P in the interior of an arbitrary triangle such that the angles formed by the sides of the triangle and the straight lines connecting P with

2. REVIVAL OF SYNTHETIC GEOMETRY

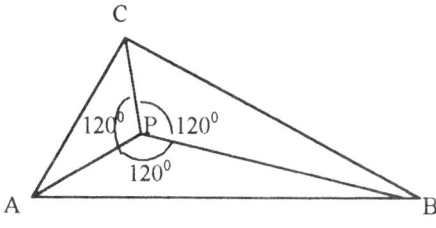

FIGURE 2.2

the vertices of the triangle are equal, i.e., $\hat{1} = \hat{2} = \hat{3}$ (Figure 2.1). In addition, there exists a point P' different from P such that $P'\hat{A}C = P'\hat{C}B = P'\hat{B}A$.

In 1822 Germinal Dandelin (1794-1847) proved the following: "If two spheres are inscribed in a circular cone in such a way that they are tangent to a plane which intersects the cone at a conic section, then the points of contact between the spheres and the plane are the foci of the conic section, and the intersections of the plane with the planes of the circles at which the spheres are tangent to the cone are the directrices of the conic section."

During that period several problems for finding maxima and minima were solved using purely geometric methods, i.e., without using the calculus of variations. For example, Steiner proved the so-called "**isoperimetric theorem**": "From all plane figures having a given perimeter, the circle surrounds the maximum area." Steiner gave more than one proof of this theorem. But, unfortunately, in his proofs he made the a priori assumption of the existence of a curve surrounding the maximum area. Dirichlet tried to convince Steiner to complete his proof, but Steiner insisted that the existence of a curve surrounding the maximum area was obvious. The proof of the existence of the maximizing curve was given in 1870 by Weierstrass using the calculus of variations. Later Constantin Carathéodory in collaboration with Study completed Steiner's result without using the calculus of variations.

Hermann Amandus Schwarz, who made contributions in partial differential equations and in analysis, gave a rigorous mathematical proof of the isoperimetric theorem in 3-dimensional space.

Steiner also proved that among all triangles having a given perimeter, the equilateral triangle surrounds the maximum area. Another result of Steiner is the following: "If A, B, C are three given points and if every angle of the triangle ABC is less than 120°, then the point P for which $PA + PB + PC$ is minimum is such that every angle at P is equal to 120° (Figure 2.2). But if one of the angles of the triangle, say A, is greater or equal to 120°, then P coincides with A." This result was actually proved earlier by Cavalieri but this was not known to Steiner, who also generalized this result for n points.

Schwarz solved the following problem: "Assume that an acute triangle is given. Consider all the triangles, each of which has its three vertices on the three sides of the given triangle. Among these triangles find the one with the minimum perimeter." Schwarz proved that the vertices of the triangle having the minimum perimeter are the point of intersection of the perpendiculars from the vertices to the opposite sides.

In 1899 Frank Morley, a professor of mathematics at John Hopkins University, proved a new theorem in Euclidean geometry: "If from every vertex of an arbitrary

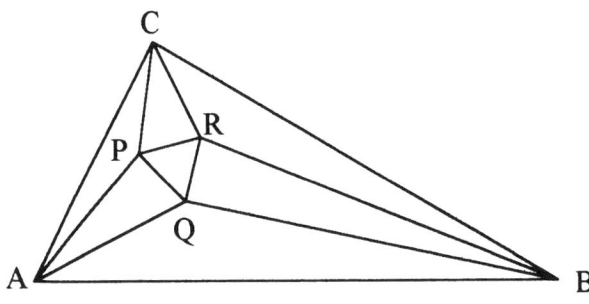

FIGURE 2.3

triangle we "draw" the trisectors of the corresponding angle, then the trisectors that are closer to the same side of the triangle intersect at the vertices of an equilateral triangle" (Figure 2.3).

The originality of this theorem lies in the consideration of trisectors of angles. Until the middle of the 19th century, no mathematician analysed trisectors of angles, because only the elements and the figures that were constructible using a ruler and a compass were "legitimately" accepted.

Later, it was understood that constructibility was not the only way to guarantee the existence of an element or a figure. This becomes particularly clear in the work for the foundation of Euclidean geometry.

Regarding constructions using the ruler and the compass, we note that in 1822 Poncelet proved that (apart from the construction of arcs of circles) all the constructions, which are carried out using a ruler and a compass only, can also be carried out using a ruler only, provided that a fixed circle and its center are given. Steiner proved again the same result in a more elegant way in his book *The geometric constructions are carried out using a straight line and a fixed circle*.

In parallel with the above works, where theorems were proved using synthetic geometry, mathematicians continued to use analytic methods to prove theorems of Euclidean geometry. For example, Gergonne gave analytic proofs of many geometric theorems (they were published in the journal *Annales de Mathématiques*).

CHAPTER 3

THE SYSTEM OF REAL NUMBERS

In Chapter 1 of Part II (in the biography of Weierstrass) we mentioned the reasons which made it necessary to give a logical and rigorous foundation of the system of real numbers. The fact that this need for the foundation of the system of real numbers was not perceived earlier, constitutes one of the most surprising events in the history of mathematics.

First of all, we present the symbols of the various number systems.

N the natural numbers 1, 2, ...
Z the integers (Z from the word "Zahlen")
Q the rational numbers (Q from the word "quotients")
R the real numbers (R from the word "reals")
C the complex numbers $a + ib$ (C from the word "complex")
H the quaternions (H from from the word "Hamilton", who introduced the quaternions)

These symbols are universally accepted today. They were first introduced by N. Bourbaki (mentioned in Part I), when he began some sixty years ago to write the series of books entitled *Éléments de Mathématique* following the contemporary spirit of axiomatizing mathematics.

The order in which the above number systems are written is the one that is considered to be correct from the logical point of view. But this order differs from the order in which these number systems appeared historically, namely:

$$\mathbf{N} \to \mathbf{Q}^+ \to \mathbf{R}^+ \to \mathbf{R} \to \mathbf{C} \to \mathbf{H},$$

where $\mathbf{Q}^+$ is the set of positive rationals (i.e., the set of fractions of positive integers) and $\mathbf{R}^+$ is the set of positive reals. $\mathbf{Q}^+$ was considered to be the set of all numbers until (as we saw in Part I) the Pythagoreans discovered that the diagonal of a square is not a rational multiple of its side.

The positive rationals and the positive reals were studied before the introduction of negative numbers.

Every set that belongs to the first of the last two tables given above and follows the set **N**, constitutes an extension of the set preceding it and, apart from the set **H**, it was invented by mathematicians in order to solve equations which could not be solved in another way. Thus, the systems **Z**, **Q**, **R**, **C** are necessary for the solution of equations such as $x + 2 = 0$, $2x = 3$, $x^2 = 2$, $x^2 + 1 = 0$ respectively.

Mathematicians often "construct" a set of the above Table from the set that immediately precedes it. As an example we will mention the construction of the rationals that makes use of the integers. This construction consists of equating the rationals with certain classes of pairs of integers. For example, the fraction 2/3 is equated with the set $\{(2,3), (4,6), ..., (-2,-3), ...\}$. But since the above table of symbols begins with the natural numbers, one poses the following question: How is the system **N** defined? According to Kronecker: "God created the natural numbers. The remaining numbers are created by humans."

But, as we will see below, people also tried to construct the natural numbers.

Let it be noted that there exist other kinds of numbers which are not included in the above table, such as the transfinite numbers and the infinitesimals.

But what are the natural numbers? In particular what is number 2 or II? We believe that it is not something perishable or variable, but what is it? There exist more than one answer to this question. We will mention only some of them.

For the "Platonics" the numbers are necessarily abstract existing objects. The number 2 is the Platonic "idea" through which objects have the property of "duality".

For the "Logicians" the numbers are objects which can be defined through logic. For example, for Bertrand Russell, 2 is the class of all unordered pairs of objects.

For the "Formalists" 2 is a class of expressions which we use on the basis of certain rules. Formalists often do not define numbers but give axioms which characterize them. Such an example is the way of constructing the system of natural numbers according to Peano, which will be presented later.

"Intuitionists" consider the natural numbers as mental objects which exist only when people think about them.

We remark that every school has a different point of view on this subject. However we think that it would be wiser not to pose the following question: Which one of the above (and of other existing) views corresponds to the "truth"?

The natural numbers (according to Peano)

Giuseppe Peano (1858-1932) published the construction of the system of natural numbers in his book *Arithmetices Principia* (1889). He gives the following:

Peano Axioms for the Natural Numbers

Let **N** be a set of elements, called the natural numbers, which satisfies the following five axioms:

(1) There exists an element of **N** denoted by 1.
(2) For any $n \in \mathbf{N}$, there exists a unique element $n^* \in \mathbf{N}$ called the successor of n in **N**.
(3) For any $n \in \mathbf{N}$, we have $n^* \neq 1$.
(4) If $m, n \in \mathbf{N}$ and $m^* = n^*$, then $m = n$.
(5) If S is a subset of **N** such that $1 \in S$, and $n \in S$ implies that $n^* \in S$, then $S = \mathbf{N}$.

It is more convenient to write $n + 1$ for n^* and we write
$$2 = 1 + 1, \ 3 = 2 + 1, \ 4 = 3 + 1, \ ...$$

Axiom 5 is called **mathematical induction**.

Addition of natural numbers is defined by induction:

(1) $1 + y = y^*$
(2) $x^* + y = (x + y)^*$

Using (1) and (2), and mathematical induction, the following basic properties of addition of natural numbers can be proved:

(3) $x + (y + z) = (x + y) + z$ \hfill (associative law)
(4) $x + y = y + x$ \hfill (commutative law)
(5) $x + z = y + z$ implies $x = y$ \hfill (cancellation law)

We only prove (3).

Proof (3) Let y, z be any two members of **N**. We first show that
$$1 + (y + z) = (1 + y) + z.$$
We have
$$\begin{aligned} 1 + (y + z) &= (y + z)^* &&\text{by (1)} \\ &= y^* + z &&\text{by (2)} \\ &= (1 + y) + z &&\text{by (1)} \end{aligned}$$

Hence (3) holds for $x = 0$. Next we show that the validity of
$$k + (y + z) = (k + y) + z \qquad \text{(induction hypothesis)}$$
implies the validity of
$$k^* + (y + z) = (k^* + y) + z.$$
We have
$$\begin{aligned} k^* + (y + z) &= [k + (y + z)]^* &&\text{by (2)} \\ &= [(k + y) + z]^* &&\text{by the induction hypothesis} \\ &= (k + y)^* + z &&\text{by (2)} \\ &= (k^* + y) + z &&\text{by (2)}. \end{aligned}$$

This proves (3) by mathematical induction.

Multiplication of natural numbers is defined by induction:

(6) $1y = y$,
(7) $x^* \cdot y = x \cdot y + y$.

Using (6) and (7), and mathematical induction, the following properties can be derived:

(8) $x \cdot (y \cdot z) = (x \cdot y) \cdot z$ \hfill (associative law)
(9) $x \cdot y = y \cdot x$ \hfill (commutative law)
(10) $x \cdot z = y \cdot z$ implies $x = y$ \hfill (cancellation law)
(11) $x \cdot (y + z) = x \cdot y + x \cdot z$ \hfill (distributive law).

We only prove (11).

Proof (11) Let y, z be any two members of **N**. We will use mathematical induction on x. For $x = 1$, (6) implies that
$$1 \cdot (y + z) = 1 \cdot y + 1 \cdot z.$$
Assume that
$$k \cdot (y + z) = k \cdot y + k \cdot z$$

is true for some $k \in \mathbf{N}$. Then

$$\begin{aligned}
k^* \cdot (y+z) &= k \cdot (y+z) + (y+z) & \text{by (7)} \\
&= (k \cdot y + k \cdot z) + (y+z) & \text{by the induction hypothesis} \\
&= k \cdot y + [k \cdot z + (y+z)] & \text{by (3)} \\
&= k \cdot y + [(k \cdot z + y) + z] & \text{by (3)} \\
&= k \cdot y + [(y + k \cdot z) + z] & \text{by (4)} \\
&= k \cdot y + [y + (k \cdot z + z)] & \text{by (3)} \\
&= (k \cdot y + y) + (k \cdot z + z) & \text{by (3)} \\
&= k^* \cdot y + k^* \cdot z & \text{by (7).}
\end{aligned}$$

This proves (11).

The concept of **order** in $\mathbf{N}$ can be introduced as follows:

(12) $x > y$ if and only if there exists $z \in \mathbf{N}$ such that $x = y + z$.

From this definition of order, it follows that

(13) $x > y$ and $y > z$ imply $x > z$ \hfill (transitive law)

(14) for any x and y in $\mathbf{N}$, one and only one of the following holds:

$$x > y, \quad x = y, \quad y > x \quad \text{(trichotomy law)}.$$

All known properties of natural numbers may be derived from the Peano axioms.

We state the following theorem (without proof) which asserts that the system of natural numbers is unique.

Theorem. Let $\mathbf{N}$ and $\mathbf{N}'$ be two sets satisfying all the Peano axioms. Then there exists a one-to-one correspondence (called an isomorphism) $f : \mathbf{N} \to \mathbf{N}'$ such that $f(1) = 1'$ and $f(n+1) = f(n) + 1'$, where $1'$ denotes the special element of $\mathbf{N}'$ satisfying 1-5 and $1'$ is the successor of $0'$ in $\mathbf{N}'$.

The integers

A **ring** $(R, 0, -, +, 1, \cdot)$ is a set R with specified elements 0 and 1, and with operations $-$, $+$ and $\cdot$, where $+$ and $\cdot$ are binary operations, and $-$ is a unary operation, which satisfy the following axioms:

(1) $(x+y) + z = x + (y+z)$ \hfill (associativity)

(2) $x + 0 = x$

(3) $x + (-x) = 0$

(4) $x + y = y + x$ \hfill (commutativity)

(5) $x \cdot 1 = x = 1 \cdot x$

(6) $(x \cdot y) \cdot z = x \cdot (y \cdot z)$ \hfill (associativity)

(7) $(x+y) \cdot z = (x \cdot z) + (y \cdot z)$, $x \cdot (y+z) = (x \cdot y) + (x \cdot z)$ \hfill (distributivity)

Recall that functions such as $u : X \to X$, $b : X \times X \to X$ and $t : X \times X \times X \to X$ are respectively called unary, binary and ternary operations on the set X.

A **commutative ring** is a ring in which multiplication is commutative (i.e., for any $x, y \in R$, we have $x \cdot y = y \cdot x$).

The reader knows how the integers $\mathbf{Z}$ are constructed from the natural numbers. We will show below that the rationals $\mathbf{Q}$ are constructed from the integers in the same way.

Familiar systems of numbers are commutative rings under the usual operations of addition and multiplication. Examples are:

Z the ring of all integers
Q the ring of all rational numbers
R the ring of all real numbers
C the ring of all complex numbers.

The ring of 2×2 matrices with entries in **Z** is not commutative.

Axioms (1), (2) and (3) of rings make $(R, 0, -, +)$ a group.

If in a group the law of commutativity $(x + y = y + x)$ holds, then the group is called Abelian.

A commutative ring is called an **integral domain** if

(8) $0 \neq 1$ and $x \cdot y = 0$ implies $x = 0$ or $y = 0$.

A commutative ring is said to be a **field** if

(9) $0 \neq 1$ and for any $x \neq 0$, there exists an element y such that $x \cdot y = 1 = y \cdot x$.

Z is an integral domain but not a field, while every field is an integral domain.

Construction of the field Q of rational numbers from the ring Z

On the set of pairs (a, b) of integers with $b \neq 0$ we define the relation $\equiv$ as follows:

$$(a, b) \equiv (c, d) \iff ad = bc.$$

Each rational number $x = \frac{a}{b}$ with $a, b \in \mathbf{Z}$ and $b \neq 0$ may be described as the solution x in **Q** of the equation $bx = a$ with coefficients a, b in **Z**.

It is easily verified that the relation $\equiv$ is an equivalence relation. We define the ratio $\frac{a}{b}$ to be the equivalence class of (a, b), i.e., the set of all pairs (c, d) where $d \neq 0$ and $ad = bc$. Observe that

$$\frac{a}{b} = \frac{c}{d} \iff (a, b) \equiv (c, d) \iff ad = bc.$$

If we call **Q** the set of all these ratios, then we obtain a field $(\mathbf{Q}, \underline{0}, -, +, \underline{1}, \cdot)$, where we have underlined zero and one in order to distinguish them from the integers 0 and 1 respectively. The operations on **Q** are defined as follows:

$\underline{0} = \frac{0}{1}$
$-\frac{a}{b} = \frac{-a}{b}$
$\frac{a}{b} + \frac{c}{d} = \frac{ad+bc}{bd}$
$\underline{1} = \frac{1}{1}$
$\frac{a}{b} \cdot \frac{c}{d} = \frac{ac}{bd}$.

It is not hard to prove that with respect to the operations $-, +, \cdot$ and the elements $\underline{0}, \underline{1}$ this set of ratios is a field, that is the field **Q** of rational numbers.

So the field **Q** is constructed starting from the ring **Z**. In general, we may start from an arbitrary integral domain and construct in the same way a field, called the **field of quotients** of the integral domain from which we have started.

But what is the relationship between **Q** and **Z**? Strictly speaking we cannot say that **Z** is a subset of **Q**. However we can say that **Q** contains a subset consisting of the ratios (quotients) of the form $\frac{a}{1}$, which is isomorphic to **Z**. In order to see this let $h : \mathbf{Z} \to \mathbf{Q}$ be defined by $h(a) = \frac{a}{1}$. We observe that

$$h(a+b) = h(a) + h(b)$$
$$h(a \cdot b) = h(a) \cdot h(b)$$
$$h(-a) = -h(a)$$
$$h(0) = \underline{0}$$
$$h(1) = \underline{1}$$

which means that the mapping h preserves the operations on $\mathbf{Z}$. Moreover, h is one-to-one. This proves that $h(\mathbf{Z})$ is an isomorphic image of $\mathbf{Z}$. For this reason we accept to identify 2 with 2/1, etc., and consider that $\mathbf{Z}$ is a subset of $\mathbf{Q}$.

The real numbers

Starting from the rational numbers, there exist two methods of constructing the system $\mathbf{R}$ of real numbers: Dedekind's method (whose roots are in the work of Eudoxus) and Cauchy's method which uses sequences of rational numbers.

As we saw before (Part I, Chapter 7), Eudoxus was a member of Plato's Academy and he was the one who defined the notion of proportion for geometric quantities. As mentioned in Part I, this work of Eudoxus constitutes a crowning moment in the history of mathematics. Eudoxus' achievement can be interpreted today as the introduction of the notion of equality between two real numbers (i.e., of the notion of the ratio between two geometric quantities) α and β. Eudoxus defines the equality $\alpha = \beta$ as follows:

"$\alpha = \beta$ if and only if the set of all rationals which are less than α coincides with the set of all rationals which are less than β and the same happens with the sets of rationals which are greater than α and greater than β respectively."

Dedekind advanced this idea of Eudoxus and defined a real number a as the pair (L, U) of sets of rational numbers, where L is the set of rationals which are less than a and U is the set of rationals which are greater than a. These sets can be described without mentioning a. For example, L can be the set of rationals x for which $x^2 < 2$ and U can be the set of rationals y for which $y^2 > 2$. We will not develop this idea any further, since we expect that the reader has already seen it in many books on algebra.

The mathematical analysts prefer the construction of the system of real numbers which was suggested by Cauchy. First of all, a real number a is defined as the set of all sequences of rational numbers which converge to a. As in Dedekind's construction, here too the construction can be carried out without mentioning a. We call **Cauchy sequence** a sequence $\langle a_n : n \in \mathbf{N} \rangle$ of rational numbers such that the number $|a_m - a_n|$ can be made arbitrarily small, if we take m and n to be sufficiently large.

Two Cauchy sequences $\langle a_n : n \in \mathbf{N} \rangle$ and $\langle b_n : n \in \mathbf{N} \rangle$ are called **equivalent** if the number $|a_n - b_n|$ can be made arbitrarily small if n is sufficiently large. Then a real number is defined as an equivalence class of Cauchy sequences.

A construction of the real numbers which is not widely known, and which does not make use of the rational numbers (as do the constructions mentioned above), is the following.

We call a function $f : \mathbf{Z} \to \mathbf{Z}$ **almost linear** if the set of all numbers of the form $|f(m+n) - f(m) - f(n)|$, where $m, n \in \mathbf{Z}$, is bounded. Two almost linear functions f and g are called "equivalent" if the set of all numbers of the form $|f(n) - g(n)|$, where $n \in \mathbf{N}$, is bounded. A real number is defined to be an equivalence class of almost linear functions $\mathbf{Z} \to \mathbf{Z}$. The real number a will be the equivalence class of the function f for which $f(n) = [an]$, where $[x]$ is the greatest integer which is $\leq x$. The sum and the product of the real numbers that correspond to the almost linear functions f and g are defined to be the equivalence classes of $f + g$ and $f \circ g$ respectively, where $(f + g)(n) = f(n) + g(n)$ and $(f \circ g)(n) = f(g(n))$. This definition of the real numbers is due to Steve Schanuel.

In addition to the construction of the system of real numbers using one of the above it is also possible to give an axiomatic description of the field of real numbers by saying that they form a **complete ordered field**.

We already know the definition of a field. Consequently, what is left to do is to define the notions "ordered" and "complete".

A field F is called **ordered** if it contains a subset P such that

(1) $x, y \in P \Rightarrow x + y \in P$
(2) $x, y \in P \Rightarrow x \cdot y \in P$
(3) Exactly one of the following holds: $x = 0$ or $x \in P$ or $-x \in P$.

We note that (3) implies that either 1 or -1 belongs to P. Since $(-1)\cdot(-1) = 1$, (2) implies that $1 \in P$. The elements of P are the **positive elements** of the field. The existence of P permits us to define an **order relation** on the field F as follows:

$$x \leq y \iff (x = y \text{ or } y - x \in P).$$

This definition implies the following propositions:

x=x (reflexivity)
$(x \leq y \ \& \ y \leq z) \Rightarrow x \leq z$ (transitivity)
$(x \leq y \ \& \ y \leq x) \Rightarrow x = y$ (antisymmetric property)
$x \leq y$ or $y \leq x$ (property of dichotomy)

Due to these four properties the relation $\leq$ is called an **order relation**.

The system $\mathbf{Q}$ of rational numbers and the system $\mathbf{R}$ of real numbers are both ordered fields. The system $\mathbf{C}$ of complex numbers is not an ordered field. Indeed, let us consider the element i. If $i \in P$, then $-1 = i \cdot i \in P$. But this is impossible since $1 \in P$. If $-i \in P$, then again $-1 = (-i) \cdot (-i) \in P$. Therefore, neither i nor $-i$ belongs to P, which contradicts property (3) of ordered fields.

An ordered field is **complete** if every non-empty set of positive elements has an **infimum**. For example, the number $\sqrt{2}$ is the infimum of all positive real numbers r for which $2 \leq r^2$. But since $\sqrt{2}$ is not a rational number, from this example we deduce that $\mathbf{Q}$ is not a complete ordered field. Essentially **there exists a single complete ordered field** in the sense of the following theorem:

Theorem 3.1. *Two arbitrary complete ordered fields are isomorphic.*

The reader may find a proof of this theorem in the book *Algebra* (1967) of Birkhoff and MacLane, in Chapter 4.3.

CHAPTER 4

THE SYSTEM OF COMPLEX NUMBERS AND THE QUATERNIONS

The complex numbers

The complex number appeared in the process of extracting the square root of an arbitrary number. In Chapter 37 of his writing *Artis magnae sive de regulis algebraicis liber unus* (1545), Cardano poses and solves the problem of separating the number 10 into two parts whose product is equal to 40. The equation is $x(10-x) = 40$, from which he obtains the roots $5+\sqrt{-15}$, $5-\sqrt{-15}$ and says the following: "By setting aside the tortures that we meet, multiply $5 + \sqrt{-15}$ with $5 - \sqrt{-15}$, hence the product is $25 - (-15)$, i.e., 40." Then, he adds that "this is the way in which arithmetic "finesse" makes progress, the outcome of which is as delicate as useless!"

Bombelli, who also worked with complex numbers in order to solve third-order equations, and who formulated the correct four operations on complex numbers thought that complex numbers are useless.

Albert Girard was more positive about complex numbers and considered complex roots as formal solutions of equations. In his book *L'invention nouvelle en algèbre* he writes the following: "One can wonder: What is the use of these "impossible" solutions (i.e., of the complex roots)? And I answer: They are useful for three reasons. They verify the general rules, they are useful and they are the only solutions."

Descartes also rejected the complex roots, which he called "imaginary". In his book *La Géométrie* he writes the following: "Neither the true nor the false (the negative) roots are always real. Sometimes they are imaginary."

Even Newton rejected the complex roots, because in his time no physical interpretation was given to complex numbers [7]. Newton believed that problems which do not have a solution which has a physical or geometric interpretation, must have complex solutions.

The negative attitude towards complex numbers can also be deduced from the following statement of Leibniz: "The Holy Spirit found an admirable way out of the dead end that appeared in mathematical analysis: this way out is through the ideal world and the "amphibious" solution between the existing and the non-existing, which we call imaginary root of the negative unity."

We have presented the above historical notes regarding the complex numbers in order to help the young researcher appreciate the bumpy road that mathematical research sometimes follows, passing through hazy and uncertain situations, before it reaches clarity.

In previous chapters we saw that geometry was restricted to Euclidean Geometry until Lobachevsky and Bolyai introduced in 1829 and 1832 respectively the non-Euclidean Geometry, in which the axiom of parallelism does not hold. The achievement of these two mathematicians opened a new road for the creation of many other new geometries.

We may say that something similar happened with algebra. During the beginnings of the 19th century, mathematicians believed that there could not exist an algebra which would be different from the algebra that existed until then. It was considered to be preposterous to try to construct a consistent algebraic system in which multiplication is non-commutative. How could there exist an algebra, which is consistent with logic and in which $a \cdot b$ is different from $b \cdot a$?

In 1843, his research in physics led William Rowan Hamilton to invent an algebra in which multiplication is non-commutative. Hamilton made the critical decision to abandon the law of commutativity with great difficulty after many years of studying a particular problem. But we will not discuss any further how Hamilton made this critical decision.

First we will present Hamilton's elegant theory where he constructs the complex numbers as pairs of real numbers. He presented this work to the Royal Irish Academy in 1833.

As mentioned before, the mathematicians of Hamilton's time considered the complex number as a strange hybrid number having the form $a + ib$, where a and b were real numbers and i was some kind of non-real number such that $i^2 = -1$. They added and multiplied complex numbers by considering them as polynomials of degree one with respect to i and by replacing i^2 with -1, wherever it appeared. Thus, one deduces that for addition we have

$$(a + ib) + (c + id) = (a + c) + i(b + d)$$

while for multiplication we have

$$(a + ib) \cdot (c + id) = (ac - bd) + i(ad + bc).$$

If we take these last relations as the definitions of addition and multiplication for complex numbers, then it is easily recognized that these two operations are associative and commutative, and they satisfy the law of distributivity.

Since a complex number is completely determined by two real numbers a and b, Hamilton came up with the idea to represent this number by the ordered pair (a, b) of the real numbers a and b.

He defined that $(a, b) = (c, d)$ if and only if $a = c$ and $b = d$. He also defined addition and multiplication between these pairs as follows:

$(a, b) + (c, d) = (a + c, b + d),$
$(a, b) \cdot (c, d) = (ac - bd, ad + bc).$

Keeping in mind these definitions, which make use of the known operations of addition and multiplication of real numbers, and using the fact that the real numbers constitute a field, it is not difficult to prove that Hamilton's pairs of real numbers also constitute a field.

It should be noted that the system of real numbers can be embedded in the system of complex numbers. This means that every real number r can be "identified" with the corresponding pair $(r, 0)$ and that this correspondence is preserved by the operations of addition and multiplication of complex numbers, i.e., $(a, 0) + (b, 0) = (a + b, 0)$ and $(a, 0) \cdot (b, 0) = (a \cdot b, 0)$. It follows from the above that a complex number of the form $(r, 0)$ can be represented simply by its corresponding real number r.

The old form $a+ib$ of a complex number follows from the one given by Hamilton. Indeed, we remark that for an arbitrary complex number (a,b), we have

$$(a,b) = (a,0) + (0,b) = (a,0) + (0,1) \cdot (b,0) = a + ib,$$

where $(0,1)$ is represented by the symbol i and $(a,0)$, $(b,0)$ are identified with the real numbers a, b respectively. Finally, we remark that

$$i^2 = (0,1) \cdot (0,1) = (-1,0) = -1.$$

Thus, the mystery that surrounded the complex numbers no longer exists, because there is nothing mysterious about a pair of real numbers.

Without going into detail, we will mention another way of introducing the field of complex numbers. The complex numbers constitute the "quotient ring" $\mathbf{R}[x]/(x^2+1)$. The elements of this ring are equivalence classes of polynomials with real coefficients, where two such polynomials are called equivalent when their difference is a multiple of $x^2 + 1$. Here the role of i is played by the equivalence class of the polynomial x.

Another way of defining the complex numbers is to consider them as 2×2 matrices with entries in $\mathbf{R}$ which have the form

$$\begin{bmatrix} a & b \\ -b & a \end{bmatrix}.$$

(Regarding matrices see the section on quaternions.)

It is easily recognized that the set of matrices of this form is closed with respect to addition and multiplication of matrices.

This way of defining the complex numbers has the advantage that one can use the arithmetic of matrices to give a short proof that the set of complex numbers constitutes a ring. It must be verified though that multiplication is commutative, because, as it is known, matrix multiplication is in general not commutative. Moreover, it is recognized that the inverse matrix of a non-zero matrix of the form

$$\begin{bmatrix} a & b \\ -b & a \end{bmatrix}$$

is the matrix

$$\begin{bmatrix} \frac{a}{k} & -\frac{b}{k} \\ \frac{b}{k} & \frac{a}{k} \end{bmatrix},$$

where $k = a^2 + b^2$ is the determinant of the matrix

$$\begin{bmatrix} a & b \\ -b & a \end{bmatrix}.$$

Since when we talk about real numbers we do not think of a matrix, one could wonder how the real numbers are related to the complex numbers when the latter are defined as matrices. In this respect we note that the mapping $h : \mathbf{R} \to \mathbf{C}$, where

$$h(a) = \begin{bmatrix} a & 0 \\ 0 & a \end{bmatrix}$$

is a one-to-one homomorphism. Consequently, $h(\mathbf{R})$ is an isomorphic image of $\mathbf{R}$ and hence it can be identified with $\mathbf{R}$, something that permits us to say that $\mathbf{R}$ is a subset of $\mathbf{C}$.

If u and v are two complex numbers, then
$$|u + v| \leq |u| + |v|,$$
$$|u \cdot v| = |u| \cdot |v|.$$

If we consider the complex numbers as points of the Cartesian plane, then the above inequality is the triangle inequality in Euclidean Geometry (Book I, 20), while the above equality is equivalent to the algebraic identity
$$(ac - bd)^2 + (ad + bc)^2 = (a^2 + b^2)(c^2 + d^2),$$
which was known to al-Khazin in 950 A.D. (Part I, Chapter 18).

If we assign to the number $a + ib$ the point (a, b) of the Cartesian plane and we call r the distance of (a, b) from the origin and θ the argument of $a + ib$, then we know that
$$a + ib = r(\cos\theta + i\sin\theta).$$

If the number $a' + ib'$ has absolute value r' and argument θ', then by applying the known formulas regarding the sum of trigonometric functions, we have
$$(a + ib) \cdot (a' + ib') = rr'(\cos(\theta + \theta') + i\sin(\theta + \theta')).$$

Using mathematical induction, this formula leads to the equality
$$(r(\cos\theta + i\sin\theta))^m = r^m(\cos m\theta + i\sin m\theta),$$
which is known as the theorem of de Moivre (1667-1754) [7].

The quaternions

The complex number $z = a + ib$ representing a point Σ with orthogonal Cartesian coordinates (a, b), can be considered as a representation of the vector $O\Sigma$, where O is the origin. Thus complex numbers are useful for the study of vectors and of rotations in the plane.

Hamilton tried to invent an analogous system of numbers for the study of vectors and rotations in 3-dimensional space. This led him to ordered quadruples (a, b, c, d) of real numbers. By defining that two quadruples (a, b, c, d) and (e, f, g, h) are equal if and only if $a = e$, $b = f$, $c = g$ and $d = h$, Hamilton realized that the following must hold:
$$(a, 0, 0, 0) + (e, 0, 0, 0) = (a + e, 0, 0, 0)$$
$$(a, 0, 0, 0) \cdot (e, 0, 0, 0) = (ae, 0, 0, 0)$$
$$(a, b, 0, 0) + (e, f, 0, 0) = (a + e, b + f, 0, 0)$$
$$(a, b, 0, 0) \cdot (e, f, 0, 0) = (ae - bf, af + be, 0, 0).$$

Hamilton called the ordered quadruples of real numbers quaternions. Hamilton defined addition and multiplication of quadruples of real numbers as follows:
$$(a, b, c, d) + (e, f, g, h) = (a + e, b + f, c + g, d + h),$$
$$(a, b, c, d) \cdot (e, f, g, h) = (ae - bf - cg - dh,\ af + be + ch - dg,$$
$$ag + ce + df - bh,\ ah + bg + de - cf).$$

Using these definitions, it follows that the real and the complex numbers have isomorphic images inside the set of quaternions, and furthermore if the quaternion $(m, 0, 0, 0)$ is identified with the real number m, then
$$m \cdot (a, b, c, d) = (a, b, c, d) \cdot m = (ma, mb, mc, md).$$

Addition of quaternions is associative and commutative, and multiplication of quaternions is associative and satisfies the law of distributivity with respect to addition. But, **multiplication of quaternions is NOT commutative**. Indeed consider the quaternions $(0,1,0,0)$ and $(0,0,1,0)$. We have $(0,1,0,0) \cdot (0,0,1,0) = (0,0,0,1)$, but $(0,0,1,0) \cdot (0,1,0,0) = (0,0,0,-1) = -(0,0,0,1)$.

If we represent by 1, i, j, k the "quaternion units" $(1,0,0,0)$, $(0,1,0,0)$, $(0,0,1,0)$, $(0,0,0,1)$ respectively, then the following multiplication table gives the products of these units.

×	1	i	j	k
1	1	i	j	k
i	i	-1	k	$-j$
j	j	$-k$	-1	i
k	k	j	$-i$	-1

From this table we obtain that $i^2 = j^2 = k^2 = ijk = -1$. Every quaternion (a,b,c,d) can be written in the form $a+bi+cj+dk$. If two quaternions are written in this form, then we can find their product by multiplying them as we multiply two polynomials of i, j, k and by taking into account the above multiplication table.

In 1844, Hermann Günther Grassmann presented classes of algebras which are more general than Hamilton's quaternion algebra. Instead of considering ordered quadruples of real numbers he considered ordered n-tuples of real numbers. At every n-tuple $(x_1, x_2, ..., x_n)$ he assigned a "hypercomplex" number of the form $x_1e_1 + x_2e_2 + ... + x_ne_n$, where e_1, e_2, ..., e_n are the units of the algebra developed by Grassmann which are analogous to the ones of the quaternion algebra. We will not discuss any further the description of Grassmann's algebras, which indeed created new approaches for the construction and study of a huge number of algebraic structures.

Another example of an algebra with a non-commutative multiplication is the "matrix algebra". This was introduced by Arthur Cayley (1821-1895) in 1857. The notion of a matrix came from the study of linear transformations of the form

$$x' = ax + by$$
$$y' = cx + dy$$

where a, b, c, d are real numbers. These transformations map the arbitrary point (x,y) to the point (x',y') and are completely determined by the coefficients a, b, c, d. Thus, the transformation in question can be represented by the symbol

$$\begin{bmatrix} a & b \\ c & d \end{bmatrix}$$

which we call "square matrix" of order 2. Since two transformations of this kind are identical if and only if they have the same coefficients, we define that two matrices

$$\begin{bmatrix} a & b \\ c & d \end{bmatrix} \text{ and } \begin{bmatrix} e & f \\ g & h \end{bmatrix}$$

are equal if and only if

$$a = e, \quad b = f, \quad c = g, \quad d = h.$$

When the above transformation is followed by the transformation

$$x'' = ex' + fy'$$
$$y'' = gx' + hy',$$

then
$$x'' = (ea + fc)x + (eb + fd)y$$
$$y'' = (ga + hc)x + (gb + hd)y.$$
The above motivates the following definition of the product of two matrices:
$$\begin{bmatrix} e & f \\ g & h \end{bmatrix} \cdot \begin{bmatrix} a & b \\ c & d \end{bmatrix} = \begin{bmatrix} ea + fc & eb + fd \\ ga + hc & gb + hd \end{bmatrix}.$$
The sum of two matrices is defined by
$$\begin{bmatrix} a & b \\ c & d \end{bmatrix} + \begin{bmatrix} e & f \\ g & h \end{bmatrix} = \begin{bmatrix} a+e & b+f \\ c+g & d+h \end{bmatrix}.$$
If m is an arbitrary real number, then we define that
$$m \cdot \begin{bmatrix} a & b \\ c & d \end{bmatrix} = \begin{bmatrix} a & b \\ c & d \end{bmatrix} \cdot m = \begin{bmatrix} ma & mb \\ mc & md \end{bmatrix}.$$

In the matrix algebra defined above, it can be verified that addition is associative and commutative. Multiplication is associative and satisfies the law of distributivity with respect to addition, but it is not commutative.

At the end of the 19th century algebras appeared in which multiplication is associative. Examples of such algebras are the **Jordan algebras** and the **Lie algebras**. A particular Jordan algebra used in Quantum Mechanics has as its elements matrices, where equality and addition are defined as in Cayley's matrix algebra, but the product of two matrices A and B is defined to be $\frac{1}{2}(AB + BA)$, where AB is the Cayley product of the matrices A and B. In this Jordan algebra it is proved that multiplication is not associative, while it is obvious that it is commutative.

A Lie algebra differs from the Jordan algebra mentioned above in that the product of two matrices A and B is defined to be $AB - BA$, where AB is again the Cayley product of the matrices A and B. In a Lie algebra multiplication is neither associative nor commutative.

When Hamilton's quaternions first appeared, it was thought that they would become an indispensable tool for future physicists, but this did not happen. The quaternions were displaced by the much more flexible theory of "vector analysis", which is due to the American physicist and mathematician Josiah Willard Gibbs (1839-1903).

On the other hand, Cayley's theory of matrices was widely developed and today it constitutes a very important and useful tool in mathematics.

The deserving fame that Hamilton's quaternions acquired is due to the fact that they were historically very important for algebra. Indeed from the moment that mathematicians realized that they could construct a system of numbers where, in contrast to what happens with the real and the complex numbers, multiplication is not commutative, they felt free to consider structures which deviated from the usual properties of the real and the complex numbers. It was necessary to realize this fact before the development of vector algebra and analysis, since vectors violate the usual laws of algebra more than quaternions do. Hamilton's work also opened the way for the theory of the so-called linear associative algebras.

CHAPTER 5

THE FUNDAMENTAL THEOREM OF ALGEBRA

We have seen that the negative integers, fractions, irrationals and complex numbers were introduced in order to secure the existence of solutions of polynomial equations. It is then natural to wonder whether we need more numbers to supply solutions for polynomial equations. The answer to this question is in the negative and is given by the fundamental theorem of algebra.

A viewpoint (John Stillwell, 1989) exists that the proof of this theorem given by Gauss (see Gauss' biography in Chapter 1 of Part II) is incomplete, and that the first rigorous proof was given after Weierstrass established the basic properties of continuous functions.

Fundamental Theorem of Algebra. If $P(z) = z^n + a_{n-1}z^{n-1} + ... + a_0$ is a polynomial of degree $n \geq 1$ with complex coefficients, then there is a number $a \in \mathbf{C}$ such that $P(a) = 0$.

Proof. We only need the following fact:

If an entire function $f : \mathbf{C} \to \mathbf{C}$ (i.e., f is analytic everywhere on $\mathbf{C}$) is bounded on $\mathbf{C}$ (i.e., there exists $M > 0$ such that $|f(z)| < M$ for every $z \in \mathbf{C}$), then f is a constant function.

First observe that

$$\lim_{z \to \infty} P(z) = \lim_{z \to \infty} (z^n(1 + a_{n-1}z^{-1} + ... + a_0 z^{-n})) = \infty.$$

Assume that $P(z) \neq 0$ for all $z \in \mathbf{C}$. Then $\frac{1}{P}$ would be an entire function. Since $P(z) \to \infty$ for $z \to \infty$, it follows that $\frac{1}{P}$ is bounded. Therefore P is constant. But this is wrong because, since $n \geq 1$, P cannot be constant. This proves the theorem.

Conclusion. Every polynomial $P(z)$ of degree m with complex coefficients is factored into exactly m linear factors.

Proof. We saw that $P(z) = 0$ has a complex solution, say z_1. If $P(z) = Q(z)(z - z_1) + R$, then $R = P(z_1) = 0$, which means that $z - z_1$ is a factor of $P(z)$. $Q(z)$ has degree $m - 1$ and, in case $m - 1 > 0$, $Q(z)$ too has a complex root. By continuing in this way we may write

$$P(z) = c(z - z_1)(z - z_2)...(z - z_m),$$

where c is the coefficient of z^m in $P(z)$.

CHAPTER 6

SET THEORY

The main difficulty in set theory lies in the notion of a set with infinitely many elements. Now, we will call these sets infinite.

The nature of infinite sets occupied mathematicians and philosophers since the time of the ancient Greeks. The known paradoxes of Zeno (Part I) can be perhaps considered as the first evidence of the difficulties associated with such sets. Neither the assumption that the ad infinitum subdivision of a line segment is possible, nor the assumption that a line segment consists of infinitely many points, let to an understanding of the notion of motion.

Aristotle had considered examples of infinite sets (such as the set of positive integers), but he did not accept an infinite set as a unified entity. Actually, Aristotle made a distinction between the "potentially infinite" and the "actually infinite". According to him only the potentially infinite exists. An example of the former is the set of positive integers, in the sense that one can add 1 to any positive integer and obtain a new positive integer. Moreover, Aristotle claimed that many sets cannot be even potentially infinite, because if this was possible, then, by adding ad infinitum the same element, these sets would exceed the boundaries of the universe! According to him, space is potentially infinite (it can be subdivided ad infinitum) and time is also potentially infinite (it can increase ad infinitum and also can decrease ad infinitum).

Proclus argued that since every diameter of a circle divides the circle into two semicircles and since there exist infinitely many diameters, it follows that the number of semicircles is twice the number of diameters. According to him, this is not a contradiction, because we cannot talk about an (actually) infinite number of diameters and semicircles, but only about an ad infinitum increasing number of diameters and semicircles. Hence by accepting only the existence of the potentially infinite and not the existence of the actual infinite, he avoided the consideration of the problem that two times infinity is equal to infinity.

During the Middle Ages, it was understood that despite the fact that their lengths are unequal, the points of two concentric circles can be brought into a one-to-one correspondence by considering that the points of the two circles that lie on the same radius correspond to one another. Moreover, Galileo remarked that since the points of two unequal line segments AB and CD (Figure 6.1) can be brought into a one-to-one correspondence, it follows that, even though these segments are unequal, they have the same number of points.

Galileo also understood that the set of natural numbers and its proper subset consisting of the squares of the natural numbers, can be brought into a one-to-one correspondence, by assigning to every natural number its square. But then, this leads one to accept the existence of two different kinds of infinity, which Galileo could not accept, since he believed that infinite sets did not differ from each other.

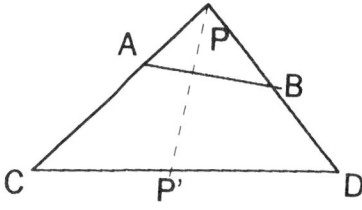

Figure 6.1

In his letter to Schumacher of the 12th of July of 1831, Gauss wrote: "I protest about the use of an infinite quantity as a self-existent unified entity. This is never allowed in mathematics. The infinite is merely a way of expressing ourselves when we talk about the limits of certain ratios, which can be approached by these ratios as close as we want, while other ratios can be made as large as we want."

Cauchy also did not accept the existence of infinite sets because he considered it to be a contradiction of the fact that an infinite set can be brought into a one-to-one correspondence with a proper subset of it.

Mathematicians of the time took the position of ignoring problems which they could not solve. They did not accept the existence of actual infinity, but they made use of series with infinitely many terms, and also used the system of real numbers. This tactic, although somewhat "hypocritical" allowed them to make progress in classical mathematical analysis. However, during the 19th century, when the problem of making mathematical analysis rigorous emerged, infinite sets had to be properly understood.

Bolzano made the first step towards the creation of set theory. In his book *Paradoxien des Unendlichen* (1851), he accepted the existence of infinite sets in the sense of actual infinity and introduced the notion of one-to-one corresondence between sets (he called such sets equivalent). His work on infinite sets was more of a philosophical rather than of a mathematical nature; he noted that such sets had paradoxical properties and hence he did not study them any further.

The creator of modern Set Theory was Georg Cantor (1845-1918), who was born in Saint Petersburg in Russia from parents of Danish-Jewish origin. His parents settled in Germany, and in 1863 he entered the University of Berlin. He intended to study to become a mechanic, but under the influence of Weierstrass he studied instead pure mathematics. He was elected professor in 1879. At the age of twenty he published his first great paper on set theory, which was considered to be brilliant and extremely original, despite the adverse critique of certain mathematicians of the old school.

Cantor's work contributed to the solution of problems, which until then remained unsolved, by introducing completely new ideas.

Leopold Kronecker fought against Cantor's ideas regarding ordinal and cardinal numbers, for more than ten years. At some point Cantor suffered a nervous breakdown, which he partly overcame and continued his work. Doubts regarding the correctness of Cantor's work remained even after Kronecker's death in 1891.

According to Cantor a "set" is a collection into a whole of definite and separate objects, which are perceived by the mind and are such that, given any object, it is possible to decide whether it belongs to this collection or not.

The reader should know that the notion of a set is a primitive notion, which admits no exact definition. What we gave above is simply an intuitive picture of this notion and not its definition.

Cantor did not accept the viewpoint of the mathematicians and the philosophers of that time, that only potential infinity exists. According to Cantor, a set has infinitely many elements when it can be brought into a one-to-one correspondence with some proper subset of it.

In his earlier work on trigonometric series, Cantor had already introduced certain set theoretic notions, such as the notion of an "accumulation point", the notion of "derived set", and the notion of "sets of the first species". If S is a set of real numbers, then the number p is called an accumulation point of S if every interval that contains p contains infinitely many points of S. The derived set S' of S consists of all accumulation points of S. There exists also a second derived set of S, which is the derived set of S', and so on. If the nth derived set of S is a set with finitely many elements, then S is called a set of rank n (and a set of the first species). A set of real numbers is called "closed" when it contains all its accumulation points. A set S of real numbers is called "open" if every point of S is an "interior point", which means that every point of S lies in an interval which contains only points of S. Cantor gave also the definitions of the notions of union and intersection of sets. Cantor was mainly interested in sets of real numbers, i.e., in subsets of the straight line, but he extended these notions to subsets of the n-dimensional Euclidean space.

The cardinal numbers of sets include the numbers 0, 1, 2, ..., beyond which are the cardinal numbers of the various kinds of infinity. In general, a cardinal number indicates how many elements a given set has. But this notion is clear when it refers to finite sets, and it must be investigated further when it refers to infinite sets [8].

We say that two sets have the same cardinal number if and only if there exists a one-to-one correspondence between their elements:

$$A \cong B \iff \text{there exists a one-to-one and onto function } f : A \to B.$$

For example, the set $\{3, 5, 7\}$ has the same cardinal number with the set $\{0, 1, 2\}$. Let it be noted that $\cong$ is an equivalence relation.

If n is a positive integer, we will say that a set A has cardinal number n if and only if $A \cong \{0, 1, 2, ..., n-1\}$, and we will write $|A| = n$. The cardinal number of the empty set is 0, i.e., $|\emptyset| = 0$.

A set A is called **finite** when $|A|$ is some natural number. If $|A|$ is not a natural number, then A is called infinite. For example, $\mathbf{N}$ is an infinite set.

A set A is said to have cardinal number $\aleph_0$ if and only if $A \cong \mathbf{N}$. These sets are called **countably infinite** or simply **countable**. For example, $f(m) = 2m$ is a one-to-one function from $\mathbf{N}$ onto the set $\{0, 2, 4, 6, ...\}$ of even numbers. This means that the cardinal number of the set of even numbers is $\aleph_0$.

The symbol $\aleph$ is the first letter of the Hebrew alphabet and is pronounced "aleph". The cardinal number of the set of real numbers is denoted by $\mathbf{c}$.

The last example that refers to the cardinal number of the set of even numbers is a particular case of the following theorem which is usually referred to as "Galileo's theorem":

"If a set S has cardinal number $\aleph_0$, then every infinite subset of S has cardinal number $\aleph_0$."

Galileo's theorem might make one believe that there exists a single infinite cardinal number. One of Cantor's achievements is that he proved the following: The power set of a set S has cardinal number greater than that of S.

We remind the reader that the power set of a set S is the set $P(S)$ whose elements are the subsets of S.

We write $|A| < |B|$ (which is read: A has a cardinal number which is smaller than that of B) if and only if $A \not\cong B$ and B has a proper subset C such that $A \cong C$. It should be noted that $P(S)$ always has a proper subset C, i.e., the sets of all singletons of S, such that $C \cong S$. We recall that a singleton $\{a\}$ of S is a set that has as its single element an element a of S. We present the following theorem of Cantor.

Theorem. If A is an arbitrary set, then the cardinal number of A is smaller than the cardinal number of the power set $P(A)$.

Proof. We argue by reductio ad absurdum. We assume that there exists a one-to-one and onto function $f : A \to P(A)$. Then every subset of A has the form $f(a)$ for some $a \in A$. Let $S = \{x \in A : x \notin f(x)\}$. S is a subset of A, hence there exists some $a \in A$ such that $S = f(a)$. If $a \in S$, then from the definition of S it follows that $a \notin f(a)$, i.e., $a \notin S$. If $a \notin S$, then again from the definition of S it follows that $a \in f(a)$, i.e., $a \in S$. Thus, $a \in S \iff a \notin S$, which is a contradiction.

If a finite set has n elements, it is easily recognized that its power set has 2^n elements. In general, if a set has cardinal number k, then the cardinal number of its power set is denoted by 2^k. For example, $|P(\mathbf{N})| = 2^{\aleph_0}$. The above theorem of Cantor implies that

$$\aleph_0 < 2^{\aleph_0} < 2^{2^{\aleph_0}} < \ldots$$

In other words, there exist infinitely many infinite cardinals (i.e., there exist infinitely many "infinities"). Conclusions of this kind caused the dispute between Cantor and Kronecker.

After the presentation of the above, the following question arises naturally:

Does there exist a subset of $P(\mathbf{N})$ whose cardinal number is greater than $\aleph_0$ and smaller than $2^{\aleph_0}$?

The hypothesis that the answer to this question is NO is called **continuum hypothesis**.

In 1940, Kurt Gödel proved that the continuum hypothesis is compatible with the usual axioms of set theory, and in 1963, Paul Cohen proved that the negation of the continuum hypothesis is also compatible with the usual axioms of set theory.

In other words, the basic notions regarding sets that we have at our disposal neither imply nor refute the continuum hypothesis. For this reason we may say that **the continuum hypothesis is independent of the axioms of set theory**, i.e., of axioms on which Ernst Zermelo (1871-1953) and Abraham Fraenkel (1891-1965) founded set theory.

Cardinal arithmetic

If k and l are two finite cardinal numbers, then their sum $k+l$ and their product $k \cdot l$ have their known meaning. We will try to generalize these operations to infinite cardinal numbers, i.e., to cardinal numbers of infinite sets, so that these operations preserve the known meaning and properties valid for cardinal numbers of finite sets.

Definition. Let $a = |A|$, $b = |B|$, $A \cap B = \emptyset$. Then the sum $a + b$ of the cardinal numbers a and b is the cardinal number of the set $A \cup B$, i.e., $a + b = |A \cup B|$.

Remark. It is easily recognized that given two (not necessarily distinct) arbitrary cardinal numbers, there exist sets X and Y such that $a = |X|$ and $b = |Y|$, where the sets X and Y are **not** necessarily disjoint. But this causes no problem, because we may choose $A = X \times \{0\}$ and $B = Y \times \{1\}$, hence we have $A \cong X$, $B \cong Y$ and $A \cap B = \emptyset$. Thus $a + b = |A \cup B|$ and the sum $a + b$ is uniquely defined in this way, because if A' and B' were another two disjoint set such that $A' \cong A$ and $B' \cong B$, then $(A' \cup B') \cong (A \cup B)$ or $|A' \cup B'| = |A \cup B|$.

It follows from this last remark that if a and b are two cardinal numbers, then the following conditions hold:

(i) There exist disjoint sets A and B such that $a = |A|$ and $b = |B|$.
(ii) If A, B, A', B' are sets such that $|A'| = |A|$, $|B'| = |B|$, $A \cap B = \emptyset$, $A' \cap B' = \emptyset$, then $|A' \cup B'| = |A \cup B|$.

If we take into account that the union of two sets is associative and commutative, then we obtain the following properties of the operation of addition of cardinal numbers.

If x, y, z are arbitrary cardinal numbers, then

(a) $x + y = y + x$ (commutativity)
(b) $(x + y) + z = x + (y + z)$ (associativity)

As an example of application of what we presented above we prove that $\aleph_0 + \aleph_0 = \aleph_0$.

Let $\mathbf{N}_\alpha$ and $\mathbf{N}_\pi$ be the sets of even and odd natural numbers respectively. $\mathbf{N}_\alpha$ and $\mathbf{N}_\pi$ are disjoint countably infinite sets and $\mathbf{N}_\alpha \cup \mathbf{N}_\pi = \mathbf{N}$. From the definition of the sum of two cardinal numbers, we have

$$\aleph_0 + \aleph_0 = |\mathbf{N}_\alpha| + |\mathbf{N}_\pi| = |\mathbf{N}_\alpha \cup \mathbf{N}_\pi| = \aleph_0.$$

The property expressed by this example obviously does not hold for finite cardinal numbers, because for them the equality $n + m = n$ holds if and only if $m = 0$.

As a second example we outline the proof of the relation $(0, 1) \cong \mathbf{R}$:

The function $f : (0, 1) \to (-1, 1)$ with $f(x) = 2x - 1$ is one-to-one and onto. Therefore $(0, 1) \cong (-1, 1)$.

The function $g : (-1, 1) \to \mathbf{R}$ with $g(x) = \tan \frac{\pi x}{2}$ is one-to-one and onto. In order to demonstrate this property of g, the reader is asked to prove that g is continuous and unbounded (both from above and from below), and that the derivative $g'(x) = \frac{\pi}{2} \sec^2 \left(\frac{\pi x}{2} \right)$ is positive, which means that g is strictly increasing.

If one takes into account that $(0, 1) \cong \mathbf{R}$, the proof of the relation $\aleph_0 + \mathbf{c} = \mathbf{c}$ is analogous to the proof of the relation $\aleph_0 + \aleph_0 = \aleph_0$.

Definition. If $a = |A|$ and $b = |B|$ are the cardinal numbers of the sets A and B respectively, then their product $a \cdot b$ is defined to be the cardinal number of the set $A \times B$, i.e., $a \cdot b = |A \times B|$.

Remark. It is proved that the definition of the product does not depend on the choice of the sets A and B. This means that if X and Y are two sets such that $A \cong X$ and $B \cong Y$, then $A \times B \cong X \times Y$ and hence $|A \times B| = |X \times Y|$.

The properties below follow from the above definition of the product of two cardinals:

If x, y, z are arbitrary cardinal numbers, then:

(a) $x \cdot y = y \cdot x$ (commutativity)
(b) $(x \cdot y) \cdot z = x \cdot (y \cdot z)$ (associativity)
(c) $x \cdot (y + z) = x \cdot y + x \cdot z$ (distributivity)

We ask the reader to prove as an exercise that if x is an arbitrary cardinal number, then:

(a) $1 \cdot x = x$
(b) $0 \cdot x = 0$
(c) $\aleph_0 \cdot \aleph_0 = \aleph_0$
(d) $\mathbf{c} \cdot \mathbf{c} = \mathbf{c}$

Definition. Let a and b be two cardinal numbers with $a \neq 0$, and let A and B be two sets such that $a = |A|$ and $b = |B|$. If we denote by B^A the set of functions from A to B, then by definition $b^a = |B^A|$.

Remark. Before accepting this definition, we must prove that it does not depend on the sets A and B. This means that we must prove the following proposition:

"If A, B, X, Y are sets such that $A \cong X$, $B \cong Y$, then $B^A \cong Y^X$."

We give an outline of the proof. Let $g : A \to X$ and $h : B \to Y$ be two functions which are one-to-one and onto. We define the function $\Psi : B^A \to Y^X$ as follows:

If $f \in B^A$ and we define $\Psi(f) : X \to Y$ by the relation $\Psi(f)(x) = h \circ f \circ g^{-1}(x)$, then we have the following diagram:

$$\begin{array}{ccc} A & \longrightarrow & B \\ \downarrow & & \downarrow \\ X & \xrightarrow{\Psi(f)} & Y \end{array}$$

Prove that the function $\Psi : B^A \to Y^X$ is one to one and onto.

Example. If A is an arbitrary set, then prove that $|P(A)| = 2^{|A|}$.

Proof. Let $B = \{0, 1\}$ be the set whose only elements are 0 and 1. At every subset D of A we assign its characteristic function χ_D, where $\chi_D : A \to B$ is defined as follows: $\chi_D(x) = 1$ if $x \in D$ and $\chi_D(x) = 0$ if $x \notin D$. The function $g : P(A) \to B^A$ with $g(D) = \chi_D$ is one-to-one and onto (prove this!). Therefore, the sets $P(A)$ and B^A have the same cardinal number, i.e., $|P(A)| = 2^{|A|}$.

We present also certain other arithmetical properties of cardinal numbers without giving the proof:

If a, b, x, y, z are arbitrary cardinal numbers, then: $a^x \cdot a^y = a^{x+y}$, $(z^y)^x = z^{y \cdot x}$ and $(a \cdot b)^x = a^x \cdot b^x$.

In one of his first papers regarding set theory, Cantor proved that two important sets are countable.

The first of these two sets is the set **Q** of rational numbers. **Q** has also the property of being **dense**. This means that between any two distinct rational numbers, there exists a rational number; in fact, there exist infinitely many rational numbers. For example, between 0 and 1, we have the following rational numbers: $\frac{1}{2}, \frac{2}{3}, \frac{3}{4}, \frac{4}{5}, \frac{5}{6}, \ldots, \frac{n}{n+1}, \ldots$

Due to this property of the rationals one could think that the infinite cardinal number of the set **Q** is greater than $\aleph_0$. Cantor proved that this is not the case and that the set of rationals is countable. We present Cantor's proof. We consider the diagram:

$$
\begin{array}{cccc}
1 \rightarrow 2 & 3 \rightarrow 4 & \cdots \\
\frac{1}{2} & \frac{2}{2} & \frac{3}{2} & \frac{4}{2} & \cdots \\
\frac{1}{3} & \frac{2}{3} & \frac{3}{3} & \frac{4}{3} & \cdots \\
\frac{1}{4} & \frac{2}{4} & \frac{3}{4} & \frac{4}{4} & \cdots \\
\cdots & \cdots & \cdots & \cdots
\end{array}
$$

Here the first horizontal line contains all the positive natural numbers in increasing order. The second line contains in increasing order all the positive fractions with denominator 2, and so on. Obviously this diagram contains all positive rational numbers. If we write them in the order shown by the arrows and if we omit every time the numbers that we have already met, we obtain an infinite sequence of numbers

$$1,\ 2,\ \frac{1}{2},\ \frac{1}{3},\ 3,\ 4,\ \frac{3}{2},\ \frac{2}{3},\ \frac{1}{4},\ \ldots$$

in which every positive rational number appears once. We denote this sequence by $\{r_1, r_2, r_3, \ldots\}$. Then the sequence

$$0,\ -r_1,\ r_1,\ -r_2,\ r_2,\ -r_3,\ r_3,\ \ldots$$

constitutes the set of rational numbers. This proves that the set **Q** is countable.

Cantor proved that the set of algebraic numbers is also countable. A complex number is said to be an **algebraic number** if it is the root of a polynomial equation:

$$f(x) = a_0 x^n + a_1 x^{n-1} + \ldots + a_{n-1} x + a_n = 0,$$

where $a_0 \neq 0$ and all the coefficients a_k are integers. The set of algebraic numbers contains the rational numbers and all their roots.

Theorem. The set of algebraic numbers is countable.

Proof. Let $f(x)$ be the above polynomial where, without loss of generality, we may assume that $a_0 > 0$. We call "height" of the polynomial $f(x)$ the number

$$h = n + a_0 + |a_1| + |a_2| + \ldots + |a_{n-1}| + |a_n|.$$

It is obvious that h is an integer ≥ 1 and that there exist only finitely many polynomials $f(x)$ with given height h. Hence the number of algebraic numbers which are roots of such polynomials of height h is also finite. We form a list of all the algebraic numbers by listing first the ones which are roots of polynomials of height 1, then the ones which are roots of polynomials of height 2, and so on. Since the algebraic numbers form an infinite sequence, it follows that the set of algebraic numbers is countable.

Theorem. The set of real numbers contained in the interval $(0, 1)$ is not countable.

Proof. The proof uses the so-called "diagonal method" of Cantor. We assume that the set in question is countable. This means that we can write the numbers of this set as a sequence, say $\{\rho_1, \rho_2, ..., \rho_n, ...\}$. Every number ρ_i of this sequence has a unique decimal expression with infinitely many non-zero digits (for example, $\frac{1}{2}$ is written as 0.4999... and not as 0.5). So we can write the above sequence as follows:

$$\rho_1 = 0.a_{11}a_{12}a_{13}...$$
$$\rho_2 = 0.a_{21}a_{22}a_{23}...$$
$$\rho_3 = 0.a_{31}a_{32}a_{33}...$$
$$...$$

where every symbol a_{ij} represents a digit among 0, 1, 2, 3, 4, 5, 6, 7, 8, 9. Although the above table contains by assumption all the real numbers that lie in the interval $(0,1)$, there exist numbers in $(0,1)$ which are not contained in this table. Such a number is $0.b_1b_2b_3...$, where for every positive integer k, $b_k = 7$ if $a_{kk} \neq 7$ and $b_k = 3$ if $a_{kk} = 7$. Indeed this number lies in the interval $(0,1)$, but it is different from every ρ_i, because it differs from ρ_1 at least at the first decimal digit, it differs from ρ_2 at least at the second decimal digit, and so on. Therefore, the original assumption led to a contradiction, and hence the set in question is not countable.

An immediate consequence of this theorem is that **the system of real numbers in not countable**.

From these last two theorems, Cantor obtained a proof of the following very important theorem.

Theorem. There exist transcendental numbers, i.e., complex numbers which are not algebraic.

Proof. Indeed, since the set of real numbers is not countable, it follows that the set of complex numbers is not countable. But since the set of algebraic numbers is countable, it follows that there exist complex numbers which are not algebraic and consequently they are transcendental.

The proofs of the last two theorems were not accepted by the entire mathematical community, because they were "non-constructive". In particular, the last theorem does not provide the construction of a concrete example of a transcendental number. In mathematics there exist many non-constructive "existence proofs" of this kind, where the existence of a certain object is proved by reductio ad absurdum. Due to the fact that non-constructive proofs were not immediately accepted, mathematicians tried to replace these proofs by other constructive proofs.

The proof that a given number is transcendental is more difficult than the proof of the existence of transcendental numbers.

In 1873 Charles Hermite (1822-1901) proved that the number e is a transcendental number, and in 1882 C. L. F. Lindemann (1852-1939) proved that π is a transcendental number.

The question of whether the number π^π is algebraic or transcendental remains open.

A useful theorem is the following:

"If a is an algebraic number which is different from 0 and 1, and b is any irrational algebraic number, then the number a^b is transcendental."

This theorem was the outcome of 30 years of efforts to prove that the so-called "Hilbert number" $2^{\sqrt{2}}$ is transcendental.

We will close this section on cardinal arithmetic with the following interesting remark:

In 1874 Cantor occupied himself with the problem of whether the set of points of a straight line and the set of points of $\mathbf{R}^n$ (i.e., the n-dimensional Euclidean space) are equinumerous or not. Three years later he proved that they are equinumerous, i.e., that there exists a one-to-one correspondence between them, and he wrote to Dedekind the following: "I see that there exists a one-to-one correspondence but I don't believe it".

The way in which Cantor obtained the correspondence in question is the same with the following, where one achieves a one-to-one correspondence between the points of the unit square and the points of the unit interval $(0,1)$.

Let (x,y) be a point of the unit square and let $z \in (0,1)$. We write x and y in their decimal form with infinitely many non-zero digits (for example, as before, the number $\frac{1}{2}$ will be written in the form 0.4999... and not in the form 0.5). Then, we separate the digits of x and y into groups so that every group has as its last digit the first non-zero digit that we meet. Thus, for example, we write

$$x = 0.3\ 002\ 03\ 04\ 6\ ...$$
$$y = 0.01\ 6\ 07\ 8\ 09\ ...$$

We form the number

$$z = 0.3\ 01\ 002\ 6\ 03\ 07\ 04\ 8\ 6\ 09\ ...$$

by selecting as groups of digits of z first the first group of digits of x, second the first group of digits of y, third the second group of digits of x, and so on.

If two numbers x or two numbers y differ at one digit, then the corresponding numbers z will also differ. This means that every (x,y) corresponds to a single z. Conversely, given a number z we separate again its decimal digits into groups in the way described above and we form the numbers x and y by following the inverse of the procedure that we followed in the formation of z. Again, two different z will give us two different pairs (x,y), which means that every z corresponds to a single pair (x,y). The one-to-one correspondence that we described above is not continuous. This essentially means that points z which lie close to each other do not necessarily correspond to points (x,y) which lie close to each other, and vice versa.

There existed mathematicians who did not accept this proof. One of them was Du Bois-Reymond (1882).

The Axiom of Choice

Let us assume that we enter a fruit store where there exist many (non-empty) baskets full of fruits. We know that it is not difficult at all to chose a (single) fruit from every basket. But the following similar question, which at first seems to be trivial, is actually complicated:

A non-empty set S is given, whose elements are pairwise disjoint non-empty sets S_a. Does there exist a set E which has as elements an element x_a from every set S_a?

The difficulty to answer this question appears when S is an infinite set. In the beginnings of the 20th century, Ernst Zermelo and others tried unsuccessfully to give an answer to this question. Zermelo had the feeling that no provable answer to this question exists, and hence the only way to overcome this difficulty was to consider the positive answer to this question as an axiom. Today we call it the "Axiom of Choice" and its statement is the following:

Axiom of Choice. Let S be a non-empty set consisting of non-empty sets. Then there exists a so called **choice function** $f : S \to \bigcup_{A \in S} A$ such that $f(A) \in A$ for every $A \in S$.

The axiom of choice is necessary for the proof of many propositions in various areas of mathematics. As an example we will prove the following proposition:

"Every infinite set has a countably infinite subset."

Proof. Let X be an infinite set. Then since $X \neq \emptyset$, we may choose an element $x_0 \in X$. Next we choose an element $x_1 \in X \setminus \{x_0\}$. Similarly we choose $x_2 \in X \setminus \{x_0, x_1\}$. After choosing in the same way x_{k-1}, we choose $x_k \in X \setminus \{x_0, x_1, ..., x_{k-1}\}$. x_k exists for every $k \in \mathbf{N}$ because X is an infinite set, which means that $X \setminus \{x_0, x_1, ..., x_{k-1}\} \neq \emptyset$ for every $k \in \mathbf{N}$. The set $\{x_k : k \in \mathbf{N}\}$ is a countably infinite subset of X.

Mathematicians invented also other axioms which are equivalent to the axiom of choice, and whose use is sometimes equally appropriate with the use of the axiom of choice. In order that the reader understands these axioms and their relation to the axiom of choice, we must first give some definitions.

If A and B are arbitrary (different or equal) sets, we remind the reader that we call relation Σ from A to B every subset of the Cartesian product $A \times B$. Instead of writing $(a, b) \in \Sigma$, we usually write $a \Sigma b$. The notation $a \Sigma b$ is read as follows: "a is in relation Σ with b".

A relation $\leq$ on a set A is called a **partial order** if and only if it is reflexive (i.e., $x \leq x$ for every $x \in A$), transitive (i.e., $(a \leq b \ \& \ b \leq c) \Rightarrow a \leq c$) and antisymmetric (i.e., $(a \leq b \ \& \ b \leq a) \Rightarrow a = b$).

We call a **partially ordered set** a pair $(A, \leq)$, where A is a set and $\leq$ is a partial order on A. A partial order $\leq$ on a set A is said to be a **linear order** (or a complete order) if and only if for every pair (a, b) of elements of A, we have $a \leq b$ or $b \leq a$. A **linearly ordered set** (or a completely ordered set) is a pair $(A, \leq)$, where A is a set and $\leq$ is a linear order on A.

Example. Let Σ be the relation on the plane $\mathbf{R} \times \mathbf{R}$ which is defined as follows: $(a_1, a_2) \Sigma (b_1, b_2) \iff (a_1 \leq b_1 \ \& \ a_2 \leq b_2)$. It is easily recognized that Σ is a partial order on $\mathbf{R} \times \mathbf{R}$. But since neither $(1, 2) \Sigma (2, 1)$ holds nor $(2, 1) \Sigma (1, 2)$ holds, it follows that Σ is not a linear order.

Example. If X is an arbitrary set, then it is easily recognized that the power set $P(X)$ is partially ordered with respect to the relation $\subset$ (i.e., the relation of "being a subset of") on $P(X)$.

In contemporary algebra and topology it is usually more appropriate to make use of the so-called "Hausdorff Maximality Principle" instead of the Axiom of Choice. In order to help the reader understand this principle we give first the following definitions.

Definition 1. Let $(A, \leq)$ be a partially ordered set.
(a) An element $u \in A$ is called an **upper bound** of a subset B of A if and only if $u \geq b$ for every $b \in B$.
(b) An upper bound u_0 of $B \subset A$ is called a **supremum** of B if and only if $u_0 \leq u$ for every upper bound u of B.
(c) An element $e \in A$ is called **maximal** if and only if for any $a \in A$, $e \leq a \Rightarrow e = a$.

Definition 1'. Let $(A, \leq)$ be a partially ordered set.

(a) An element $v \in A$ is called a **lower bound** of a subset B of A if and only if $v \leq b$ for every $b \in B$.

(b) A lower bound v_0 of $B \subset A$ is called an **infimum** of B if and only if $v_0 \geq v$ for every lower bound v of B.

(c) An element $e' \in A$ is called **minimal** if and only if for any $a \in A$, $e' \geq a \Rightarrow e' = a$.

Example. Let X be an arbitrary set and let $\mathcal{B}$ be a subset of the partially ordered set $(P(X), \subset)$. Then the supremum of $\mathcal{B}$ is $\bigcup_{B \in \mathcal{B}} B$ and the infimum of $\mathcal{B}$ is $\bigcap_{B \in \mathcal{B}} B$.

The "Hausdorff Maximality Principle", which is equivalent to the "Axiom of Choice", as mentioned above, is the following:

"Let $\mathcal{J}$ be the set of all linearly ordered subsets of a partially ordered set $(A, \leq)$. We know that $(\mathcal{J}, \subset)$ is a partially ordered set. The Hausdorff maximality principle asserts that $(\mathcal{J}, \subset)$ has a maximal element."

The following theorem, which is called "Zorn's Lemma", is also a proposition which is equivalent to the "Axiom of Choice".

"Let $(A, \leq)$ be a partially ordered set in which every linearly ordered subset has an upper bound. Then A has a maximal element."

Remark. The last theorem constitutes a typical example of an "existence theorem". This means that the proof of this theorem does not provide a constructive method of finding a maximal element, but simply asserts the existence of this element. This remark is also true in the case of the proof of the Hausdorff Maximality Principle.

An application of Zorn's Lemma is the following proposition:

"If A and B are non-empty sets, then either there exists a one to one function of A into B or there exists a one-to-one function of B into A."

From this proposition we obtain the following conclusion:

"If A and B are two sets, then $|A| \leq |B|$ or $|B| \leq |A|$."

Another application of Zorn's Lemma is the proof of the proposition that bears the name the "Well-ordering Principle" of Ernst Zermelo.

Definition. A linearly ordered set $(A, \leq)$ is called **well-ordered** if and only if every non-empty subset B of A contains a single element $b \in B$ such that $b \leq x$ for every $x \in B$, which is called "the minimum element of B". If $(A, \leq)$ is a well-ordered set, then the relation $\leq$ is called a **well-ordering relation**.

Examples.

(a) The set of natural numbers is well-ordered with respect to the known relation "less than or equal".

(b) The set of rational numbers is not well-ordered with respect to the relation "less than or equal".

We will say that a set A **can be well-ordered** if there exists a well-ordering relation on it.

Theorem (Well-ordering Principle).
Every set can be well ordered.

Note. The well-ordering principle presents another example of a non-constructive proposition, because the proof of the last theorem provides no method of constructing the well-ordering relation of a set A. The theorem simply asserts the existence of a well-ordering relation. It does not say which one is it. In particular, we don't know, for example, how the set of real numbers can be well-ordered.

It is proved that the Well-ordering Principle implies the Axiom of Choice.

If we summarize the above, we have:

Axiom of Choice $\Rightarrow$ Hausdorff Maximality Principle $\Rightarrow$ Zorn's Lemma $\Rightarrow$ Well-ordering Principle $\Rightarrow$ Axiom of Choice.

Next we discuss briefly the notion of an "ordinal number".

In the arithmetic of finite numbers, the cardinals are the numbers 0, 1, 2, 3, ... and the "ordinal" numbers are the ones that give an ordering to the objects of a set: first, second, third, and so on. There exists no distinction between the finite cardinal numbers and the finite ordinal numbers, i.e., the numbers 0, 1, 2, 3, ... are the finite cardinal and at the same time ordinal numbers.

But what is the meaning of an "infinite ordinal number"? Just like an infinite cardinal number is obtained by an infinite set, an infinite ordinal number is obtained by an infinite well-ordered set. We have the following definition:

Definition. Two well-ordered sets $(A, \leq)$ and $(B, \leq')$ are said to be order-isomorphic (or simply isomorphic) if there exists a one-to-one and onto function $f : A \to B$ such that if $a_1, a_2 \in A$ and $a_1 \leq a_2$, then $f(a_1) \leq' f(a_2)$. Such a function $f : A \to B$ is called an **order-isomorphism** (o-i).

It is easily recognized that if $f : A \to B$ is an o-i, then $f^{-1} : B \to A$ is also an o-i. Moreover, if $g : B \to C$ is an o-i, then $g \circ f : A \to C$ is an o-i. If $(A, \leq)$ and $(B, \leq')$ are isomorphic, then we write $(A, \leq) \approx (B, \leq')$ or simply $A \approx B$. The relation $\approx$ is reflexive, symmetric and transitive. Here, we call $\approx$ a "relation" in the sense that it is a relation on every set of well-ordered sets.

In general, we consider the "ordinal number" as a notion which obeys the following rules:

K-1 At every well-ordered set $(A, \leq)$ we assign an ordinal number which we denote by $ord(A, \leq)$. If α is an ordinal number, then there exists a well-ordered set $(A, \leq)$ such that $\alpha = ord(A, \leq)$.

K-2 If $(A, \leq)$ and $(B, \leq')$ are two well-ordered sets, then $ord(A, \leq) = ord(B, \leq') \iff (A, \leq) \approx (B, \leq')$.

K-3 $ord(A, \leq) = 0 \iff A = \emptyset$.

K-4 If $(A, \leq)$ is a well-ordered set such that $A \cong \{1, 2, ..., k\}$, where k is a positive integer, then $ord(A, \leq) = k$.

The ordinal number of the set $\{1, 2, 3, ...\}$ with the usual order relation "less than or equal" is denoted by the Greek letter ω. So we have

$$\omega = ord(1, 2, 3, ...).$$

Note. Even though a given set has a single cardinal number, the same set may have two different ordinal numbers if it can be well-ordered in two different ways. For example, the set $\{1, 2, 3, ...\}$ can be well-ordered in the following two different ways:

$$(\{1, 2, 3, ...\}, \leq) = \{1, 2, 3, ...\} \quad \text{and} \quad (\{1, 2, 3, ...\}, \leq') = \{1, 3, 5, ..., 2, 4, 6, ...\}.$$

It can be proved that these two well-ordered sets are not order-isomorphic.

Cantor's set theory was a daring step in an area with which mathematicians had occupied themselves from time to time, since the days of the ancient Greeks. Set theory requires a rigorous application and use of logical arguments. It asserts the existence of infinite sets with all the more larger cardinals, and in general it introduces notions which cannot be perceived through intuition. Therefore, it would have been very odd if these extremely revolutionary ideas had not met with strong opposition even from the great mathematicians of that time. Powerful arguments against set theory were stated by Kronecker, Felix Klein and Poincaré. Poincaré calls set theory a "pathological case" which will be rejected by future generations. As opposed to that, other famous mathematicians, like Adolf Hurwitz and Hadamard, appreciated the important applications to the theory of transfinite numbers. Applications were also made in Measure Theory and in Topology. Hilbert disseminated Cantor's ideas in Germany and in 1926 he said: "No one can drive us away from the heaven that Cantor created for us". Hilbert was a devout supporter of Cantor's cardinal arithmetic, which he characterized as the most extraordinary product of abstract mathematical thought. Bertrand Russell characterizes Cantor's work as the most important acquisition of that time.

This was the situation that was created during the beginnings of the 20th century as far as set theory is concerned.

CHAPTER 7

LOGIC

Logic was not always considered to be a branch of mathematics. In particular, Aristotle (384-322 B.C.), who was the first person in the West who wrote about logic, did not consider Logic part of mathematics.

Among the principles stated by Aristotle in Logic are the following:

$\neg(\neg p) \iff p$ \hfill (double negation)
$p \vee \neg p$ \hfill (the principle of the excluded middle)
$(p \Rightarrow q) \iff (\neg q \Rightarrow \neg p)$ \hfill (contraposition)

We remind the reader about the meaning of these symbols: If A and B are propositions, then the propositions (A and B), (A or B), (A implies B), (not A) are denoted by

$$A \wedge B, \quad A \vee B, \quad A \Rightarrow B, \quad \neg A$$

respectively. The proposition $(A \Rightarrow B) \wedge (B \Rightarrow A)$ is denoted by $A \iff B$ and is read "A and B are equivalent".

Aristotle was greatly occupied with a type of logical procedure which is called "syllogism", and which dominated "logical reasoning" for the next two thousand years. Aristotle analysed the following four types of basic declaratory phrases:

$S \ a \ P$ meaning: all S are P
$S \ e \ P$ meaning: no S is P
$S \ i \ P$ meaning: some S are P
$S \ o \ P$ meaning: some S are not P

Aristotle remarked that PeS is equivalent to SeP and that PiS is equivalent to SiP. He accepted that SaP implies SiP. However, by accepting SiP, he accepted the existence of some S which may not exist. Obviously, Aristotle had not conceived the notion of the empty set ($\emptyset$) used today. For example, today we accept the proposition "all unicorns have horns", but we do not accept the proposition "some unicorns have horns"!

We call "syllogism" a logical procedure by which, starting from two given propositions, allows us to obtain as a conclusion a third proposition. We present the following four first kinds of syllogisms:

(1) MaP \quad (2) MeP \quad (3) MaP \quad (4) MeP
    $\underline{SaM}$ \quad\quad $\underline{SaM}$ \quad\quad $\underline{SiM}$ \quad\quad $\underline{SiM}$
    SaP \quad\quad SeP \quad\quad SiP \quad\quad SoP

An example of a syllogism of type (4) is the following:

No irresponsible person is wise
<u>Some members of parliament are irresponsible</u>
Some members of parliament are not wise

The Stoical philosophers (200 B.C.) were essentially the first who introduced what we call today truth-tables in logic, which were introduced again after many centuries by Ludwig Wittgenstein (1889-1951). In particular, the Stoical philosophers examined the problem whether proposition $p \vee q$ is true whenever p or q is true, as well as the problem of whether proposition $p \Rightarrow q$ is true whenever p is false. They obtained the conclusions that we present below using truth-tables, where the letters T and F mean "true" and "false" respectively.

p	q	$p \vee q$
T	T	T
T	F	T
F	T	T
F	F	F

p	q	$p \Rightarrow q$
T	T	T
T	F	F
F	T	T
F	F	T

According to the Stoical philosophers, a phrase has only two truth-values, T or F, i.e., it is either true or false. In particular, they believed that a phrase like "An earthquake will take place tomorrow" is either true or false. It is believed that their view that even phrases of this kind have only two truth-values, is due to the fact that, in their minds, the future was already predetermined by fate. In other words, whether an earthquake will take place tomorrow or not is already predetermined by fate, and for this reason the phrase in question has only two truth-values.

The ideas of Gottfried W. Leibniz (1646-1716) seemed to constitute a progress in the area of logic. Leibniz alleged that a universal symbolic language would be a most useful one not only for mathematicians but for science in general. Unfortunately, Leibniz published nothing regarding these ideas, perhaps because he was preoccupied with the dispute with Newton about who was the first to discover infinitesimal calculus, or perhaps because he was preoccupied with his (successful) efforts to help George I to occupy the throne of England.

The so-called **symbolic logic** came into the limelight in 1847. Symbolic logic is a branch of logic, where one investigates, using mathematical symbols, the logical conclusions that are usually used in mathematics. During that year, George Boole (1815-1864) (*An investigation of the laws of thought*, Walton and Maberly, 1854) and Augustus de Morgan (*Formal logic, or the calculus of inference*, Taylor and Walton, 1847) introduced the so-called **algebra of logic**, an algebra of sets or relations, which was developed further but without reaching the level of contemporary symbolic logic.

The following two laws are due to De Morgan:

$$\neg(p \wedge q) \iff (\neg p \vee \neg q) \qquad \neg(p \vee q) \iff (\neg p \wedge \neg q).$$

The next great step was made by Gottlob Frege (1848-1925), who was the first to conceive the contemporary picture of the universal quantifier and the existential quantifier, but he did not make use of their contemporary notation, i.e., of $\forall$ and $\exists$ respectively.

Frege too tried to express the entire content of mathematical science using logical symbols, thus reducing mathematics to logic and adopting the philosophic thesis that bears the name "Logicism".

A central point of Frege's theory is the "hypothesis" (comprehension scheme) that every property expressed by a predicate of the language that we use corresponds to a unique and clearly determined set, whose elements are exactly those that have the property in question. Frege expresses this hypothesis briefly as follows:

$$\exists y \forall x (x \in y \iff P(x)), \qquad (\star)$$

where $P(x)$ is an arbitrary proposition that expresses the property that was mentioned above and is written using symbols, which possibly include the free variable x. This "hypothesis" of Frege is accompanied by the following axiom that guarantees the uniqueness of the set y, whose existence was assumed in $(\star)$:

$$\forall y \forall z (\forall x (x \in y \iff x \in z) \Rightarrow y = z). \qquad (\star\star)$$

In particular, axiom $(\star\star)$ implies the existence of at most one y which has no elements, i.e., of at most one empty set. Axiom $(\star\star)$ includes implicitly the hypothesis that all objects included in the language presented by Frege are sets.

Immediately after Frege completed the writing of his book, in which he developed the above ideas, Bertrand Russell (1872-1970) sent him a letter, in which he wrote to him that a serious problem exists regarding the case in which $P(x)$ is the proposition $\neg(x \in x)$. Indeed, if y is such that

$$\forall x(x \in y \iff \neg(x \in x)),$$

then one obtains the following particular case

$$y \in y \iff \neg(y \in y)$$

which is a contradiction.

This last argument is known as "Russell's paradox". A way of eliminating this contradiction is not to allow expressions like $x \in x$.

In order to avoid this paradox, Russell and Whitehead suggest the so-called "theory of types", according to which every symbol that represents an entity must correspond to some natural number which is the type of the symbol, and the expression $a \in b$ is allowed only if the type of b is by one unit larger than the type of a. Despite the fact that there exist today various writings on the "theory of types", mathematicians, in their majority, prefer to use other methods in order to avoid Russell's paradox and other paradoxes as well. People who study problems-paradoxes of this kind follow the theory of sets of Gödel and Bernays, where a distinction is made between "sets" and "classes", and where only sets can be members of other sets. Unfortunately though, one has to also add a number of axioms to this theory in order to be able to determine which classes are sets and which classes are not sets. Besides, Logicians prefer to use the theory of sets of Ernst Zermelo (1871-1953) and Abraham Fraenkel (1891-1965), in which $(\star)$ is modified and has the following form:

$$\forall z \exists y \forall x (x \in y \iff (x \in z \wedge P(x))).$$

But even here the founders of this theory were obliged to introduce additional axioms which make $(\star)$ lose its simplicity.

The fact that $(\star)$ can lead to contradictions, influenced many mathematicians and made them more cautious when they face procedures of this kind apart from the most basic ones. Among these mathematicians are the so-called "Intuitionists", whose ideas are presented immediately below.

The School of the Intuitionists

This school appeared near the end of the 19th century, when an effort was made towards a more rigorous foundation of both the number system and geometry. The discovery of the existence of the paradoxes mentioned above, strengthened further the development of this school.

L. Kronecker is considered as the first intuitionist. According to him, Weierstrass's rigorous treatment of mathematics introduced principles which cannot be acceptable, and Cantor's work regarding transfinite numbers and set theory is not mathematics but mysticism. He said that God created the natural numbers while all the rest were a "questionable" work of humans. He also accepted the fractional numbers, because they can be defined with the help of the natural numbers.

Kronecker rejected the theory of irrational numbers as well as the theory of continuous functions. After learning that Lindemann had proved the transcendentality of it, he asked him: "Why should we occupy ourselves with such problems since irrational numbers do not exist?" Kronecker said that "every theorem of mathematical analysis must be such that one can interpret it as something that provides relations between integers only."

Another objection of Kronecker concerns the fact that in many areas of mathematics, there do not exist constructive methods and criteria for the determination (after a procedure consisting of finitely many steps) of the objects considered. He believed that the definitions, must at the same time provide a way or ways of calculating the object defined (after a procedure consisting of finitely many steps), and that proofs of "existence" must be such that they provide a way of calculating, as accurately as we please, the quantity whose existence is being proved. It should be noted that Kronecker tried very little to promote the Theory of Intuitionism which he supported. His work in number theory was very successful, and equally successful was his work in algebra; in these works he did not conform to what he had suggested. In his time, his philosophy had no supporters and for twenty years no one followed his ideas. But when the paradoxes mentioned above appeared, the Intuitionistic Philosophy came into the limelight.

H. Poincaré (1854-1912) was a major supporter of this philosophy He claimed that it is ridiculous to try to found mathematics on logic, because then mathematics would become an endless tautology. Poincaré alleged that if Logicism was scrutinized carefully, then it is doubtful whether anything would be left from it. He also predicted that the mathematics that serves useful purposes will continue to develop following its own principles and will continue to make its usual achievements, which are definite and which will never need to be reconsidered.

Poincaré did not accept principles whose definition requires the use of infinitely many words. Thus, according to Poincaré, a set obtained using the axiom of choice is essentially not defined when the choice is made from every set that belongs to an infinite family of sets. In addition, Poincaré believed that arithmetic theory cannot be built axiomatically, because such a structure is intuitively obvious. Also he claimed that mathematical induction is a fundamental intuition and not simply an axiom. He insisted that all laws and all proofs must be constructive.

Poincaré agreed with Russell that the origin of the paradoxes mentioned earlier, are that we define collections or sets which include the object that is being defined. Thus, the set A of all sets includes A. But A cannot be defined until every element of A is defined.

Discussions related to the logical foundation of mathematics took place between Emile Borel (1871-1956), René Baire (1874-1932), Jacques Hadamard (1865-1963) and Henri Lebesgue (1875-1941). Borel agreed with Poincaré that the set of integers should not be axiomatized. Borel criticised the axiom of choice because it requires uncountably many choices, which our intuition cannot conceive. Hadamard and Lebesgue went further by saying that even a sequence of countably many arbitrary successive choices is not intuitively perceived, because it requires infinitely many operations which we cannot conceive.

According to Lebesgue, all difficulties are reduced to the fact that one must know what it means to say that a mathematical object "exists". In the case of the axiom of choice Lebesgue wonders: "If simply I have in mind some way of selection, can't I change my choices during the time that I think?" According to him, even the choice of a single object from some set presents the same difficulties. Hence, we must know that the selected object exists, which means that we must name it explicitly at the moment we select it. Thus, Lebesgue rejected Cantor's proof of the existence of transcendental numbers. But Hadamard and Borel noted that if we follow Lebesgue's objections then we must also reject the existence of the set of real numbers.

The above sporadic objections were put into a systematic philosophy by L. E. J. Brouwer, who is the contemporary founder of Intuitionism (1881-1966). But just like Kronecker, his work on Topology was not consistent with the intuitionistic philosophy that he professed.

According to Brouwer, the fundamental intuition is the appearance of events in a time series. "Mathematics is obtained when the notion of duality that is created as time elapses takes an abstract form, i.e., when it is relieved of the events that caused it. The resulting void type [the relation of n to $n+1$], which is common to all these dualities, becomes the initial intuition in Mathematics and, as it repeats itself ad infinitum, it creates new mathematical objects." Thus, from the repetition that takes place ad infinitum, our mind forms the notion of successive natural numbers. This idea was also supported by Kant, by W. R. Hamilton, and by the philosopher A. Schopenhauer.

According to Brouwer, "mathematical reasoning" is a process of construction which is independent of the world of our experience. This process is based only on the fundamental mathematical intuition, which must not be considered as an undefinable notion, but rather as a tool through which all undefinable notions that we meet in the various mathematical systems become perceived.

Brouwer believed that the only possible process of foundation of mathematics is the "constructive" process, which has to distinguish which notions are acceptable by intuition and which ones are not. Mathematical ideas exist in our mind before "language", "logic" and "experience". Neither experience nor logic, but intuition is the one that determines whether ideas are correct. It should be noted that when he talks about experience he talks about its philosophic sense and not about its historic sense.

Herman Weyl (1885-1955) also belonged to the School of the Intuitionists.

We will not discuss any further the Theory of Intuitionism. By summarizing we would say that an intuitionist believes that the proof of a proposition must be "constructive". For example, a proposition, where one is asked to prove the existence of some object, the proof must consist in providing (via construction and display) such an object.

In addition, in order to prove a proposition of the form $p \vee q$ we must prove that p is true or q is true. As an example we consider the following two propositions:

A: Every even natural number ≥ 4 is the sum of two prime numbers (Goldbach's conjecture).
$\neg A$: Not all even natural numbers ≥ 4 are the sum of two prime numbers.

According to the classical viewpoint, which accepts the principle of the excluded middle, the proposition $A \vee (\neg A)$ is true. But an intuitionist cannot assert that $A \vee (\neg A)$ is true, because both propositions A and $\neg A$ remain unproved up to this day.

Obviously the mathematics followed by an intuitionist (which is sometimes called "intuitionistic mathematics") differs greatly from classical mathematics, which was accepted by the great majority of mathematicians even before 1900. Intuitionists investigate "mental mathematical constructions without making any reference to matters regarding the nature of the constructed objects, as for example, whether these objects exist independently of whether we have knowledge of their existence or not". As opposed to that, other mathematicians adopt the classical viewpoint because they believe that, "in mathematics, objects and mathematical spaces "exist" just like platonic ideas exist". For example, they believe that the propositions that refer to these objects and describe their properties either hold or do not hold, and hence they are either true or false. The reader may find more about the Intuitionistic Philosophy in the book: *P. Benacerraf and H. Putnam (Editors), Philosophy of Mathematics, Selected Readings, Cambridge*, 1983.

The School of the Formalists

Hilbert founded the school of formalists, which after the classical and the intuitionistic school is the third school that studied the philosophy of mathematics.

Hilbert's work in this direction began with his efforts to found the arithmetic system without using set theory. Hilbert had already proven that the axiomatic foundation of geometry is consistent if the number system is consistent. So it was crucial for him to prove the consistency of the number system.

Hilbert rejected Kronecker's claims that the irrational numbers must be excluded. He accepted the existence of the infinite and he was a supporter of Cantor's work. Hilbert presented these ideas at the 1904 International Congress of Mathematicians, but then for fifteen years he did not study these problems any further. He later returned to this subject and continued his work on the foundation of mathematics until the end of his life. He published some very basic papers during the 1920s and gradually a group of mathematicians agreed with his views.

The formalists allege that logic must be studied in parallel with mathematics, and that each of the various branches of mathematics has its own axiomatic foundation consisting of logical and mathematical notions and principles. Logic is a language of signs (symbols) which uses formulas to express mathematical propositions and expresses the mathematical arguments following certain axioms. These axioms express the rules through which formulas are derived from other formulas. All signs and symbols are relieved of their meaning in relation to their content. The various formulas may have intuitive meaning, but these meanings do not form part of mathematics.

Hilbert accepts the principle of the excluded middle and remarks that prohibiting a mathematician from using this principle is like prohibiting a boxer from using his fists.

Since mathematics involves symbolic expressions, all laws of Aristotelian Logic can be applied to these expressions. In this new sense, the mathematics of infinite sets is acceptable. In addition, by avoiding the explicit use of the word "all", Hilbert was hoping to avoid the presence of "paradoxes".

In order to state the logical axioms, Hilbert introduced symbols for the notions and relations "and", "or", "negation", "exists", and the like. These symbols constituted the building blocks for the construction of the various formulas.

In order to deal with the notion of infinity, apart from the known axioms, Hilbert uses the so-called "transfinite axiom"

$$A(\tau A) \to A(a)$$

According to Hilbert, this means that if property A holds for an object τA, while τA is taken to be a point of reference, i.e., which we believe to be true (something that we trust), then this property A holds for any other object. For example, if we assume that A means "can be corrupted", then if Aristides the Just (whom we trust) can be corrupted, then every person can be corrupted.

According to Hilbert every mathematical proof consists of the following procedure: Verify that some formula holds, verify that this formula implies some other formula, and establish the validity of the second formula. The proof of an arbitrary theorem consists of a sequence of such steps. In this process, the substitution of a symbol by another symbol or by a group of symbols is an allowed operation. Hence, formulas are produced by applying the rules of manipulation of the symbols that belong to already derived formulas.

It follows from the above that, for the formalists, mathematics is a collection of axiomatic systems, each of which has its own logic, its own principles, its own axioms, its own rules for obtaining theorems, and its own theorems. Thus, mathematics ends up being a collection of formal systems in which formal expressions are derived from other formal expressions, and the work of mathematical science is to develop these systems.

Of course, an important question remains: Are the derived conclusions free of contradictions? According to a theorem in Logic, the existence of any contradictory proposition implies the proposition "$1 = 2$". Thus the basic question is: Are we sure that in a given system we cannot derive the proposition "$1 = 2$"?

During the years 1920-1930, Hilbert and his students Wilhelm Ackermann (1896-1962), Paul Bernays (1888-1977), and J. von Neumann (1903-1957), gradually formulated the so-called Beweistheorie (proof theory) or Hilbert's "Metamathematics". This is a method for proving the consistency (non-existence of contradictions) of an arbitrary axiomatic system. In metamathematics, Hilbert suggests the use of a special logic which must be basic and in which any objections must be eliminated. We will not discuss any further this logic.

Today the question of whether a large part of classical mathematics is consistent is reduced to the question of whether the system of natural numbers is consistent.

Hilbert's school had proved the consistency of very simple formal systems and it was thought at that time that the consistency of the arithmetic system and the consistency of set theory would be also proved in the near future. But in 1931 Kurt Gödel (1906-1978) an Austrian mathematician from the University of Vienna, proved a remarkable result: He proved that the consistency of a system that follows ordinary logic, which includes the system of natural numbers, cannot be proved if we restrict ourselves to the use of notions and methods of this system. This means that such a system contains propositions S, such that neither S nor $\neg S$ can be proved inside the system. In other words, the "consistency" of number theory cannot be proved using the formal logic of metamathematics. Commenting on this result, Weyl said the following: "God exists because mathematics is consistent and Devil exists because we cannot prove the consistency of mathematics".

The above result of Gödel follows from a more general result, the so-called "Gödel's incompleteness theorem": "If an arbitrary formal (axiomatic) theory T, which includes number theory, is consistent and if the axioms of the formal system of arithmetic are axioms or theorems of T, then T is incomplete". Here the phrase "T is incomplete" means that there exists a proposition S of number theory such that neither S not $\neg S$ can be derived from the axioms of T. But since either S or $\neg S$ is true, it follows that there exists a true proposition in number theory whose validity cannot be proved inside T.

This last result can be applied to the Russell-Whitehead system, to the Zermelo-Fraenkel system, and to Hilbert's axiomatic number theory.

The incompleteness of a system is a drawback of the system, because it is impossible to derive in it all propositions which are true in it. Moreover, something even worse happens, namely there exist propositions for which we cannot decide whether they are true or false, even though intuitively they seem to be true. The incompleteness of a system cannot be removed by accepting proposition S or proposition $\neg S$ as an axiom of this system, because Gödel proved that any system that contains number theory must contain a proposition which is not provable inside this system.

One of the consequences of Gödel's theorem is that no axiomatic system is appropriate to derive even some branch of mathematics, because every such axiomatic system is incomplete. There exist propositions, in which the notions involved belong to the system and which cannot be proved inside the system. It is possible though to prove that these propositions are true using informal reasoning, as, for example, using the logic of metamathematics.

The above theorem shows that there exist limitations in what can be achieved by axiomatization: This is in contrast to the viewpoint that was formed during the end of the 19th century, i.e., that the power of mathematics can be extended if by increasing the number of axioms in every axiomatic branch of mathematics. Gödel's result ruled out the complete axiomatization of mathematics. Of course, this does not mean that it is impossible to devise new methods of "proof" which will go beyond Hilbert's metamathematics. Besides, this is something that Hilbert himself also believed.

As was expected, the program of the Formalists was not accepted by the Intuitionists and the disputes regarding these matters continued to exist.

Weyl's viewpoint is the following: "The question of whether we will obtain in the future a final foundation of mathematics and a final picture of "what is mathematics" remains open. We don't know what direction will be followed in order to find the final solution, and we don't even know if we must hope that there exists some final objective answer to this question. The fact that humans occupy themselves with mathematics may perhaps be a creative and extremely original activity, just like language and music, but the historic evolution of this activity is logically and objectively completely unpredictable."

CHAPTER 8

FUNCTIONAL ANALYSIS

One of the main characteristics of contemporary mathematics is the separate study of basic notions of a known mathematical system, their reconnection (just like it happens with the parts of an engine) and formation of new combinations, and the subsequent study of these new combinations. As an example we will mention the system **R** of real numbers. The system **R** becomes better understood if we study its structure separately and from different points of view. For example:

The system **R** has an "algebraic" structure: It is a set where several algebraic operations have been defined (the most important of which are addition and multiplication), which have certain properties, as are **associativity**, **commutativity**, and others.

The system **R** has an "order" structure: It is a set where an order relation (denoted by $<$) has been defined, which has certain basic properties, as is transitivity (i.e., $(a < b \ \& \ b < c) \Rightarrow a < c$).

The system **R** has the structure of a "metric space": It is a set such that to every pair of its elements, which are called points, corresponds a number called their "distance", and this distance has certain properties, as, for example, it is zero if and only if the two points coincide.

The importance of the system **R** lies not only in the existence of these three or possibly even other structures, but also in the relations that exist between these structures. The fact that given three real numbers a, b and c, the inequality $a < b$ implies the inequality $a + c < b + c$, is a property that follows from the fact that **R** has both an algebraic and an order structure.

In general, the separate study of the various structures of a mathematical system provides general theories which are useful both in obtaining a better understanding of the system where these structures were originally observed, as well as in studying other systems where the same structures exist.

In particular, let us restrict ourselves to the area of Mathematical Analysis. The part of mathematics that began with infinitesimal calculus and was greatly developed in various directions under the same spirit is called **Classical Analysis**. Differential equations, series (and in particular the power series and the Fourier series), and functions of a complex variable constitute some of the branches of classical analysis. Due to its development, classical analysis had become so enormous that even experts could hardly follow its entire development.

Under the influence of this situation, certain mathematicians tried to isolate and study separately the basic principles (structures) on which classical analysis was based. Among these mathematicians are Riemann, Weierstrass, Cantor, Lebesgue, Hilbert, Riesz, and others. This effort led to the birth of topology, of contemporary algebra, and of the theory of measure and integration. When mathematicians began to apply these new theories to classical analysis, they initiated what we call

Contemporary Analysis. During its development, contemporary analysis led to the simplification of the proofs of many theorems of classical analysis. In addition, the study of the structures of the systems of real and complex numbers shed light and simplified greatly many problems of classical analysis.

During the end of the 19th century, use of transformations or operators acting on functions was made in many areas of mathematics. Even the usual operations of differentiation and anti-differentiation can be considered as operations that act on functions and produce new functions.

In the Calculus of Variations, one considers integrals of the type

$$J = \int_a^b F(x, y(x), y'(x))dx,$$

acting on a class of functions $y(x)$, among which one seeks the function that maximizes or minimizes the integral.

In differential equations one defines a differential operator as

$$L = \frac{d^2}{dx^2} + p(x)\frac{d}{dx} + q(x),$$

which acts on a class of functions $y(x)$. One seeks a concrete function $y(x)$ such that when L acts on $y(x)$ it gives 0 and also $y(x)$ satisfies some initial or boundary conditions.

In the area of integral equations, one considers operators such as

$$\int_a^b K(x,y)y(x)dx,$$

which act on a class of functions $y(x)$ and produce new functions.

In Functional Analysis of all these operators can be considered as acting on a class of functions which are considered to be points of a suitable space. Then an operator transforms points into points and in this sense it **constitutes** a generalization of the usual geometrical transformations.

We note that some of the above operators transform functions into real numbers and not into functions; such operators are called today **functionals**, while the term operator usually refers to transformations that transform functions to functions.

Functional Analysis unifies and generalizes the theory of operators, thus it is consistent with the general trend observed in the 20th century.

Functionals

A function of "lines" is a real function F whose values depend on all the values of a function $y(x)$ defined on some interval $[a, b]$. The functions $y(x)$ are considered as points of a space where both the neighborhood of a point and the limit of a sequence of points could be defined. Volterra defined for a functional $F(y(x))$ the notions of continuity, of the derivative and of the differential. But these definitions turned out to be inappropriate for the abstract theory of the calculus of variations, so they were later replaced.

The idea that a set of functions defined on the same interval can be considered as a set of points of a space, was stated before Volterra by Riemann, Giulio Ascoli (1843-1896), Cesare Arzela (1847-1912). Hadamard, and Emile Borel.

The foundation of an abstract theory of function spaces begins with the Doctoral Dissertation (1906) of the French mathematician Maurice Fréchet (1878-1973).

In his "functional calculus", he tried to unify ideas that already existed in the works of Cantor, Volterra, Arzela, Hadamard, and others. Fréchet introduced the basic notions of Cantor's set theory in function spaces, where points are functions. He also introduced in these spaces the notion of the limit of a set of points, but he did not give a clear definition of this last notion. However, he presented the characteristic properties of this limit which were general enough to unify the various kinds of limits that already existed in the theories that were developed until then. In this connection he introduced a class of L-spaces, where L indicates that the notion of a limit exists in everyone of these spaces: A_1, A_2,... are arbitrary elements of any of these spaces, then one must be able to determine whether or not there exists a unique element A, called the limit of the sequence $\langle A_n \rangle$, when it exists such that

(a) If $A_i = A$ for every i, then $\lim \langle A_i \rangle = A$.
(b) If A is the limit of $\langle A_n \rangle$, then A is the limit of every infinite subsequence of $\langle A_n \rangle$.

For every space of the class L, he incorporated the following notions which were already known from set theory:

- The derived set E' of a set E consists of those points of E which are limits of sequences of points of E.
- The set E is closed if E' is contained in E.
- The set E is perfect if $E' = E$.
- A point A of E is an interior point of E if A is not the limit of sequence of points that do not belong to E.
- The set E is compact either if it has finitely many points or if every infinite subset of E has at least one limit point.

When E is closed and compact it is called extremal (what he calls "compact" is now called "relatively sequentially compact" and what he calls "extremal" is now called "sequentially compact").

The first important theorem of Fréchet is the following: "If $\langle E_n \rangle$ is a descending sequence (i.e., $E_{n+1} \subset E_n$) of closed subsets of an extremal set, then their intersection is non-empty."

He introduced the following definitions: 1. "A functional U is called continuous at a point A of E, if $\lim_{n \to \infty} U(A_n) = U(A)$ for every sequence $\langle A_n \rangle$ of points of E which converges to A."

2. "U is upper semicontinuous at A, if $U(A) \geq \limsup U(A_n)$ for every sequence $\langle A_n \rangle$ of points of E which converges to A. U is lower semicontinuous at A if $U(A) \leq \liminf U(A_n)$ for every sequence $\langle A_n \rangle$ of points of E which converges to A."

Using the above definitions, Fréchet proved the following theorem: "Every continuous functional defined on some extremal set E is bounded and takes a maximum and a minimum value on E."

Fréchet introduced the notion of a metric space. In a metric space he defines a function which plays the role of a distance; he called this function écart and denotes it by (A, B). Today écart is called metric. The écart is defined for every two points A and B of the space and satisfies the following conditions:

(a) $(A, B) = (B, A) \geq 0$
(b) $(A, B) = 0$ if and only if $A = B$
(c) $(A, B) + (B, C) \geq (A, C)$ \hfill (the triangle inequality)

He calls $\mathcal{E}$ the class of metric spaces.

Fréchet gives examples of function spaces:

- The set of all continuous real functions defined on the same interval I. The distance of two functions f and g is defined to be equal to $\max_{x \in I} |f(x) - g(x)|$. It is a space of the class $\mathcal{E}$.
- Another example is the set of all sequences of real numbers. If $x = \langle x_1, x_2, ... \rangle$ and $y = \langle y_1, y_2, ... \rangle$ are any such sequences, then the écart of x and y is defined to be

$$(x, y) = \sum_{p=1}^{\infty} \frac{1}{p!} \frac{|x_p - y_p|}{1 + |x_p - y_p|}.$$

This is a space with infinitely many dimensions.

Using the spaces of the class $\mathcal{E}$, Fréchet was able to define for a functional the notions of continuity, of differentiability, and of the differential. But again these definitions were not appropriate for the calculus of variations.

Charles Albert Fischer (1884-1922) modified the definition of Volterra for the derivative of a functional in such a way that the improved definition is appropriate for the calculus of variations. The differential of a functional was defined in terms of the derivative. Moreover, Elizabeth Le Stourgeon (1881-1971) gave additional necessary definitions. The basic notion of the differential was a variation of the notion of differential given by Fréchet.

An important work in the theory of functionals is *Fondamenti di calcolo delle variazioni* (2 volumes, 1922, 1924), by L. Tonelli (1885-1946) where he treats the calculus of variations from the point of view of functionals.

Linear Functional Analysis

The calculus of variations studies functionals with rather special properties, which do not hold for functionals in general. Furthermore, these functionals are nonlinear and the nonlinearity creates difficulties which are not relevant to the functionals and the operators which appear in the study of linear integral equations.

The theory of Hilbert spaces was motivated by the study of linear integral equations. David Hilbert (1862-1943) remarked that a linear integral equation can be transformed into a system of infinitely many linear equations with respect to the Fourier coefficients of the unknown function. He considered the linear space l_2, which consists of all sequences $\langle x_n \rangle$ of complex numbers for which $\sum_{n=1}^{\infty} |x_n|^2 < +\infty$, and he defined for every pair of elements $x = \langle x_n \rangle \in l_2$ and $y = \langle y_n \rangle \in l_2$ their inner product $(x, y) = \sum_{n=1}^{\infty} x_n \overline{y_n}$. The space l_2 can be considered as an infinite-dimensional space which is an extension of the notion of Euclidean space. Later, F. Riesz (1880-1956) considered the functional spaces that we call today L^2-spaces and was able to give a satisfactory answer to the problem of expanding a function into a Fourier series. Below, we will give an axiomatic foundation of Hilbert spaces.

Schmidt, Fischer and Riesz, who contributed to the further development of the theory of solving linear integral equations, began also to work for the creation of an abstract theory of linear functionals and operators.

The American mathematician E. H. Moore (1862-1932) realized that there exist common characteristics between the theory of finitely many linear equations with finitely many unknowns, the theory of infinitely many linear equations with infinitely many unknowns, and the theory of linear integral equations. He developed his "General Analysis", which was an effort to establish an abstract theory containing the above concrete theories as particular cases. But this effort was of limited success.

We now present some background material regarding the theory of integral equations.

The equations that include the integrals of unknown functions are called **integral equations**. Among all integral equations, the ones that were studied the most are **linear integral equations**, i.e., the ones which are linear with respect to the unknown functions.

Let D be a domain in the n-dimensional Euclidean space and let $A(x)$, $f(x)$, $k(x,y)$ be functions that are defined for $x = (x_1, x_2, ..., x_n) \in D$, $y = (y_1, y_2, ..., y_n) \in D$. The integral equations of Fredholm type are of one of the following forms:

$$\int_D k(x,y)\phi(y)dy = f(x) \qquad (1)$$
$$\phi(x) - \int_D k(x,y)\phi(y)dy = f(x) \qquad (2)$$
$$A(x)\phi(x) - \int_D k(x,y)\phi(y)dy = f(x) \qquad (3)$$

where $\phi(x)$ stands for the unknown function and $\int_D dy$ denotes the n-fold integral $\int ... \int_D dy_1...dy_n$.

The equations of the form (1), (2) and (3) are called respectively equations of the **first**, the **second** and the **third kind**. The equations of the second kind have been studied in detail, while in many cases equations of the third kind can be reduced to equations of the second kind. The function $k(x,y)$ is called the **kernel** of the integral equation.

The integral equations of Volterra type are of one of the following forms:

$$\int_a^x k(x,y)\phi(y)dy = f(x) \qquad (1')$$
$$\phi(x) - \int_a^x k(x,y)\phi(y)dy = f(x) \qquad (2')$$
$$A(x)\phi(x) - \int_a^x k(x,y)\phi(y)dy = f(x) \qquad (3')$$

where $\phi(x)$ stands for the unknown function.

The equations of the form (1'), (2') and (3') are also called respectively equations of the **first**, the **second** and the **third kind**.

The kernels in the equations (1)-(3) and (1')-(3') are often written in the form $\lambda k(x,y)$, where λ is a parameter. This form is preferred when the equations are related to problems regarding eigenvalues that we will discuss below.

The theory of integral equations began in 1823 with N. H. Abel, who studied the relation between time and the orbit of a body that falls in the gravitational field. Let $\phi(t)$ be a quantity which changes with time and which is related, according to a certain law, to the values that it takes in a time interval of either the past or the future. Then the law according to which $\phi(t)$ changes can be described mathematically with an integral equation. The situation is similar when the variable t is not time but some other parameter. In this way, various problems in physics can be reduced to the solution of integral equations.

Let us consider now the homogeneous integral equation of the second kind:

$$\phi(x) - \lambda \int_D k(x,y)\phi(y)dy = 0$$

where D is a bounded and closed domain and $k(x,y)$ is continuous on $D \times D$. If this equation has a non-trivial solution $\phi(x)$ for some λ, then λ is called an **eigenvalue** or a **characteristic value** that corresponds to the kernel $k(x,y)$ and the non-trivial solution $\phi(x)$ is called an **eigenfunction** that corresponds to the kernel $k(x,y)$.

For an arbitrary eigenvalue λ, there exists a set of linearly independent eigenfunctions that correspond to λ such that every eigenfunction that corresponds to λ can be written as a linear combination of eigenfunctions that belong to this set. Such a set of linearly independent eigenfunctions that correspond to the eigenvalue λ is called a **fundamental system** that corresponds to λ and the number of elements of the fundamental system is called the **index of the eigenvalue** λ. The index of an eigenvalue is always finite.

An important contribution to the theory of integral equations is the following famous theorem of Hilbert, which was later called the **Hilbert-Schmidt Theorem**:

Let $k(x,y)$ be a continuous and symmetric kernel and let $h(x)$ be a function such that $|h(x)|^2$ is integrable on a bounded and closed domain D. Then a function $f(x)$ such that

$$f(x) = \int_D k(x,y)h(y)dy$$

can be expanded into a series of the form

$$f(x) = \sum_{n=1}^{\infty} c_n \phi_n(x),$$

where $\langle \phi_n(x) \rangle$ is a complete orthonormal system of fundamental functions that correspond to $k(x,y)$ and

$$c_n = \int_D f(x)\phi_n(x)dx \qquad (n=1,2,...).$$

The above series converges uniformly.

Note. A kernel $k(x,y)$ is called **symmetric** if the function $k(x,y)$ is real and $k(x,y) = k(y,x)$.

By making use of the above theorem, it is proved that if λ is not an eigenvalue, then we can find a solution $\phi(x)$ of the Fredholm equation

$$\phi(x) - \lambda \int_D k(x,y)\phi(y)dy = f(x),$$

where the kernel is symmetric.

The first essential step towards an abstract theory was made in 1907 by E. Schmidt and Fréchet.

The elements of Schmidt's function spaces are infinite sequences $z = \langle z_p \rangle$ of complex numbers such that

$$\sum_{p=1}^{\infty} |z_p|^2 < +\infty.$$

Schmidt introduced the notion of the **norm** $\|z\|$ of an element z, where

$$\|z\| = \left(\sum_{p=1}^{\infty} z_p \overline{z_p} \right)^{1/2}.$$

Following Hilbert, Schmidt used the notation (z, w) for the sum $\sum_{p=1}^{\infty} z_p w_p$ so that $\|z\| = \sqrt{(z, \overline{z})}$. Today we define $(z, w) = \sum_{p=1}^{\infty} z_p \overline{w_p}$. The elements z and w of the space are called orthogonal if and only if $(z, \overline{w}) = 0$. Later, Schmidt proves the following theorem which constitutes a generalization of the Pythagorean Theorem:

If $z_1, z_2, ..., z_n$ are n pairwise orthogonal elements of the space, then the relation

$$w = \sum_{p=1}^{n} z_p$$

implies that

$$\|w\|^2 = \sum_{p=1}^{n} \|z_p\|^2.$$

This theorem implies that n pairwise orthogonal elements are linearly independent.

Schmidt extended Bessel's inquality to this general space as follows:

If $\langle z_n \rangle$ is an infinite sequence of orthonormal elements such that $(z_p, \overline{z_q}) = \delta_{pq}$ (where $\delta_{pp} = 1$ and $\delta_{pq} = 0$ when $p \neq q$) and w is an arbitrary element, then

$$\sum_{p=1}^{\infty} |(w, \overline{z_p})|^2 \leq \|w\|^2.$$

It is also proved that the norm satisfies Schwarz's inequality and the triangle inequality.

A sequence $\langle z_n \rangle$ of elements of the space is said to **converge strongly** to z if $\|z_n - z\|$ converges to zero. If every strong Cauchy sequence $\langle z_n \rangle$ (i.e., every sequence for which $\|z_p - z_q\|$ converges to zero as p and q tend to ∞) converges to some element z of the space, then we say that the space is complete. The property of completeness is a very important property of a space.

Then, Schmidt introduces the notion of a (strongly) closed subspace. A subset A of the space is called a closed subspace if it is a closed set in the sense of convergence defined above and if it is algebraically closed, which means that if w_1 and w_2 are elements of A, then $a_1 w_1 + a_2 w_2$ (where a_1 and a_2 are arbitrary complex numbers) is an element of A. It is proved that such closed subspaces exist. To this end, it is enough to consider an arbitrary sequence $\langle z_n \rangle$ of linearly independent elements and then consider all the linear combinations of finitely many terms of $\langle z_n \rangle$. The closure of the set of these elements is an algebraically closed subspace.

So let A be a given closed subspace. Schmidt proves first that if z is any element of the space, then there exist two unique elements w_1 and w_2 such that $z = w_1 + w_2$, where w_1 belongs to A and w_2 is orthogonal to A, which means that w_2 is orthogonal to every element of A. (This result is known today as the "projection theorem", because w_1 is the projection of z on A.)

Moreover, $\|w_2\| = \min \|y - z\|$, where y ranges over arbitrary elements of A, and the minimum is attained only at $y = w_1$. The number $\|w_2\|$ is called the distance of z from A.

In 1907, Schmidt and Fréchet remarked that the space L^2 of functions f having the property that $|f|^2$ is Lebesgue integrable, has geometric properties which are completely analogous to the ones of the Hilbert space whose elements are sequences of numbers. This analogy was clarified further a few months later when Riesz proved that a notion of distance can be defined in the space L^2, and based on this notion one can build a geometry in L^2 (this proof was based on the so-called "Riesz-Fischer theorem", which states that there exists a one-to-one correspondence between the set of functions in L^2 and the set of sequences $\langle x_n \rangle$ of real numbers for which $\sum_{n=1}^{\infty} x_n^2 < +\infty$). In particular, it is proved that the space L^2 can be identified with the space of sequences, each of which is the sequence of Fourier coefficients of a function in L^2 with respect to a given complete orthonormal system of functions.

The problem of finding a function $f(x)$, whose Fourier coefficients $\langle a_n \rangle$ with respect to a given orthonormal system $\langle g_n \rangle$ are given and which satisfies the relations

$$\int_a^b g_n(x) f(x) dx = a_n \qquad n = 1, 2, ...,$$

constitutes the so-called "moment problem" (where the integration is in the sense of Lebesgue). This problem came up in the work of Riesz in 1907. In 1910, Riesz tried to generalize this problem, and in this effort Riesz made also use of Hölder's inequalities

$$\sum_{i=1}^n |a_i b_i| \leq \left(\sum_{i=1}^n |a_i|^p \right)^{1/p} \left(\sum_{i=1}^n |b_i|^q \right)^{1/q}$$

and

$$\left| \int_M f(x) g(x) dx \right| \leq \left(\int_M |f(x)|^p dx \right)^{1/p} \left(\int_M |g(x)|^q dx \right)^{1/q}$$

where $\frac{1}{p} + \frac{1}{q} = 1$. This motivated him to introduce the set L^p of functions f which are measurable on a set M and have the property that $|f|^p$ is Lebesgue integrable on M. The first important theorem of Riesz in this direction is that if a function $h(x)$ is such that the product $f(x)h(x)$ is integrable for every $f \in L^p$, then h belongs to L^q, and, conversely, the product of a function in L^p and a function in L^q is always a Lebesgue integrable function. It is assumed, of course, that $1 < p < \infty$ and $\frac{1}{p} + \frac{1}{q} = 1$. The L^p-spaces are complete with respect to the metric

$$d(f_1, f_2) = \left(\int_a^b |f_1(x) - f_2(x)|^p dx \right)^{1/p}.$$

Fréchet proved that to every continuous linear functional $U(f)$ defined on L^2, there corresponds a unique function $u(x)$ that belongs to L^2 such that

$$U(f) = \int_a^b f(x) u(x) dx$$

for every $f \in L^2$.

In 1909, Riesz generalized this result by expressing $U(f)$ as a Stieltjes integral, namely

$$U(f) = \int_a^b f(x) du(x).$$

Then, Riesz generalized this last result to linear functionals A which satisfy the condition that for all $f \in L^p$,

$$A(f) \leq M \left(\int_a^b |f(x)|^p dx \right)^{1/p}$$

where M depends only on A. Then there exists a function $a(x)$ in L^q, which is unique (in the sense that one can add to it only a function whose integral is equal to zero) and such that

$$U(f) = \int_a^b a(x) f(x) dx$$

for every $f \in L^p$. This last result is known as the Riesz representation theorem.

Note. Regarding the subjects discussed in this and in other chapters, the reader may consult the book *Functions of Real Variables*, S. Athanasopoulos - S. Papadamis, Athens, 1998, [8].

The main part of functional analysis concerns the abstract theory of operators, which appear in differential and integral equations.

This abstract theory connects the theory of eigenvalues of differential and integral equations with linear transformations, which take place in the n-dimensional space.

An example of such an operator is

$$g(x) = \int_a^b k(x,y) f(y) dy$$

where k is given. This operator assigns g to f and satisfies certain additional conditions. If we denote by A the abstract operator and we write $g = Af$, then in this case linearity means that

$$A(\lambda_1 f_1 + \lambda_2 f_2) = \lambda_1 A f_1 + \lambda_2 A f_2 \tag{1}$$

where the λ_i ($i = 1, 2$) are arbitrary real or complex constants.

The indefinite integral $g(x) = \int_a^x f(t) dt$ and the derivative $f'(x) = Df(x)$ are linear operators which act on appropriate classes of functions.

We say that an operator A is continuous if and only if whenever the sequence of functions $\langle f_n \rangle$ converges (in the sense of the space of functions considered) to the function f, the sequence $\langle A f_n \rangle$ converges to Af.

In the case of an abstract operator A, the property that corresponds to the property of the symmetry of the kernels $k(x, y)$ in the particular case of integral equations is, in general, the property of being "self-adjoint".

If for arbitrary functions f_1 and f_2, we have

$$(Af_1, f_2) = (f_1, Af_2)$$

where (f_1, f_2) denotes the scalar product of two functions of the space, then the operator A is called **self-adjoint**. In the case of integral equations, if

$$Af_1(x) = \int_a^b k(x,y) f(y) dy,$$

then
$$(Af_1, f_2) = \int_a^b \int_a^b k(x,y) f_1(y) f_2(x) dy dx$$
$$(f_1, Af_2) = \int_a^b \int_a^b k(x,y) f_2(y) f_1(x) dy dx$$

If the kernel k is symmetric, then $(Af_1, f_2) = (f_1, Af_2)$.

It is proved that the eigenvalues of arbitrary self-adjoint operators are real numbers and the eigenfunctions that correspond to distinct eigenvalues are pairwise orthogonal.

Riesz considered the solutions of the integral equation
$$\phi(x) - \lambda \int_a^b k(x,t) \phi(t) dt = f(x)$$
which belong to L^p-spaces (i.e., to spaces that he had introduced earlier).

Riesz considered the expression
$$\int_a^b k(s,t) \phi(t) dt$$
as a transformation acting on $\phi(t)$. Moreover, since the functions $\phi(t)$ belonged to the space L^p, the above transformation maps functions that belong to L^p to functions that belong to the same space or to some other space. In particular, a transformation or an operator T that transforms functions that belong to L^p into functions that belong to L^p is called linear on L^p if (1) is satisfied and T is bounded. This means that there exists a constant M such that for all f in L^p for which
$$\int_a^b |f(x)|^p dx \leq 1$$
we have
$$\int_a^b |T(f(x))|^p dx \leq M^p.$$
The infimum of all such M is called the norm of T and is denoted by $\|T\|$.

Riesz introduced also the notion of the adjoint of the operator T. For an arbitrary $g \in L^q$ and for T acting on L^p, the expression
$$\int_a^b T(f(x)) g(x) dx$$
defines, for a given g and for f ranging over L^p, a functional on L^p. Consequently, it follows from the Riesz representation theorem that there exists a function $\psi(x)$ in L^q, which is unique and such that
$$\int_a^b T(f(x)) g(x) dx = \int_a^b f(x) \psi(x) dx. \tag{2}$$
The adjoint or transpose operator of T is denoted by T^* and is defined to be the operator on L^q which assigns ψ to g via (2), i.e., $T^* g = \psi$. T^* satisfies the equality $(Tf, g) = (f, T^* g)$. T^* is a linear transformation which is defined on L^q and $\|T^*\| = \|T\|$.

Riesz was able to prove that the equation
$$T(\phi(x)) = f(x) \tag{3}$$

where T is a linear transformation defined on L^p, f is a known function, and ϕ is the unknown function, has a solution if and only if

$$\left| \int_a^b f(x)g(x)dx \right| \leq M \left(\int_a^b |T^*(g(x))|^q dx \right)^{1/q}$$

for all $g \in L^q$. In this way he was led to the notion of the inverse transformation or inverse operator T^{-1}, and in the same way he was led to the inverse operator $(T^*)^{-1}$. With the help of the adjoint operator he proves the existence of the inverses.

In 1910 Riesz introduced the notation

$$\phi(x) - \lambda K(\phi(x)) = f(x) \tag{4}$$

where K now represents $\int_a^b k(x,t) \star dt$ and $\star$ represents the function on which K acts.

In order to deal with the problem of eigenvalues in the case of integral equations, Riesz generalized to abstract operators the notion of complete continuity that was already introduced by Hilbert.

An operator K on L^2 is called **completely continuous** if K maps every weakly convergent sequence of functions to a strongly convergent sequence of functions, i.e., if $\langle f_n \rangle$ is weakly convergent, then $\langle K(f_n) \rangle$ is strongly convergent. (A sequence $\langle f_n \rangle$ is said to converge weakly to f in L^p, if $\int_a^b |f_n(x)|^p dx < M$, where M is independent of n, and if for every $x \in [a,b]$, $\lim_{n \to \infty} \int_a^x (f_n(t) - f(t))dt = 0$.) He also proves that when K is symmetric, its spectrum is discrete and not continuous, while the eigenfunctions that correspond to distinct eigenvalues are pairwise orthogonal.

The general definition of normed spaces appeared in the period 1920-1922 in the works of Stefan Banach (1892-1945), Hans Hahn (1879-1934), Eduard Helly (1884-1943) and Norbert Wiener (1894-1964).

Banach, in his efforts to generalize integral equations, made fundamental contributions to functional analysis.

In particular, he wanted to construct a normed space in which the norm is not defined through a scalar product. Even though in L^2 we have $\|f\| = (f,f)^{1/2}$, in a Banach space we cannot define the norm in this way because this space has no scalar product.

Banach defined what was later called a Banach space: This space E, whose elements are denoted by $x, y, z, \ldots$, is a space whose elements satisfy the following three groups of axioms:

The first group contains 13 axioms, which assert that E is a commutative group with respect to the operation of addition, that E is closed with respect to the operation of multiplication by a real number, and that the known properties of associativity and distributivity hold for the operations between real numbers and elements of E.

The second group of axioms characterizes the notion of the norm of an element, denoted by $\|x\|$. If a is an arbitrary real number and $x, y \in E$, then the norm has the following properties:

(a) $\|x\| \geq 0$
(b) $\|x\| = 0$ if and only if x is the zero vector.
(c) $\|ax\| = |a| \cdot \|x\|$
(d) $\|x+y\| \leq \|x\| + \|y\|$

The third group contains the axiom of completeness which asserts that if $\langle x_n \rangle$ is a Cauchy sequence with respect to the norm, meaning that $\lim_{n,p \to \infty} \|x_n - x_p\| = 0$, then there exists $x \in E$ such that $\lim_{n \to \infty} \|x_n - x\| = 0$.

The notion of orthogonality which is valid in Hilbert spaces, does not exist in Banach spaces. The first and the third group of axioms also hold in Hilbert spaces, but the conditions imposed by the second group are weaker than the ones that are satisfied by the norm in Hilbert spaces. Banach spaces include the L^p-spaces, the space of continuous functions and the space of essentially bounded measurable functions.

Banach was able to prove a number of known propositions regarding his spaces. One of his basic theorems is the following: "Let $\{x_n\}$ be a set of elements of E such that $\sum_{p=1}^{\infty} \|x_p\| < \infty$. Then the series $\sum_{p=1}^{\infty} x_p$ converges with respect to the norm to an element x of E."

Banach introduced the following definitions:

1. An operator F is said to be continuous at x_0 in relation to a set A, if the following conditions are satisfied: $F(x)$ is defined for all $x \in A$, x_0 belongs to both A and A' (the derived set of A), and whenever $\langle x_n \rangle$ is a sequence of elements of A which converges to x_0, the sequence $\langle F(x_n) \rangle$ converges to $F(x_0)$.

2. A sequence of operators $\langle F_n \rangle$ is said to converge with respect to the norm to an operator F on a set A, if for every $x \in A$, $\lim_{n \to \infty} F_n(x) = F(x)$.

3. An operator F is called "additive" if for any x and y, $F(x+y) = F(x)+F(y)$.

It can be proved that a continuous additive operator has the property that $F(ax) = aF(x)$ for every real number a. It is also proved that if F is additive and continuous at a single element (point) of E, then F is continuous everywhere and F is also bounded. This means that there exists a constant M which depends only on F such that $\|F(x)\| \leq M\|x\|$ for all $x \in E$. Furthermore if $\langle F_n \rangle$ is a sequence of continuous additive operators and if F is an additive operator such that $\lim_{n \to \infty} F_n(x) = F(x)$ for every x, then F is continuous and there exists M such that $\|F_n(x)\| \leq M\|x\|$ for all x and n.

Banach proved several important theorems for abstract equations. If F is a continuous operator which is defined on E and takes values in E, while there exists a number $M \in (0,1)$ such that for all x' and x'' in E,

$$\|F(x') - F(x'')\| \leq M\|x' - x''\|$$

then there exists a unique $x \in E$ such that $F(x) = x$.

An even more important theorem is the following: Let

$$x + hF(x) = y \tag{5}$$

where y is an unknown function in E, F is a continuous additive operator which is defined on E and takes values in E, and h is a real number. Let M be the infimum of all numbers M' for which $\|F(x)\| \leq M'\|x\|$ for all x. Then for any y and for any value of h, which satisfies the inequality $|hM| < 1$, there exists a function x, which satisfies (5). Moreover,

$$x = y + \sum_{n=1}^{\infty} (-1)^n h^n F^{(n)}(y)$$

where $F^{(n)}(y) = F(F^{(n-1)}(y))$. This result generalizes the method of of Volterra for solving integral equations.

In 1929 Banach introduced another important notion in functional analysis, namely the notion of the conjugate space of a Banach space. This space consists of all continuous bounded linear functionals which are defined on the given space. The norm of every functional that belongs to the conjugate space is taken to be equal to its least upper bound. The conjugate space is a Banach space. This work generalizes the work of Riesz regarding the spaces L^p and L^q, where $q = \frac{p}{p-1}$, because the space L^q is isomorphic to the conjugate space of the space L^p. Recall the Riesz representation theorem,

$$U(f) = \int_a^b a(x)f(x)dx.$$

The relation between L^p and L^q is the same with the relation between the conjugate space and the given Banach space.

An important theorem, which generalizes a theorem of Hahn and which is known as the Hahn-Banach theorem, is the following:

"Let p be a real function which is defined on some complete normed linear space R and let p satisfy, for all x and y in R, the relations:

(a) $p(x+y) \leq p(x) + p(y)$,
(b) $p(\lambda x) = \lambda p(x)$ for all $\lambda > 0$.

Then there exists an additive functional f on R such that

$$-p(-x) \leq f(x) \leq p(x)$$

for all x in R."

Let R and S be two Banach spaces and let U be a continuous linear operator which is defined on R and which takes values in S. Let R^* and S^* denote the sets of continuous linear functionals on R and S respectively. Then U induces a mapping from S^* to R^*, as follows: If $g \in S^*$, then $g(U(x))$ is defined for all $x \in R$. This definition implies that $g(U(x))$ is a continuous linear functional on R, which means that $g(U(x))$ is an element of R^*. Thus if $U(x) = y$, then $g(y) = f(x)$, where f is an element of R^*. The mapping $U^* : S^* \to R^*$ which is induced by U is called the adjoint operator of U.

Banach was able to prove that if U^* has a continuous inverse, then the equation $y = U(x)$ is solvable for any $y \in S$. Moreover, if $f = U^*(g)$ is solvable for every $f \in R^*$, then U^{-1} exists and is continuous on the range of U, where the range of U in S is the set of all y for which there exists $g \in S^*$ with the property that $g(y) = 0$ when $U^*(g) = 0$.

Banach applied this theory to operators of the form $U = I - \lambda V$, where I is the identity operator and V is a completely continuous operator. The abstract theory can be applied to the L^2-space of functions defined on $[0,1]$ and to the operators

$$U_\lambda(x) = x - \lambda \int_0^1 K(\cdot, t) x(t) dt$$

where $\int_0^1 \int_0^1 K^2(s,t) ds dt < \infty$. When applied to the equation

$$y(s) = x(s) - \lambda \int_0^1 K(s,t) x(t) dt \qquad (6)$$

and to its conjugate equation

$$f(s) = g(s) - \lambda \int_0^1 K(t,s) g(t) dt, \tag{7}$$

the general theory implies that if λ_0 is an eigenvalue of (6), then λ_0 is also an eigenvalue of (7) and vice versa. Moreover, (6) has a finite number of linearly independent eigenfunctions that correspond to λ_0 and the same is true also for (7). In addition, if $\lambda = \lambda_0$, then (6) is not solvable for every y. Indeed, a necessary and sufficient condition for (6) to have a solution is that the following n "orthogonality" conditions are satisfied:

$$\int_0^1 y(t) g_0^{(p)}(t) dt = 0 \qquad p = 1, 2, ..., n$$

where $g_0^{(1)}, ..., g_0^{(n)}$ is a set of linearly independent solutions of the equation

$$0 = g(s) - \lambda_0 \int_0^1 K(t,s) g(t) dt.$$

The axiomatization of the Hilbert space

It was realized in 1923 that the appropriate mathematical tool for Quantum Mechanics is the so-called spectral theory of Hilbert spaces. This was a powerful motivation for studying Hilbert spaces. This research began with John von Neumann (1903-1957) in 1927, who studied the space of sequences having the property that the sum of the squares of their terms is finite,

In two of his publications von Neumann presented an axiomatic approach to the Hilbert spaces and to the operators which act on Hilbert spaces. His main motivation was the formulation of a general theory of eigenvalues of a class of operators called Hermitian operators.

He introduced the L^2-space which consists of all measurable complex functions f which are defined on some measurable subset E of the complex plane and are such that $|f|^2$ is integrable. He also introduced the corresponding space of sequences a_1, a_2, ... of complex numbers having the property that $\sum_{p=1}^{\infty} |a_p|^2 < +\infty$. The Riesz-Fischer theorem implies that it is possible to construct a one-to-one correspondence between the functions and sequences of this space as follows:

Choose a complete orthonormal system of functions $\langle \phi_n \rangle$. If f is an element of the function space, then the Fourier coefficients of f with respect to the system $\langle \phi_n \rangle$ constitute a sequence that belongs to the above space of sequences. Conversely, if we start from a sequence, then there exists a unique function in L^2 (in the sense that one can add to it any function whose integral is zero) whose Fourier coefficients with respect to the system $\langle \phi_n \rangle$ are the terms of this sequence.

The scalar product among functions is

$$(f, g) = \int_E f(z) \overline{g(z)} dz$$

where $\overline{g(z)}$ is the conjugate complex number of the complex number $g(z)$, and the scalar product among sequences is

$$(a, b) = \sum_{p=1}^{\infty} a_p \overline{b_p}.$$

Then for f that corresponds to a, and for g that corresponds to b, $(f,g) = (a,b)$.

A Hermitian operator R on the space of functions is a linear operator such that for all f and g in its domain of definition, $(Rf, g) = (f, Rg)$. Similarly a Hermitian operator on the space of sequences is a linear operator having the property that for all a and b in its domain of definition, we have $(Ra, b) = (a, Rb)$.

Von Neumann's theory provides an axiomatic foundation for both the function space and the space of sequences:

(A) H is a complex vector space. Let us call its elements vectors. This means that H is equipped with an operation of addition of its elements, and an operation of multiplication of a vector with a complex number so that if $f_1, f_2 \in H$ and a_1, a_2 are arbitrary complex numbers, then $a_1 f_1 + a_2 f_2 \in H$.

(B) H is equipped with a scalar product denoted by (f, g) i.e., with a complex function of two variables $f, g \in H$, which has the following properties:
(a) $(af, g) = a(f, g)$
(b) $(f_1 + f_2, g) = (f_1, g) + (f_2, g)$
(c) $(f, g) = \overline{(g, f)}$
(d) $(f, f) \geq 0$
(e) $(f, f) = 0$ if and only if $f = 0$

Two elements f and g are called orthogonal if $(f, g) = 0$. The norm of f is denoted by $\|f\|$ and is equal to $\sqrt{(f, f)}$; the quantity $\|f - g\|$ defines a metric on H.

(C) H constitutes a separable space with respect to the above metric, i.e., H contains a countable dense subset with respect to the metric $\|f - g\|$.

(D) For any positive integer n, there exists a set of n linearly independent elements of H.

(E) H is complete. This means that if $\langle f_n \rangle$ is any sequence of elements of H such that $\|f_n - f_m\| \to 0$ as m, n tend to ∞, then there exists $f \in H$ such that $\|f - f_n\| \to 0$ as $n \to \infty$.

We will not discuss any further von Neumann's axiomatic foundation and development of the Hilbert space.

Functional analysis has many applications in the generalized moment problem, in statistical mechanics, in the theorems of existence and uniqueness, in partial differential equations, in fixed-point theorems, etc. Furthermore functional analysis plays an important role in the calculus of variations and in the theory of representation of continuous compact groups. It also has applications in algebra, in approximation methods, in topology and in the theory of functions of a real variable.

Here, we make a special reference to the **Theory of Distributions**, which also constitutes a subject that lies in the area of functional analysis. The most appropriate introduction to the theory of distributions is the one given by the founder of this theory, Laurent Schwartz (1915-2002), in his book *La théorie des distributions, Hermann, Paris*, 1973.

The rules of Symbolic Calculus were introduced by Heaviside. Although these rules were not at all rigorous, they were achieving their aim. When Dirac introduced the famous function $\delta(x)$, whose properties are that $\delta(x) = 0$ for $x \neq 0$, $\delta(0) = +\infty$, and $\int_{-\infty}^{+\infty} \delta(x) dx = 1$, the formulas of Symbolic Calculus became even more unacceptable from the mathematical point of view. Claiming that the so-called Heaviside function $y(x)$, which is equal to 1 for $x \geq 0$ and equal to 0 for $x < 0$, has as derivative the function of Dirac, that was apparently violating every permissible limit.

Most of the time, when such a situation full of contradictions and "antinomies" is presented, a mathematical theory is later created which in a sense makes legitimate (mathematicizes) the existing situation. The birth of such a mathematical theory contributes to the progress of mathematics and physics. Regarding the case at hand, it is interesting to note that at that time there were some related theories, such as the theories of Carson and Van der Pol. Despite the fact that they these theories were rigorous, they did not satisfy the physicists, either because they were based on the Laplace transform (which altered the nature of the problem), or because they omitted completely the function δ and its derivatives, thus ignoring methods whose success was already an unquestionable fact.

Schwartz's theory of distributions was able to bridge the gap between mathematicians and physicists.

Below, we will limit ourselves to giving the definition of vector distributions, the definition of scalar distributions, some examples and some related properties and definitions.

1. Vector distributions and scalar distributions

1) Let $\mathbf{R}^m$ be the m-dimensional Euclidean space. If $t = (t_1, t_2, ..., t_m)$ is a point in $\mathbf{R}^m$, we set
$$|t| = (t_1^2 + t_2^2 + ... + t_m^2)^{1/2}.$$
We denote by $D(\mathbf{R}^m)$ the set of complex functions $\phi : \mathbf{R}^m \to \mathbf{C}$ having the following properties:

(a) ϕ has derivatives of any order, i.e., all the derivatives
$$\frac{\partial^{k_1+k_2+...+k_m} \phi}{\partial t_1^{k_1} \partial t_2^{k_2} ... \partial t_m^{k_m}}$$
exist and are continuous for every $0 \leq k_i < +\infty$ ($i = 1, 2, ..., m$).

(b) There exists a number $r > 0$ (which depends on ϕ) such that $\phi(t) = 0$ for $|t| \geq r$.

Examples of such functions are given by the following formula, where a is an arbitrary positive number:
$$P_a(t) = \begin{cases} exp\left(\frac{-a^2}{a^2-|t|^2}\right) & \text{for } |t| < a \\ 0 & \text{for } |t| \geq a \end{cases}$$

The set $D(\mathbf{R}^m)$ obviously constitutes a vector space with respect to the usual addition of functions and the multiplication of functions with complex numbers. The zero of this space is the function that takes the value zero everywhere. In what follows, we consider m to be fixed and we will write simply D instead of writing $D(\mathbf{R}^m)$.

Definition. A vector space E is said to be a **space with convergence** if and only if we can single out a subset of the set of sequences of elements of E, whose elements are called **convergent sequences** and satisfy the following conditions:

(i) At every convergent sequence $\langle x_n \rangle$ corresponds a unique element $x \in E$ called the limit of the sequence $\langle x_n \rangle$. In this case we say that the sequence $\langle x_n \rangle$ converges to x and we write $\lim_{n \to \infty} x_n = x$ or $x_n \to x$.

(ii) If $x_n = x$ for every $n \in \mathbf{N}$, then $\langle x_n \rangle$ converges to x.

(iii) If $\langle x_n \rangle$ converges to x, then every subsequence of $\langle x_n \rangle$ converges to x.

(iv) If $x_n \to x$ and $y_n \to y$, while $a, b \in \mathbf{C}$, then $ax_n + by_n \to ax + by$.

(v) If a sequence $\langle a_n \rangle$ of complex numbers converges to a complex number a, then $a_n x \to ax$.

Remark. The above conditions do **not** imply that E is necessarily a topological space. The notion of a space with convergence is more general than the notion of a topological space.

Definition. Let $\langle \phi_n \rangle$ be a sequence of functions in D and let $\phi \in D$. We say that $\langle \phi_n \rangle$ converges to ϕ if all the functions ϕ_n vanish outside a ball which is independent of n (i.e., $\phi_n(t) = 0$ for $|t| \geq r > 0$, where r is independent of n) and all the derivatives of ϕ_n converge uniformly on $\mathbf{R}^m$ to the corresponding derivatives of ϕ as $n \to \infty$. (Attention: uniform convergence does not refer to the order of differentiation.)

It is obvious that this definition of convergence makes D a space with convergence.

2) Let E, F be two vector spaces with convergence and let $T : E \to F$ be a mapping from E to F. T is called continuous at a point $x \in E$ if and only if for any sequence $\langle x_n \rangle$ of points of E that converges to x, the sequence $\langle T(x_n) \rangle$ converges to $T(x)$.

If T is also a linear mapping, then it is enough to consider only the point $0 \in E$. That is, if T is continuous at 0, then it is continuous at every $x \in E$. Indeed, let $\langle x_n \rangle$ be a sequence that converges to x. Then $\langle x_n - x \rangle$ converges to 0. Hence $T(x_n - x) \to 0$ and, since T is linear, we deduce that $Tx_n \to Tx$.

2. **Definitions.** Every continuous linear mapping $D \to E$, where E is a vector space with convergence, is called a E-**valued distribution in** m **dimensions**, while for $E = \mathbf{C}$ we obtain the **scalar distributions**.

If f is a scalar distribution and $\phi \in D$, we will write $\langle \phi, f \rangle$ instead of $f(\phi)$. The **conjugate distribution** $\overline{f}$ of a scalar distribution f is defined by the formula

$$\langle \phi, \overline{f} \rangle = \overline{\langle \overline{\phi}, f \rangle}.$$

As mentioned earlier, the theory of distributions was introduced in order to mathematicize existing "antinomies". The notion of a distribution generalizes the notion of a function in such a way that operations that hold only for a restricted class of functions in classical analysis, hold now for very wider classes of functions.

Let us consider the case of functions $f : \mathbf{R} \to \mathbf{C}$. If f is integrable on every bounded interval of $\mathbf{R}$, then the formula

$$\langle \phi, f \rangle = \int_{-\infty}^{+\infty} \phi(t) \overline{f(t)} dt \qquad (\star)$$

defines a scalar distribution of a single variable. Indeed, the integral in the second side of the equality exists for every $\phi \in D$ and constitutes a linear operation. In addition, if $\langle \phi_n \rangle$ is a sequence of elements of D such that the ϕ_n vanish outside a bounded interval and $\langle \phi_n \rangle$ converges uniformly to 0 on $\mathbf{R}$ (something that obviously holds in case $\langle\phi_n\rangle$ converges to 0 in the sense of the notion of convergence in D), then the integrals $\int_{-\infty}^{+\infty}\phi_n(t)\overline{f(t)}dt$ converge to zero. This proves that the linear functional f is continuous and satisfies the definition of a distribution.

The above can be extended to functions that take values in vector spaces and in particular in Banach spaces.

The distributions that correspond to locally integrable functions do not determine uniquely the functions from which they are obtained, because the integral $\int_{-\infty}^{+\infty}\phi(t)\overline{f(t)}dt$ that determines them does not change if we modify f on a set of measure zero. In relation to that, we mention without proof the following proposition.

Proposition. The distributions
$$T_{f_1}(\phi) = \int_{-\infty}^{+\infty}\phi(t)\overline{f_1(t)}dt$$
$$T_{f_2}(\phi) = \int_{-\infty}^{+\infty}\phi(t)\overline{f_2(t)}dt$$
coincide if and only if $f_1 = f_2$ almost everywhere.

Note. If f is continuous, then f is uniquely determined by the corresponding distribution T_f, i.e., if $\langle\phi, T_f\rangle = \int_{-\infty}^{+\infty}\phi(t)\overline{f(t)}dt = 0$ for every $\phi \in D$, then $f = 0$ everywhere.

3) A particular distribution defined by $(\star)$ is Heaviside's distribution. To this end we consider the function
$$\theta(t) = \begin{cases} 0 & \text{for } -\infty < t < 0 \\ 1 & \text{for } 0 \leq t < +\infty \end{cases}$$
which defines the distribution
$$\langle\phi,\theta\rangle = \int_0^\infty \phi(t)dt.$$

4) The distribution δ_a $(a \in \mathbf{R})$ of Dirac is defined by the relation
$$\langle\phi,\delta_a\rangle = \phi(a).$$
If $a = 0$, then we write $\delta_0 = \delta$ and we have $\langle\phi,\delta\rangle = \phi(0)$.

5) We will consider now a distribution which is not obtained by $(\star)$ and which is useful in quantum mechanics.

We consider the function $f(t) = \frac{1}{t}$ $(t \neq 0)$, which is not integrable on any open interval that contains 0. Let us consider the functional F which is defined as follows:
$$\langle\phi,F\rangle = P.V.\int_{-\infty}^{+\infty}\frac{\phi(t)}{t}dt = \lim_{\epsilon\to 0}\int_{|t|\geq\epsilon}\frac{\phi(t)}{t}dt.$$
(This limit exists and is equal to $-\int_{-\infty}^{+\infty}\phi'(t)\log|t|dt$.)

Obviously F is linear. We will prove that it is also continuous. Let $\phi_n \to 0$ in D and let $[-a,a]$ be an interval that contains the supports of all the ϕ_n. (If ϕ is continuous, then we define the support of ϕ to be the closure of the set $\{x : \phi(x) \neq 0\}$.) We have

$$\begin{aligned}\langle\phi_n, F\rangle &= P.V.\int_{-\infty}^{+\infty}\frac{\phi_n(t)}{t}dt \\ &= P.V.\int_{-\infty}^{+\infty}\frac{\phi_n(t)-\phi_n(0)+\phi_n(0)}{t}dt \\ &= \phi_n(0)\cdot P.V.\int_{-a}^{a}\frac{dt}{t} + P.V.\int_{-a}^{a}\frac{\phi_n(t)-\phi_n(0)}{t}dt \\ &= \phi_n(0)\cdot 0 + \int_{-a}^{a}\frac{\phi_n(t)-\phi_n(0)}{t}dt\end{aligned}$$

where in the last relation we omitted P.V. because the function $\frac{\phi_n(t)-\phi_n(0)}{t}$ is integrable on $[-a,a]$. The mean value theorem gives

$$\left|\frac{\phi_n(t)-\phi_n(0)}{t}\right| \leq \max_{-a\leq t\leq a} |\phi'_n(t)|$$

that is

$$|\langle\phi_n,F\rangle| \leq 2a \cdot \max_{-a\leq t\leq a} |\phi'_n(t)|$$

from which we obtain that $\langle\phi_n,F\rangle \to 0$ as $n\to\infty$.

The distribution F is denoted by $P.V.\frac{1}{t}$ (or simply by $\frac{1}{t}$). Consequently

$$\langle\phi, P.V.\frac{1}{t}\rangle = P.V.\int_{-\infty}^{+\infty}\frac{\phi(t)}{t}dt.$$

In addition, the distribution $\log|t|$ is defined by the formula

$$\langle\phi,\log|t|\rangle = P.V.\int_{-\infty}^{+\infty}\phi(t)\log|t|dt.$$

6) Then we will consider distributions which take values in some Banach space E. An example of such a distribution T is given by the formula

$$T(\phi) = \langle\phi,f\rangle x$$

where f is a scalar distribution and $x \in E$.

We denote by D'_E the set of distributions which take values in E. For $T_1, T_2 \in D'_E$, $\phi \in D$ and $a \in \mathbf{C}$, we put $(T_1+T_2)\phi = T_1\phi + T_2\phi$ and $(aT)\phi = aT\phi$. In this way D'_E becomes a vector space. We shall say that a sequence $\langle T_n\rangle$ of elements of D'_E converges to the element T of D'_E if and only if for any $\phi \in D$, the sequence $\langle T_n\phi\rangle$ converges to $T\phi$ in the space E.

The notion of the support of a distribution is very useful in the theory of distributions. We shall say that a distribution T vanishes on an open set A if $T\phi = 0$ for every function $\phi \in D$ whose support is contained in A. Let U be the union of all the open sets on which the distribution T vanishes. Then the complement of U, i.e., U^c, is called the **support** of T. Obviously the support of a distribution is a closed set.

It is proved that if a distribution T and a function ϕ have disjoint supports, then $T\phi = 0$.

When a set A contains the support of a distribution T, we shall say that T is **concentrated** on A.

The class of distributions having compact support plays an important role in the theory of distributions. In particular, if we consider the class of scalar distributions with compact support and we consider these distributions as continuous linear functions $D \to E$, then this class can be uniquely extended to the space $\mathcal{E}$ of all complex functions having derivatives of any order.

In the space $\mathcal{E}$ the convergence of a sequence is defined as follows: $\langle\phi_n\rangle$ converges to ϕ if and only if every derivative of ϕ_n converges uniformly on every compact set to the corresponding derivative of ϕ as $n\to\infty$.

Obviously the space D constitutes a subspace of $\mathcal{E}$. Let us consider a sequence of elements of D which converges to some element of D in the space D. Then this sequence converges to the same element in the sense of convergence in $\mathcal{E}$. This follows by a simple comparison of the two notions of convergence. For this reason we say that convergence in D is stronger than convergence in $\mathcal{E}$. Besides there exist sequences in D which converge in the sense of convergence in $\mathcal{E}$ to elements of $\mathcal{E}$ which do not belong to D.

Moreover, it is proved that every element of $\mathcal{E}$ is the limit in the sense of convergence in $\mathcal{E}$ of a sequence of elements of D.

Convergence in the space $\mathcal{E}'_E$ is defined as it was defined in D'_E: A sequence $\langle T_n \rangle$ of elements of $\mathcal{E}'_E$ converges to an element T of $\mathcal{E}'_E$ if and only if for any $\phi \in \mathcal{E}$, the sequence $\langle T_n \phi \rangle$ converges to $T\phi$ in E. This means that convergence in $\mathcal{E}'_E$ is stronger than convergence in D'_E. An example of a scalar distribution with compact support is the distribution δ_a.

3. Let S be the set of infinitely differentiable functions having the property that

$$\lim_{|t| \to \infty} \left((1 + |t|^2)^{\frac{j}{2}} \left| \frac{\partial^{k_1+k_2+\ldots+k_m} \phi(t)}{\partial t_1^{k_1} \partial t_2^{k_2} \ldots \partial t_m^{k_m}} \right| \right) = 0$$

for every $j = 0, 1, 2, \ldots$ and for every index of differentiation $(k_1, k_2, \ldots, k_m)$. Every function $\phi \in S$ is called **rapidly decreasing** as $|t| \to \infty$.

The subspace of D'_E which consists of continuous linear mappings $S \to E$ is called the subspace of **tempered distributions**.

A sequence $\langle \phi_n \rangle$ of elements of S is said to converge to 0 in S if and only if every expression of the form

$$P(t) \frac{\partial^{k_1+\ldots+k_m} \phi_n(t)}{\partial t_1^{k_1} \ldots \partial t_m^{k_m}},$$

where $P(t)$ is an arbitrary polynomial, converges uniformly to 0 on $\mathbf{R}^m$. This notion of convergence is weaker than the notion of convergence in D.

In addition, we have $S \subset \mathcal{E}$ and convergence in S is stronger than convergence in $\mathcal{E}$.

The space of all continuous linear mappings $S \to E$ is denoted by S'_E and contains the distributions that correspond to functions with polynomial growth.

It is easily recognized that the space S'_E contains the distributions with compact support.

For any $k = 0, 1, \ldots$, we obtain an important subspace of D'_E by considering the continuous linear mappings $C_0^{(k)} \to E$, where $C_0^{(k)}$ consists of functions which have compact support and continuous derivatives of order k.

A sequence $\langle \phi_n \rangle$ of elements of $C_0^{(k)}$ converges to 0 in $C_0^{(k)}$ if and only if all the ϕ_n have their supports contained in the same interval and their derivatives of order i ($i = 0, 1, \ldots, k$) converge uniformly to 0.

A continuous mapping $C_0^{(k)} \to E$ is called **distribution of order** k.

If the range space E constitutes a Banach space and the distributions considered are scalar, then it is proved that the definitions of convergence above introduced are equivalent with certain topological or pseudo-topological structures. Moreover, this equivalence implies that the spaces of distributions considered are complete in the following sense: If for every element ϕ that belongs to the space where the distributions considered are defined, the sequence $\langle T_n \phi \rangle$ converges, then the associated limits define a distribution on the space considered.

4. We said before that a distribution T vanishes on an open set A if and only if $T\phi = 0$ for every $\phi \in D$ whose support is contained in A.

If T_1, T_2 are two distributions such that their difference vanishes on the open set A, then we say that T_1, T_2 are equal on the set A. In particular, a scalar distribution

f can be equal to a locally integrable function on an open set A. For example, the distribution δ_a is equal to the zero function on the set $\{t \in \mathbf{R}^m : t \neq a\}$ and the distribution $P.V.\frac{1}{t}$ is equal to the function $f(t) = \frac{1}{t}$ on the set $\{t \in \mathbf{R} : t \neq 0\}$.

In general, let f be a function which is locally integrable on $\mathbf{R}^m \setminus \{t_0\}$. The construction of a distribution (called also f) which is equal to $f(t)$ on every $t \neq t_0$ plays an important role in applications. If such a distribution is constructed, then for any $\phi \in D$ that vanishes in a neighborhood of t_0, we have

$$\langle \phi, f \rangle = \int_{\mathbf{R}^m} \phi(t) \overline{f(t)} dt.$$

The distribution f is called a **regularization** of the function f. It is easily recognized that if f_1, f_2 are two regularizations of f, then $f_1 - f_2$ is a distribution which is concentrated at the point t_0.

In Appendix II we are very pleased to be able to quote the entire lecture given by Schwartz when he was invited to speak at the General Mathematics Seminar (G.M.S.) of the Mathematics Department of the University of Patras (this lecture is published in Volume 9 of the G.M.S.).

CHAPTER 9

TOPOLOGY

Topology which was originally known as "analysis situs" was developed in the 19th century. Topology can be thought of as a branch of geometry which studies the properties of figures which are preserved when the figures are bent, stretched, contracted or deformed in an such a way that no new points are created and no existing points are identified with each other, thus the transformations allowed are one-to-one correspondences between the points of the initial and the transformed figure. These transformation map points which are "close" together to points which are also "close" together. This last property is called continuity and it is required that both the transformation and its inverse are continuous. Such a transformation is called a homeomorphism. In that sense topology is the "Geometry of the elastic membrane", because if the figures were made of an elastic material, many of them could be transformed into homeomorphic figures. Since an elastic strip can be transformed homeomorphically into a circle or a square, it is topologically equivalent to these figures. But it is not topologically equivalent to the figure of the number 8, because this would require the identification of two points of the strip.

Topology studies objects independently of their surrounding space. Topologically, two figures may be homeomorphic even when it is impossible to transform topologically the entire surrounding space of the first figure into the surrounding space of the second figure. Hence if we take a rectangular strip of paper and we connect its two farthest extremities, then we obtain a cylindrical strip. But if we first rotate by 360° one of the extremities and then we connect it with the other, then the new figure is topologically equivalent to the initial one, but there exists no topological transformation of the 3-dimensional space into itself which transforms the first figure into the second one.

As it happens with many branches of mathematics, many results in topology were obtained by mathematicians without realizing that they were creating part of a theory, whose independent existence was only appreciated in the course of time. In the 20th century, topology is divided into "pointset topology", or general topology, and "combinatorial topology" or algebraic topology.

Pointset Topology

A set of points is called a "space" when the points of this set are related to each other in a certain way. For example, in Euclidean space the distance between two points "measures" how close they are to each other and this notion of distance permits us to define the limit of a sequence of points. In 1906 Fréchet introduced the notion of abstract spaces, in an effort to unify Cantor's theory of pointsets; he considered functions as points of an abstract space. The birth of functional analysis with its Hilbert and Banach spaces, made even more important the study of sets of points considered as spaces. In functional analysis the limit of a sequence plays a crucial role, thus the topological properties of the associated abstract space become very important. Furthermore the operators in functional analysis are transformations which map one space to another.

Fréchet understood that the main principle used to relate the points in a Euclidean space is not necessarily the function that gives the distance between two points of the set. He introduced various other notions generalizing the notion of distance by introducing the class of metric spaces.

In a metric space, a neighborhood of a given point, is the set of all points whose distance from the given point is less than some positive number, say ϵ. These neighborhoods are circular. We may also have square neighborhoods. But it is also possible to assume that the neighborhoods are subsets of a given set which are defined in some specific way without making use of a metric. Actually, Felix Hausdorff (1868–1947) used the notion of a neighborhood of a point to define a generalization of metric spaces.

Hausdorff defines a topological space as a set of points x such that to every x there corresponds a family of subsets U_x. These subsets, which are called neighborhoods, must satisfy the following conditions:

(a) To every point x corresponds at least one neighborhood U_x that contains x.
(b) The intersection of two neighborhoods of x contains a neighborhood of x.
(c) If y is a point of U_x, then there exists a U_y such that $U_y \subset U_x$.
(d) If $x \neq y$, then there exist U_x and U_y such that $U_x \cap U_y = \emptyset$.

Hausdorff also introduced countability axioms:

(a) For any x, the set of the U_x is at most countable.
(b) The set of all distinct neighborhoods is countable.

The following definitions are used in pointset topology:

We call *limit point* of a set of points, a point a such that every neighborhood of a contains points of the set which are different from a. A set is called *open* if every point of this set belongs to a neighborhood which contains only points of this set. A set which contains all its limit points is called *closed*.

A space or a subset of a space is called **compact** if every infinite subset of it has a limit point. For example, the set of points of a straight line in the Euclidean space is not compact because the infinite set $\{1, 2, 3, ...\}$ has no limit point. A set is called *connected* if no matter how we divide it into two disjoint sets, at least one of these sets contains limit points of the other. For example, the set consisting of the points of the curve $y = \sin \frac{1}{x}$ and of the points of the interval $(-1, 1)$ of the y-axis is connected.

A space is called *separable* [8] if it contains a countable subset whose *closure* (i.e., the union of this set with the set of its limit points) is equal to the space itself.

We now introduce the notions of a "continuous transformation" and of a "homeomorphism": Let X and Y be topological spaces. A transformation $f : X \to Y$ is called continuous when to every point $x \in X$ corresponds a single point $f(x) = y \in Y$, and when given a neighborhood of y, there exists a neighborhood of x (or of every x for which $f(x) = y$), such that the images of the points in this neighborhood belong to the given neighborhood of y (this is a generalization of the ϵ-δ definition of a continuous function, where ϵ determines the neighborhood of a point in the space of images (Y) and δ determines a neighborhood of a point in the initial space (X)).

A **homeomorphism** between two topological spaces X and Y is a one-to-one correspondence between X and Y which is a continuous transformation of X to Y and a continuous transformation of Y to X.

Pointset topology studies properties which remain invariant under continuous transformations and under homeomorphisms between topological spaces.

Hausdorff contributed new results on the notion of completeness of a metric space (this notion was introduced by Fréchet in his dissertation in 1906). A metric space is called *complete* when every sequence $\langle a_n \rangle$, that satisfies the condition that given $\epsilon > 0$, there exists a positive integer N such that $|a_n - a_m| < \epsilon$ for all integers n and m which are greater than N, has a limit point. Hausdorff's fundamental result is that every metric space can be extended to a complete metric space in a unique way.

The development of abstract spaces gave rise to new questions. For example, such a question was asked by Fréchet: "If a topological space is defined by a system of neighborhoods, is this space necessarily metrizable?" That is, is it possible to introduce in this space a metric that preserves the structure of the space so that the limit points remain limit points? A result, which is related to this question and is due to Paul S. Urysohn (1898-1924), asserts that every completely separable normal space is metrizable. We call a topological space **normal** when it has the property that given two arbitrary disjoint closed subsets of it, there exist two disjoint open subsets of it, each containing exactly one of the given closed subsets. Another related result, also due to Urysohn is that every separable metric space is homeomorphic to a subset of the "Hilbert cube". We remind the reader that the Hilbert cube consists of all infinite sequences $\langle x_i \rangle$ with the property that $0 \leq x_i \leq \frac{1}{i}$, where the distance is defined by the relation:

$$d = \sqrt{\sum_{i=1}^{\infty} (x_i - y_i)^2}.$$

A very important notion in topology is the notion of the dimension of a topological space. We recall that at the end of the 19th century, G. Cantor discovered that there exists a one-to-one correspondence between the set of points of a line segment and the set of points of the surface of a square. Moreover, G. Peano had discovered the existence of a continuous mapping of a line segment onto the set of points of a square (this is the so-called Peano curve). Following the progress achieved in pointset topology, it became necessary to consider sets which are more complicated than the sets that were known until then. This made it necessary to give an exact definition of "dimension". In 1913, L. E. J. Brouwer gave a definition of "dimension" which was based on an idea of Poincaré. In 1922, K. Menger and P. Urysohn began the theory of dimension for separable metric spaces.

Later, P. S. Alexandrov and W. Hurewicz made important contributions to the development of this theory. The development of the theory of dimension for general metric spaces was carried out by M. Katetov and K. Morita, who worked independently of each other. In addition, a more general theory regarding normal topological spaces was established, even though the results that hold in metric spaces cannot always be generalized to normal topological spaces.

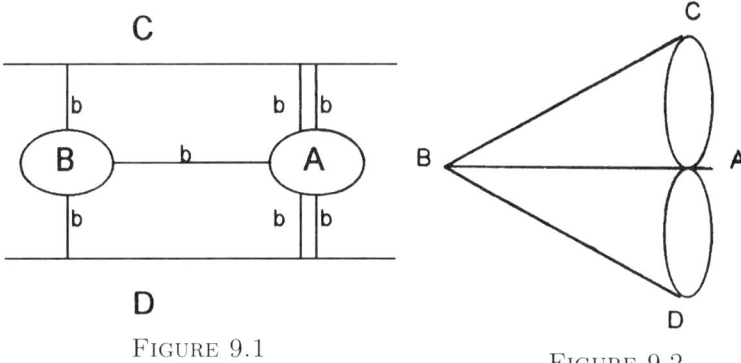

FIGURE 9.1   FIGURE 9.2

Combinatorial Topology

Combinatorial topology is based on pointset topology, but it considers figures as groups consisting of smaller constituents, just like a wall consists of bricks. The first efforts in this direction were made be Leibniz who tried to formulate basic geometric properties of geometric figures, and to combine these properties following certain operations in order to obtain new properties. In his letter addressed to Huygens (1679) he mentioned that he was not satisfied with the study of geometric figures through analytic geometry, because, apart from the fact that this method was not direct (elegant), it was also of a quantitative nature. "I believe", he wrote in that letter, "what we lack is a purely geometric method which expresses directly the position (situs) in the same way that algebra expresses the quantity".

A very well known problem of a topological nature is the "problem of the bridges of Koenigsberg". On the Pregel River that runs through the small town Koenigsberg there exist two islets and seven bridges denoted by b in Figure 9.1. The problem is the following: Is it possible to walk on the seven bridges and traverse every bridge exactly once?

Euler, who was living in Saint Petersburg at that time, learned about this problem and solved it in 1735. Euler simplified the statement of the problem by replacing the lands with the points A, B, C, D and the bridges that connect these lands with line segments or arcs, see Figure 9.2. Hence the problem is equivalent to the following: "Is it possible to draw Figure 9.2 with one stroke of the pencil, i.e., without lifting the pencil from the paper and without drawing the same edge more than once?" Euler proved that this is impossible.

In general Euler proved that in order to be able to draw a linear figure, as for example Figure 9.3, with one stroke of the pencil and without drawing the same edge more than once, it is necessary that the figure has an even number of odd vertices (a vertex is called even or odd if it is the end of an even or odd number of edges respectively). A connected linear figure can be drawn with one stroke of the pencil and without drawing the same edge more than once if and only if the figure has either no odd vertices or only two odd vertices. Since Figure 9.2 has 4 odd vertices, it cannot be drawn with one stroke of the pencil. Figure 9.3 has two odd vertices, A and B, and hence it can be drawn with one stroke of the pencil without drawing the same edge more than once. This can be done if one starts from A and ends at B or if one starts from B and ends at A. Later, the German mathematician Johann Benedikt Listing (1808-1882) proved that a connected linear figure with $2n$

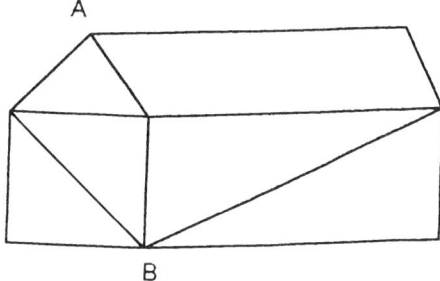

FIGURE 9.3

odd vertices can be drawn with n strokes of the pencil, where one starts from an odd vertex and ends at another odd vertex.

Historically the first theorem in topology is the theorem expressed by the following formula of Euler, which seems to have been known even to Descartes (1639). Let us consider the surface of an arbitrary convex polyhedron. If V, E and F are the number of vertices, the number of edges and the number of faces of the polyhedron respectively, then

$$V - E + F = 2.$$

This theorem belongs to topology because the formula still holds when a topological transformation is applied to the polyhedron in question. After such a transformation, the edges, in general, will no longer be straight, the faces will no longer be plane and the surface of the polyhedron will become a curvilinear surface; but Euler's formula that relates the vertices, the edges and the faces will still hold. The most interesting case is when all the faces are triangles, hence we have a so-called "triangulation" (a separation of the surface into linear or curvilinear triangles). It is easy to reduce the general case, where the faces are polygons, to this particular case: It is enough to divide the faces into triangles (for example, by drawing the diagonals from some vertex of the given face). Consequently, we may restrict ourselves to the case in which a triangulation exists. The combinatorial method in the topology of surfaces consists of studying one of the triangulations of the given surface. Of course we are interested in the properties of the triangulation which are independent of the triangulation that was selected at random, i.e., we are interested in the properties which are common to all triangulations of the given surface and therefore express a property of the surface under study.

Euler's formula provides such a property which we will examine in more detail. The expression $V - E + F$, where V, E and F are the number of vertices, the number of edges and the number of faces in the given triangulation respectively, is called the Euler characteristic of the triangulation. Euler's theorem asserts that for all triangulations of a surface which is homeomorphic to the surface of a sphere, the Euler characteristic is equal to 2. It can be proved that for every surface (and not only for surfaces which are homeomorphic to the surface of a sphere), all triangulations of this surface have the same Euler characteristic.

It is not difficult to calculate the Euler characteristic of the various surfaces. First of all the Euler characteristic of a torus is equal to zero. Indeed, given a triangulation of the surface of a sphere, if we remove from it two triangles which have no point in common and we glue the sides of these triangles, we obtain a

FIGURE 9.4

triangulation of a surface which is homeomorphic to the surface of a torus. Here the number of vertices and the number of edges remains unchanged, while the number of triangles is reduced by two, which means that the Euler characteristic of the obtained triangulation is equal to zero.

Now let us consider the surface that is obtained from a triangulation of the surface of a sphere by removing the interior of $2p$ triangles which have pairwise no point in common. (This can be achieved if we subdivide appropriately an arbitrary triangulation in such a way that the resulting triangulation has enough triangles.) Then the Euler characteristic is reduced by $2p$. In addition, it is easily recognized that the Euler characteristic does not change when cylindrical tubes are attached to every pair of holes that were obtained on the surface of the sphere after the removal of the triangles (Figure 9.4). This conclusion follows from the fact that the Euler characteristic of every attached tube is (as we saw above) equal to zero. This means that the number of vertices equals the number of edges on the surface of every tube. Hence on a closed two-sided surface of genus p, the Euler characteristic is equal to $2 - 2p$.

Later we will give an important property of triangulations which satisfies the so-called condition of topological invariance. This means that every triangulation of a given surface has this property if at least one triangulation of the given surface has it. It is the property that informs us whether the triangulation is orientable or not. But let us explain first what is the orientation of a triangulation.

Every triangle can be oriented, i.e., one can determine a specific direction according to which its perimeter is traced by giving one of the two orderings of its vertices (i.e., A, B, C or A, C, B). Let us assume that on an arbitrary surface, two triangles are given which have a common side and no other point in common (Figure 9.5). Two orientations of these triangles are said to be compatible when they produce opposite directions on their common side. (In the plane or in any other two-sided surface, this means that when the two triangles lie on the same side of the surface, then they are traced both either clockwise or counterclockwise.) A triangulation of a given closed surface is called orientable if the orientations of all the triangles that belong to it are chosen in such a way that the orientations of any two triangles having a common side and no other point in common are compatible. If

9. TOPOLOGY

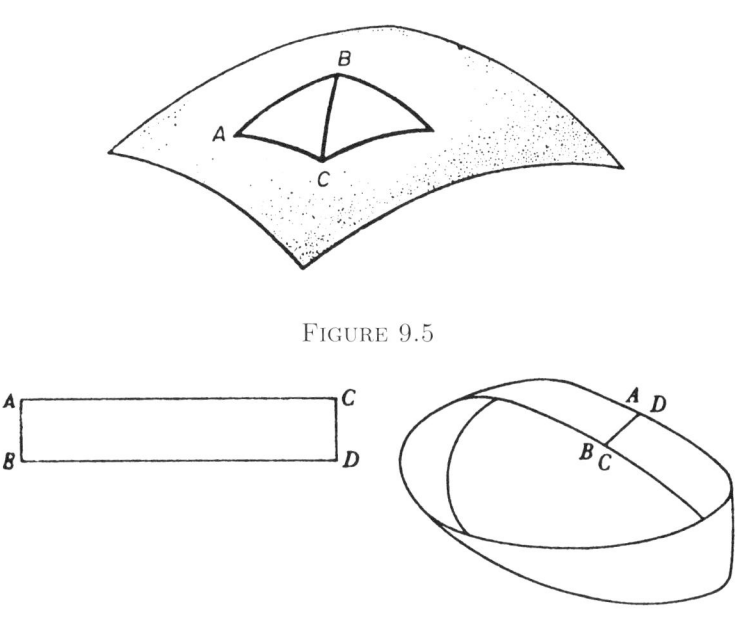

FIGURE 9.5

FIGURE 9.6

this happens, then: **Every triangulation of a two-sided surface is orientable and every triangulation of a one-sided surface is non-orientable**. This is the reason that the two-sided surfaces are called orientable, while the one-sided surfaces are called non-orientable.

If we choose an arbitrary triangulation of a Möbius strip (Figure 9.6 on the right) which is a one-sided surface, indeed we will note that this surface is non-orientable. The Möbius strip is constructed as follows: We take a rectangular strip (Figure 9.6 on the left) and we connect its two ends after having rotated the most narrow edge by 180° (Figure 9.6). The fact that the obtained surface has a single side can be characterized by the fact that one can paint the entire surface with a continuous movement of the brush. But if we connect the two ends of the strip without having rotated first one of them, then, in order to paint the entire obtained surface, one must lift at some moment the brush in order to paint also the other side of the surface.

One-sided and two-sided surfaces can also be defined via a normal to the surface. So let this normal have a "given" direction. If by moving this normal in an arbitrary way on the surface we return to the point of departure and the normal still has the same given direction, then the surface is two-sided, while if the direction of the normal can be made opposite of the initial one, then we say that the surface is one-sided.

In order to be able to obtain the simplest possible triangulation of the projective plane (Part I, Chapter 14) we need only draw in the projective plane three arbitrary straight lines which do not pass through the same point (Figure 9.7).

These straight lines separate the projective plane into four triangles, from which one lies in the finite part of the plane, while every other triangle is divided by the straight line at infinity into two parts. In Figure 9.7, one of the triangles that extend to infinity is shaded. It follows also from Figure 9.7 that if we try to give

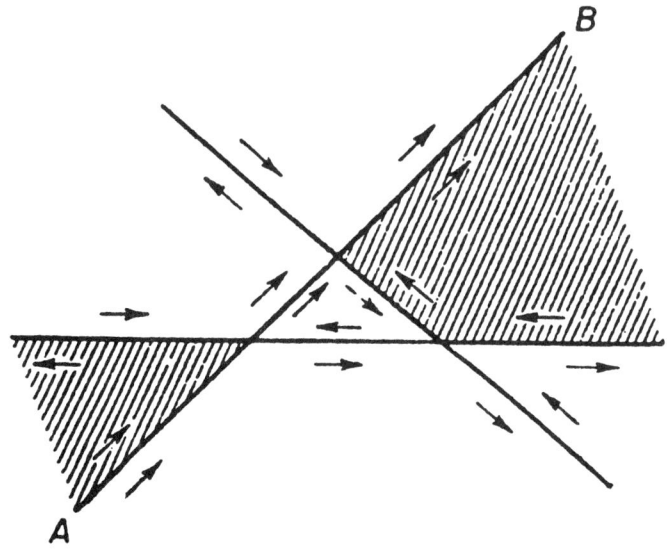

FIGURE 9.7

to all four triangles compatible orientations, we will not succeed. In particular, if we choose the orientations of the four triangles as they are marked in Figure 9.7, then the algebraic sum of the sides is different from zero, while it should have been equal to zero if the orientations were compatible. In this case the straight line AB is taken twice in the summation.

The Euler characteristic and the property of being orientable or non-orientable provide a complete system of topological invariants for closed surfaces. This means that two closed surfaces are homeomorphic (i.e., there exists a homeomorphism between them) if and only if their triangulations have the same Euler characteristic and they are both either orientable or non-orientable.

In order to clarify further the triangulation of the projective plane, we present the following.

The projective plane. The set of straight lines and planes of the 3-dimensional Euclidean space which pass through a given point S of the space is said to be a projecting pencil of straight lines and planes with center S. If we intersect this pencil with a plane P which does not pass through the center S, then to every point of the plane P corresponds a straight line of the pencil which intersects the plane P at this point, and to every straight line of the plane P corresponds a plane of the pencil which intersects the plane P along this straight line.

But in this way we do not obtain a one-to-one mapping of the set of straight lines and planes of the pencil onto the set of points and straight lines of the plane P, because the straight lines and the plane of the pencil which are parallel to the plane P do not intersect the plane P and hence, in this way, they correspond to no point and no straight line of the plane P. For this reason we "agree" to say that the straight lines of the pencil which are parallel to P intersect P at ideal points (at infinity) which lie in the directions of these straight lines, and that the plane of the pencil which is parallel to P intersects P along an ideal straight line (at infinity).

The plane P with the addition of these ideal points and this ideal line is said to be a projective plane and is denoted by P^*. Hence the set of straight lines and planes of the pencil with center S is mapped, in a one-to-one fashion, onto the set of (real or ideal) points and (real or ideal) straight lines of the projective plane P^*.

So we agree to say that a (real or ideal) point lies on a (real or ideal) straight line of the projective plane P^* if the corresponding straight line of the pencil lies on the corresponding plane of the pencil. In this sense, two arbitrary straight lines of the projective plane intersect (at a real or ideal point), while two arbitrary planes of the pencil intersect along a straight line of the pencil. The above imply, among other things, that the ideal straight line is the set of ideal points.

Essentially, the completion of the plane with its ideal elements is meant to be used as a device for the study of the pencil of straight lines and planes which pass through the same point.

The combinatorial method is not only applied to the study of surfaces (i.e., of 2-dimensional manifolds) but also to the study of manifolds of arbitrary dimensions. For example, in the case of 3-dimensional manifolds the role of triangulations is played by the separations of manifolds into tetrahedra. Such separations are called 3-dimensional triangulations of the manifolds. The Euler characteristic of a 3-dimensional triangulation is, by definition, the number $a_0 - a_1 + a_2 - a_3$, where a_i, $i = 0, 1, 2, 3$, is the number of i-dimensional elements of the triangulation (i.e., a_0 is the number of vertices, a_1 is the number of edges, a_2 is the number of faces and a_3 is the number of tetrahedra). The study of manifolds of dimensions greater than three can be made in an analogous way. But the further discussion of the subject goes beyond the scope of the present book.

Before closing this chapter we will mention one more problem, whose topological nature was also recognized later.

The four color problem

Let us assume that a country is divided into areas (for example, prefectures) and we want to construct a map of this country in such a way that when two areas have a common boundary, they have different colors.

The problem posed is whether four colors are always sufficient to color any map. This is the four color problem.

It is easily proved that there exist maps where three colors are not enough. For example, four colors are necessary to color the areas A, B, C, D in Figure 9.8.

It is also proved that five colors are sufficient to color any map.

The statement of the four color problem given above is not absolutely accurate. There exist certain additional conditions that the coloring of the map must satisfy, but we will not occupy ourselves with them.

The conjecture that four colors are sufficient to color any map was stated in 1852 by Francis Guthrie, and it was announced to A. De Morgan by his brother Frederick Guthrie in 1852. In 1878, A. Cayley became interested in the problem. But the more general interest of the mathematical community arose when J. P. Heawood discovered a mistake in the "proof" given by A. B. Kempe (1879).

Map coloring problems exist also for other surfaces, not only for plane ones. The problem is essentially the same for all simply connected surfaces, while the situation is completely different when the surfaces are not simply connected, as is, for example, the torus (inner tube of a car's wheel). It has been proved that

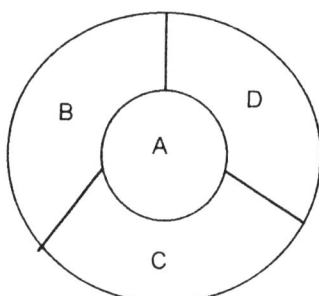

FIGURE 9.8

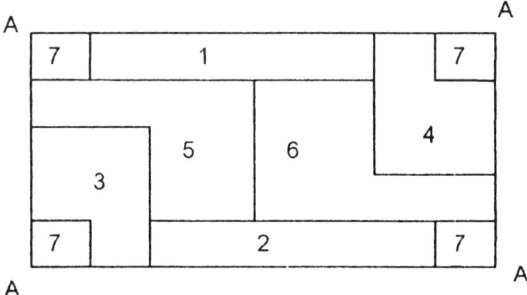

FIGURE 9.9

seven colors are sufficient to color a map on the surface of a torus and that for certain maps on the torus, seven colors are necessary, as it happens in Figure 9.9, where every area is neighboring with every other area. Map coloring is a topological problem because it is about relations between areas, which are preserved under a topological transformation. The least number of colors which are necessary to color a given map is called the **chromatic number** of the given surface of the map and constitutes another example of a topological invariant.

In 1976, K. Appel and W. Haken solved the four color problem. Their main achievement is that they reduced the problem to the study of a finite number of cases: they proved that if we know the answer for a finite number of certain maps (a number which is indeed very large), then we will know the answer for all maps. This reduction of the problem to a specific finite number of cases was the first step towards the proof. The second step was to carry out a calculation.

The reduction of the problem to a finite number of cases constitutes a usual way of mathematical reasoning, which was in general (but not in its details) known long before. That is, it was known that a certain reduction has to be made.

The calculation, i.e., the second step towards the solution, was carried out with the help of a computer, which was running for approximately 1200 hours during $4\frac{1}{2}$ years in order to give the answer: Yes, four colors are sufficient.

After the first enthusiasm that followed the above result, negative comments and complaints made their appearance:

- The program must be checked again.
- It is impossible to check the program.

- The first part of the proof seems to contain mistakes.
- How beneficial for us and for mathematics was this result?

We will not occupy ourselves with these questions. It is possible that, in the future, mathematicians will find a "better" solution for the four color problem.

Mathematicians working hard for decades or even for centuries enrich mathematics. In this process, general proofs of certain propositions become shorter, new principles are discovered, and older ones are unified in more general theories. The technical terms (such as "Banach spaces", "Hilbert spaces", etc.) may frighten and estrange the non-specialists, but they are merely short phrases which clarify the effort, the work, and the thought of the scientists who lived before us.

CHAPTER 10

FUNCTIONS OF REAL VARIABLES

During the 19th century, a number of "strange discoveries" were made. Examples are:

- Continuous functions which are not differentiable.
- Series of continuous functions whose sum is a discontinuous function.
- Continuous functions which are not piecewise monotone (i.e., increasing or decreasing)
- Functions which have bounded derivatives but they are not Riemann integrable.
- Non-integrable functions which are limits of integrable functions

All these discoveries were highly non-intuitive. The need to understand these results led to the development of the theory of real functions. Another motivation for the further study of the behavior of functions came from the analysis of Fourier series. The theory of Fourier series, as it was formulated by Dirichlet, Riemann, Cantor, Dini (1845-1918), Jordan and other mathematicians of the 19th century, was a satisfactory tool for the needs of applied mathematics, but it was not rigorous.

The rigorous theory of functions of real variables can be considered as a direct continuation of the work of Riemann, Darboux, Du Bois-Reymond, Cantor and others.

In Newton's work, it was proved that the area of a domain can be calculated using the antiderivative, i.e., the function whose derivative is the given function.

Cauchy (Part II, Chapter 1) defined the integral as the limit of a sum and not as an operation which is the inverse of differentiation. There was at least an important reason for this new definition of the integral: Fourier was working with discontinuous functions and hence the known formulas for the Fourier coefficients were given by integrals of discontinuous functions.

In his work on the definite integral, Cauchy begins with continuous functions. Let $f(x)$ be a continuous function on the interval $[x_0, X]$. If $[x_0, X]$ is divided into subintervals by the values $x_1, x_2, ..., x_n$, where $x_n = X$, then the integral of $f(x)$ is by definition equal to

$$\lim_{n \to \infty} \sum_{i=1}^{n} f(\xi_i)(x_i - x_{i-1}),$$

where ξ_i is an arbitrary value of x in the interval $[x_{i-1}, x_i]$. This definition presupposes that $f(x)$ is continuous on the interval $[x_0, X]$ and that the length of the greatest subinterval tends to zero.

Cauchy proves that this integral exists no matter how one chooses the x_i and the ξ_i. But his proof was not complete because he did not have at his disposal the notion of uniform convergence.

Later, following the notation that was introduced by Fourier about the integral, Cauchy put
$$F(x) = \int_{x_0}^{x} f(t)dt$$
and proved that $F(x)$ is continuous on the interval $[x_0, X]$. He considered the expression
$$\frac{F(x+h) - F(x)}{h} = \frac{1}{h}\int_{x}^{x+h} f(t)dt$$
and using the mean value theorem for integrals, he proved that
$$F'(x) = f(x).$$
This last relation is the so-called fundamental theorem of infinitesimal calculus.

After proving that all the antiderivatives of $f(x)$ differ by a constant, he defined the indefinite integral by the relation
$$\int f(x)dx = \int_{a}^{x} f(t)dt + C.$$
He also remarked that if $f'(x)$ is continuous, then
$$\int_{a}^{b} f'(x)dx = f(b) - f(a).$$

Cauchy also studied the so-called improper integrals, where $f(x)$ becomes infinite for some value of x in the interval of integration or the interval of integration extends to infinity. In the case where $f(x)$ is discontinuous at $x = c$, no matter whether $f(x)$ takes a finite value at $x = c$ or not, Cauchy defined the integral of $f(x)$ as follows:
$$\int_{a}^{b} f(x)dx = \lim_{\epsilon_1 \to 0^+} \int_{a}^{c-\epsilon_1} f(x)dx + \lim_{\epsilon_2 \to 0^+} \int_{c+\epsilon_2}^{b} f(x)dx$$
whenever these two limits exist. For $\epsilon_1 = \epsilon_2$, one obtains what Cauchy named "principal value" of the integral.

Notions as the area of a domain enclosed by a curve, the length of a curve, the volume of a solid enclosed by surfaces, and the area of a surface were intuitively acceptable. The fact that mathematicians were able to calculate these quantities with the help of the integral was considered as an important achievement of infinitesimal calculus.

Cauchy, whose aim was the so-called arithmetization of mathematical analysis, defined the geometric quantities that we mentioned above via the integrals that were used for the calculation of these quantities.

In this way Cauchy imposed automatically certain restrictions on these notions. Hence the formula
$$S = \int_{a}^{b} \sqrt{1 + (f'(x))^2}\, dx$$
giving the arc length of a curve $y = f(x)$ holds for the curves for which the derivative of $f(x)$ exists, so that the integral in question has a meaning.

But the way in which the arc length of non-differentiable curves is defined and in general the definition of the area and the volume remained an open problem. Many of these questions were answered during the next century by the researchers who were working in topology and analysis.

Riemann extended the notion of the integral for functions $f(x)$ which are defined and bounded on some interval $[a,b]$. He divided this interval into subintervals Δx_1, Δx_2, ..., Δx_n (where Δx_i also denotes the length of the interval Δx_i) and defined the oscillation of f on Δx_i as the difference between the greatest and the smallest value of $f(x)$ on Δx_i.

Then he proved that a necessary and sufficient condition for the sums

$$S = \sum_{i=1}^{n} f(x_i)\Delta x_i$$

to tend to a unique limit (which means then that the integral exists) as the maximum of the Δx_i tends to zero, is that "the sum of the intervals Δx_i on which the oscillation of $f(x)$ is greater than any given number λ tends to zero as the length of the intervals tends to zero".

Riemann remarked that the above condition, which refers to the oscillations, permited him to replace continuous functions with functions having isolated singularities and with functions having an everywhere dense set of points of discontinuity as well. In this way Riemann eliminated from the definition of the integral the restrictions of "continuity" and "piecewise continuity", which were imposed on the function under the integral.

In 1854 Riemann gave a new necessary and sufficient condition for a bounded function $f(x)$ to be integrable on $[a,b]$. He worked as follows: Let

$$S = M_1 \Delta x_1 + ... + M_n \Delta x_n$$
$$s = m_1 \Delta x_1 + ... + m_n \Delta x_n$$

where m_i and M_i are the minimum and the maximum value of $f(x)$ in Δx_i respectively. Next, let $D_i = M_i - m_i$. Without further explanation, Riemann said that the integral of $f(x)$ on $[a,b]$ exists if and only if

$$\lim_{\max \Delta x_i \to 0} (D_1 \Delta x_1 + D_2 \Delta x_2 + ... + D_n \Delta x_n) = 0$$

for all choices of the Δx_i which subdivide $[a,b]$.

The proof that this last condition is indeed necessary and sufficient for the existence of the integral was given by Gaston Darboux (1842-1917).

Darboux proved also that a bounded function $f(x)$ is integrable on the interval $[a,b]$ if and only if the set of points of discontinuity of $f(x)$ has measure zero, which means that these points can be covered by a finite number of intervals, the sum of the lengths of which can be made as small as we want. In 1875 Darboux proved that the fundamental theorem of infinitesimal calculus is true for functions which are integrable in the above sense.

O. Bonnet gave a proof of the mean value theorem of differential calculus without using the assumption that $f'(x)$ is continuous. Darboux made use of this proof and showed that

$$\int_a^b f'(x)dx = f(b) - f(a)$$

where f' is assumed only to be integrable in the sense of Riemann-Darboux.

The Stieltjes integral

The first extension of the notion of the integral actually came from the study of a class of problems different from those described above. In 1894 Thomas Jan

Stieltjes (1856-1894) in his extremely original paper *Recherches sur les fractions continues*, introduced a generalization of the Riemann-Daboux integral. This paper introduced problems of a new nature, both in the theory of analytic functions and in the theory of functions of a real variable.

Stieltjes begins by considering a positive distribution of mass along a straight line. He remarks that such a distribution is given by an increasing function $\phi(x)$, which specifies the total mass in the interval $[0, x]$ for $x > 0$. Here the discontinuities of ϕ correspond to masses which are concentrated at certain points. For such a distribution of mass in the interval $[a, b]$, Stieltjes forms the Riemann sums,

$$\sum_{i=0}^{n-1} f(\xi_i)(\phi(x_{i+1}) - \phi(x_i))$$

where the points $x_0, x_1, ..., x_n$ form a subdivision of $[a, b]$, and ξ_i lies in the interval $[x_i, x_{i+1}]$. Then he proves that if f is continuous on $[a, b]$ and the maxima of the subintervals of the subdivisions tend to zero, then the above sums tend to a limit, which he denotes by $\int_a^b f(x)d\phi(x)$. Stieltjes only used this integral in the specific problem that he considered and he did not try to give a general definition. However, he did define the relevant integral in the case that the interval is of the form $(0, \infty)$:

$$\int_0^\infty f(x)d\phi(x) = \lim_{b \to \infty} \int_0^b f(x)d\phi(x).$$

It was later realized that the Stieltjes integral has many other applications, and was studied by several mathematicians.

The notions of "content" and "measure"

The study of the set of points of discontinuity of a function led to the question of how one can measure the "size" of such a set (because this size determines whether the function is integrable or not).

The notion of the "content" of a set and later the notion of the "measure" of a set were introduced in order to extend the known notion of length to sets which are not intervals of a straight line. The notion of the "content" of a set is based on the following idea: Let E be a set of points distributed in some way in an interval $[a, b]$. Let us assume that we cover the points of E by small subintervals of $[a, b]$ in such a way that the points of E are interior points or, in the worst case, extremities of some of these subintervals. Next we reduce further the lengths of the subintervals and, if necessary, we add other subintervals so that we continue to cover the points of E, while at the same time we reduce the sum of the lengths of these subintervals. The infimum of these sums is called the (exterior) content of E. This somewhat sketchy formulation does not constitute the final notion of "content" that became accepted, but it gives us a picture of what mathematicians were trying to do at that time.

Even though the above notion of "content" was not completely satisfactory, its use revealed the existence of sets of positive content which are nowhere dense. Functions which are discontinuous on such sets are not Riemann integrable (we recall that a set contained in an interval is called nowhere dense when it is not dense in every subinterval of the interval in question). In addition, the existence of functions was revealed which had bounded but non-integrable derivatives. Despite this situation, the mathematicians of the 1880s continued to believe that the Riemann integral could not be generalized.

Several efforts were made to overcome the obstacles in the "theory of content" and to make rigorous the notion of the area of a domain. Peano (1887) introduced an improved notion of "content". During the 19th century, Jordan made another decisive step in the theory of content (étendue) in *Cours d' analyse* (Vol. I, 1893). Despite this progress, the definition of "content" was still not entirely satisfactory.

The next step in the theory of content was made by Borel in *Leçons sur la théorie des fonctions* (1898), where he was able to isolate and bypass the weak points of the previous theories regarding the notion of "content".

Cantor had proved that every open set U of the real line is the union of **denumerably** many open intervals which are pairwise disjoint. Borel used this result and defined the measure of a bounded open set U of the real line as the sum of the lengths of the intervals that comprise the set U. Later he defined the measure of the union of denumerably (i.e., finitely or countably) many pairwise disjoint measurable sets as the sum of the measures of these sets, and he defined the measure of the set $A \setminus B$, where A and B are measurable sets and $B \subset A$, as the difference of the measure of A from the measure of B. With these definitions he was able to assign a measure to every set, which is formed via the union of denumerably many pairwise disjoint measurable sets, and to the difference of two arbitrary measurable sets A and B, where $B \subset A$. Then he considered sets of measure zero and he proved that a set of positive measure is uncountable.

Borel's "measure theory" improved greatly the theory of content of Peano and Jordan, but Borel did not give any applications in integration theory.

The Lebesgue integral

The generalization of the notions of the measure of a set and of the integral is due to a student of Borel, namely Henri Lebesgue (1875-1941). This work is considered the definitive work on this subject.

Although Lebesgue's theory of integration is applied to sets in the n-dimensional space, for simplicity we will restrict ourselves to the case of one dimension.

Let E be a set of points in the interval $[a, b]$. The points of E can be covered as interior points by finitely or countably many subintervals $d_1, d_2, \ldots$ of the interval $[a, b]$. (The ends of $[a, b]$ can be ends of some d_i.) It is proved that the intervals $d_1, d_2, \ldots$ can be replaced by a set of pairwise disjoint intervals $\delta_1, \delta_2, \ldots$ so that every point of E is an interior point of some δ_i or the common extremity of two adjacent of these intervals. Let $\sum \delta_i$ denote the sum of the lengths of the δ_i. The infimum of the set of $\sum \delta_i$ for all possible sequences $\langle \delta_i \rangle$ is called the exterior measure of E and is denoted by $m_e(E)$. The interior measure $m_i(E)$ of E is by definition the exterior measure of the set E^c, i.e., of the complement of E with respect to $[a, b]$, or, in other words, the exterior measure of the set of points $a \leq x \leq b$ which do not belong to E.

One can prove various results, which include the inequality $m_i(E) \leq m_e(E)$. The set E is defined to be measurable if $m_i(E) = m_e(E)$ and the measure $m(E)$ of E is defined to be this common value. Lebesgue proved that the measure of the union of countably many pairwise disjoint measurable sets is equal to the sum of the measures of these sets. It can be shown that every Jordan measurable set is Lebesgue measurable and its measure remains the same. Lebesgue's notion of measure differs from Borel's by a part of a set of measure zero in the sense of Borel. We mention that the cardinality of the set of Lebesgue measurable subsets of the real line $\mathbf{R}$ is equal to $2^\mathbf{c}$, while the cardinality of the set of Borel measurable subsets of $\mathbf{R}$ is equal to $\mathbf{c}$. This means that there exist Lebesgue measurable sets which are not Borel measurable. We recall that $\mathbf{c}$ is the cardinality of the set $\mathbf{R}$ of real numbers. Lebesgue also noted that there exist sets which are not Lebesgue measurable.

The next important notion introduced by Lebesgue is the notion of a measurable function. Let E be a bounded measurable subset of the x-axis. A function $f(x)$, which is defined for every $x \in E$, is called measurable on E if for any real number A, the set $\{x \in E : f(x) > A\}$ is measurable.

The Lebesgue integral is defined as follows: Let $f(x)$ be a bounded and measurable function defined on the measurable subset E of the interval $[a, b]$. Let A and B be the infimum and the supremum of $f(x)$ on E respectively.

The interval $[A, B]$ (on the y-axis) can be divided into n subintervals

$$[A, l_1], \ [l_1, l_2], \ ..., \ [l_{n-1}, B]$$

where $A = l_0$ and $B = l_n$. Let e_r be the set of points $x \in E$ for which

$$l_{r-1} \leq f(x) \leq l_r, \qquad r = 1, 2, ..., n.$$

Then the sets $e_1, e_2, ..., e_n$ are measurable. Let the sums S and s be defined as follows

$$S = \sum_{r=1}^{n} l_r m(e_r), \qquad s = \sum_{r=1}^{n} l_{r-1} m(e_r).$$

The sums S have an infimum J and the sums s have a supremum I. Lebesgue proved that for bounded measurable functions, $I = J$. This common value is called the Lebesgue integral of $f(x)$ on E and is denoted by

$$I = \int_E f(x) dx.$$

If E happens to be $[a, b]$, then the integral is denoted by $\int_a^b f(x) dx$, but it is understood that the integration is in the sense of Lebesgue. If $f(x)$ is Lebesgue integrable and the value of its integral is finite, then $f(x)$ is called "summable".

If $f(x)$ is Riemann integrable on $[a, b]$, then it is also Lebesgue integrable on $[a, b]$. But there exist Lebesgue integrable functions which are not Riemann integrable. If $f(x)$ is integrable both in the sense of Riemann and in the sense of Lebesgue, then the two integrals have the same value.

The Lebesgue integral makes sense for a function which is not necessarily continuous almost everywhere (i.e., continuous on every point apart from the points of a set of measure zero). For example, the Dirichlet function, which is equal to 1 for every rational value of x and equal to 0 for every irrational value of x in $[a, b]$, is not Riemann integrable, but it is Lebesgue integrable and its integral is equal to zero.

The notion of Lebesgue integration can also be extended to unbounded functions. If $f(x)$ is Lebesgue integrable and unbounded in the interval of integration, then $|f(x)|$ is also Lebesgue integrable. Unbounded functions can be Lebesgue integrable but not Riemann integrable.

For practical purposes the Riemann integral is sufficient. Indeed, in his writing *Leçons sur l'intégration et la recherche des fonctions primitives* (1904), Lebesgue proved that a bounded function is Riemann integrable if and only if the set of its points of discontinuity has measure zero.

However, for works of a theoretical nature the Lebesgue integral is much more powerful. In this connection we note the Lebesgue measure has the property of being countably additive as opposed to Jordan's notion of "content" which is only finitely additive.

An example of the simplicity that characterizes the theorems that use the Lebesgue integral is provided by the following result which was given by Lebesgue himself in his doctoral dissertation.

Let us assume that the functions $u_1(x), u_2(x), \ldots$ are summable (integrable) on a measurable set E and that the series $\sum_{n=1}^{\infty} u_n(x)$ converges to a certain function $f(x)$. Then $f(x)$ is measurable. If in addition $S_n(x) = \sum_{i=1}^{n} u_i(x)$ is uniformly bounded (i.e., $|S_n(x)| \leq B$ for all $x \in E$ and for all n), then a theorem of Lebesgue asserts that $f(x)$ is Lebesgue integrable on E and

$$\int_E f(x) dx = \lim_{n \to \infty} \int_E S_n(x) dx.$$

But if instead of working with the Lebesgue integral, we worked with the Riemann integral, then, in order to reach the above result, we would need to make the additional assumption that the sum of the series $\sum_{n=1}^{\infty} u_n(x)$ is an integrable function.

The Lebesgue integral is particularly useful in the theory of Fourier series. If we restrict ourselves to the Riemann integral, then we know that the Fourier coefficients a_n and b_n of a bounded integrable function tend to zero as n tends to ∞. But if we use the Lebesgue integral, then

$$\lim_{n \to \infty} \int_a^b f(x) \left\{ \begin{array}{c} \sin nx \\ \cos nx \end{array} \right\} dx = 0$$

where $f(x)$ is an arbitrary Lebesgue integrable function, which may be bounded or unbounded. This last result is the well-known Riemann-Lebesgue Lemma.

In his work of 1903, Lebesgue also proved that if a bounded function f is given by a trigonometric series, i.e.,

$$f(x) = \frac{a_0}{2} + \sum_{n=1}^{\infty} (a_n \cos nx + b_n \sin nx),$$

then the a_n and the b_n are the Fourier coefficients. In 1905 Lebesgue gave a new sufficient condition for the convergence of a Fourier series of a function $f(x)$.

Lebesgue also proved (*Leçons sur les séries trigonométriques*, 1906, p. 102) that it is possible to integrate "termwise" a Fourier series even if it does not converge uniformly to $f(x)$. Actually, if $f(x)$ is Lebesgue integrable, then

$$\int_{-\pi}^{x} f(x) dx = a_0(x + \pi) + \sum_{n=1}^{\infty} \frac{1}{n}(a_n \sin nx + b_n(\cos n\pi - \cos nx)),$$

where x is an arbitrary point of the interval $[-\pi, \pi]$, independently of whether the initial series that corresponds to $f(x)$ is convergent or not. Moreover, this new series on the left-hand side of the last equality converges uniformly on the interval $[-\pi, \pi]$.

Lebesgue also proved that Parseval's equality

$$\frac{1}{\pi} \int_{-\pi}^{\pi} (f(x))^2 dx = 2a_0^2 + \sum_{n=1}^{\infty} (a_n^2 + b_n^2),$$

holds for every function $f(x)$ for which the function $|f(x)|^2$ is Lebesgue integrable on the interval $[-\pi, \pi]$. Later, Pierre Fatou (1878-1929) proved that

$$\frac{1}{\pi} \int_{-\pi}^{\pi} f(x)g(x) dx = 2a_0 c_0 + \sum_{n=1}^{\infty} (a_n c_n + b_n d_n),$$

where a_n, b_n and c_n, d_n are the Fourier coefficients of the functions $f(x)$ and $g(x)$ respectively for which $|f(x)|^2$ and $|g(x)|^2$ are Lebesgue integrable on the interval $[-\pi, \pi]$.

We note that we still don't know a necessary and sufficient condition for a Lebesgue integrable on the interval $[-\pi, \pi]$ function $f(x)$ to have a Fourier series which converges to $f(x)$ almost everywhere.

Lebesgue also studied the relation between the notions of the integral and the antiderivatives (indefinite integrals). There exist examples of functions f which are Riemann integrable but such that the function $\int_a^x f(t) dt$ has no derivative (and in particular neither left nor right derivative) at some points. On the other hand, Volterra proved in 1881 that a function $F(x)$ can have a bounded derivative in some interval I, which is not Riemann integrable on I.

Lebesgue proved that if f is Lebesgue integrable on $[a, b]$, then the function $F(x) = \int_a^x f(t) dt$ has a derivative equal to $f(x)$ almost everywhere. Conversely, if a function g is Lebesgue integrable on $[a, b]$ and its derivative $g' = f$ is bounded, then f is Lebesgue integrable and the formula $g(x) - g(a) = \int_a^x f(t) dt$ is valid.

The situation is much more complicated if g' is not bounded. In this case g' is not necessarily integrable, hence one must first characterize the functions g for which g' exists almost everywhere and is integrable.

If one of the following two limits

$$\limsup_{h \to 0, h > 0} \frac{g(x+h) - g(x)}{h}, \qquad \liminf_{h \to 0, h > 0} \frac{g(x+h) - g(x)}{h}$$

is finite everywhere, Lebesgue proved that g is necessarily a function of bounded variation (see the Note below).

Lebesgue also proved (1904) the following result: "A function g of bounded variation has a derivative g' almost everywhere and g' is integrable." Note that this doesn't necessarily imply

$$g(x) - g(a) = \int_a^x g'(t) dt$$

because the difference between the quantities in this equality is not zero but a non-constant function of bounded variation whose derivative is equal to zero almost everywhere. The functions g of bounded variation for which the last relation holds must be absolutely continuous, i.e., they must have the following property: "The total variation of g in an open set U (i.e., the sum of the total variations of g in every connected component of U) tends to 0 when the measure of U tends to zero". Such functions were studied by Guiseppe Vitali.

Note. Let $f(x)$ be a bounded function on $[a,b]$ and let $a = x_0, x_1, ..., x_{n-1}, x_n = b$ be an arbitrary subdivision of $[a,b]$. Let $y_0, y_1, ..., y_{n-1}, y_n$ be the values of $f(x)$ at the points $x_0, x_1, ..., x_{n-1}, x_n$ respectively. Then for every subdivision, we have

$$\sum_{r=0}^{n-1}(y_{r+1} - y_r) = f(b) - f(a).$$

We put

$$t = \sum_{r=0}^{n-1}|y_{r+1} - y_r|.$$

To every subdivision of $[a,b]$ corresponds a number t. When the supremum of all the numbers t (which is called the total variation of f) is finite, f is called a function of **bounded variation** on $[a,b]$.

With his work Lebesgue also contributed to the theory of multiple integrals. Some results in this direction are contained in his doctoral dissertation, but the best result for multiple integrals were obtained by Guido Fubini (1879-1943).

Lebesgue's work constitutes one of the most important contributions in the progress of mathematics in the 20th century. Still, there existed famous mathematicians who objected to it.

Hermite was one of the mathematicians who considered the strange functions mentioned in the introduction as "illnesses" or as part of a sad pathological situation of functions, which are irrelevant to the important problems of pure and applied mathematics. Actually, Hermite tried to oppose the publication of Lebesgue's paper *Note on Non-Ruled Surfaces Applicable to the Plane*.

Using Lebesgue's theory of integration, previous results about integrals are generalized and furthermore one obtains simple and clear theorems about series. Later, other extensions of the notion of the integral were introduced. One of them is due to Johann Radon (1887-1956). This extension includes the Stieltjes and the Lebesgue integral, and is known as the Lebesgue-Stieltjes integral.

These extended integrals have applications in probability, spectral theory, ergodic theory and harmonic analysis.

CHAPTER 11

ABSTRACT ALGEBRA

The purpose of the present chapter is to give to the reader a general picture of the new algebraic theories, which began to appear during the 19th century, but reached their complete development during the 20th century. These theories have had a wide impact on contemporary mathematical research.

The object of Abstract Algebra (or what we call Contemporary Algebra), as well as of Classical Algebra, is the study of operations and of the various rules of calculation. It is not restricted though only to the study of the properties of numerical operations, but it also studies operations of a much more general nature. This trend of generalization is also dictated by practical needs; for example, in mechanics we add forces, velocities, etc.

In linear algebra, where various ideas and methods of algebra have wide applications in practical calculations, the operations are applied to various classes of objects, such as matrices, linear transformations, vectors in n-dimensional spaces, etc.

These classes of objects differ from each other according to the properties of the operations on their elements. Examples of such classes that we have already met are the groups, the rings, the ideals, and the fields.

During the 19th century, almost all works regarding these various types of algebras refer to specific systems whose elements were known objects, as numbers, vectors, and other. Only during the last decades of the 19th century mathematicians understood that certain algebras can be unified by making their content abstract. Hence the groups of permutations, the groups of transformations, as well as other particular groups, can be studied together simultaneously as sets of elements or objects which obey an operation whose nature is determined by certain abstract properties that we have already met in Chapter 9 of Part I.

We recall that all one-to-one mappings of a set M onto itself, i.e., all permutations of M, constitute a group under the operation of composition, which is defined by the relation $(f \circ g)(x) = f(g(x))$ $(x \in M)$. A group G is called a group of permutations on M if every element of G is a permutation on M.

In this way, the unification of the various structures led to the creation of an abstract algebra as the study of many classes of algebras, each of which has as elements concrete objects and served various purposes in various areas of mathematics; for example, the groups of permutations were used in the theory of algebraic equations. In the course of time, the application of the abstract theory to particular areas of mathematics became gradually less important, so that the study of abstract structures and the finding of their various properties became an end in itself.

Today abstract algebra constitutes a wide area of research. We will present only certain initial activities and we will indicate the multiple paths of research that one can follow. Apart from the usual difficulties that arise due to the fact that various authors use different terms and the meaning of the terms changes from time to time, algebra has that additional difficulty of using a large number of new terms. A comprehensive vocabulary of the algebraic terms would require a separate many-paged book.

The first abstract structure was the structure of a group. Many basic ideas of the abstract theory of groups had made their appearance at least since 1800 in the works of Gauss, Abel, Galois, Cauchy and others. After the foundation of the abstract theory, it was not difficult for the founders to incorporate from works of the past the ideas and theorems which are related to the subject.

The formulation of the notion of the abstract group and its associated properties followed a slow pace. In 1849, Cayley had proposed the generalization of the notion of the group of permutations, but his proposition did not attract the attention of his contemporaries. In 1858, starting from the notion of the group of permutations, Dedekind gave the definition of an abstract group with finitely many elements. This work can be considered as one of the basic contributions in the theory of abstract groups.

In addition, Kronecker gave the abstract definition of a finite Abelian group, which was analogous to the definition given by Cayley in 1849.

In 1878, Cayley published four more works on finite abstract groups. In these works, as well as in his earlier works of 1849 and 1854, he stresses that a group can be considered as a more general notion and that we should not restrict ourselves only to groups of permutations, but he claims that every (finite) group can be represented in the form of a group of permutations. The impact of these works of Cayley was larger than the impact of his previous works, because the time had become ripe for the creation of an abstract notion of a group.

In his book *Die Substitutionentheorie und ihre Anwendung auf die Algebra* (1882), Eugen Netto restricts himself to the study of groups of permutations, but the way in which he introduces the various notions and the way in which he proves the various theorems indicate clearly that he recognizes their abstract character.

Netto uses the results of the mathematicians before him and also occupies himself with the notions of "isomorphism" and "homomorphism". We call isomorphism a "one-to-one" correspondence between the elements of two groups such that if $a \cdot b = c$, where a, b, c are elements of the first group, then $a' \cdot b' = c'$, where a', b', c' are the corresponding elements of the second group. A homomorphism is a mapping of one group into another with the property that if $a \cdot b = c$, where a, b, c are elements of the first group, then $a' \cdot b' = c'$, where a', b', c' are the images of a, b, c in the second group.

During 1880, new ideas on groups came into the limelight. Influenced by the work of Jordan, Klein proves as part of the well-known Erlangen Program, that groups of transformations (with infinitely many elements) can be used for the classification of geometries. We remind the reader that Klein's basic idea was that every geometry can be characterized by a certain group of transformations and that the basic study of every geometry is the characterization of the properties that remain invariant under the action of its group of transformations.

In addition these groups are continuous, which means that the parameters contained in these transformations can take all real values. Hence in the transformations which express rotation of axes, the angle θ can take all real values. In their work on automorphic functions, Klein and Poincaré used another kind of group with infinitely many elements, the so-called discrete group (see Poincaré's biography in Part II).

Sophus Lie, who collaborated with Klein in 1870, occupied himself with continuous groups of transformations for reasons independent of the problem of the classification of geometries. Indeed, he had observed that many ordinary differential equations, which were solved with older methods, remained invariant under the action of various classes of continuous groups of transformations. This observation led him to the idea that he could shed more light on the problem of solving differential equations and to their classification by making use of groups.

In 1874 Lie introduced his general theory regarding groups of transformations. Such a group is represented by

$$x'_i = f_i(x_1, x_2, ..., x_n, a_1, a_2, ..., a_n), \qquad i = 1, 2, ..., n \qquad (1)$$

where the f_i are analytic functions of the x_i and the a_i. The a_i are parameters, while the x_i are variables and $(x_1, x_2, ..., x_n)$ is a point of the n-dimensional space. Both the parameters and the variables can take real or complex values. Hence if the space is 1-dimensional, the class of transformations

$$x' = \frac{ax+b}{cx+d}, \qquad (2)$$

where the a, b, c, d take all real values such that $ad - bc \neq 0$, constitutes a continuous group.

The groups given by (1) are called finite because the number of the parameters is finite. Of course, this does not mean that the number of these transformations is finite; clearly, this number is infinite.

In case (2), we have a group which has only three parameters, because what is important is the value of the ratios $\frac{a}{d}$, $\frac{b}{d}$, $\frac{c}{d}$ and not the value of the parameters a, b, c, d. In the general case, the product of two transformations

$$x'_i = f_i(x_1, ..., x_n, a_1, a_2, ..., a_n)$$
$$x''_i = f_i(x'_1, ..., x'_n, b_1, b_2, ..., b_n)$$

is the transformation

$$x'''_i = f_i(x''_1, ..., x''_n, c_1, c_2, ..., c_n)$$

where the c_i are functions of the a_i and the b_i.

In his work, which was published in 1883 and which refers to continuous groups, Lie introduced continuous groups of transformations with infinitely many parameters. We will not discuss this subject.

During 1880, four types of groups were known:
- Discrete groups of finite order, examples of which are the groups of permutations.
- Infinite discrete groups, examples of which were given in the theory of automorphic functions.
- Finite continuous groups of Lie, examples of which are the groups of transformations of Klein and the more general analytic transformations of Lie.
- Infinite continuous groups of Lie, which are defined through differential equations.

The work of Walter von Dyck (1856-1934) unified under the notion of the abstract group the three main roots of group theory, namely theory of equations, number theory, and infinite groups of transformations. Dyck was influenced by Cayley and was a student of Felix Klein. During the years 1882 and 1888 he published a series of works on abstract groups, which included both discrete and continuous groups. Dyck calls a group a set of elements equipped with an operation (for example, multiplication) which has the property that the product of two elements of the set is an element of the set, while this operation is associative but not commutative and every element has its inverse.

Dyck gave particular emphasis to the research regarding the notion of "generators" of a group. The generators of a group are a particular set consisting of independent elements of the group such that every element of the group can be expressed as the product of powers of the generators and of their inverses.

When there exist no restrictive conditions on the generators, the group is called "free". If $A_1, A_2, \ldots$ are the generators, then an expression of the form

$$A_1^{\mu_1} A_2^{\mu_2} \ldots$$

where the μ_i are positive or negative integers, is called a "word".

The generators may satisfy various relations of the form

$$F_i(A_i) = 1$$

where we have a word or several words equal to the unit element of the group.

Dyck proved that the presence of such relations implies that the group has certain properties. In a paper published in 1883, Dyck applied the theory of abstract groups to groups of permutations, to groups that appear in number theory, as well as to groups of transformations of differential equations.

After the establishment of the "abstract notion of a group", mathematicians began to extend to abstract groups the theorems that were proved about particular cases of groups. For example, Frobenius extended to finite abstract groups a theorem of Sylow, which states that every finite group whose order (i.e., the number of its elements) is divided by the nth power of a prime number p always contains a subgroup of order p^n.

Besides, many researchers introduced directly new notions in the abstract theory of groups. For example, Dedekind and Miller introduced the notion of the "commutator". If s and t are two arbitrary elements of a group G, then the element $s^{-1}t^{-1}st = [s,t]$ is called the commutator of s and t. This name is due to the fact that the element $s^{-1}t^{-1}st$ has the property that $ts \cdot (s^{-1}t^{-1}st) = st$. That is, s commutes with t if and only if their commutator is the unit element of the group. The subgroup of G that is generated by all its commutators (i.e., the intersection of all the subgroups of G that contain all the commutators of G) is called the "commutator subgroup" of G. The automorphisms of an abstract group, i.e., the one-to-one transformations of the elements of the group to elements of the same group with the property that if $a \cdot b = c$, then $a' \cdot b' = c'$, were studied by Hölder and E. H. Moore.

The development of the abstract theory of groups followed many directions. One of these directions was the finding of all groups having a given order, i.e., a given number of elements. This problem is not yet generally solved. But certain particular cases have been solved, as, for example, when the order of the group is equal to p^2q^2, where p and q are prime numbers.

Another direction of research was the effort to determine the so-called solvable groups and the simple groups, i.e., the groups which have no non-trivial proper normal subgroups.

For the convenience of the reader we give the following definitions and notions.

(α) If H is a subgroup of a group G and $Hx = xH$ for every $x \in G$, then H is called a "normal (or invariant) subgroup" of G. Here Hx is the set of elements of the form hx, where $h \in H$. The meaning of xH is analogous.

If H is a normal subgroup of a group G, then it can be proved that the set $(Ha)(Hb)$ is equal to the set Hab. This means that the sets Ha ($a \in G$) constitute a group which is called the "quotient group" of G modulo H and is denoted by G/H. (If G is an additive group, then G/H is denoted by $G - H$ and is called the "difference group".)

The commutator subgroup C of a group G is a normal subgroup of G and the quotient group G/C is Abelian.

If A, B are two subsets of a group G, then the subgroup of G that is generated by the commutators $[a, b]$, where $a \in A$ and $b \in B$, is called the commutator of A and B and is denoted by $[A, B]$.

(β) A finite sequence of subgroups

$$G = G_0 \supset G_1 \supset G_2 \supset ... \supset G_r = \{e\}$$

of a group G is called a "normal chain" if G_i is a normal subgroup of G_{i-1} for $i = 1, 2, ..., r$ and e is the unit element of the group G. If $a \in G$, then we denote by $\langle a \rangle$ the group whose elements are all the powers of a. The sequence G_0/G_1, G_1/G_2, ..., G_{r-1}/G_r is called a sequence of quotient groups.

(γ) Let us assume that we are given a sequence of subgroups G_i ($i = 0, 1, 2, ...$) of a group G such that $G = G_0$ and $[G_i, G_i] = G_{i+1}$ for every i. Then we have a normal chain

$$G = G_0 \supset G_1 \supset ...$$

If $G_r = \{e\}$ for some r (where e is the unit element of G), then G is called a "solvable group". Frobenius proved that all groups whose order is not divisible by the square of some prime number are solvable. The problem of finding all solvable groups forms part of the more general problem of determining the structure of a given group.

A problem in group theory which awaits its solution is the so-called "Burnside problem": We know that every finite group is generated by a finite number of elements and every element of the group has a finite order which divides (and therefore is bounded by) the order of the group. (Order of an element a is called the smallest natural number n (if it exists) for which $a^n = e$, while if such a natural number does not exist, then the order of a is equal to ∞.) In 1902, William Burnside (1852-1927) posed the question of whether the converse of this last proposition holds. That is, if a group G is generated by a finite number of elements and if every element of this group has a finite order which is bounded by some fixed positive integer, is G then a finite group? Only partial solutions have been given so far to this problem.

Another problem that also awaits its solution is the so-called "isomorphism problem": Determine when two groups, each of which is defined via generators and other relations, are isomorphic.

A surprising phenomenon that is observed in group theory is that soon after the appearance of abstract group theory mathematicians turned their efforts to "representations" (we will explain immediately below the meaning of this term) in more concrete algebraic structures in order to obtain new results regarding the abstract groups.

Representations of groups

The abstract theory of groups presents a certain similarity with elementary geometry; indeed they are both founded on certain systems of axioms which constitute the starting points for the derivation of the entire content of each theory. The example of analytic geometry illustrates the fact that analytic and arithmetic methods can be useful in the investigation of geometric problems.

An application of the methods of analysis and of classical algebra to group theory constitutes the so-called "theory of representation of groups". Just like analytic geometry provides much more than just methods for solving geometric problems via the use of mathematical analysis, the theory of group representations is not only used as an auxiliary mechanism for the study of the properties of groups, but also constitutes a link between fundamental notions and problems of analysis and group theory. Furthermore it allows us to find expressions regarding group theory which are given through arithmetic relations, and to find representations in groups involving various analytic relations. A large part of the important applications of group theory in physics is directly related to the theory of group representations.

From linear algebra, which is taught to us during the first years of our studies, we know how to multiply two matrices. We know that this operation of multiplication is associative but in general not commutative. The non-singular square matrices of a given order form a group with respect to the operation of multiplication since the product of two non-singular matrices is a non-singular matrix and the role of the unit element is played by the identity matrix, while every non-singular matrix has an inverse non-singular matrix.

We remind the reader of the inverse of a matrix: Let A be a square matrix whose entries belong to a ring or to a field (as are, the field $\mathbf{R}$ of real numbers and the field $\mathbf{C}$ of complex numbers). If there exists a matrix A^{-1} such that $AA^{-1} = A^{-1}A = I$, then A^{-1} is called the inverse matrix of A and A is called non-singular.

The matrix I is the identity matrix of the same order with A and the (i,j)-element of I (i.e., the element that lies in the ith row and in the jth column of I) is equal to Kronecker's δ-symbol δ_{ij}, where $\delta_{ii} = 1$ and $\delta_{ij} = 0$ for $i \neq j$.

Let us assume that we are given a group G and to every $g \in G$ we assign a certain non-singular matrix A_g of order n with entries in $\mathbf{C}$ in such a way that if $g, h \in G$, then $A_{g \cdot h} = A_g \cdot A_h$. In this case we say that we have a representation of the group G by matrices of order n. We usually omit the words "by matrices" and we simply talk about a representation of G of order n. A representation of order n of a given group G is simply a homomorphic mapping (a homomorphism) of G into the group of non-singular matrices of order n. It follows from the general properties of homomorphic mappings that in every representation of G the unit element of G is mapped to the identity matrix and elements of G, which are inverse to each other, are mapped to matrices, which are inverse to each other.

Matrices of order 1 are simply complex numbers. Consequently, a representation of order 1 of G is a relation that makes every element of G to correspond to a complex number in such a way that a product of two arbitrary elements of G corresponds to the product of the complex numbers that correspond to these elements of G.

When we know one representation of a group G, we can then obtain infinitely many representations of G. Indeed, let $g \mapsto A_g$ be a representation of G by matrices of order n. We choose an arbitrary non-singular matrix P of order n and we set $B_g = P^{-1}A_gP$ for every $g \in G$. Then the correspondence $g \mapsto B_g$ also constitutes a representation of G, because:

$$B_{gh} = P^{-1}A_{gh}P = P^{-1}A_gA_hP = P^{-1}A_gPP^{-1}A_hP = B_gB_h.$$

The representations that we obtain in this way are called equivalent and are not considered to be different from the one given initially.

Another notion which is useful in the study of representations of groups is the notion of "character of a group". A character of a group is a function $x(s)$ which is defined for all elements s of the group and takes values in some field (as is the field $\mathbf{C}$ of complex numbers), and which is such that it never takes the value zero, and it satisfies the relation $x(ss') = x(s)x(s')$ for all elements s, s' of the group. Two characters x, x' of a group are different if $x(s) \neq x'(s)$ for at least one element s of the group. We will not discuss though any further the usefulness of the notion of character in group theory.

Abstract Field Theory

The notion of the field R which is generated by n quantities $a_1, a_2, ..., a_n$, i.e., the field of all the quantities which are formed by adding, subtracting, multiplying and dividing these quantities (division by zero being excluded), and the notion of the extension of a field R by adjoining in it a new element λ which does not belong to R, are assumed to be known to the reader.

The abstract theory of fields was introduced by Heinrich Weber who had already accepted the theory of abstract groups. According to Weber a field is a set of objects equipped with two operations, "addition" and "multiplication", which satisfy the condition that the sum and the product of two elements of the set is an element of the set. These operations also satisfy the laws of associativity, commutativity and distributivity. In addition, every element must have a unique inverse with respect to every operation, apart from division by zero.

To the following fields, which were known since the 19th century,

- The field of real numbers
- The field of complex numbers
- The field of algebraic numbers
- The field of rational functions (of one or more than one variables),

Kurt Hensel added the field of p-adic numbers. The introduction of this new field led to further research about algebraic numbers (*Theorie der Algebraischen Zahlen*, 1908). Hensel first of all remarked that an arbitrary integer D can be uniquely expressed as a sum of powers of a prime number p. Namely:

$$D = d_0 + d_1p + ... + d_kp^k$$

where the d_i are some integers which take values in $\{0, 1, ..., p-1\}$. For example,

$14 = 2 + 3 + 3^2$
$216 = 2 \cdot 3^3 + 2 \cdot 3^4$

He also remarked that every non-zero rational number r can be written in the form

$$r = \frac{a}{b}p^n$$

where a, b are integers which are not divisible by p, and n is an integer.

Based on these remarks, Hensel introduced the so-called p-adic numbers. These numbers are expressions of the form

$$\sum_{i=-p}^{\infty} c_i p^i$$

where p is a prime number, and the c_i are rational numbers of the form $\frac{\lambda}{\mu}$, where λ, μ are co-prime integers and μ is not divisible by p.

Hensel defined the four basic operations on these numbers and proved that they constitute a field. A subset of the p-adic numbers can be brought into a one-to-one correspondence with the rational numbers and it can be proved that this subset is isomorphic with the set of rational numbers (in the sense of isomorphism between two fields). In the field of p-adic numbers, Hensel defined units, p-adic integers, and a notion of order which is analogous to the notion of order of rational numbers. We will not discuss this subject any further.

The existing great variety of "fields" led Ernst Steinitz (1871-1928) to the effort to attempt a global study of the theory of abstract fields. This study is contained in his work entitled *Algebraische Theorie der Körper*, which is of basic importance. According to Steinitz, the class of fields can be divided into two types. Let K be a field and let us consider all the subfields of K (for example, the set of rational numbers is a subfield of the field of real numbers). The elements that belong to all the subfields of K also constitute a subfield of K, which is called the "prime subfield" P of K.

There exist only two types of prime subfields. The unit element e is obviously contained in P, hence the elements

$$e,\ 2e,\ ...,\ ne,\ ...$$

are also contained in P.

Either these elements are all distinct or there exists a positive integer p such that

$$pe = \underbrace{e + e + ... + e}_{p \text{ times}} = 0$$

In the first case P must contain all fractions $\frac{ne}{me}$, and since all these elements from a field, P is isomorphic to the field of rational numbers. In this case K is said to have "characteristic 0".

But if $pe = 0$, then it is proved that the smallest such p is a prime number and it is called the characteristic of K.

It is proved that if we start from the prime subfield of one of the two types described above, we can obtain the field K through the method of successive extensions. This method consists of considering an element a of K, that does not belong to P, and forming the field $P(a)$ of all rational functions of a whose coefficients belong to P. Then, if necessary, we take an element b of K, that does not belong to $P(a)$, and we work with b and $P(a)$ in the same way we worked with a and P. This process continues as long as necessary.

Rings

The structures of a "ring" and of an "ideal" were known to Dedekind and Kronecker who used them in their works on algebraic numbers.

The contemporary meanings of these notions are the following.

An abstract ring is a set of elements equipped with two operations. With respect to one of these operations this set constitutes a commutative group. The second operation has the property that the application of this operation on any two elements of the set gives an element of the set, and this operation satisfies the law of associativity but not necessarily the law of commutativity. In addition the unit element may or may not exist, and the law of distributivity, as expressed by the relations $a(b+c) = ab + ac$ and $(b+c)a = ba + ca$, also holds.

An ideal of a ring R is a subring M of R having the property that if $a \in M$ and r is an arbitrary element of R, then the elements ar, ra belong to M. If for any $a \in M$ and for any $r \in R$, only ar belongs to M, then M is called a right ideal of R, while if for any $a \in M$ and for any $r \in R$, only $ra \in M$, then M is called a left ideal of R.

The entire ring R is called the unit ideal. The ideal (a) generated by a single element a consists of all elements of the form $ra + na$, where $r \in R$ and n is an integer. If R has a unit element e, then $ra + na = ra + nea = (r + ne)a = r'a$, where r' is an element of R. An ideal that is generated by a single element is called a principal ideal. Every ideal which is different from $\{0\}$ and R is called a proper ideal. Similarly, if a_1, a_2, ..., a_m are m given elements of a ring R that has a unit element, then the set whose elements are the sums $r_1 a_2 + r_2 a_2 + ... + r_m a_m$, where the coefficients r_i belong to R, is an ideal of R, which is denoted by $(a_1, a_2, ..., a_m)$ and is the smallest ideal that contains a_1, a_2, ..., a_m. A ring R is called Noetherian (Emmy Noether, 1882-1935) if every ideal of R is of the form $(a_1, a_2, ..., a_m)$, where $a_1, a_2, ..., a_m \in R$.

Since an ideal M of a ring R is an additive subgroup of R, it separates R into equivalence classes. Two elements a, b of R are said to be equivalent with respect to M if the element $a - b$ belongs to M, in symbols $a \equiv b \pmod{M}$.

If T is an epimorphism of a ring R onto a ring R', in the sense that T is a surjective map and $T(ab) = (Ta)(Tb)$, $T(a+b) = (Ta) + (Tb)$ and $T1 = 1'$, then the elements of R whose image under T is the zero element of R' constitute an ideal of R which is called the kernel of T, while R' is isomorphic to the ring of equivalence classes of elements of R modulo the kernel of T. Conversely, if L is an ideal of R, we can form the ring R *modulo* L and an epimorphism of R onto R *modulo* L whose kernel is L. R *modulo* L is also denoted by R/L and is called a quotient ring.

The definition of a ring does not require that every non-zero element has an inverse element with respect to the operation of multiplication. If it happens that the ring has a unit element and every non-zero element has an inverse, then the ring is said to be a "division ring" (division algebra) and constitutes a non-commutative field.

In 1905 Wedderburn proved that a finite division ring is a commutative field. Until 1905, the only known division algebras were the commutative fields and the quaternions (Part II, Chapter 4). Later other commutative and non-commutative algebras were constructed. Then the theory of linear associative algebras followed and the object of study of abstract algebras was advanced further when Wedderburn published his work *On Hypercomplex Numbers*, where he generalized results of Elie Cartan (1869-1951).

Even during the first decades of the 19th century, various successful applications of complex numbers led mathematicians to the study of problems of constructing numbers which are represented by triples, quadruples, etc. of real numbers in a way which is analogous to the way in which the complex numbers were constructed as pairs of real numbers.

Since the middle of the 19th century, mathematicians began to construct many special systems of this kind, the so-called systems of hypercomplex numbers, while during the end of the 19th century and the first quarter of the 20th century, a general theory of hypercomplex numbers was developed, which had interesting applications in certain areas of mathematics and physics.

We call hypercomplex number of order n a "number" which can be represented by a n-tuple $(a_1, a_2, ..., a_n)$ of real numbers which we call temporarily coordinates of this number. The hypercomplex numbers $(a_1, a_2, ..., a_n)$ and $(b_1, b_2, ..., b_n)$ are said to be equal if $a_i = b_i$ for $i = 1, 2, ..., n$, while their sum is defined by the relation $(a_1, a_2, ..., a_n) + (b_1, b_2, ..., b_n) = (a_1 + b_1, a_2 + b_2, ..., a_n + b_n)$. In addition, if a is a real number, then by definition $a(a_1, a_2, ..., a_n) = (aa_1, aa_2, ..., aa_n)$.

The further completion of the system of hypercomplex numbers requires the definition of the operation of multiplication and the proof of other properties of this system. But we will not extend the discussion any further in this direction. Historically, the first example of a hypercomplex system is the system of quaternions.

The theory of rings and ideals was systematized by E. Noether, one of the few great women mathematicians, who was given in 1922 a lecturer position at the University of Göttingen. Many results regarding rings and ideals were already known when Noether began her work, but by appropriately formulating the abstract notions she was able to incorporate them in an abstract theory.

Associative algebras

Hypercomplex numbers were defined above as objects whose description requires real numbers, which means that the hypercomplex numbers were considered as systems of real numbers. Since this viewpoint was considered as somewhat restricted, the following definition became gradually accepted for reasons that make theoretical research easier.

A set S of quantities is called an algebra (or a hypercomplex system) over a field P if the following conditions as satisfied:

(a) For any $a \in P$ and for any element x of the system S, an element of S is defined which is called the product of a and x, and is denoted by ax.

(b) To every pair of elements x, y of S, corresponds by definition a unique element of S which is called the sum of x and y, and is denoted by $x + y$.

(c) To every pair of elements x, y of S, corresponds by definition a unique element of S which is called the product of x and y, and is denoted by xy.

In addition these three operations must have the following properties, where a, b, c, ... denote elements of P and x, y, z, ... denote elements of S:

(1) $x + y = y + x$.
(2) $(x + y) + z = x + (y + z)$.
(3) The system S has a zero element θ which has the property $x + \theta = x$.

(4) $a(x+y) = ax + ay$.
(5) $(x+y)a = xa + ya$.
(6) $(ab)x = a(bx)$.
(7) $\theta x = \theta$ and $1 \cdot x = x$, where 1 is the unit element of P.
(8) There exist quantities $e_1, e_2, ..., e_n$ in S such that every element in S can be uniquely expressed in the form $a_1 e_1 + a_2 e_2 + ... + a_n e_n$.
(9) $(ax)y = x(ay) = a(xy)$.
(10) $x(y+z) = xy + xz$ and $(y+z)x = yx + zx$.

In the above definition the elements of P play the role that was played by the real numbers.

The first eight conditions mean that S constitutes a finite-dimensional linear space over the field P.

A combination of conditions 9 and 10 gives the equalities

$$(ay + bz)x = a(yx) + b(zx)$$
$$x(ay + bz) = a(xy) + b(xz)$$

Today the name "algebra" is preferred over the name "hypercomplex system". The name "hypercomplex system" is used for simpler systems such as the "quaternions".

It follows from properties 1-10 that multiplication in algebras is not necessarily associative and commutative, and algebras are not assumed to have a unit element nor the possibility of "dividing" one element by another.

Every algebra has a base, i.e., a system of elements $e_1, e_2, ..., e_n$ with the property that every element of the algebra can be uniquely represented by an expression of the form $a_1 e_1 + a_2 e_2 + ... + a_n e_n$, where the a_i belong to the field P. An algebra can have infinitely many bases, but the number of elements of every base is the same and is called the rank of the algebra.

The system of complex numbers considered as an algebra over the field of real numbers has a base consisting of the numbers 1 and i. But the sets $\{2, 3i\}$ and $\{1, a + bi\}$, where a, b are real numbers and $b \neq 0$, also constitute bases of this algebra.

Among all algebras, associative algebras play a very important role. These are the algebras in which the operation of multiplication has the property $x(yz) = (xy)z$.

The most interesting algebras among the non-associative algebras are Lie algebras, in which multiplication satisfies the following properties:

$$xy = -yx, \qquad x(yz) + y(zx) + z(xy) = 0.$$

Lie algebras are interesting because there exists a close relationship between Lie algebras and Lie groups.

Everything that we mentioned above in relation to the achievements in the area of abstract algebra, does not provide, of course, the complete picture of what was achieved, even during the first half of the 20th century.

Until 1900, the various algebraic objects that were studied, such as the matrices, the hypercomplex systems, etc., were based on the systems of real and complex numbers. What followed is a trend towards the study of abstract structures, as are the abstract groups, the abstract rings, the ideals, etc.

We can say that abstract algebra in a sense "undermined" its own role in mathematics. The various notions and principles were introduced in it, in order to unify the apparently different situations. This was achieved by group theory. But after the formulation of the abstract theories, mathematicians gradually distanced themselves from the concrete structures and concentrated their research on these abstract structures. Hence, with the introduction of hundreds of particular notions, the object of study was divided into other more specific activities, which were more or less independent from one another and were not related to the concrete areas that were considered initially. In other words, the unification mentioned above was followed by diversification and specialization. Hence we have reached the point where many who work in the area of abstract algebra ignore the tools of the abstract structures that they study and furthermore they are not interested whether their results have any applications in concrete areas.

CHAPTER 12

CATEGORIES AND FUNCTORS

During the 20th century, there was a great trend towards creating more abstract theories. The study of the Euclidean plane was replaced by the study of vector and topological spaces, where certain properties of the plane take an abstract form. The notion of a "category" constitutes the culmination of this trend towards the construction of abstract structures.

The Theory of Categories began with the article *General Theory of Natural Equivalences* of S. Eilenberg and S. Mac Lane (*Trans. Amer. Math. Soc.* (1945), 231-294). Here we will restrict ourselves in defining the basic notions of the theory and giving examples in order to help the reader to understand these notions. The further development of the subject goes beyond the scope of the present book.

Today, when mathematicians study vector spaces, they are led to the study of linear transformations. Similarly, when they study groups (which are the abstract form of groups of permutations), they are led to the study of homomorphisms. The common properties that exist in the structures of the examples that we give below can be placed under a single abstract form with the introduction of the notion of a "concrete category".

When we say "concrete category" we mean the following two things:

(a) A class of sets, where every set is equipped with a certain structure.
(b) The class of all functions that map one of these sets to another and at the same time preserve their structures.

Example 1

We consider the class of all sets and the class of all functions between these sets. In this example, condition (b) above is satisfied automatically because the sets that we consider have no structures. This category is called the "category of sets".

Example 2

A "monoid" is a set which has a fixed element 1, called the unit element, and is equipped with a binary operation $\cdot$ such that

$$(a \cdot b) \cdot c = a \cdot (b \cdot c) \quad \text{and} \quad 1 \cdot a = a = a \cdot 1.$$

A "monoid homomorphism" is a function f from a monoid A to another monoid A', which preserves the structure of a monoid, namely:

$$f(a \cdot b) = f(a) \cdot f(b) \quad \text{and} \quad f(1) = 1'$$

(where $1'$ is the unit element of A'). For example, if we take 0 as the concrete element and $+$ as the binary operation, then the set of natural numbers becomes a monoid.

The singleton $\{1\}$, where $1 \cdot 1 = 1$, also constitutes an example of a monoid. The mapping of the set of natural numbers onto the singleton $\{1\}$ constitutes a monoid homomorphism.

We note that group homomorphisms form a particular case of monoid homomorphisms. The class of monoids together with the class of monoid homomorphisms form a "concrete category".

Example 3

A set is called "preordered" when it is equipped with a binary relation $\leq$ which is reflexive and transitive. Namely, $a \leq a$, while if $a \leq b$ and $b \leq c$, then $a \leq c$. We call "monotone mapping" a function f that maps a preordered set A to a preordered set A' and preserves the order. Namely, if $a \leq b$, then $f(a) \leq f(b)$. For example, the set of natural numbers and the set of even numbers equipped with $\leq$ are preordered sets. The function $f(a) = 2a$ maps the natural numbers onto the even numbers and preserves the order, i.e., it is a monotone mapping. The class of all preordered sets together with the class of all monotone mappings between these sets constitute a "concrete category".

Remark. We note that the identity mapping ($f(x) = x$) which is defined on a set always preserves the structure of the set. In addition, the composition $f \circ g$ of two functions that preserves the structure of a set also constitutes a function that preserves the structure of the set.

The reader is asked to prove that the class of groups together with the class of group homomorphisms constitute a "concrete category", the so-called "category of groups".

And now we move on to the definition of a "category".

A category consists of two things: a collection of **objects** and for each pair of objects a **morphism** (called also an arrow). An object may be some mathematical structure, for example, a group or a vector space, and a morphism may be a map between objects. The morphisms are required to satisfy some fairly natural conditions. For example, the identity map between an object and itself is always a morphism, and the composition of two morphisms (if defined) is always a morphism.

Usually we want the morphisms to preserve the mathematical structure of the objects. For example, if the objects are groups, a morphism could be a group homomorphism. If the object is a vector space, the linear maps would be a good choice for morphisms.

Category theory is the branch of mathematics which formalizes a number of algebraic properties of collections of transformations between mathematical objects (groups, topological spaces, etc.) of the same type, which contain the identity mapping and are closed with respect to compositions of mappings.

We call categories the objects studied in category theory.

Let us now be more specific.

Definition of a "Graph"

A **graph** consists of a class of "arrows", of a class of "objects", and of two mappings from the class of arrows to the class of objects.

The first of these mappings is called S (the first letter of the English word source) and the second T (the first letter of the English word target).

$$\{arrows\} \xrightarrow[target]{source} \{objects\}$$

If f is an array, $S(f) = A$, and $T(f) = B$, then we write

$$f : A \to B \quad \text{or} \quad A \xrightarrow{f} B.$$

We call **category** a graph that satisfies the following conditions:

(1) To every pair of arrows $f : A \to B$ and $g : B \to C$ (where $T(f) = S(g)$) corresponds an arrow $g \circ f : A \to C$.
(2) If $f : A \to B$, $g : B \to C$, and $h : C \to D$, then $(h \circ g) \circ f = h \circ (g \circ f)$.
(3) To every object A corresponds an identity arrow 1_A whose source is A and whose target is A.
(4) If $f : A \to B$, then $f \circ 1_A = f$, while if $g : B \to A$, then $1_A \circ g = g$.

Note. A "concrete category" is a category whose objects are sets equipped with a certain structure and whose arrows are the mappings that map any of these sets to any other of these sets and they preserve the underlying structures. In "concrete categories", 1_A represents the identity mapping on A, while the symbol $\circ$ denotes the composition of two mappings. We give below examples of categories.

Example 4

Let A be an arbitrary set. Assume that A is the class of objects and A is also the class of arrows. Let $S(a) = T(a) = a$ for every $a \in A$.

Define 1_a to be a and define $a \circ a$ to be a. Then the conditions required by the definition of a category are satisfied and we have what is called the **discrete** category that corresponds to A. In other words, an arbitrary set can be considered as a "category".

Example 5

Let $(A, 1, \bullet)$ be a monoid (with the unit element 1 and binary operation $\bullet$). As class of objects we take the singleton $\{*\}$ and as class of arrows we take A, while we define that $S(a) = T(a) = *$ for all $a \in A$.

Let 1_* be the unit element of the monoid and let $a \circ b$ be $a \bullet b$. The graph that we obtain in this way constitutes a category, due to the monoid structure. We remark that in this way a monoid can be considered as a category. If we consider the monoid of the natural numbers with $+$ as the binary operation as a category in this way, then one could ask the question: What is number 2? The answer is that number 2 is a certain uniquely determined arrow of the monoid of natural numbers when this monoid is considered as a category.

Example 6

Let $(A, \leq)$ be a preordered set. We take A to be the class of objects and we take $\{(a, b) : a \leq b\}$ to be the class of arrows. (Here a and b are assumed to be elements of A.) Let $S((a, b)) = a$ and $T((a, b)) = b$. The reflexivity of the relation $\leq$ implies that (a, a) constitutes an arrow for every $a \in A$. Let (a, a) be the identity arrow that corresponds to a so that $1_a = (a, a)$. We define the composition of two arrows as follows: $(b, c) \circ (a, b) = (a, c)$. The transitivity of the relation $\leq$ implies that (a, c) constitutes an arrow, whenever (a, b) and (b, c) constitute arrows. Hence again we obtain a category.

Remark
  (a) Examples 1, 2, 3 show that many interesting mathematical objects constitute categories, whereas examples 4, 5, 6 show that many important mathematical entities can be considered as categories.
  (b) It can be proved that there exists a category with exactly two objects and with exactly one non-identity arrow. (This category is sometimes called "the number 2".)

Functors

If A and B are two categories, a functor F from A to B is a mapping that maps the objects of A to objects of B and at the same time it maps the arrows of A to arrows of B in such a way that the following conditions are satisfied:

  (1) If g is an arrow of A with source a and target a', then $F(g)$ is an arrow of B with source $F(a)$ and target $F(a')$.
  (2) $F(1_a) = 1_{F(a)}$.
  (3) $F(g \circ h) = F(g) \circ F(h)$.

We saw above that sets, monoids and preordered sets can be considered as categories.

Question: How are the mappings that preserve the structure between such entities compared to the functors between these entities, when these mappings are considered as categories?

The answer to this question is given as follows:

Example 7

Let us assume that A and B are sets, each of which is considered as a category (Example 4).

Let F be a mapping from A to B. If $a \in A$, then F maps a to $F(a)$. But a coincides with 1_a and $F(a)$ coincides with $1_{F(a)}$. Consequently $F(1_a) = 1_{F(a)}$.

If a and b are arbitrary elements of A, then they also constitute arrows of A. If $a \neq b$, then $S(a) = T(a) = a \neq b = S(b) = T(b)$, which means that they cannot be composed. If $a = b$, then $F(a \circ a) = F(a) = F(a) \circ F(a)$. Therefore F constitutes a functor.

Example 8

We assume that A and B are monoids, each of which is considered as a category. Let F be a monoid homomorphism of A to B. Without loss of generality we may assume that $\{*\}$ is the class of objects for both categories A and B (see Example 2).

We assume that $F(*) = *$. Since F is a homomorphism, $F(1_*) = 1_{F(*)}$. Moreover, $F(a \circ a') = F(a \cdot a') = F(a) \cdot F(a') = F(a) \circ F(a')$. Therefore, F constitutes a functor.

Example 9

A monotone mapping between two preordered sets can be considered as a functor. The details of the proof are left to the reader as an exercise.

The next examples show us that many interesting mathematical entities can be considered as functors.

Example 10

As we saw in Example 4, a set can be considered as a category. Let A be a discrete singleton-category and let B be the category of sets. For every set S, there exists a unique functor from A to B such that $F(1_A)$ is the identity function on S. This functor can be considered as being the set S.

Example 11

Let A be a category with two objects, $\mathbf{a}$ and $\mathbf{o}$, and four arrows, $1_\mathbf{a}$, $1_\mathbf{o}$, $\mathbf{s} : \mathbf{a} \to \mathbf{o}$, and $\mathbf{t} : \mathbf{a} \to \mathbf{o}$. This category can be pictured as follows:

$$\mathbf{a} \rightrightarrows \mathbf{o}.$$

We assume that F maps $\mathbf{a}$ to a set X and $\mathbf{o}$ to a set Y. We assume that $F(\mathbf{s})$ and $F(\mathbf{t})$ are functions with domains equal to X and ranges equal to Y. Then F constitutes a functor from A to the category of sets. We can consider this functor as a graph with Y being its class of objects, X being its class of arrows, $F(\mathbf{s})$ being the mapping source, and $F(\mathbf{t})$ being the mapping target.

The notion of a natural transformation is one of the basic notions in the theory of categories. We will introduce this notion and we will give an example of a natural transformation.

Let A, B be two categories and let F, G be two functors from A to B. A **natural transformation** from F to G is a mapping t which assigns to every object a of A an arrow $t(a)$ of B from $F(a)$ to $G(a)$ in such a way that for any arrow f of A from a to b, we have $G(f) \circ t(a) = t(b) \circ F(f)$. This definition is represented by the following diagram:

$$\begin{array}{ccc} F(a) & \xrightarrow{t(a)} & G(a) \\ F(f) \downarrow & & \downarrow G(f) \\ F(b) & \xrightarrow[t(b)]{} & G(b) \end{array}$$

Let it be noted that a, b are objects of the category A, while all other objects and arrows in the above diagram belong to B.

Example 12

We saw in Example 10 that a set S can be considered as a functor F from the discrete category (having a single object) to the category of sets such that F maps the single object to S and its arrow to the identity function on S. Let S, S' be two sets and let F, F' be the corresponding functors. Let f be a function from S to S' and let t be a mapping that assigns the arrow f from S to S' to the unique object $*$ of the discrete category having a single object. Then

$$F'(1_*) \circ t(*) = 1_{S'} \circ f = f = f \circ 1_S = t(*) \circ F(1_*).$$

From this it follows that t is a natural transformation from F to F'.

Conversely, it is recognized that every natural transformation from F to F' must have this form.

The theory of categories is considered by many people as the appropriate language for the description of many phenomena. Others criticised this as a complicated way of expressing simple ideas. The researchers who work in this area of mathematics, which is not among the research interests of the author of this book, say that we must not forget that the abstract definitions of this theory summarize the ideas and methods of many branches of mathematics, something that can be used in an effort to unify both proofs and results.

CHAPTER 13

RECENT DISCOVERIES AND ACHIEVEMENTS

The Fast Fourier Transform

When we talk about the estimation of a function $f(x)$, we mean the application of certain algorithms (a sequence of operations) for obtaining approximate values of this function.

The appearance of high speed computers brought about drastic changes in the way in which mathematicians estimate functions. Before the appearance of computers, during the 1940s, the primary role was played by mathematical tables. The first volume (1943) of the journal *Mathematical Tables and Other Aids to Computation* (MTAC), which is the forerunner of the journal *Mathematics of Computation*, was mainly publishing tables of mathematical functions. One of the main aims of the journal was to facilitate the exchange of information regarding the mistakes contained in mathematical tables.

In the past, mathematical tables were constructed by making tedious and tiring calculations by hand, while today they are created by high speed computers. The high speed computers brought about a real revolution (1950) in the area of numerical analysis, many researchers stopped using numerical tables, and turn to the finding of more effective approximation methods for the estimation of functions.

We will not discuss extensively the multiple spectrum of these methods. We will only discuss briefly the method of the fast Fourier transform.

The Fourier transform of a function f defined on the real line is usually defined to be the function g that is given by the relation

$$g(x) = \frac{1}{\sqrt{2\pi}} \int_{-\infty}^{+\infty} f(y) e^{-ixy} dy$$

where the integral is assumed to exist in some sense. The Fourier transform is a powerful tool of contemporary and classical analysis, and of its applications. In applications it is always desirable to minimize the number of operations used in a certain procedure. In this connection the so-called discrete Fourier transform comes into the limelight. It is the result of the transformation, that assigns to every n-tuple $(x_0, x_1, ..., x_{n-1})$ of complex numbers a n-tuple $(y_0, y_1, ..., y_{n-1})$ of complex numbers, which is given by the relations

$$y_q = \frac{1}{n} \sum_{p=0}^{n-1} x_p e^{-\frac{2\pi i p q}{n}} \qquad q = 0, 1, ..., n-1.$$

The so-called fast Fourier transform is the consequence of a simple remark, whose results are very significant. Its discovery is attributed to J. W. Cooley and J. W. Tukey (1965), and to C. F. Gauss (1805).

This ingenious remark is that the above sum, i.e., the discrete Fourier transform, can be calculated if one expresses it appropriately as a sum of subsums and if one makes use of algebraic properties of the roots of unity (i.e., of the numbers of the form $e^{-\frac{2\pi ipq}{n}}$) that appear in the calculations. In this respect we note that for certain values of n the Vandermonde matrix that is formed by the powers of a root of unity can be written appropriately as a product whose factors have many entries equal to 0, which reduces the number of calculations regarding the subsums.

We remind the reader that a matrix of the form

$$\begin{bmatrix} 1 & 1 & 1 & \ldots & 1 \\ x_1 & x_2 & x_3 & \ldots & x_n \\ x_1^2 & x_2^2 & x_3^2 & \ldots & x_n^2 \\ \vdots & \vdots & \vdots & & \vdots \\ x_1^{n-1} & x_2^{n-1} & x_3^{n-1} & \ldots & x_n^{n-1} \end{bmatrix}$$

is called the Vandermonde matrix (A. T. Vandermonde, 1735-1796), and its determinant is equal to $\prod_{i>k}(x_i - x_k)$.

It follows from the definition of the discrete Fourier transform that the calculation of every y_q requires first the calculation of $n - 1$ products (i.e., x_p times $e^{-\frac{2\pi ipq}{n}}$ for $p = 1, \ldots, n - 1$) and then the calculation of $n - 1$ summations (the partial sums).

Since there exist n values of y_q, the entire effort involves $n(n-1)$ summations and the same number of multiplications.

The simplest case of the fast Fourier transform occurs when n can be written as a product of two factors, i.e., $n = hk$. In this case the fast Fourier transform may present $n(h + k - 1)$ summations and $n(h + k)$ multiplications. For example, if $n = 100 = 10 \cdot 10$, then the known method requires 9900 summations and 9900 multiplications, while the method of the fast Fourier transform requires only 1900 summations and 2000 multiplications. We remark that this large difference between this number and the previous number of operations becomes even larger as n increases and the process is repeated. The method of the fast Fourier transform constitutes a useful tool of numerical analysis. (For more details see Appendix I.)

Chaos Theory

The laws that govern most of the phenomena that can be studied by physical sciences, engineering, and social sciences are, of course, nonlinear. These laws, when written in mathematical language, often take the form of systems of nonlinear equations. The derivation and study of such systems form the core of what could be called nonlinear science. Because nonlinear phenomena occur in all the disciplines, nonlinear science is by nature interdisciplinary.

The physical and, therefore, the mathematical phenomenology associated with nonlinear systems is incredibly rich. Thus, it is not surprising that in addition to using techniques from a wide range of mathematical branches (perturbation theory, real and complex analysis, group theory, differential and algebraic geometry, topology, etc.), the analytical study of nonlinear problems has also led to the discovery of new mathematical notions.

One of the beauties of nature is that diversity is strongly coupled to unity (or unification). Indeed, many apparently disparate nonlinear systems have the commonality that they exhibit a similar behavior. "Chaotic" and "integrable" behaviors are among the best studied such behaviors arising in a variety of physically important equations. In what follows I will give a brief overview of some of the significant notions in the theory of chaos. An overview of some of the important developments in the theory of integrable systems can be found in [50].

We know that even certain small and very insignificant events can change drastically the course of history. Examples of such events are the assassination of some person that becomes the cause of a revolution or a war, and the spontaneous and sudden decision of a person to run to catch a flight which results in a tragic accident.

There exist numerous examples which show the complexity of nature. Approximately over the last thirty years, mathematicians, physicists, biologists, astronomers, and people who occupy themselves with economic sciences created a new path that leads us to realize the size of the complexity that nature presents. This new science, which is called Chaos, offers us a way of seeing the existence of "order" and "patterns", where before we were only observing something random, irregular, and unpredictable, in other words a situation that was called "chaotic".

Chaos theory is part of mathematical science and has attracted the interest not only of the scientific community but also of the public.

The bibliography that refers to this subject and the bibliography that refers to the associated theory of Fractals and the theory of Dynamical Systems is already very large ([10], [18], [41]).

We will discuss briefly chaos theory from the mathematical point of view.

In mathematics the word "point" is always synonymous with the phrase "element of a set". Let us consider the following set which is somewhat complex.

Let $I = [0, 1]$ be the unit interval in the real line. Let X be the set whose elements are all finite unions of pairwise disjoint closed subintervals of I. In other words, every point of X is of the form $[x_1, x_2] \cup [x_3, x_4] \cup ... \cup [x_k, x_{k+1}]$, where the $[x_i, x_{i+1}]$ are closed subintervals of I which are pairwise disjoint. Hence, I is an element of X. We assume also that the singletons of elements of I are not considered to be elements of X.

Let T_c be the mapping from X into itself $(T_c : X \to X)$ that acts on every $P \in X$ by subtracting the open middle thirds of the subintervals of I that comprise P.

Hence, since I is a point of X, its image $T_c(I)$ is equal to $\left[0, \frac{1}{3}\right] \cup \left[\frac{2}{3}, 1\right]$ (i.e., we have subtracted from I the open interval $\left(\frac{1}{3}, \frac{2}{3}\right)$), which is a point of X. What will happen now if the mapping $T_c : X \to X$ is repeated? For example, let us start with I and form the sequence $T_c(I), T_c^2(I), T_c^3(I), ...$, where $T_c^2(I) = T_c(T_c(I))$, $T_c^3(I) = T_c(T_c^2(I)) = T_c(T_c(T_c(I)))$, and so on.

This descending sequence of finite unions of pairwise disjoint closed intervals is known to us from the theory of sets of real numbers: the intersection of all the sets in this sequence is the classic set of Cantor. (The index c reminds us of Cantor's name.)

The transformation just described is an interesting example of a dynamical system. In general, we call "dynamical system" every mapping of a set into itself. Of course, this definition is so general that it is essentially useless.

The notion of a dynamical system becomes useful when the set on which the mapping acts is equipped with a certain interesting mathematical structure (algebraic, analytic or geometric). Indeed, such structures exist in all studies on dynamical systems.

The notion of chaos depends on the notion of a dynamical system. A first definition is the following: Chaos theory is the study of the behaviour of dynamical systems in the neighborhood of infinity. In particular, let T be an arbitrary transformation of an arbitrary set X into itself. We form the successive mappings T, T^2, T^3, ... and then we pose some "reasonable" question regarding this sequence of mappings. For example, we may take X to be the set of real numbers, we may define $Tx = \cos x$ for $x \in X$, and we may pose the (easy) question what happens with the sequence x, Tx, $T^2 x$, ... when x takes various values.

A second example is the dynamical system T_c that we mentioned earlier.

A third example is the following:

Let X be the unit interval with its extremities identified. Hence a point of X is a real number between 0 and 1, including 0 and 1, under the assumption that 0 and 1 are considered to be the same point. We consider on X the mapping T_2 which assigns the number $2x \pmod 1$ to every x. Hence, we have

$$T_2\left(\tfrac{1}{3}\right) = \tfrac{2}{3}$$
$$T_2\left(\tfrac{3}{4}\right) = \tfrac{3}{2} \pmod 1 = \tfrac{3}{2} - 1 = \tfrac{1}{2}$$
$$T_2\left(\tfrac{1}{2}\right) = 0$$
$$T_2(\sqrt{2}-1) = T_2(1.4142... - 1) = 2\sqrt{2} - 2 \pmod 1 = 2.8284... - 2 = 0.8284...$$

How does the sequence x, $T_2 x$, $T_2^2 x$, $T_2^3 x$, ... behave for the various values of x? Can it converge to a certain limit? Can the closure of the set $\{T_2^n x\}_{n=0}^{\infty}$ be a certain open set? Even though this example seems to be simple, it is actually complicated.

An example of a dynamical system, that lies closer to the area of classical analysis is the following: Let a and b be two arbitrary real numbers and let us define the mapping $T_{ab} : \mathbf{R}^2 \to \mathbf{R}^2$ by the relation $T_{ab}(x,y) = (y + 1 - ax^2, bx)$.

Transformations of this kind were introduced and studied by Hénon, and the study of their asymptotic properties is interesting and delicate. These properties obviously depend on the choice of the parameters a and b. For example, if we take $a = 1.3$ and $b = 0.3$, then there exist seven points in the plane which are "periodic attractors". This means that the repeated images of a point of the plane either tend to infinity or first they approach some of these seven points, then they approach some other of these seven points, and so on, until at the eighth step they start again to approach the first point, then the second, and so on.

But if we take $a = 1.4$ and $b = 0.3$, then the repeated images of certain points will tend to infinity, while for other points these images will form a complicated set of curves, which is said to be a "strange attractor" or a "Hénon attractor". The main property of a strange attractor is that it is an infinite set whose dependence on the initial conditions is very "delicate", which means that very small changes of the initial conditions imply huge changes of the system. For example, strange attractors are observed in the study of turbulence.

As it appears, we may consider that the reason for which the word "chaos" was used to describe this theory is due to a subjective (and not mathematical) reaction to the appearance of an unexpected and abrupt change in the behavior of the system. This may appear in two different parts of the theory. On the one hand, a dynamical system depends on certain parameters (just like the transformation T_{ab} of Hénon depends on a and b), and on the other hand, the sequence x, Tx, $T^2 x$, $T^3 x$, ... of a dynamical system T applied to some x always depends on the initial point x.

The abrupt change that is observed in the behavior of the transformations of Hénon's type (i.e., the point at which the periodic attractors are transformed into strange attractors) is considered to constitute "chaos" (unpredictability), and the existence of the strange attractor of Hénon (which obviously cannot be seen immediately from the definition of the dynamical system), is also considered to constitute "chaos" (unpredictability).

It should be noted that also other results in the science of mathematics cause surprise, especially when they involve the notion of infinity. By simply observing a transformation, it is not possible, of course, to say how the infinite sequence of repeated images, mentioned above, will behave. But is the fact that a small change of the parameters causes an abrupt change in the behaviour of the system a sufficient reason for us to name the phenomenon "chaotic"?

Chaos theory is closely related to the Theory of Fractals. An idiomatic term which gives the correct translation of the word fractal in Greek is "$θρύμμα$"; it was introduced and justified by the author of the present book in one of his talks at the Academy of Athens [20].

The dynamical system T_c, which was discussed above, led us to the Cantor set which is often met in chaos theory. An important property of the Cantor set is that the part of the set that lies in the interval $\left[0, \frac{1}{3}\right]$ is similar to the entire Cantor set. In order to see this we need only multiply by 3 every point x of the Cantor set that lies in $\left[0, \frac{1}{3}\right]$, and in this way we obtain the entire Cantor set. Similarly, the part of the Cantor set that lies in the interval $\left[\frac{2}{9}, \frac{7}{27}\right]$ is similar to the entire Cantor set. In general, every neighborhood of every point of the Cantor set contains a subset which is similar to the entire Cantor set. The ad infinitum repetition of transformations tends to produce sets which preserve the property of being similar to parts of themselves, i.e., sets which remain unchanged under a change of the drawing scale. The so-called Hénon attractor constitutes a second example where the phenomenon of "self-similarity" is observed.

This last phenomenon was known earlier to the mathematical community. The Cantor set constitutes the model for the construction of a large family of sets which are obtained by making various architectural changes. Example: if in the construction of the Cantor set, instead of removing the open middle third of an interval, we remove the open middle fifth of an interval, then we obtain a set which is analogous to the Cantor set. In addition, we may remove the open middle third not of an interval but of a rectangle or of a disc or of other interesting plane figures. All these sets present typical properties which are pleasant and interesting.

The Cantor set is a fractal and all the variants of this set mentioned above also constitute fractals.

Another fractal is the well-known set of Mandelbrot. Benoit B. Mandelbrot (1924) is considered to be the founder of the theory of fractals, which he published in his very important writing *The fractal geometry of nature* (1977, 1982, 1983).

The Mandelbrot set is a subset of the complex plane which is obtained as follows. We consider the function $f(z) = z^2 + c$, where the parameter c is some

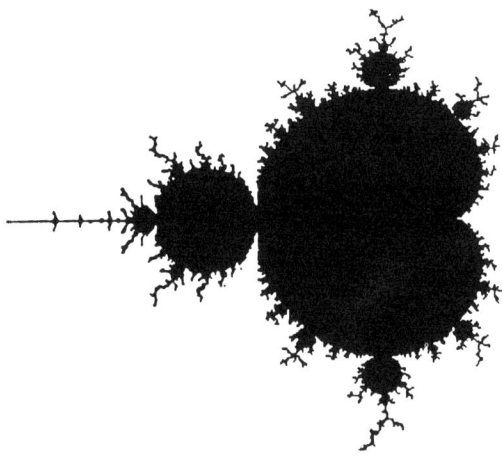

FIGURE 13.1. The Mandelbrot set

complex number. The Mandelbrot set (Figure 13.1) consists of the values of c for which the orbit of 0, i.e., the sequence 0, $f(0)$, $f^2(0)$, $f^3(0)$, ... is bounded.

But let us examine in somewhat more detail the notion of a fractal. A characteristic property of fractals is that if one examines them with all the more powerful magnifying glasses, they always reveal more details.

In addition, as we also saw in the case of the Cantor set, the structure that fractals present when we observe them under a small scale, is the same with one that they present when we observe them under a larger scale.

The rigorous mathematical definition of the notion of a fractal makes use of the Hausdorff-Besicovitch dimension of a set.

It is known that to every set of the Euclidean space (of the natural space that surrounds us), no matter how "pathological" this set is, corresponds a number called the "topological dimension" of the set and is denoted by D_T. For example, the topological dimension of a point is 0, the topological dimension of a line segment is 1, the topological dimension of a triangle is 2, the topological dimension of a sphere is 3. To the same set corresponds also a second number, which is called the Hausdorff-Besicovitch (1919) dimension of the set and is denoted by D.

Of course, we cannot develop here the theory of Hausdorff-Besicovitch dimension. We limit ourselves only to some remarks. The first idea that the general notion of volume and, more generally, the notion of size is necessary in order to investigate the dimensions of continuous sets is due to Cantor. This idea was advanced greatly by the ingenious Greek mathematician Constantin Carathéodory (1914). "Carathéodory's measure" is well known in mathematical analysis. Hausdorff went one step further than Carathéodory and introduced the well-known and widely used "Hausdorff measure". Finally, the notion of Hausdorff-Besicovitch dimension was introduced with the significant contribution of Besicovitch, and plays a fundamental role in the Geometry of Fractals.

Both dimensions D_T and D are non-negative and they do not exceed the Euclidean dimension of the space where the set considered belongs.

The dimension D_T is always an integer, while the dimension D, which is also called the "fractal dimension", is not necessarily an integer; furthermore we always have the inequality $D \geq D_T$.

We now give the following definition:

"Every set whose **Hausdorff-Besicovitch** dimension is strictly greater than its topological dimension, i.e., $D > D_T$, is called a **Fractal**."

Let us consider a part of the coastline of some country. It is a fractal. It is a very irregular curve whose length is obviously greater than the length of the line segment connecting its two endpoints. Now if we try to calculate the "exact" length of the coastline considered by using some of the many existing methods, we will see that it is so large and indeterminate that it is preferable that we consider it as infinite! Here we deal with a non-rectifiable curve. This remark implies that the classical notion of length is not appropriate for a quantitative comparison of the various coastlines. But this comparison, as well as others of a similar nature, must be made, and at this point the need for the introduction of the notion of fractal dimension becomes evident. The truly ingenious idea is due to Hausdorff. Let us follow him in his first thoughts: In order to calculate the length of the perimeter of a closed polygon we add the first powers of the lengths of its sides (where 1 is the dimension of a straight line in Euclidean space). This exponentiation to power 1 does not change the length of a side and hence this process of exponentiation to power 1 seems completely redundant. However, we will see in a while that it is indicative of the path that we will follow. We also note that in order to calculate the area of the surface surrounded by this closed polygon, we divide it into small squares and we add the second powers of the lengths of the sides of these squares (where 2 is the Euclidean dimension of the plane). And now let us make the following basic remark: if, during the processes that we described above, we took different powers of the lengths than 1 and 2 respectively, then we would not obtain the desired results regarding the length of the perimeter and the area of the polygon. In a while we will see how and where this idea will be used in the calculation of the fractal dimension.

Let us return to the calculation of the length of the coastline using one of the usual methods. Imagine that we run over the coastline by making steps of constant length, say ϵ, and assume that we needed B steps to cover the entire coastline. Then the product $B\epsilon$ is an approximate length of the coastline. Now if we run again over the coastline by making smaller steps, say of length ϵ_1 ($\epsilon_1 < \epsilon$), we will need B_1 steps to cover the entire coastline, hence $B_1\epsilon_1$ constitutes a second approximate length of the coastline and $B_1\epsilon_1 > B\epsilon$. If we continue to reduce the length of our step, then we obtain always larger numbers which tend to infinity. The notion of fractal dimension is hidden in a remark made by Lewis Fry Richardson (1881-1953) at an unsuspected time, which was found in his notes in 1961, i.e., after his death. This remark in combination with Hausdorff's idea of measure lead us to the desired result. The measurements made by Richardson assert that, in order to approximate the length of the coastline considered, $F\epsilon^{-D}$ steps are needed, where F, D are two constants and ϵ is the length of the step. Now if we multiply the number of steps by ϵ, then we obtain the approximate value $F\epsilon^{-D} \cdot \epsilon = F\epsilon^{1-D}$, which leads us, as before, to a dead end because $F\epsilon^{1-D}$ tends to infinity as ϵ tends to zero, due to the fact that D is less than 1. But if we follow Hausdorff's idea mentioned above and instead of multiplying the number of steps $F\epsilon^{-D}$ by ϵ we multiply it by ϵ^D (as we did in the case of the polygon), then we will obtain the number $F\epsilon^{-D} \cdot \epsilon^D = F$, which obviously does not depend on the length ϵ of our step. It is proved that D is the fractal dimension of the curve formed by the coastline and is equal approximately to $\frac{3}{2}$, i.e., it is larger than its topological dimension which is equal to 1. It is natural to say that the length of the coastline in dimension D is approximately equal to F.

Similarly, if we want to calculate approximately the length of the coastline in some other dimension d, which is smaller or greater than D, then the distance that we will find will tend to zero or to infinity respectively as ϵ tends to zero. The desired behaviour of the distance is obtained if and only if $d = D$.

Let us consider again the Cantor set, which we denote by C. Since C is a totally disconnected set, its topological dimension is known to be zero, i.e., we have $D_T = 0$. In order to find the fractal dimension of C, we work as follows: We consider a line segment of length 1 and we divide it into $N = b$ equal subintervals of length $r = \frac{1}{b}$. Similarly we consider a square of side 1 and we divide it into $N = b^2$ equal squares of side $r = \frac{1}{b}$. We remark that in both cases the ratio $\frac{\log N}{\log(1/r)}$ is equal to the topological dimension of the corresponding set, i.e., it is equal to 1 or 2 respectively. Since in the case of the set C, after every removal of the open middle interval, $N = 2$ intervals are left, each of which is the $1/3$ of the interval from which it was obtained, i.e., $r = \frac{1}{3}$, based on the preceding remark we take as fractal dimension of the set C the number $D = \frac{\log 2}{\log 3} = 0.6309...$ Hence we have $D > D_T$.

A way to make intuitively clear the notion of non-integer dimension of a set is the following. We remark that in the case of the Cantor set the fractal dimension is greater than 0 and less than 1. We may easily explain this fact if we consider the Cantor set as a natural structure with mass, as follows. We assume that the interval $[0, 1]$ is a bar which is made of some material. After the removal of the open middle interval we distribute the removed mass in the two remaining intervals, and so on. At the end, the entire mass of the bar is distributed in the set C, which is something less than a continuous line segment and something more than a vanishing pointset, when we consider it in this way as a natural entity. Hence it is natural to accept that the fractal dimension of C is some number between 0 and 1.

Now we give another example which also will help to clarify the notion of non-integer dimension of a set.

We consider a distribution of "point-masses" in a spherical volume. With the term point-mass we mean an indivisible unit of natural mass located at some point in space. It is a conventional notion. Now let us imagine that we approach this set of point-masses from a very large distance. At first the mass seems like a single uniform formation. But as we get closer, we see that this formation consists of smaller formations. If we approach even closer, we see that each of these smaller formations consists of even smaller formations, and so on. This example, which at first seems to be artificial and prefabricated, is the one that describes the distribution of stars in the Universe. The Hausdorff dimension of the distribution of stars was calculated using astronomical methods and was found approximately equal to 1.23.

It is interesting to see how the total mass of such a formation is related to the Hausdorff dimension. It is known that the total mass of a distribution is proportional to the number r^D, where r is a unit of length and D is the dimension of the space in which the object under investigation lives. If $M(r)$ is the quotient of the mass over the mass of unit volume, then we have $M(r) = r^D$. For example, the mass of a ball of radius r which is filled with sand is proportional to r^3. This means that when we have no other information about the content of the ball, we usually assume that the matter that is contained in it is uniformly distributed, hence $D = 3$. Now let us assume that by examining the ball from a closer distance we see that the sand is not uniformly distributed, but it is contained in smaller balls of radius r/b, each of which has mass $1/a$ times smaller than the total mass contained in the initial ball. Now assume that we approach even closer and by examining each of the small balls we find that they consist of even smaller balls of radius r/b^2 and that each of these balls has mass $1/a^2$ smaller than the total mass. Now if we assume that we proceed in the same way and we find always smaller balls, then at the nth step, i.e., after n successive observations, we find that $M(r) = M\left(\frac{r}{b^n}\right) a^n = r^D \frac{a^n}{b^{nD}}$. If we continue this process ad infinitum, i.e., if n tends to infinity, then the last relation provides the total mass of the ball only if $D = \frac{\log a}{\log b}$. This value of D, which is not necessarily an integer, is the Hausdorff-Besicovitch dimension or the fractal dimension of the distribution of mass that is contained in the ball of radius r, and this distribution is another example of a fractal.

We return to the chaotic phenomena in order to add that these phenomena are observed everywhere in nature, even in heart beats. Under certain conditions the human heart becomes a chaotic system. It is well known that this situation can be controlled by placing a pacemaker, but even then one cannot exclude chaotic phenomena because the initial conditions may very easily change. The understanding of the presence of chaos in a system helps us to describe not only the irregular heart beats or the evolution of similar phenomena, but also many other aspects of our complex Universe at a small or large scale.

The astronomers use chaos theory in order to give models of the Universe at its Birth, of the motion of stars inside the galaxies, and of the motions of planets, satellites, and comets inside the solar system.

Chaos theory helps in the study of charged particles trapped by the magnetism of the Earth, and when they enter the atmosphere, they generate phenomena such as the Dawn, the Aurora Borealis, and others.

The biologists observe chaotic phenomena in the changes of the populations of insects and birds, in the propagation of epidemics, in the metabolism of cells, and in the transmission of impulses through our nervous system [49].

The physicists use chaos theory in the study of the motion of electrons inside atoms and of atoms inside molecules or gases, as well as in the theory of elementary particles and elsewhere.

Some of the most marvelous examples of chaos are met in mathematics, where the solutions of apparently simple problems present a tremendously complex behaviour.

Usually when we refer to fractals as pictures or shapes or structures, we consider them as static objects, as, for example, in the case of a tree or a cloud, etc. But such a consideration provides minimal information about the evolution or the birth of a given structure.

Very often, as it happens, for example, in Botany, apart from the complicated geometric picture that a completely grown plant presents, we are very much interested in knowing the way it grew, the dynamic growth of this plant. The same thing happens with mountains, with mountain chains whose geometric form is the result of geological transformations of the crust of the earth and of phenomena of erosion. These are causes that continue today and will continue in the future to influence the shape of the mountains.

We can make analogous remarks about the phenomenon of accumulation of zinc during an experiment of electrolysis or for the creation of the human lung. In other words, to simply talk about Fractals is insufficient when one ignores the dynamic process that is followed for their formation.

Even though it would be historically fair to name the persons who were the forerunners of chaos theory, as, for example, the French mathematician Henri Poincaré, we will not attempt this here.

Chaos theory and the theory of integrable systems are still relatively young; they are only 30 years old and we are not sure whether they will change the picture that we have today about the Universe. What is certain though is that these theories have entered the research programs of many branches of science, where the leading role is played by pure and applied mathematics, and by computers. They also give researchers the courage and the inspiration to face the constantly changing reality of our natural world.

The Bieberbach Conjecture

It is one of the most famous problems of mathematical science which remained unsolved for many decades until it was solved by Professor de Branges of Purdue University of the USA. The solution was given in August of 1984.

The Bieberbach conjecture is about the class of one-to-one analytic (i.e. conformal) mappings f, which are defined on the unit disc $\{z \in \mathbf{C} : |z| < 1\}$ of the complex plane and have the property that $f(0) = 0$ and $f'(0) = 1$. Every such function f can be written as a convergent power series: $z + a_2 z^2 + a_3 z^3 + ...$, where the a_n are complex numbers. These functions are known as univalent.

In 1916 the German mathematician Ludwig Bieberbach formulated the conjecture that for $n = 2, 3, 4, ...$, we have $|a_n| \leq n$ and that the equality $|a_n| = n$ holds only for the Koebe function, which is given by the series

$$z + 2az^2 + 3a^2 z^3 + ...$$

where a is a complex number such that $|a| = 1$.

We note that this last function is the "rotation" of the function

$$z(1-z)^{-2} = z + 2z^2 + 3z^3 + ...$$

The most important property of a one-to-one analytic function was given by Riemann in 1851 in his famous "representation theorem" [7], whose statement is as follows: "Every simply connected proper domain in the complex plane is the image of the unit disc under a conformal mapping." Here when we say "proper" domain we mean a domain which does not coincide with the entire complex plane.

An important fact in the proof of this theorem is that one follows the principle that the function that maps the unit disc onto a simply connected proper domain can be expressed via its value at $z = 0$ and the value of its derivative at $z = 0$. The Bieberbach conjecture improves this principle because it provides a further estimation of the coefficients (and hence an estimation of all the values of the derivatives at $z = 0$) of the function that maps the unit disc onto a simply connected proper domain. Hence we can say that the Bieberbach conjecture constituted an important problem in the theory of complex functions from the geometric vantage point.

The Bieberbach conjecture can also be considered as an infinity of problems, where for any n, we seek to choose, among all conformal mappings f of the unit disc with the property that $f(0) = 0$ and $f'(0) = 1$, the function that maximizes the value $|a_n|$. What de Branges was able to prove is that the Koebe function constitutes the unique solution of each of these problems.

The Bieberbach conjecture remained unproved for 68 years. Bieberbach himself had proved the inequality $|a_2| \leq 2$. The inequality $|a_3| \leq 3$ was proved in 1923 by the Czech mathematician Karl Loewner, whose method makes use of a concrete differential equation, which is also used by de Branges in his proof.

The case $n = 4$ was proved in 1955 by Paul Garabedian and Menachem Schiffer, who had moved to the USA in 1940. In 1968 R. Pederson and M. Ozawa proved the case $n = 6$, and in 1972 Pederson and Schiffer proved the case $n = 5$.

In 1923 J. E. Littlewood, who was working in some other direction, proved that for all n, the inequality $|a_n| < en$ holds, where e is the base of the natural logarithms. For many years, many efforts were made to replace e in the last inequality by a smaller number. In 1978 David Horowitz proved that $|a_n| < 1.0657n$ using inequalities which are due to Carl Fitzgerald. The inequality $|a_n| \leq n$ was also proved for various particular classes of conformal mappings, as is the class of star-like functions (R. Nevanlinna, 1920), the class of functions for which the a_n are real numbers (Dieudonné, 1931), and the class of functions which are in "some" sense close to the function of Koebe (1931). But despite all these efforts, no correct proof of the Bieberbach conjecture was given until de Branges gave his proof in 1984.

In his proof, de Branges combines the differential equation of Loewner with an ingenious construction of a system of special functions and reduces the problem to a system of polynomial inequalities. By coincidence the same inequalities were studied by Richard Askey and George Gasper in 1976 for completely different reasons and their results lead to the proof of the conjecture.

Even though the final proof given by de Branges is independent of these results, de Branges claims that the powerful computer of Purdue University played an important role in his discovery.

Without knowing the inequalities of Askey and Gasper, de Branges had worked with Walter Gautschi in the computer department of Purdue University in order to verify certain particular cases. The satisfactory results given by this effort of his with computers encouraged him in his final effort to prove the Bieberbach conjecture.

De Branges essentially was able to prove a conjecture of the Russian mathematician I. M. Milin, which implies the validity of the Bieberbach conjecture.

The solution of this famous problem leads to other problems in the geometry of conformal mappings. Perhaps the greatest benefit that comes out of de Branges' work is that it leads us to perceive the chain that connects complex analysis, optimization theory, functional analysis and special function theory.

Achievements in Number Theory

During the 20th century, one observes many times that mathematicians obtained results regarding abstract structures, which were then used to prove again results regarding abstract structures, and so on. This phenomenon is similar to carpenters using their tools to construct again new tools without using any of their tools to construct a house! We will mention certain cases which constitute exceptions to the above observation.

The Diophantine equation $x^2 + k = y^3$, where k is an integer different from zero, has its roots in the work of Diophantus (see Problem 17 of Book VI of his writing *Arithmetica*). This equation occupied Claude-Gaspar Bachet (1581-1638) during the 17th century and it is known since then as the Bachet equation. A complete solution of this equation was given in 1968 by Alan Baker for any given k. At first, the solution given by Baker consisted of providing a very large bound $M(k)$ for the quantities x and y. But soon thereafter Baker himself and other mathematicians, such as H. Davenport, obtained a practical method which provided a solution of the equation for a given k. Later, it was also proved that for $k = 28$, the Bachet equation has only three solutions, i.e., $x = 6, 22, 225$. In addition, it was proved that for $k = 999$, the solutions of the Bachet equation are $x = 1, 27, 251, 1782, 2295, 3370501$.

In 1900 in Paris, David Hilbert, who was one of the most eminent mathematical figures of the 20th century, presented at the second International Congress of Mathematicians a list consisting of 23 problems, which he considered to be the aims of mathematical science during the 20th century. Among these problems were also the following:

(1st) To prove or disprove the "continuum hypothesis".

(2nd) To determine whether the system of natural numbers is consistent or not.

(8th) To study problems regarding the distribution of prime numbers, and in particular to examine whether the non-trivial roots of the function

$$\zeta(z) = \sum_{n=1}^{\infty} \frac{1}{n^z}$$

where z is a complex number, lie on the straight line $x = \frac{1}{2}$.

(10th) To find an algorithm (a computer program) which decides whether a given polynomial Diophantine equation (whose integer coefficients and exponents are known) has a solution or not.

Today we know that some of Hilbert's problems are not solvable in the way that Hilbert had requested. For example, it has been proved that the known axioms of set theory (under the assumption that they are consistent) imply neither the "continuum hypothesis" nor its negation. The 10th problem of Hilbert also belongs to this category.

The 8th problem of Hilbert still remains unsolved and is one of the most important problems of contemporary number theory.

Among the achievements of the 20th century that we have already discussed quite extensively (Part I, Chapter 24) is the Last Theorem of Fermat (1637), i.e., that the equation

$$x^n + y^n = z^n$$

has no positive integer solutions for $n > 2$. We point out that this conjecture was studied by G. Frey, K. Ribet and J.-P. Serre, and it was finally proved in 1994 by A. Wiles with the help of R. Taylor.

Other achievements of the 20th century in number theory, whose success is due to computers, are the following.

Twenty new perfect numbers were discovered and one hundred new pairs of friendly numbers.

In addition, programs were constructed which are capable of calculating products with 100 integer factors in only a few hours.

The solution of the cattle-problem of Archimedes (200 B.C.) that we already discussed in Part I (Chapter 17) was also solved in a remarkable way: With the help of a computer it was proved (1965) that this problem is equivalent to the Diophantine equation

$$x^2 - (8 \cdot 2471 \cdot 957 \cdot 4657^2)y^2 = 1.$$

Partition of a positive integer

We call "partition" of a positive integer every way in which this number can be written as the sum of non-increasing positive integers. For example, the number 5 has 7 partitions, namely, $5, 4+1, 3+2, 3+1+1, 2+2+1, 2+1+1+1, 1+1+1+1+1$.

$p(n)$ denotes the number of partitions of a positive integer n. For example, $p(5) = 7$.

In 1918, Ramanujan (1887-1920) and Godefrey Harold Hardy (1877-1947) gave the first fast way of calculating $p(n)$ for an arbitrary n, and in 1937 Hans Rademacher complemented their work by giving the first known formula for $p(n)$, which is the following:

$$p(n) = \frac{1}{\pi\sqrt{2}} \sum_{k=1}^{\infty} A_k(n) \sqrt{k} \frac{d}{dn} \left(\frac{\sinh\left(\frac{\pi}{k}\sqrt{\frac{2}{3}\left(n - \frac{1}{24}\right)}\right)}{\sqrt{n - \frac{1}{24}}} \right)$$

where

$$A_k(n) = \sum_{0 \leq h \leq \left[\frac{k}{2}\right], \gcd(h,k)=1} 2\cos\left(\pi s(h,k) - \frac{2\pi nh}{k}\right)$$

and the so-called Dedekind sums are defined as follows

$$s(h,k) = \sum_{r=1}^{k-1} \frac{r}{k}\left(\frac{hr}{k} - \left[\frac{hr}{k}\right] - \frac{1}{2}\right)$$

($s(0,1)$ is taken to be equal to 0.)

Hardy and Rademacher contributed greatly to the discovery of the last formula, but the "intuitive spark" came from Ramanujan, who was an exceptional genius born near Madras in India. Ramanujan, who is considered to be one of the greatest number theorists of the 20th century, sometimes attributed his discoveries to the Divine Providence. He had once said the following:

"For me an equation means nothing
unless it expresses a thought of God."

Ergodic Theory

The origin of ergodic theory is the so-called **ergodic hypothesis**, which constituted the cornerstone of classical statistical mechanics, as it was created by L. Boltzmann and J. Gibbs during the end of the 19th century. The results of H. Poincaré, C. Carathéodory, G. D. Birkhoff, and J. von Neumann, marked the beginnings of ergodic theory.

We present a passage from an article of P. R. Halmos (*Amer. Math. Monthly, Vol. 97, No.* 7), where he explains with remarkable simplicity and elegance the origin of this theory.

If we hit a billiard ball perpendicularly with respect to a side of the table, then it will hit the opposite side, it will return to the point of departure, and it will continue to move back and forth between the two sides. Now if we hit the ball from the midpoint of one side towards the midpoint of an adjacent side, then it will hit the midpoint of the side lying opposite the side from which the ball began to move, it will hit the midpoint of the fourth side, it will return to the point of departure, and it will continue to trace the same path again and again. Of course, there exist also other ways of hitting the ball in order to create periodic orbits just like the ones we described above.

But what will happen if we hit the ball under such an angle that no periodic orbit occurs? Boltzmann, who was one of the founders of ergodic theory, studied this phenomenon during the 19th century and formulated the so-called **ergodic hypothesis**. According to this hypothesis, (in case the orbit is not periodic) the billiard ball follows an orbit which passes through all the points of the billiard table. But if one takes into account the topological properties of a plane rectangle and the topological properties of a curve in this rectangle, it is not difficult to prove that the ergodic hypothesis does not hold as it was formulated above. Later, a second ergodic hypothesis was formulated, which constitutes a variation of the first and is the following: Even though the orbit of the billiard ball will not pass through any point of the billiard table, it will pass as close as we want from any point of this table. In other words, the set consisting of the points of the orbit of the billiard ball is everywhere dense in the billiard table.

Statistical mechanics was treating subjects of such nature until G. D. Birkhoff published a basic work on this subject in 1931. Birkhoff proved that if we consider an arbitrary subset of the billiard table, as is, for example, its right half or the interior of a circle whose center is the center of the table, then the average time that the ball will remain in this subset is proportional to the area of this subset, independently of where is the point of departure of the ball. The technical difficulty that this last result presents is that we talk about the average time interval which means that the limit involved in the interpretation of this phrase exists, but what is surprising in this conclusion is that this average time is constant and independent of the point of departure of the ball. Among other things, this conclusion implies that if one observes the interior of an arbitrary circle on the billiard table, no matter where this circle is located and no matter how small this circle is, it is certain that the billiard ball will enter repeatedly the interior of this circle, which means that the orbit of the ball (i.e., the set of points that belong to the orbit of the ball) is indeed everywhere dense.

This work of Birkhoff was published more than 50 years ago and since then ergodic theory constitutes an important part of the science of mathematics and is connected with classical and contemporary analysis.

Simulation

In its most general meaning, simulation is a method according to which we use models for studying the nature of certain phenomena.

This method is used in cases in which the experimental study of phenomena is time-consuming, difficult and expensive.

Among the various types of simulation techniques, we will mention only the so-called "Monte Carlo Method".

On a smooth horizontal surface we draw a system of parallel straight lines such that the distance of two neighboring straight lines is, for example, two centimeters and we consider a needle (which is idealized mathematically, i.e., it is a line segment) of length equal to one centimeter. We throw randomly the needle on this surface and we pose the following question: What is the probability that the needle will not fall between two of the parallel straight lines, but it will intersect some of them? This is the famous **problem of Buffon** and the method of solving it presents no particular difficulties. This problem constitutes an exercise in many books on elementary probability theory. The answer to the above question is that the requested probability is equal to $\frac{1}{\pi}$.

This result is surprising. Why does π appear in the answer to our question? Even though this question is not the main subject with which we are concerned here, since we have posed it we will discuss it. The answer to a problem that lies in the area of probability theory depends on the distribution of our various data. In the case at hand these are the data that we obtain during the throwing of the needle on the surface. Various possible probabilistic distributions were studied, which led, as it was expected, to different answers. One of these possibilities is to assume that the probabilistic distribution of the midpoint of the needle lies uniformly in a line segment of length say 3 centimeters which is perpendicular to the parallel straight lines. In addition, independently of this distribution we will assume that the probabilistic distribution of the angle (of the slope) formed by the needle and this perpendicular line segment is uniform between $0°$ and $180°$. It is obvious that from the moment that angles are involved in the study of the problem, the presence of π in the answer given above is natural.

If we accept the answer given above to our question, i.e., if we accept that the requested probability is equal to $\frac{1}{\pi}$, then we are led to the following strange result. If we repeat the experiment many times (i.e., if we throw the needle many times) and we consider the quotient of the number of successes over the number of trials, then the "law of large numbers" informs us that this quotient is very close to the number $\frac{1}{\pi}$. Hence we are led to the conclusion that the number π can be calculated by performing a natural (probabilistic) experiment, without carrying out many calculations.

This last conclusion, which was known approximately two centuries ago, was the forerunner of the contemporary technique which is known as the "Monte Carlo Method" and was introduced in 1945 by Ulam and von Neumann.

The Monte Carlo method has many applications for problems where the required calculations are complicated, lengthy and difficult. In these cases the problem is replaced by a problem in probability theory, which has the same answer with the original problem and whose solution is achieved either experimentally or with the help of a computer.

An easy example of application of this method is the calculation of a definite integral $\int_0^1 f(x)dx$, where f is a bounded function (for example, $0 \leq f(x) \leq 1$, whenever $0 \leq x \leq 1$). We work as follows: We choose (possibly with the help of a computer) a large set of pairs (x, y) of random numbers, where $0 \leq x \leq 1$ and

$0 \leq y \leq 1$, and we calculate the quotient of the number of pairs for which we have $y \leq f(x)$ over the total number of pairs selected. This quotient is approximately equal to the value of the integral. Even though one might consider this example to be "naive" for this theory, the usual applications follow a similar spirit.

The method took the name Monte Carlo because it is reminiscent of one of the most important centers of games of chance. What the name seems to suggest is that if you cannot solve a certain problem try to "leave it to chance". That is, replace it by a certain problem regarding a game of chance, whose probability of solution is larger.

CHAPTER 14

LANGUAGE

Perhaps it is unfortunate that ideas can be transmitted from person to person only (or almost only) through a language. The realization of this fact and of the difficulties implied by it, naturally leads to the need for making at least a brief presentation regarding language. Indeed the extent of clarity achieved may depend on the position that one has regarding language and on the determination with which one preserves this position.

Following the rules of good behaviour, when we withdraw from a reception, we take leave of our hosts by saying "Thank you very much, everything was perfect, good night". And the hosts reply "We were very pleased and we are very happy that you could come". Here we have a case where the use of the language means almost nothing. What we say has almost nothing to do with what we think; we simply comply with the rules of good behaviour. Such a use of a language is called pre-symbolic.

The pre-election speeches of politicians and the reviews made by music critics (in their majority) use a less pre-symbolic language. When a politician cries out that "Mr so-and-so drove the country to an unspeakable moral miasma", the politician does not, of course, believe what he/she says, but hopes to make the audience believe it. A music critic may write the following: "The tunes presented on the screen of conscience in order to form a dynamic model of such a deep perspective were such that it was impossible for the audience not to feel the presence of a third dimension." Now if the reader tries to interpret this phrase, word for word, it is obvious that it means absolutely nothing. In such cases the exact choice of words presents no interest regarding the transmission of "ideas".

In everyday discussions people make use of a language whose exactness varies between that of pre-symbolism and that demanded by mathematical science.

A further specialization in the use of language is observed in financial contracts, in official documents, and elsewhere. Here, the ideas that we want to transmit are not extremely difficult to comprehend, but the existence of clarity in the text is far more important than the one observed in our everyday discussions.

The reader may easily find many other examples which are similar to those presented above. It is obvious that the more specialized the notions involved in the discussion, the greater the confusion that dominates the discussions of an everyday nature. Therefore, it must not come as a surprise the fact that the difficulties that stem from the use of language in mathematics are the most complex. In order to overcome these difficulties we must make great efforts.

The transmission of ideas may fail in two ways. Namely, either the other person fails to understand what we want to transmit to him/her or the other person understands something different from what we want to transmit to him/her.

Of course, to fail to understand something is less dangerous than "understanding" something different, because in the second case this person will act with the false belief that the idea transmitted to him/her is the one that we wanted to transmit.

The reader may naturally think that apart from the cases where use of a pre-symbolic language is made, the number of cases where misunderstandings are observed can be drastically reduced, if we agree to define a priori the notions that present an ambiguous meaning.

Indeed, such an agreement is necessary. Almost all of us have heard during a discussion the question: "Can you give the definitions of the notions that you use?" But what do we mean by the word "definition"? If, by saying "definition", we mean what is written in dictionaries, then this barely helps us, because dictionaries try to achieve something that cannot be achieved. **They try to define all the words**.

The fact that it is impossible to define all the words (notions) can be easily proved. Indeed, a useful definition of a word must give the meaning of this word with the help of other words, which are already defined, i.e., whose meanings are known to us. From this it follows that one is then able to avoid the use of a word which is defined. In every discussion the use of the defined word may be avoided simply by making use of the phrase that defines it. In other words, since the defining phrase has the same meaning with the word that it defines, the phrase can very well replace the word.

Consequently, a defined word is an unnecessary word. We remark that we have reached a dead end: If all the words could be defined, then no word would be necessary. That is, we would not have a language. The contradiction that we obtained leads us to the conclusion that certain words cannot be defined.

The effort to define a word that belongs to a set of words in terms of words of this set leads to paradoxes, as is the following well-known paradox in which the Barber of Seville is "defined" as follows: He is the one who shaves all the men in Seville who don't shave themselves. (It is assumed that the Barber of Seville has a beard and needs to shave.)

Now the following question is posed: "Who shaves the Barber of Seville?" Since he shaves only those who don't shave themselves, it follows that it is not possible that he shaves himself. But if he doesn't shave himself, then, according to the definition we gave, he must be one of the persons that he shaves. Thus:

"If the Barber of Seville shaves himself, then he does not shave himself, and if he does not shave himself, then he shaves himself."

Another way of expressing the same conclusion is the following:

"It is neither true nor false that the Barber of Seville shaves himself."

The problem posed by the paradox of the Barber of Seville attracts the interest because the "definition of the Barber of Seville" seems to be analogous to the definitions given in mathematics. Of course, it is easy to determine the source of this difficulty: The definition of the "Barber of Seville" involves the set of men that live in Seville. If we consider that the Barber of Seville lives in Seville, then, as we saw, the definition becomes "circular", which means that the Barber of Seville is defined partly via himself. However, if we consider that he does not live in Seville, but he is some other human being (which is introduced by the definition itself), then no paradox exists and the definition is acceptable.

Another paradox is the following: We call an adjective "self-descriptive" if it describes itself. If this does not happen, then the adjective is called "non-self-descriptive". Hence an adjective is self-descriptive if, whenever we place it in the empty places of the phrase below, the phrase becomes true:

.... is a .... word

The adjectives "polysyllabic" and "English" are two examples of self-descriptive words. There exist many adjectives which make the above phrase "false" or meaningless when they are placed in the empty places, as, for example, are the words "long", "Chinese" or "bald", "ungrateful" respectively.

Now let us consider the adjective "non-self-descriptive" and let us pose the question whether this adjective is self-descriptive or not. If we assume that it is self-descriptive, then we obtain a true proposition when we place it in the empty places of the above phrase. Therefore, this adjective is non-self-descriptive, as the adjective itself says. But if we assume that it is non-self-descriptive, then this assumption coincides with the statement that we obtain if we place in the empty places of the above phrase the adjective "non-self-descriptive". Consequently, it is by definition a "self-descriptive" adjective. In other words:

If it is non-self-descriptive, then it is self-descriptive, and if it is self-descriptive, then it is non-self-descriptive.

In this case too the paradox appears because the term "non-self-descriptive" is defined with the help of all the adjectives to which this term belongs. The definition is acceptable if the term "non-self-descriptive" is considered as a kind of word which does not belong to the adjectives; otherwise it is not acceptable.

The above examples show that circular definitions must be avoided. We will see below how this can be achieved.

We proved above that all the words (notions) cannot be defined. This means that a basic table of words must exist, which are not definable and whose notions are captured through internal procedures, as it happens during our childhood. From the moment that we have formed this basic table, the linguistic basis, and we have captured the meanings of the notions of the words contained in it, all the remaining (definable) words acquire their meaning through the words contained in the basic table.

The need for the existence of a linguistic basis seems not to have been recognized as it should, and perhaps no effort has been made in this direction.

One of the words that could belong to the basic table is the word "truth". In mathematics, we know that the word "set" is also considered to be undefinable. Obviously, the undefinable words form the basic table and must be the ones that admittedly present the greatest difficulties in our effort to define them.

So probably the best answer to the question "What is truth?" posed by Pontius Pilate is the following phrase: "Truth is a word that belongs to the basic table and hence it cannot be defined."

Of course, we must not confuse the meaning of the phrase "undefinable word" with the meaning of the phrase "meaningless word". The words of the linguistic basis are verbal symbols of primitive notions, while the words which are defined through them are placed on a secondary footing, as they express in stenographic form expressions which are constructed by words of the basic table. But what is the origin of the "meaning" of the undefinable words? This is a difficult question which we cannot probably answer completely. As we mentioned above, we perceive the notions of the words of the basic table through internal psychological procedures, by observing and completing the various correlations.

Notions differ from person to person because they depend on the corresponding correlations, on our experiences with words, on the abilities of each one of us to complete the correlations, etc.

For certain particular and simple notions, we have a unanimous agreement. But for other words, which are usually abstract, we might have a complete disagreement. These extreme cases correspond to what we usually call "objective" and "subjective" respectively. Of course, many words, such as the word "truth", are placed between these two extreme cases.

We note that the words such as "truth" and "set", which we said belong to the basic table, are simply examples of words which *could* belong to the basic table. Perhaps we could find other more basic and more difficult words through which we could define these words. We always have a wide possibility of choice when we select words which will be considered as the basic ones.

We will close this chapter with comments regarding the notion of "logic", making an effort to answer the question whether it belongs to the linguistic basis or not.

Perhaps it is expected that the word "logic" indeed belongs to the linguistic basis. In order to see if there is a reason for that, let us try to give a certain definition of the notion of "logic". For example,

"Logic is a particular mental procedure which leads the person to assert in full conviction that a particular set of circumstances implies necessarily another specific situation under coexisting conditions. Logic is based on the ability of the person to recognize a certain analogy (coincidence of essential factors) between the given conditions and other conditions in which the outcome is known from experience."

The above definition indeed tells us certain things about "logic", as we perceive and accept it, but it is vague and above all incomplete. The words **analogy**, **mental procedure**, **conviction**, as well as others, are equally difficult to perceive and equally basic as is the word "logic".

In the past, and in particular during the 20th century, many efforts have been made to analyze the principles of logic and to formulate some of them with the help of other, perhaps more elementary and more primitive terms. Of course, it is obvious that in every such analysis a part of the notion of "logic" must necessarily be assumed to belong to the linguistic basis, something that makes the entire effort even more difficult.

Today the matter of the analysis of the notion of "logic" has become so vast that its study constitutes a particular branch.

For the reasons that we presented above as well as for many other reasons the word "logic" is considered to belong to the linguistic basis.

APPENDIX I

The Fast Fourier Transform (FFT)

An **algorithm** is a method ("recipe") of performing calculations. During our school years, when we multiplied two integers or when we subtracted one integer from another by saying "one to be carried over, etc.", we were making use of algorithms. Computers use much more complicated algorithms in order to perform calculations, which could not be done without the use of a computer.

The Fast Fourier Transform (henceforth FFT) is an algorithm, which had a drastic impact on our society. This purely mathematical idea speeded up greatly production in various industries.

The great advantage of the FFT is that it reduces drastically the number of calculations needed in order to calculate the Fourier transform. In general, if we know n values of a function, then, in order to calculate its Fourier transform without using the FFT, we need n^2 calculations, while if we use the FFT, only $n \log n$ calculations are required, i.e., much less.

We recall that the logarithm with base b of a number n, i.e., $\log_b n$, is the number m for which $b^m = n$. For example, we have $\log_2 4 = 2$, $\log_2 8 = 3$, $\log_{10} 100 = 2$. In other words, $\log_b n$ is approximately equal to the number of digits of n, when the latter is written in the system with base b. For example, $\log_{10} 1000 = 3$, $\log_{10} 374113 \approx 5.57$, $\log_{10} 1000000 = 6$.

The larger the number n, the more impressive the reduction of the number of calculations, when one makes use of the FFT. For example, if $n = 2^{10} = 1024$, then $n^2 = 1048576$, while $n \log_2 n = 1024 \cdot 10 = 10240$ a number that differs from n^2 approximately by a factor of 100. If $n = 2^{20} = 1048576$, then $n^2 = 1099511627776$, while $n \log_2 n = 20971520$ is a number that differs from n^2 approximately by a factor of 50000.

People who work with computers told us that if we have at our disposal a powerful computer, then by making use of the FFT we can calculate approximately one billion decimal digits of the number π in less than an hour, while if we use the same computer and we don't use the FFT, the same result is achieved almost after ten thousand years!

The central idea on which the FFT is based is due to Carl Friedrich Gauss in 1805, i.e., two years before the presentation by Fourier of his famous memorandum to the Académie des Sciences in Paris. This idea of Gauss was published after his death. But the algorithm introduced by the FFT was rediscovered and used in some computer program by James Cooley and John Tukey in 1965. We will try to explain the FFT as simply as possible.

α) We will begin first with the calculation of certain Fourier coefficients of a function without making use of the FFT.

Let $f(x)$ be a function defined on $[0, 1]$. Then we know that the Fourier coefficient of order k of f is given by the formula

$$c_k = \int_0^1 f(x) e^{2\pi i k x} dx.$$

Here we have to calculate an integral whose calculation usually does not constitute a simple procedure. Apart from certain cases, where there exists a formula providing the exact value of the integral, we usually make an approximate calculation.

In order to obtain these approximations we divide the interval $[0, 1]$ into 2^N equal subintervals. We calculate the values of f at the points $\frac{j}{2^N}$, where $j = 0, 1, ..., 2^N - 1$, and we write

$$c_k = \frac{1}{2^N} \sum_{j=0}^{2^N-1} f\left(\frac{j}{2^N}\right) e^{2\pi i k \frac{j}{2^N}} \qquad (1)$$

for $k = 0, 1, ..., 2^N - 1$.

Example. Let $N = 2$ and $f(x) = x^2$, whenever $0 \le x \le 1$. We have to calculate four coefficients c_k ($k = 0, 1, 2, 3$). We consider the values of the function at the points $0, \frac{1}{4}, \frac{1}{2}, \frac{3}{4}$, i.e., at the values of $\frac{j}{2^N}$ for $j = 0, 1, 2, 3$. Then for every k, we have:

For $j = 0$, $\quad 0^2 e^{2\pi i k \frac{0}{4}} = 0.$
For $j = 1$, $\quad \left(\frac{1}{4}\right)^2 e^{2\pi i k \frac{1}{4}} = \frac{1}{16} e^{\pi i k \frac{1}{2}}.$
For $j = 2$, $\quad \left(\frac{2}{4}\right)^2 e^{2\pi i k \frac{2}{4}} = \frac{1}{4} e^{\pi i k}.$
For $j = 3$, $\quad \left(\frac{3}{4}\right)^2 e^{2\pi i k \frac{3}{4}} = \frac{9}{16} e^{3\pi i k \frac{1}{2}}.$

Now we may calculate the coefficients by taking also into account that $e^{\frac{\pi i}{2}} = i$. We find

$$c_0 = \frac{14}{64}, \quad c_1 = -\frac{1}{8}\left(\frac{1}{2} + i\right), \quad c_2 = -\frac{6}{64}, \quad c_3 = \frac{1}{4}\left(-\frac{1}{4} + \frac{1}{2}\right).$$

The above way of calculating the Fourier coefficients is time-consuming. The subdivision of the interval $[0, 1]$ into four equal subintervals is obviously insufficient. If we assume that we subdivide $[0, 1]$ into $2^{10} = 1024$ equal subintervals, then we must add up 1024 products, which means that in order to calculate the Fourier transform, more than one million calculations are required.

β) Now let us try to shorten our work by making use of matrices.

We know that if A and B are two matrices, then their product AB is defined only if the number of columns of A equals the number of rows of B. We also know that it is possible that the product AB is defined while the product BA is not. And even if A and B are square matrices of the same order, it is not always the case that $AB = BA$.

We assume that the reader knows how to multiply two matrices.

Now we write relation (1) by making use of matrices.

Matrix A (which we call F_{2^N}) has as entries the numbers $e^{2\pi i k \frac{j}{2^N}}$, where j varies with the column and k varies with the row. Matrix B (which we call vector because it consists of a single column) has as entries the values of the function f at the points $\frac{j}{2^N}$, where $j = 0, 1, ..., 2^N - 1$. If we divide the product of these two matrices by 2^N, we obtain the coefficients $c_0, c_1, ..., c_{2^N-1}$. We have:

$$\begin{bmatrix} e^{2\pi i 0 \frac{0}{2^N}} & \cdots & e^{2\pi i 0 \frac{2^N-1}{2^N}} \\ \vdots & & \vdots \\ e^{2\pi i (2^N-1) \frac{0}{2^N}} & \cdots & e^{2\pi i (2^N-1) \frac{2^N-1}{2^N}} \end{bmatrix} \cdot \begin{bmatrix} f\left(\frac{0}{2^N}\right) \\ \vdots \\ f\left(\frac{2^N-1}{2^N}\right) \end{bmatrix} = \begin{bmatrix} 2^N c_0 \\ \vdots \\ 2^N c_{2^N-1} \end{bmatrix}$$

$$A \qquad\qquad\qquad B \qquad\qquad AB.$$

γ) Unfortunately, despite the use of matrices, we achieved no significant shortening of the work required for the calculation of the Fourier coefficients. The matrix F_{2^N} has 2^{2N} entries, i.e., 2^{20} in case $N = 10$, and 2^{20} multiplications are required, i.e., more than a million, in order to calculate the coefficients.

The trick introduced by the FFT consists in decomposing (appropriately) the matrix F_{2^N} into a product of three other matrices.

The key idea consists in reordering the numbers that we have to multiply so that we avoid making the same multiplication twice. An elementary school pupil, who has to multiply (without using a calculator) 9996496×8426735, makes his work easier by placing the second number above the first so that he/she multiplies 8426735 by 9 only once and each time he/she moves the obtained product by one position to the left. Something like that also happens in the case of the FFT.

When we calculate a Fourier transform and we multiply the term $e^{2\pi i k \frac{j}{2^N}}$ with the corresponding value $f\left(\frac{j}{2^N}\right)$ of the function, then we come to the same product $e^{2\pi i \frac{kj}{2^N}} \cdot f\left(\frac{j}{2^N}\right)$ more than once, i.e., as many times as we can write the number kj as a product of two numbers. For example, if $kj = 24$, then

$$kj = 1 \cdot 24 = 24 \cdot 1 = 2 \cdot 12 = 12 \cdot 2 = 3 \cdot 8 = 8 \cdot 3 = 4 \cdot 6 = 6 \cdot 4.$$

Consequently, it would be very useful to find a way to calculate the product that corresponds to the number $kj = 24$ only once instead of eight times.

Gauss invented a way to take advantage of the existence of the coincidences mentioned above, and he achieved that in his effort to determine orbits of asteroids.

Of course, Gauss did not use the above terminology, i.e., he did not work with matrices because matrices were discovered in 1858. But the algorithm that Gauss employed is written in matrix form as follows:

$$[F_{2^N}] = \begin{bmatrix} I_{2^{N-1}} & D_{2^{N-1}} \\ I_{2^{N-1}} & -D_{2^{N-1}} \end{bmatrix} \cdot \begin{bmatrix} F_{2^{N-1}} & 0 \\ 0 & F_{2^{N-1}} \end{bmatrix} \cdot [\text{Mixing Matrix}]$$

$$[1] \qquad\qquad\qquad [2] \qquad\qquad\qquad [3].$$

The first of these new matrices denoted by [1] includes four submatrices (which we denote by I, D, I, $-D$). All the entries of these submatrices, apart from the entries in the main diagonals, are equal to 0.

In the submatrices denoted by I the entries in the main diagonals are equal to 1. For example, if $N = 3$ and we put $\omega = e^{\frac{2\pi i}{2^N}}$, then we have:

$$I_{2^{N-1}} = \begin{bmatrix} 1 & 0 & 0 & 0 \\ 0 & 1 & 0 & 0 \\ 0 & 0 & 1 & 0 \\ 0 & 0 & 0 & 1 \end{bmatrix}, \qquad D_{2^{N-1}} = \begin{bmatrix} \omega^0 & 0 & 0 & 0 \\ 0 & \omega^1 & 0 & 0 \\ 0 & 0 & \omega^2 & 0 \\ 0 & 0 & 0 & \omega^3 \end{bmatrix}$$

The half of the submatrix denoted by [2] is filled with zero entries and the number of its non-zero entries is equal to half the number of the non-zero entries of the matrix F_{2^N}.

The third submatrix denoted by [3] is the mixing matrix whose job is to mix (shuffle) the elements of the vector (i.e., the values of the function that correspond to the subdivision points that we mentioned) without changing their values, as it happens when we shuffle a pack of cards. We give an example of an 8×8 mixing matrix, which mixes 8 values:

$$\begin{bmatrix} 1 & 0 & 0 & 0 & 0 & 0 & 0 & 0 \\ 0 & 0 & 1 & 0 & 0 & 0 & 0 & 0 \\ 0 & 0 & 0 & 0 & 1 & 0 & 0 & 0 \\ 0 & 0 & 0 & 0 & 0 & 0 & 1 & 0 \\ 0 & 1 & 0 & 0 & 0 & 0 & 0 & 0 \\ 0 & 0 & 0 & 1 & 0 & 0 & 0 & 0 \\ 0 & 0 & 0 & 0 & 0 & 1 & 0 & 0 \\ 0 & 0 & 0 & 0 & 0 & 0 & 0 & 1 \end{bmatrix} \cdot \begin{bmatrix} \mu_0 \\ \mu_1 \\ \mu_2 \\ \mu_3 \\ \mu_4 \\ \mu_5 \\ \mu_6 \\ \mu_7 \end{bmatrix} = \begin{bmatrix} \mu_0 \\ \mu_2 \\ \mu_4 \\ \mu_6 \\ \mu_1 \\ \mu_3 \\ \mu_5 \\ \mu_7 \end{bmatrix}$$

The first matrix, the mixing matrix, "shuffles" the entries $\mu_0, ..., \mu_7$ of the second matrix by placing the entries with even index ($\mu_0, \mu_2, \mu_4, \mu_6$) in the upper part and the entries with odd index ($\mu_1, \mu_3, \mu_5, \mu_7$) in the lower part.

The above decomposition of the matrix $[F_{2^N}]$ into a product of three other matrices reduces approximately by half the work needed for the calculation of the Fourier transform.

So instead of having to perform one million calculations (when $N = 10$) we have to perform now about half a million calculations: $2 \cdot (512 \cdot 512)$ calculations involving the two submatrices denoted by [2] and approximately 1000 calculations involving the submatrices denoted by [1]. Usually the work performed by the mixing matrix is not taken into consideration. But, in any case, when we say that $n \log n$ calculations are required in order to calculate a Fourier transform by making use of the FFT, what we mean essentially is that $cn \log n$ calculations are required, where the constant c depends on the details that appear during the process of performing the calculations.

Let it be noted that the process of calculating the Fourier transform via the method of the FFT can be continued. The two submatrices of the matrix [2] can be decomposed into products along the way followed above and give new submatrices, each of which has $2^{N-2} = 2^8 = 256$ entries; then each of these submatrices can be decomposed into a product, and so on. Hence the number of calculations is drastically reduced in 10 (because $N = 10$) successive steps. We have

1st step: From 1 million, it is reduced to $\frac{1}{2}$ of a million + (approximately) 1000

2nd step: ... it is reduced to $\frac{1}{4}$ of a million + (approximately) $1000 = \frac{1}{4}$ of a million + (approximately) 2000

$\vdots$

10th step: ... it is reduced to $\frac{1}{1024}$ of a million + (approximately) 10000 = (approximately) 11000

Of course, if we try to carry out the above calculations by hand, we will have to carry out an enormous and extremely exhausting work. But with the help of a computer, the FFT has become a powerful tool.

We will close this subject by making a few comments.

Due to the fact that the use of the FFT is indeed very effective, many people use it also in problems, where it is not necessarily useful. The use of Fourier analysis is not appropriate for all kinds of functions and for all kinds of problems. Sometimes the scientists who use Fourier analysis look like a person who is trying to find a lost coin under the light of a lamp-post, not because he/she lost the coin there but because light exists there.

The use of Fourier analysis is appropriate for the solution of linear problems. Non-linear problems present greater difficulties, and it is harder to predict the behaviour of non-linear systems than the behaviour of linear systems. Even a small change in the initial data of a non-linear system can cause a great change in the final result.

The law of gravitation is non-linear and its use in order to make long-term predictions is very difficult and probably impossible, even in the case of the famous three body problem. The system is very unstable. In his book *Fourier Analysis* (*Cambridge University Press, UK*, 1988) T. W. Körner ironically writes the following: "The great discovery of the 19th century was that the equations that refer to nature are linear, and the great discovery of the 20th century was that they are non-linear."

Appendix II

Historical Roots and Basic Notions in the Theory of Distributions
by L. Schwartz
October 1982

Note. A tape recorder was used to record Professor Schwartz's lecture. Then some members of the Mathematics Department had to work hard to come up with this final text which was approved by the speaker.

The same lecture was given in several other foreign Universities but it is published for the first time, now, in the *Proceedings of the General Mathematics Seminar* of the University of Patras.

N. Artemiadis

It is a great pleasure for me to give this lecture here. It is a difficult lecture, though, because you ask me to speak about the history of Distributions, in which I am involved, so that I cannot avoid speaking about myself, which is always a difficult problem. So I decided to do exactly that: to show you in what way I could discover, or I discovered, the Distributions, seeing some things quite soon and some things very late; in what way my strengths and my weaknesses can be compared with those of other people.

Because, when one examines this discovery, as well as many others, one can see that there were many people before who had something, even maybe a large part of the subject, but could not go far enough. And then, at some time, something, imposed by the circumstances, comes out, which was not recognized before.

I shall try to give other examples, because I want to show you, in a way, the strength and the weakness of the human spirit for discovery.

Let me give a very famous example: Einstein discovered Relativity, but Lorentz had already completely written, what is called the "Lorentz Group", namely the group of transformations of space-time which leave invariant Maxwell's Equation of Electromagnetism. So he wrote this result! But it was more or less ignored, because it was not assumed that electromagnetism and the velocity of light had to be invariant. Why should they be invariant? When one has light as something propagating in a matter like ether, then it does not have to be invariant with respect to the various observers, moving with respect to each other.

It was only a negative physical experiment Michelson made trying to measure the velocity of light with respect to various places moving one with respect to another, which indicated that there was no observable difference. After Michelson's experiment had been repeated, it was admitted that the velocity of light seemed to be constant. Again nobody considered that as a fundamental fact. Then along came Einstein[1]. It was already known from experiments that the velocity of light is invariant and that there was a group leaving invariant Maxwell's Equation and the velocity of light, i.e., the Lorentz group. But he had to introduce the new concept of space-time, and in the same way many other new formulas and concepts on Mechanics (force, energy, mass) and Physics; nobody would believe it in the beginning. And there was a very strong contravening current. There are still people who want to fight Relativity, but they are really crazy. At that time it was the opposite: it was Einstein who wasn't believed immediately. Yet afterwards, he was completely recognized and the system became universal. But it was considered to be so difficult, that it was said in many circles that only two or three people could understand relativity. Nowadays Relativity is taught in all Universities in the world; and even High School pupils know something about Relativity.

Usually, when a new concept is established, one can find that this concept had already existed but was not recognized as something useful and fundamental.

This is also true for the Distributions. It is difficult to imagine how many people, who lived much earlier, already knew something about them. Even I knew many things which I did not bring together. And when I did, I found very oblique ways to get the results and not immediately the definition which exists now.

So one has to overcome external and internal difficulties in order to establish new theories. This is what I want to show.

Let us return to the past. You find some traces of future distributions in Riemann, in Gauss, in Dirichlet, in the Theory of Harmonic Functions.

For instance it is known that a harmonic function f ($\Delta f = 0$) in $\mathbf{R}^n$ (in order to be harmonic it has to be twice differentiable) is as a matter of fact infinitely differentiable. Then people observed that in order to define a harmonic function it was not even necessary to have derivatives of the second order and a harmonic function could be defined as a continuous function such that for every sphere $S(R)$ the value $f(0)$ at the center was equal to the mean value of f on the sphere $S(R)$. This could be taken as a definition:

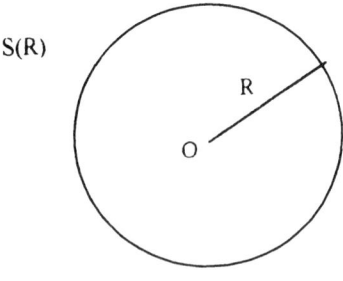

FIGURE 1

If a continuous function has the property that the mean value on every sphere is equal to the value at the center, then it is necessarily twice differentiable and harmonic. So, one has a definition of harmonic functions without derivatives a priori.

---

[1] I don't want to compare myself with Einstein, but show how discoveries usually occur through various steps and various people.

The same happened with holomorphic functions. If you have a holomorphic function $f(z)$ of one complex variable in a complex plane, Cauchy proved that it is indefinitely differentiable and if C is a closed rectifiable curve, then

$$\int_C f(z)dz = 0.$$

There is also a converse theorem by Morera saying that if a continuous function has integral zero over every closed rectifiable curve, then it is holomorphic, so it has derivatives. Therefore one can also define a holomorphic function without using derivatives. Of course this does not lead immediately to distributions but it implies that sometimes it is not necessary to assume the derivatives in the usual sense to get a property which finally comes back to derivatives.

At the end of the 19th and the beginning of the 20th century, people considered various integral equations, convolution integral equations, Volterra convolution, Mercer convolution of the type:

$$h(t) = \int_0^t f(t-\tau)g(t)d\tau.$$

Convolution played an important role at that time in the studies of electricity, in differential equations, in partial differential equations and, in particular, workers in electricity were led to consider phenomena which started at time zero (one switches on the current, for instance) and then they observe the reactions. So there are functions which are zero for negative t and just different from zero for non-negative t. This produces an integral from 0 to t.

It was known that if one had a complex electrical system which contained resistances, self-inductions, capacities, it had an impedance $Z(t)$ which was a function of t, so that if you had put electromotive force $e(t)$ you obtained a current with intensity $i(t)$. Then one has the relationship

$$e(t) = \int_0^t Z(t-\tau)i(\tau)d\tau,$$

i.e., the convolution of the intensity and the impedance of the electromotive force. This was known at the end of the 19th and the beginning of the 20th century.

Then the British engineer and mathematician Heaviside introduced **symbolic calculus** (1893): instead of writing $e(t)$, $Z(t)$, $i(t)$ he wrote $i(p)$, $Z(p)$, $e(p)$ and he wrote that in the following algebraic rule $e(p) = Z(p) \cdot i(p)$.

This was justified later by using the notion of Laplace transform, which was not systematically used at that time[2]. It was purely symbolic. He introduced a δ impulsion (it was not called δ, it was called a "unity impulsion") much before Dirac. This δ impulsion represented for him an imaginary, very strange, electromotive force acting for a very short time, so that the integral was equal to one. It was exactly the future notion of Dirac function.

Then he introduced the so-called unit scale y which was a current equal to zero for negative t and equal to one for positive t.

He observed that the derivative in some sense of this unit scale should be equal to δ: $y' = \delta$, (in what sense was not very precise) but he observed that δ must be

---

[2] Carson 1926, van der Pol 1932.

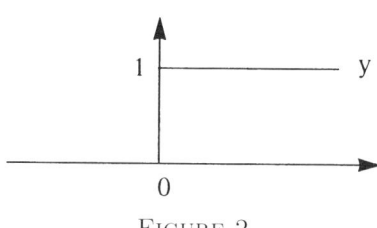

FIGURE 2

zero except at the origin, and that

$$\int_{-\epsilon}^{\epsilon} \delta(x)dx = y(\epsilon) - y(-\epsilon) = 1.$$

So he stated everything he knew about Dirac's δ-function at that time: δ is the derivative of the function y above, it is zero everywhere except at the origin and its integral equals one. But this cannot be possible because δ is almost everywhere equal to zero, so that its integral is zero, according to Lebesgue theory. So δ didn't really exist, but he wrote that. He called it 1; δ' he called p; and y he called $\frac{1}{p}$. The formula $\delta' * y = \delta$ (here is the modern notation with distributions) Heaviside wrote as $p \cdot \frac{1}{p} = 1$. He made a complete symbolic calculus with these functions of p. He constructed a symbolic algebra in which he used polynomial multiplications, decomposition of rational fractions into simple elements and it became a fantastic construction with asymptotic developments and so on; but no mathematician had accepted that! It was completely rejected by the community of mathematicians who said he was just crazy. It was a little strange because it led him mostly to perfectly true results; he also found theorems which were absolutely precise (for instance the extension of the fundamental solution E of a differential equation on **R** with constant coefficients, $P(D)E = \delta$, written as $P(p)E(p) = 1$ or $E(p) = \frac{1}{P(p)}$, computed by decomposition of the rational fraction $\frac{1}{P(p)}$ into simple elements), and which nobody could "reasonably" explain. One could prove these theorems, but he found them using this symbolic computation and nobody would admit that there might be some truth in it, even if it led to true results.

He was rejected and considered insane, at the end of his life he lost partly his equilibrium. It may be that to be a mathematician is as dangerous as to be a man of politics! You may be rejected if you are not in the normal current!

This was a first experience and it is now part of the Theory of Distributions, but it was then rejected.

Later, Dirac introduced in 1927 the δ-function, which is called the "Dirac function", with the same definition: δ is a function which is zero everywhere except at the origin and whose integral has to be one. Then he said another thing about δ: We have an approximation of δ as shown in the following picture:

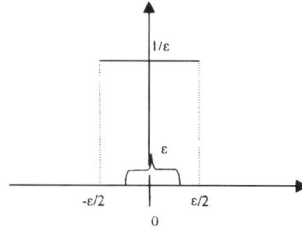

FIGURE 3

This is a discontinuous approximation of δ with the value $\frac{1}{\epsilon}$ on an ϵ-interval. He also gave another, this time continuous, approximation which was a Gauss curve, very concentrated, called a Gauss curve with a small parameter. In both cases the integrals are equal to one (see Figure 4).

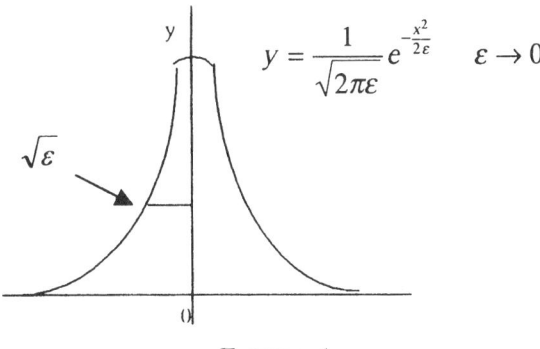

FIGURE 4

He then introduced the derivative δ', which was completely unacceptable for mathematicians, but his explanation was that: if the function in Figure 4 is an approximation of δ, you just differentiate it and you obtain a function as follows:

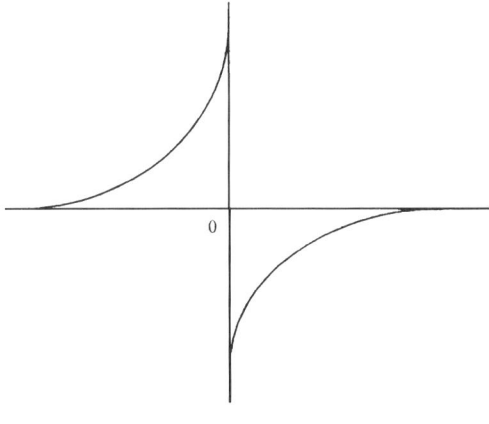

FIGURE 5

which should be an approximation of δ'. He then differentiated the δ-function many times. But not only did he introduce the δ-function, he also made changes of variables; for instance, he considered $\delta(t^2 - x^2 - y^2 - z^2)$ which introduces a distance in terms of Relativity. This was a change of variables in Distributions before they actually existed! Then he manipulated all these in such a way that in 1940 in all books of Theoretical Physics there were fantastic computations with distributions prior to the existence of distributions in Mathematics! These computations were sometimes wrong, since they were based on intuition which often is misleading. Still, there were a lot of true formulas which are now well proved. So Dirac was not only the inventor of the Dirac function, but he and many physicists also did a lot of computations so that in every book of Physics there were many δ-functions and distributions with no justification at all from the mathematical point of view (which came at least 15 years later). The mathematicians felt only contempt for that and they rejected the whole method. Yet Dirac did not become insane, but the Physicists formed a world outside of Mathematics.

I followed a course, when I was a student, in the year 1935, in which I heard about the δ-function and its derivatives, I discussed it with some friends, so it remained in my mind that this existed in Physics and had absolutely no sense in mathematics. I discussed it with the mathematicians of that time, my professors, my classmates and we came to the conclusion that it was nothing, we had nothing to do with it. It was absolutely impossible to find any kind of justification, especially for δ' which was the first true obstacle. Because δ could be a measure, but then a measure has no derivative. But I kept that open in my mind.

One of the first completely coherent things on the subject was introduced by Sobolev just before the war (1936). Sobolev introduced some kinds of generalized functions which were very near my definition of distributions. Sobolev instead of considering the space C_{comp}^{∞} (of infinitely differentiable functions with compact support) (that I called $D^{\infty} = D$), considered C_{comp}^{m}, which he called D^m, introduced continuous linear functionals on D^m and manipulated some derivatives, so that he had a correct definition, possible to use. Something completely lacking in his theory was the support, the carrier of a distribution; and the carrier of a distribution depends on the Theorem of Unity Partition which was discovered in France during the war by Dieudonné. Sobolev had no convolution, no Fourier transform, no strong topological properties, but he had many formulas and, in particular, he was able to see that a partial differential equation with a boundary value could be solved in this way by putting the boundary value in the right-hand side, a decisive fact. It was quite a good approximation of distributions. However in the world of mathematicians it remained a curiosity in the sense that it was just a publication among others, for a particular aim and not presented as a general tool used for many things. Most of the mathematicians ignored it (including myself). Of course this appeared just before the war and Sobolev himself did not continue much afterwards in this direction. He had it published and then went on to other subjects. However he introduced very interesting objects which still carry his name, the spaces H^m, $m \geq 0$ an integer, called the Sobolev spaces. With his generalized derivative he defines H^m, for integer m, as the space of L^2 functions whose derivatives, in the weak sense of generalized functions, belong to L^2; they need not be continuous, they are differentiable in the sense of distributions and not in the usual sense. That was a really considerable work, and the Sobolev spaces remained because they are an essential tool for the study of partial differential equations. The beginning was a little forgotten; it is difficult to know exactly why, but this is the case (as often happens) and it is comparable with the Lorentz group. Lorentz found the group, but he forgot it and he continued in other directions. In order that a new **essential** idea becomes **universally accepted**, it is not sufficient that somebody introduces it casually, it is necessary that one or several people introduce it on a **massive** scale, and be strongly persuasive.

Bochner made another very good approach. The case of Bochner is extremely interesting because in some way he was at the least possible distance from distributions, although he did not go further. So he published at the end of his book about Fourier Integrals (1932) a chapter in which he introduces "formal functions" on the

real line **R**. A formal function was defined as follows: Consider a square integrable function f, multiply it by a polynomial Q and apply $P(D)$, a partial differential operator with constant coefficients. So it is the derivative of a slowly increasing function:
$$P(D)(Q \cdot f).$$

These beings, introduced by Bochner, are exactly the same as temperate distributions which I introduced later on, on $\mathbf{R}^N$, N arbitrary. But nobody understood it, and I ignored it then myself. So it remained also a curiosity and so difficult to explain, that he put that chapter at the end of the book (24 pages), as a curiosity. He did make some computations with that; he did know that the Fourier operation transforms convolution into multiplication and multiplication into convolution. So he knew everything in some way, without topology. Yet, even he did not recognize it as a new fundamental tool.

For instance, if you want to define δ in this way (which he did not do) you have to define it as
$$\delta = D^2 \left(\frac{1}{2}|x|\right)$$
(the second derivative of the function one-half of modulus of x).

But the second derivative of $\frac{1}{2}|x|$ is not usable as δ! You cannot handle δ in this way. So he did not define the δ of the physicists. Thus there was no relationship with the experiments and with the physicists, and there was no relationship with Heaviside. It is a quite exceptional scope on what is the introduction of a theory, because in some way everything was contained in that chapter and nobody recognized it; even Bochner didn't estimate it as its correct value. He considered it as a curiosity and he never spoke of it again. You see how things advanced exactly as in the case of Lorentz. Lorentz found the Lorentz group but there was no conclusion about that.

A second time Bochner considered a partial differential equation with constant coefficients: $P(D)f = 0$ in $\mathbf{R}^N$, and he had to know what was a "generalized" solution of the partial differential equation. This could be done by using his Fourier Transform, but he did not do it this way! He said: f is a generalized solution of $P(D)f = 0$ if there exist infinitely or several times differentiable functions f_n which converge uniformly to f on every compact subset of $\mathbf{R}^N$, and such that $P(D)f_n = 0$, in the usual sense. This could be done only for an operator with constant coefficients, by regularization: the f_n are of the form $f * \phi_n$, ϕ_n smooth functions with compact support, ϕ_n was differentiable enough, so that $P(D)(f*\phi_n)$ was meaningful, but not infinitely differentiable. So there is a second article in which he discovers in a way the derivative in the sense of distributions, but he establishes no relationship between both articles!

The main defect is that one could define f satisfying $P(D)f = 0$, but one could not compute the values of the derivatives. For instance in the hyperbolic equation of a vibrating string
$$\frac{1}{v^2}\frac{\partial^2 f}{\partial t^2} - \frac{\partial^2 f}{\partial x^2} = 0,$$
it was known a long time ago that the general solution is
$$f = u(x + vt) + w(x - vt).$$

In order to be a solution, u and w must be twice differentiable. What about the case of a function which is a sum of this kind, where u and w are just continuous functions without derivatives? That was one of my obsessions since 1934: in what sense is it possible to say that a $u(x + vt)$ and a $w(x - vt)$, where u and w are continuous functions, can form a solution of the wave equation, although they have no derivatives in the usual sense? So one cannot actually compute either $\frac{\partial^2 f}{\partial t^2}$ or $\frac{\partial^2 f}{\partial x^2}$, but in some sense their combination is zero. I ignored also this definition of Bochner. I found it again myself in 1944. I shall describe that below. Anyway it didn't solve my problem. Namely f can be approximated by $f_n = u_n + w_n$, where

$$f_n = u_n(x + vt) + w_n(x - vt).$$

Then one could say that f was a generalized solution, but one could compute neither $\frac{\partial^2 f}{\partial t^2}$ nor $\frac{\partial^2 f}{\partial x^2}$. Just the combination was zero by a limiting procedure, and one could not compute intermediate terms. Jean Leray had introduced, early in 1934, the weak derivatives, and the weak solutions of partial differential equations, by integrating by parts (as it is now in distributions), and I attended his lectures in 1935. He defined the weak derivative **when it was a function**, and also a weak **solution** of $\frac{1}{v^2}\frac{\partial^2 f}{\partial t^2} - \frac{\partial^2 f}{\partial x^2} = 0$, **without each term having any meaning!** Always the same difficulty! I now had this objective which remained always in my mind for the future: how to define a generalized solution of a partial differential equation so that one can say that f is a generalized solution, but also that $\frac{\partial^2 f}{\partial t^2}$ has some meaning and $\frac{\partial^2 f}{\partial x^2}$ also, in such a way that if you combine them you find zero?

Hadamard introduced finite parts of divergent integrals. In his very remarkable book on Hyperbolic Partial Differential Equations he studied fundamental solutions of these equations. The very definition of a fundamental solution had never been written in its full generality. The following definition comes with distributions: if $P(D)$ is a partial differential operator with constant coefficients, E is called a fundamental or an elementary solution if $P(D)E = \delta$ in the sense of distributions. So that now it is very easy, but because δ did not exist at that time, the elementary solution was a solution of $P(D)E = 0$ in some open set, and with some kind of singularity at the origin or on the wave cone. With these singularities in the way, it was difficult to define the fundamental solution. A lot of computations have been done by Hadamard himself and by other scientists on fundamental solutions of PDE's by Fourier transform, or some other method, without a fundamental solution being anywhere correctly defined. They have now a meaning with the theory of distributions. Hadamard also defined the finite parts of divergent integrals, which played an enormous role in the theory of hyperbolic partial differential equations (1932). So I already knew about the finite parts of divergent integrals of Hadamard, and fundamental solutions. I knew them well and I knew that something had to be done also in this domain. Now, I remember that George de Rham came to Clermont-Ferrand during the war (1942) and gave a lecture about this notion of "currents". He had a notion of currents, which was quite fascinating, according to which a current of degree p, on a manifold of dimension N could be either a differential form of degree p, or a submanifold of dimension $N - p$ with boundary, or a sum of a differential form w and a manifold with boundary: $\Gamma = w + v$. And then there was a notion of coboundary $d\Gamma$ of this current, which was the coboundary of the differential form plus or minus the boundary of the manifold:

$$d\Gamma = dw \pm \beta v, \quad d\Gamma = dw + (-1)^{P \pm 1}\beta v.$$

But this was just formal and he wanted to have something more than that, and we had a discussion. He then told me: "I would like to find a kind of generalization of Lebesgue integration". In Lebesgue integration on $\mathbf{R}^N$ there are absolutely continuous measures with respect to dx and singular measures as a unit mass. But there are many intermediates, as measures on surfaces, and you have a space of measures, which is a complete vector space with well-known properties. I would like more generalized currents in which one may have differential forms corresponding to absolutely continuous functions; I would also like some manifolds corresponding to δ (which was not called δ, but ϵ) or measures on surfaces; so one may have also many intermediates and this could be a true complete vector space with nice properties. But it seems to be so considerably more difficult than Lebesgue theory, that probably we are very far from that. "It's not for our generation".

I was also impregnated with this. This was in 1942 and I said: "I think it is probably impossible, since we have no good tools to do it". Two and a half years later, I found the distributions, although at that time I considered it as an impossible task! I had therefore many things by that time, not all related to each other. I worried about PDE's with generalized solutions; I worried about δ, δ' of the physicists and whether they can be related to de Rham's currents. I had also functional analysis and duality. It was known after André Weil's book of integration on topological groups (1940) that instead of considering abstract measures on a sigma field it was possible to define a Radon measure μ on a locally compact topological space X as a continuous linear functional on the space $C(X)$ of continuous functions vanishing outside compact subsets of X, μ was a functional, so that for every continuous function ϕ with compact support, $\mu(\phi)$ was a number; μ was a linear form on $C(X)$, continuous on each $C_K(X)$ (the space of continuous functions on X with support K, K a compact subset of X) equipped with the sup-norm.

Frederic Riesz's theorem is that every measure gives birth to such a functional, but A. Weil reverted the process and defined in his book a Radon measure as being a continuous linear functional on the space of continuous functions with compact support. He had the $\mu(\phi)$ and he had the carrier or support of μ, introduced by Henri Cartan in potential theory. Henri Cartan called "noyau fermé des masses" what is now called the support of μ. I learned that in his book in 1940. I had these functionals and I knew they were very important!

So it appeared to me that the dual of a nice functional space might be a very important space too. That was a fundamental step! I kept this in mind and tried to study other cases. I thus made (just for myself), in the year 1943, during the worst of the war, a theory of duality not for Banach spaces but for Fréchet spaces or for other topological spaces. I did not know the work of Mackey who had done many things in the United States but they were not known in France at that time.

I had a topological vector space E, for instance, $E = C([0,1])$. I had the dual E' which in this case is $M([0,1])$, the space of measures: $E' = M$. It could be very interesting to consider E' for non-Banach spaces E, but the theory of duality for non-Banach spaces, at that time, at least in France, practically did not exist.

I did that for myself. I took E, a topological vector space, considered its dual E' and tried to define the usual things.

The Hahn-Banach theorem and the Banach-Steinhaus theorem were known. I studied the bounded subsets, the weak and strong topology on the dual E' (the

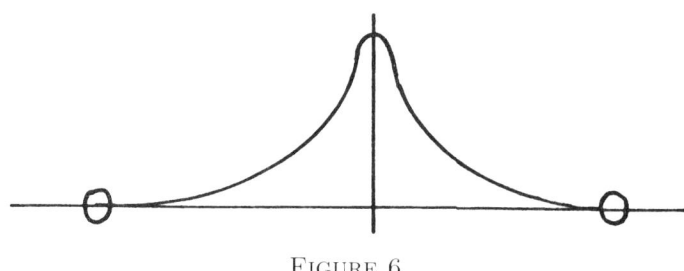

FIGURE 6

strong one being the topology of uniform convergence on the bounded subsets of E), the bidual, and the reflexivity. I took the particular case $E = D([0,1]) = C^\infty([0,1])$, and considered its dual E', which is, in the distributions of today, the space $D'([0,1])$ of distributions on $[0,1]$ (more difficult than the space $D'(\mathbf{R}^N)$ of distributions on $\mathbf{R}^N$, because of the singularities at the boundary $\{0,1\}$; but $D(\mathbf{R}^N) = C^\infty_{comp}(\mathbf{R}^N)$ is much more complicated than $D([0,1]) = C^\infty([0,1])$, that was my reason!). I didn't consider the possibility of introducing differentiation on $D'([0,1])$ by transposing the differentiation on $D([0,1])$! I just considered the topological properties (for instance $D([0,1])$ is reflexive). "Well, I said, I have a dual, I have reflexivity, it's nice. But it's probably completely useless." This was in 1943, one and a half years before the distributions! I forgot it, I didn't think of it anymore. But the war prevented me from doing mathematics!

This proves that when people find various new concepts (as did, not only Heaviside, who was rejected, but Sobolev and Bochner, and myself) they may often take certain steps without **realizing** that they are finding something very useful. Some further work is necessary in order to see that it may be used, and that it is important.

Now comes an article by Deny and Choquet (October 1944), in which they prove something about polyharmonic functions (they appeared as functions which are solutions of an iterated Laplacian). But the theory is easy with the iterated Laplacian because polyharmonic functions, just as harmonic functions, can be defined without derivatives and they are immediately infinitely differentiable functions. Then I made some generalization, a small article of 3 pages which was partly the start of the whole thing, called "Sur certaines familles non fundamentales de fonctions continues" (November 1944). In this, instead of finding polyharmonic functions, I found **generalized** solutions of a PDE with constant coefficients. And the generalized solutions which I used were in the exact sense of Bochner, as the limit of usual solutions, by regularization (I invented them myself, ignoring Bochner's article).

Immediately this trick upset me: I had generalized solutions of a PDE but I could just say that a function f is a generalized solution of the wave equation $\frac{\partial^2 f}{\partial t^2} - \Delta f = 0$ without $\frac{\partial^2 f}{\partial t^2}$ and Δf having any particular sense, once again!

But in order to define that, I used convolution with a C^∞ function ϕ with compact support: $f * \phi$, and took the limit, if f is a function, of $f_n = f * \phi_n$. The ϕ_n's were ≥ 0, C^∞ functions as in Figure 6, converging to δ in the sense that their support converged to zero and their integral was equal to one. And I said that f is a generalized solution of $P(D)f = 0$, if the $P(D)(f * \phi_n)$ are 0.

Then I showed my work to Cartan, but he told me: "C^∞ functions with compact support are a bit monstruous beings because they are not analytic. If you take a C^∞ function with compact support it is not analytic at the points marked in Figure 6 (with all the terms being zero, and all the derivatives equal to zero) and therefore it has no Taylor expansion. These functions will be horrible, maybe you should not consider them."

I said, "yes of course, I know, I must be careful but they give the good trick here".

So actually he was a little sceptical about the use of them, but I was not sceptical at all because they were giving here the shortest proof. In fact, I found only a few years ago an article by N. Wiener in 1926, in which he says that although the C^∞_{comp} functions are monstrous beings, they are very useful indeed; and he proves a theorem with this, which can be considered as a theorem of Distributions.

Then, in November 1944, suddenly I had a good complete theory for everything, a theory which gave derivatives to all beings and made unification between many things. They were the Distributions.

I could then justify δ and δ', the generalized solutions of the PDE's with separate meaning for the terms. Now when I have

$$\left(\frac{\partial^2}{\partial t^2} - \Delta\right) f = 0$$

each term $\left(\frac{\partial^2 f}{\partial t^2} \text{ and } \Delta f\right)$ separately can be a distribution and the sum of the two distributions gives zero.

I had exactly what I wanted and that for me was the beginning of the revelation. During a long night of thinking I found most of the theorems. Just in one night! Cartan was now immediately enthusiastic, also the other members of Bourbaki, whose support helped me very much.

However my definition **was not** the definition which is adopted now, it was a different one, **much more complicated!**

It was not yet $T(\phi)$, it was something else. I did not call them distributions, but operators, because of their origin in Choquet and Deny's problem. I considered the space $D(\mathbf{R}^N)$ of C^∞_{comp} functions and the space $\mathcal{E}(\mathbf{R}^N)$ of C^∞ functions with arbitrary support; I put good topologies on the D_K's and on $\mathcal{E}$; then an operator T was not a distribution, but a linear map from D into $\mathcal{E}$, continuous on each D_K, and such that $T : D \to \mathcal{E}$ is commuting with convolution with D.

I called this map: $\phi \mapsto T.\phi$. This was a convolution operator: to every test function $\phi \in D$ it assigned a function $T.\phi \in \mathcal{E}$, with the commutation $T.(\phi * y) = (T.\phi) * y$. So I called them operators. These operators, if T is a function, gave exactly the usual convolution:

$$(f.\phi)(x) = \int f(x - \xi)\phi(\xi)d\xi;$$

a locally integrable function was an operator. I put on the space D' of operators the topology of uniform convergence on bounded subsets. One could differentiate the operators infinitely many times. The derivative of an operator, DT, is defined by $DT.\phi = T.D\phi$ (DT convoluted by ϕ equals T convoluted by $D\phi$) (**no minus sign**). So every operator, in particular every continuous function, is infinitely differentiable in the space of operators. A function which has no usual derivative has a derivative operator.

I immediately saw δ: the convolution with δ is just the identity operator, namely $\delta : \phi \mapsto \phi$. δ' is just the derivative, namely $\delta' : \phi \mapsto \phi'$.

So I had all that concerned the Physicists. I had the convolution of two distributions: $(S*T).\phi = T.(S.\phi)$, if S has a compact support: if S has a compact support, ϕ is indefinitely differentiable with compact support, $S.\phi$ is also infinitely differentiable with compact support, and $S*T$ is an operator. So that, apart from the restriction of compact support, you could convolute arbitrary operators.

I remember that, sometime later on, after working long into the night, I had a dream. I was so enthusiastic about the possibility of having convolution of two arbitrary operators (but of course the result of convolution is an operator), that I dreamed I was explaining to somebody that you could convolute whatever you wanted, for instance, you could say $Mozart * Beethoven$, but of course it would not be a musician, but an operator; and you could convolute $Nancy$ and $Strasbourg$, but of course it would not be a city, but an operator. It may seem extremely strange that, having the $\mu(\phi)$ of Weil to define a Radon measure, having studied the duality E-E' in 1943, I defined an operator $\phi \mapsto T.\phi$ from D into $\mathcal{E}$, instead of a linear form $\phi \mapsto T(\phi)$, a distribution in the current meaning. It comes from the origin of the inversion of Choquet and Deny's problem. I didn't realize that a duality formula with $T(\phi)$ a number, would have given also a differentiation, with $T'(\phi) = -T(\phi')$, $(P(D)T)(\phi) = (-1)^t T(P(D)\phi)$. This came only several months later! My $T.\phi$ of November 1944 is the $T*\phi$ of today. This proves the important negative role of the inhibitions; to discover is sometimes to break inhibitions!

A strong obstacle was to define a product αT, the multiplicative product of a distribution by a C^∞ function α. It is very difficult to define a product because multiplication does not commute with convolution! But I had also a theorem at this time: that every distribution T, locally speaking, is a derivative of a continuous function f (maybe without usual derivative), i.e., $T = D^p f$, so that all you have to do is to define $\alpha D^p f$ formally and then you can use Leibniz's formula to get it. But it was necessary to have this finiteness theorem, that every operator, locally, is a derivative of a continuous function (without usual derivative). Of course one had to prove that αT is independent of the representation $T = D^p f$; this was done by a limiting procedure, it was true for a C^∞ function, and the C^∞ functions were dense in D'. I had the topology, and all my theorems of 1943 about duality and about the usual topological vector spaces came back. However I was unable to find the Fourier Transform. Here again the definition as convolution operators made things awkward.

I couldn't find it because I tried to define $T.\phi$ as a function but not $T(\phi)$, a complex number. Moreover, it was necessary to introduce this space $J(\mathbf{R}^N)$ of C^∞ functions, rapidly decreasing at infinity as well as their derivatives and its dual $J'(\mathbf{R}^N)$, the space of temperate distributions. It was unusual, at this time to introduce **many** functional spaces. But it came five months later. I was in Grenoble in 1944-1945. Cartan was in Paris and we corresponded and met often. During several months I was stubborn with this $T.\phi$; manipulated it better and better but I wasn't able to get the result. Eventually one day I said, well why not take $T(\phi)$? It was in February or March 1945 (about four or five months after beginning); I wrote immediately a letter to Cartan, saying I had the trick! One must consider distributions as linear forms, and not as operators. Why the name distributions? Because $\mu(\phi)$, when μ is a measure, is a distribution of charges in the universe; electric charges for instance; with distributions you have the dipoles, you have the magnets, double layers, you have distributions which are exactly physical distributions of masses, magnetic or electric distributions with positive and negative charges; and if you take the space, introduced by Deny later, of all distributions which are of finite energy and therefore may intervene in physics, they are distributions, not measures. As soon as de Rham learned about my article on distributions he found immediately the general notion of currents, which he was looking for, for many years! It was still difficult at the time of the operators, but with the distributions it worked perfectly. Although I had that also in my mind for years, I was essentially preoccupied at that time with $\mathbf{R}^N$, multiplication, convolution, Fourier transform (with the functional spaces S, S', O_M, O_C); I knew more or less how to manage to get currents, but really didn't think of them seriously; but he considered immediately these currents, their coboundary, their Laplacian on a Riemannian manifold, their cohomology, etc. He published his book *Variétés différentiables. Formes, courants, formes harmoniques* in 1955. I found also distributional sections of vector bundles, but published them only much later.

After two articles in the *Annales de l'Institut Fourier* (1945-1947), I slowly wrote the book; it was achieved in 1950-1951. Then, of course, everything was arranged. I think I had the first elements when I was a student, in 1934 (Dirac introduced his function in 1926-1927), I had the definite form in 1945 and I published the book in 1950-1951.

I also want to mention that in doing research, one can waste a lot of time on easy things and find immediately very difficult ones. For instance, there was a theorem which I was thought about for one whole week, eight hours every day, and it became more and more difficult to prove it (I believed the theorem was true). Each day, at the end of the day, I believed I had the proof, but I was too tired and went to sleep and in the morning I found the proof was false. I tried to prove that a distribution carried by a compact subset K can be expressed as a finite sum of derivatives of measures carried by K. It's false, K has to be "Whitney regular". This lasted for one entire week of suffering. Research is not only enjoyment; it is enjoyment when you find the results, but you may suffer for a long time and the enjoyment is the result of this suffering!

I became more and more upset about that theorem and in the morning of the last day I found a counterexample of four or five lines. The theorem was false! This I wrote in my book using some kind of irony against myself. It goes as follows: "One could believe that the following theorem is true (and I wrote the theorem); it is not true though, as proved by the following very trivial counterexample." And I gave the very trivial counterexample and nobody knew I had suffered a whole week to find it.

So that was a little of this history. I gave a course in Paris in 1945 about Distributions (cours Peccot) which was attended by about 25 people; half of them were Physicists and in particular workers in electricity.

I also gave a lecture about that in the Symposium of Harmonic Analysis in Nancy in 1947. There were many foreign mathematicians present. Among them Harald Bohr was very enthusiastic, and he invited me to Copenhagen where I gave several lectures on distributions, to a large audience; that was my first travel abroad in my professional life, the beginning of a long series! At the same occasion, I went to Lund, where I made acquaintance with Marcel Riesz and Gårding. I lectured also in Oxford and London in 1947, and I was invited to the Canadian Mathematical Congress in 1949 at Vancouver, also to speak about distributions.

At the International Congress of Mathematicians, Cambridge, Mass., USA, 1950, I received the Fields Metal for my work on distributions (before the publication of the book!), and I gave a talk on the theorem of kernels, which was published only later on. So the resonance of the distributions grew reasonably!

However, when the book appeared in 1950, it was not yet quite accepted.

I had to fight a battle against two quite opposite categories of critics; some people said it was so simple that it could not really be useful, and some others said it was such a complicated definition of a generalized function that it could not be handled and could not be used.

So some people found it too simple to give useful results while others found it too complicated to be usable. Sometimes I thought myself one or the other of these contradictory things! So a battle had to be fought which I remember very well. Also Gårding told me he kept it in his memory, after I gave my lecture in Lund, in 1947. Hörmander was about sixteen years old then, I guess. He only entered the University a little later so he learned distributions in the first years of his studies, and rapidly found applications.

I remember the battle I fought to make the distributions universally accepted, which is part of the researcher's world. Some of my younger students, such as Lions, or Malgrange, may say that it was immediately accepted, but this is not quite true. They came to the distributions as students, in Nancy in 1945-1950, they accepted them immediately; other young people started research in the years 1953-1955, when the battle was over. The work of the younger generation has been an essential part of the success.

In the years after 1950 came many articles or books; Gelfand-Shilov's books on generalized functions are very complete and self-contained. Now every mathematician introduces easily new vector spaces; and later, distributions were generalized by Sato's hyperfunctions.

Remark. Distributions can be considered as a generalization of functions in the sense in which real numbers are a generalization of rational numbers. How do we generalize the rationals? Using cuts defined by Dedekind on the straight line. After all, at some time, it became necessary to define correctly the irrational numbers! And Dedekind introduced cuts on the real line, a rational number defined a cut, so that the reals are a generalization of the rationals.

There is an article by Peano, in 1912, in which he considers Heaviside's computations and he says: "I am sure that something is to be found now; there must be a notion of generalized functions which are to functions what the reals are to the rationals," so very much earlier than 1944!

Reading that text again, I think I exaggerate the convergence of many various questions to the same solution: distributions. It has been so, and this explains the strength of my own enthusiasm for distributions, because it solved together many problems I had before. But it is sure that until the final collective solution came, these problems were not bound together in my mind, they were not thought of as having the same solution: partial differential equations (wave equation), Dirac's δ of the physicists, de Rham's currents, duality (the dual of $C^\infty([0,1])$) were discovered by me as **independent** things. Their unification by the distributions causes my enthusiasm!

APPENDIX III

The Theory of Wavelets

During the last decade intensive mathematical research activity was observed in the so-called "theory of wavelets", which we will also call briefly "wavelets".

Usually applications of a mathematical theory appear, when the theory becomes well understood. In the case of wavelets, what happened is the opposite. The theory was suggested by the various problems that were studied in physics. The mathematicians stepped in and "cleared" up the dominating vague situation by giving to it the necessary structure and order.

One can find roots of the theory of wavelets in works of mathematicians and physicists even before 1930.

In a writing regarding history, one cannot give an extensive account of a theory. We will limit ourselves in presenting certain generalities. The reader may find an extensive bibliography on wavelets in S. G. Krantz's book *A panorama of harmonic analysis*.

The theory of wavelets constitutes an extension of Fourier analysis. We recall that the Fourier transform is the method (the process) according to which we decompose a given function into the frequencies that comprise it, just like a prism resolves light into various colors. The Fourier transform transforms a function f, which depends on time, into a new function $\hat{f}$, which depends on the frequency. This new function is called the Fourier transform of the initial function (or the Fourier series of the initial function, when the initial function is periodic). When the independent variable of the functions (or the "signals", as they are sometimes called) is time — as it happens in music or in the fluctuations of the stock-market — the frequency is usually measured in hertzs, i.e., in circles per second, as one can see in Figure 1.

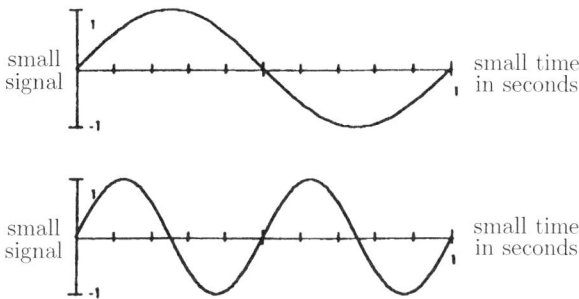

FIGURE 1. The frequency of $\sin 2\pi x$ is 1 circle per second (1 hertz). The frequency of $\sin 2\pi 2x$ is 2 circles per second (2 hertzs)

A function and its Fourier transform constitute two sides of the same information. The function reveals to us the information regarding time and hides from us the information regarding frequencies. For example, the function that corresponds to the recording of a piece of music shows us how the air pressure (caused by the music waves) varies with time, but it does not inform us which are the frequencies — the notes — that create the music. In contrast to that, the Fourier transform informs us about the frequencies, but it hides from us the information regarding time: in music the Fourier transform tells us which are the notes played, but it does not tell us when they are played.

Anyway, the function and its Fourier transform provide complete information regarding the time and the frequency of the signal considered. We can calculate the Fourier transform of the function once the function is given to us, and conversely we can obtain again the function from its Fourier transform.

As it is known, the Fourier series of a periodic function f with period 1 is written as follows:

$$f(t) = \frac{1}{2}a_0 + (a_1 \cos 2\pi t + b_1 \sin 2\pi t) + (a_2 \cos 2\pi 2t + b_2 \sin 2\pi 2t) + \ldots$$

or

$$f(t) = \frac{1}{2}a_0 + \sum_{k=1}^{\infty}(a_k \cos 2\pi kt + b_k \sin 2\pi kt). \qquad (1)$$

The Fourier coefficients $a_1, a_2, a_3,\ldots$ inform us how much of the functions $\cos 2\pi t$, $\cos 2\pi 2t$, $\cos 2\pi 3t,\ldots$ are included in the function f (i.e., how much cosines of frequency 1 hertz, 2 hertzs, 3 hertzs,...). The coefficients $b_1, b_2, b_3,\ldots$ inform us how much of the functions $\sin 2\pi t$, $\sin 2\pi 2t$, $\sin 2\pi 3t,\ldots$ are included in the function f (i.e., how much sines of frequency 1 hertz, 2 hertzs, 3 hertzs,...). The Fourier series refers only to the sines and the cosines which are integral multiples of the basic frequency. k in the above series represents the frequency.

We know that the Fourier coefficients of a periodic function f with period 1 are given by the equations

$$a_k = 2\int_0^1 f(t)\cos 2\pi kt\, dt, \qquad b_k = 2\int_0^1 f(t)\sin 2\pi kt\, dt. \qquad (2)$$

When we write the Fourier series of a function as in (1), we measure the frequency in hertzs, i.e., in circles per second. If we want to measure the frequency in radians per second, we take the integrals in (2) from 0 to 2π.

When the Fourier coefficients of a function are known to us, we can construct the function by making use of (1).

The only frequencies used in the case of a Fourier series of a periodic function are the integral multiples of the basic frequency of the function (the basic frequency being the inverse of the period of the function). If our function is not periodic, but it has the property that rapidly decreases as the independent variable tends to infinity so that the area of the domain that lies below its graph is finite, then again it is possible to express this function as a sum of sines and cosines, i.e., to decompose this function by making use of frequencies. In this case though, in order to calculate the Fourier transform, we must calculate the coefficients for all possible frequencies. Then the requested relations are the following:

$$a(\tau) = \int_{-\infty}^{+\infty} f(t)\cos 2\pi\tau t\, dt, \qquad b(\tau) = \int_{-\infty}^{+\infty} f(t)\sin 2\pi\tau t\, dt.$$

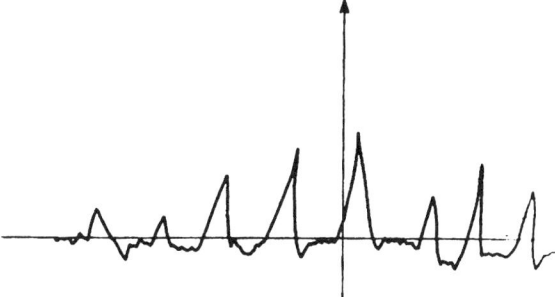

FIGURE 2. Electrocardiogram

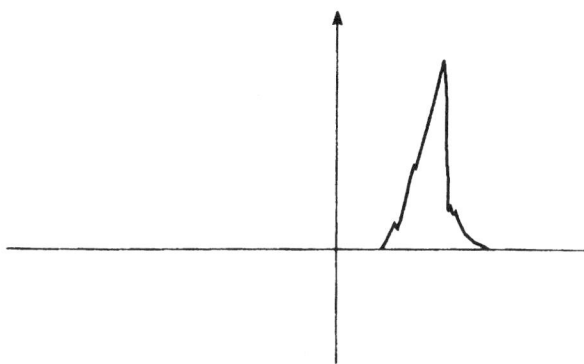

FIGURE 3. "Suspicious heart beat"

The reader understands that the aim of the above outline of Fourier analysis is to remind him/her what this theory is aiming at so that he/she will be able to compare this theory with the theory of wavelets that we will discuss also briefly below.

The aim of the theory of wavelets is similar but broader than the aim of Fourier analysis. Let us give a clear explanation of what we mean.

The theory of wavelets starts from the realization that the functions of sine and cosine do not constitute the most appropriate mean for the study of certain phenomena.

For example, let us assume that we want to find a certain way, a "mechanism", through which we will be able to trace certain "suspicious" heart beats in an electrocardiogram. (The presentation that we make here is intentionally oversimplified in order to make ourselves better understood.) First of all, we want this mechanism to be able to decompose the electrocardiogram into its components (i.e., into more simple waves). If it is observed that one of these components is the "suspicious" heart beat that we want to trace, then the mechanism must be in position to inform us accordingly.

In particular, imagine that we have an electrocardiogram, as is the one shown in Figure 2, and assume that the "suspicious" heart beat is the one shown in Figure 3.

FIGURE 4. Vibrating chord

What we are trying to achieve is to decompose the wave (the function) shown in Figure 1 into fundamental components and to be able to decide whether one of these components is the wave shown in Figure 3. One might pose the question "what is the use of Fourier analysis in this problem?". The answer to this question is that Fourier analysis allows us to decompose the wave (the function) shown in Figure 2, as well as the one shown in Figure 3, into sines and cosines, and then it allows us to compare the Fourier coefficients that we obtain from these two decompositions.

But this method (mechanism) may prove to be extremely insufficient for the very simple reason that in certain cases the functions of sine and cosine have nothing to do with the waves that we are trying to analyze, as it happens with the example of the electrocardiogram.

Historically, the introduction of sine and cosine, came from the fact that these functions are eigenvalues of a differential equation, to which mathematicians were led during the middle of the 18th century in the effort to determine the mathematical laws that govern the motion of a vibrating chord whose extremities are fixed at the points 0 and π. Indeed, an elementary analysis of the tension of the chord shows that if $y(x,t)$ denotes the ordinate of the chord at time t above point x, then $y(x,t)$ satisfies the differential equation $\frac{\partial^2 y}{\partial t^2} = a^2 \frac{\partial^2 y}{\partial x^2}$, where a is a certain parameter.

The next step in contemporary Fourier analysis (in wavelets) is to replace the classical functions of sine and cosine by other more flexible units which are appropriate for the study of problems like the ones regarding the electrocardiogram. Generally speaking, these are the reasons that dictated the introduction of the theory of wavelets.

Let ψ be a function whose graph looks like a small wavy curve (wavelet) whose integral is equal to zero. The function ψ generates a family of "wavelets" of the form $\psi(at + b)$, where a and b are real numbers. By changing a, we obtain either a contraction or a dilation of the function ψ, and by changing b, we obtain a translation of the function ψ. This 2-parametric family of functions of the form $\psi(at + b)$ is the one that replaces the functions of sine and cosine used in classical Fourier analysis.

The so-called "Continuous Wavelet Transform" of a function $f(t)$ transforms this function into a function $c(a, b)$ of two variables, where

$$c(a,b) = \int f(t)\psi(at+b)dt.$$

The "Discrete Wavelet Transform" is the one in which the parameters a and b take only discrete values. This means that in this case we use only wavelets of the form $\psi(2^k t + l)$, where k and l are integers.

The basic process in wavelets is similar to the one followed in classical Fourier analysis, as we described it above. The coefficients inform us in which way the functions used in the decomposition of a given function (i.e., the functions of sine and cosine or the wavelets) must be changed so that we will be able to construct the initial function (or signal). One can construct the function by adding wavelets of different sizes and of different positions in the same way that one constructs a function by adding functions of sine and cosine. In addition, the method of calculating the coefficients is the same with the one followed in classical Fourier analysis: we multiply the function by a wavelet and we calculate the integral of this product. By contracting and dilating the wavelets in order to change their frequency, everything gets changed. The wavelets automatically adapt to the various components of the function (i.e., of the signal). This last process is known in English terminology as multiresolution. A scholarly translation of this term in Greek would be "$\pi o \lambda \lambda \alpha \pi \lambda \acute{\eta}$ $\alpha \nu \acute{\iota} \chi \nu \epsilon \upsilon \sigma \eta$".

The wavelets are also called "mathematical microscope" because by contracting wavelets at will we increase the ability of the "microscope" to magnify, which in turn allows us to examine the function (signal) in more detail. This cannot be achieved if one uses classical Fourier analysis, i.e., by modifying the functions of sine and cosine. Due to the theory of wavelets we are no longer forced to reduce every problem to waves produced by sine and cosine functions. On the contrary, we are now in a position to "invent" an appropriate Fourier analysis for every given problem.

We have at our disposal powerful techniques which allow us to locate the problem and study it both with respect to the variable regarding space and with respect to the variable regarding phase. Many branches of applied mathematics owe their rapid progress to the theory of wavelets.

An algorithm similar to the FFT algorithm of classical Fourier analysis that we presented in Appendix I also exists in wavelets.

The theory of wavelets is still in its youth and the field of research in this theory is very wide. This is the reason that I have decided to include certain facts about wavelets in the present book, which is mainly addressed to young researchers of mathematical science.

A SHORT CURRICULUM VITAE OF THE AUTHOR

Nicolaos Artemiadis son of Kyriakos and Despina was born in Constantinople in 1917. In 1922, when the Downfall of the Greeks in Asia Minor began, his family moved to Greece. He graduated from the Department of Mathematics of the Aristotle University of Thessaloniki (1939) and from the Institut Français d'Athènes (1951). He studied in France (Sorbonne) from 1951 until 1957, where he obtained the following degrees:

A) Diplôme d'Etudes Supérieures en Analyse Supérieure (1954).
B) Diplôme d'Etudes Supérieures en Algèbre et Théorie des Nombres (1956).
Γ) Docteur ès Sciences Mathématiques (Doctorat d'Etat) avec la mention "Très honorable" (1957).

He taught as an Assistant Professor (1958-1960) and as an Associate Professor (1961-1966) at the University of Wisconsin in the USA. He also taught as a Full Professor at Southern Illinois University in Carbondale, USA (1966-1977).

He taught at the Aristotle University of Thessaloniki (1960-1961) as a tenured Professor of an extraordinary independent chair, and at the University of Patras (1975-1984) as a regular Professor of the A' Chair of Mathematics.

> Since 1984 he is an Emeritus Professor of the University of Patras.
> He is a regular member of the Academy of Athens (November 20, 1986).
> He is a regular member of the Academia Tiberina (Italy - June 24, 1987).
> He is a regular member of the New York Academy of Sciences (1992).
> He served as the Vice-President of the Academy of Athens during the year 1999 and as the President during the year 2000.

His research interests lie in the areas of Real, Harmonic and Complex Analysis, in which he carried out the main body of his original research works, which are published in reputable, American and European scientific journals.

He also wrote the following books and monographs:

(1) *A Course in Harmonic Analysis*, Volumes I, II, "Democritus" Center of Nuclear Research, 1972, 1973.
(2) *Real Analysis*, Southern Illinois University Press - Feffer and Simons, Inc., London - Amsterdam, 1976.
(3) *An Introduction to Contemporary Mathematical Analysis*, Lychnos, Athens, 1976, 1979. (This book is reprinted every two to three years by the University of Patras.)
(4) *Complex Analysis*, Lychnos, Athens, 1976, 1980, 1992.
(5) *Functions of Real Variables*, Papadamis Publications, Athens, 1977, 1982.

(6) *Elementary Geometry from an Advanced Vantage Point*, a publication of the Greek Mathematical Society, 1978.
(7) *The instruction of Mathematics in Higher and other Educational Institutions in Greece*, (Monograph) Published by the University of Patras, 1979.
(8) *An Introduction to Functional Analysis*, Lychnos, Athens, 1990, a publication of the Academy of Athens.
(9) *History of Mathematics*, Athens, 2000, a publication of the Research Committee of the Academy of Athens.
(10) More than 30 articles on various scientific topics published in the Proceedings of the Academy of Athens, 1987-2000.

MILITARY SERVICE

(1) He was a student at the School of Reserve Officers in Artillery in Thessaloniki from March 3, 1940 until October 28, 1940.
(2) During the second world war he served in the line of fire as a second lieutenant and later as a lieutenant in anti-aircraft artillery from October 28, 1940 until May 31, 1941.
(3) He served as a lieutenant in various units from January 6, 1945 until December 10, 1948.

Total military service: Sixty two (62) months.

BIBLIOGRAPHY

(1) Aaboe, A.: *Episodes from the Early History of Mathematics*, Random House, 1964.
(2) Anglin, W. S.: *Mathematics: A Concise History and Philosophy*, Springer-Verlag, 1994.
(3) Anglin, W. S. and Lambek, J.: *The Heritage of Thales*, Springer-Verlag, 1995.
(4) Apostol, H. G.: *Aristotle's Philosophy of Mathematics*, Univ. of Chicago Press, 1952.
(5) Artemiadis, N. K.: *Real Analysis*, Southern Illinois University Press, Feller and Simmons, Inc., 1976.
(6) Artemiadis, N. K.: *Introduction to Contemporary Mathematical Analysis*, First Edition, Publications of the University of Patras, 1976.
(7) Artemiadis, N. K.: *Complex Analysis*, Lychnos Publications, Athens, 1976, 1980, 1992, 1998.
(8) Artemiadis, N. K.: *Functions of Real Variables*, S. Athanosopoulos - S. Papadamis Publications, Athens, 1977, 1982, 1992, 1998.
(9) Artemiadis, N. K.: *Harmony in Nature: The Role of Mathematics in its Understanding*, Proceedings of the Academy of Athens, Volume 62, 1987.
(10) Artemiadis, N. K.: *The Geometry of Fractals*, Proceedings of the Academy of Athens, Volume 63, 1988.
(11) Artemiadis, N. K.: *Crowning Moments in Mathematics*, Proceedings of the Academy of Athens, Volume 63, 1988.
(12) Artemiadis, N. K.: *A Short Account of the work of A. N. Kolmogorov*, Proceedings of the Academy of Athens, Volume 63, 1988.
(13) Artemiadis, N. K.: *The Life and Work of I. Newton*, Proceedings of the Academy of Athens, Volume 63, 1988.
(14) Artemiadis, N. K.: *On some New Major Developments in Mathematics*, Proceedings of the 3rd Congress of Geometry, Thessaloniki, 1991.
(15) Artemiadis, N. K.: *From the History of Mathematics of Older Times*, Proceedings of the Academy of Athens, Volume 67, 1992.
(16) Artemiadis, N. K.: *Recent Discoveries in Number Theory*, Proceedings of the Academy of Athens, Volume 68, 1993.
(17) Artemiadis, N. K.: *Fermat's Last Theorem has been proved*, Proceedings of the Academy of Athens, Volume 68, 1993.
(18) Artemiadis, N. K.: *Chaos - Fractals - Dynamical Systems*, Proceedings of the Academy of Athens, Volume 69, 1994.
(19) Artemiadis, N. K.: *A remark on the space $(H(X), h)$ of fractals*, Proceedings of the Academy of Athens, Volume 70, 1995.

(20) Artemiadis, N. K.: *The Neologism Fractal in Mathematics and its translation into Greek*, Proceedings of the Academy of Athens, Volume 70, 1995.

(21) Artemiadis, N. K.: *Elementary Geometry from an Advanced Vantage Point*, Second Edition, Publications of the Greek Mathematical Society, Athens, 1998.

(22) Balaguer, M.: *Platonism and Anti-Platonism in Mathematics*, Oxford University Press, 1998.

(23) Ball, W. W. R.: *History of Mathematics*, Macmillan and Co., 1888.

(24) Bell, E. T.: *Les Grands Mathématiciens*, Payot, Paris, 1950.

(25) Boyer, C. and Merbach, B.: *A History of Mathematics*, John Wiley and Sons, Inc., 1989.

(26) G. Contopoulos, *Order and Chaos in Dynamical Astronomy*, Springer-Verlag, 2002.

(27) Georgakopoulos, K.: *Ancient Greek Exact Scientists*, L. Georgiadis Publications, Athens, 1995.

(28) Eves, H.: *Great Moments in Mathematics*, The Dolciani Mathematical Expositions (Published by the M. A. A., 1983, in two volumes).

(29) Fermat, Pierre de: *Oeuvres* (4 Volumes and Supplement), Gauthier Villars, Paris 1891-1912, 1922.

(30) A. S. Fokas and V. E. Zakharov, Eds., *Important developments in soliton theory*, Springer-Verlag, 1993.

(31) Heath, T. L.: *A History of Greek Mathematics*, Volume 1, Oxford University Press, 1921.

(32) Heath, T. L.: *Greek Astronomy*, J. M. Dent and Sons, 1932.

(33) Heath, T. L.: *Mathematics in Aristotle*, Oxford University Press, 1949.

(34) Heath, T. L.: *Aristarchos of Samos*, Oxford University Press, 1952.

(35) Heath, T. L.: *The Works of Archimedes*, Dover, 1953.

(36) Heath, T. L.: *The Thirteen Books of Euclid's Elements*, Three Volumes, Dover, 1956.

(37) Heath, T. L.: *Apollonius of Perga*, Barnes and Nobles, 1961.

(38) Heath, T. L.: *A Manual of Greek Mathematics*, Dover, 1963.

(39) Heath, T. L.: *Diophantus of Alexandria*, Dover, 1964.

(40) Kershner, R. B. and Wilcox, L. R.: *The Anatomy of Mathematics*, Ronald Press Co., 1950.

(41) Kline, M.: *Mathematical Thought from Ancient to Modern Times*, Oxford University Press, 1972. (This three-volume writing contains an extensive bibliography.)

(42) Morgan, F.: *Riemannian Geometry*, A. K. Peters, 1998.

(43) Parsons, E. A.: *The Alexandrian Library*, The Elsevier Press, 1952.

(44) Peitgen, H.-O., Jüngens, H. and Saupe, D.: *Chaos and Fractals*, Springer-Verlag, 1992.

(45) Penrose, R.: *The Emperor's New Mind*, Oxford University Press, 1990.

(46) Penrose, R.: *Shadows of the Mind*, Oxford University Press, 1994.

(47) Smith, D. E.: *History of Mathematics*, Two Volumes, Dover Publications, 1958.

(48) Verriest, G.: *Les Nombres et les Espaces*, Arman Colin, 1951.

(49) *Mathematics*, The MIT Press, 1956.

(50) *The Encyclopedic Dictionary of Mathematics*, Second English Edition, The MIT Press, 1986.

BIBLIOGRAPHY (For Appendices I, III)

(1) Holschneider, M.: *Wavelets*, Clarenton Press, Oxford, 1995.
(2) Hubbard, B. B.: *The world according to wavelets*, A. K. Peters, Natick, Mass., USA, 1998.
(3) Kahane, J. P.: *Lemarié - Rieusset: Séries de Fourier et ondelettes*, Paris, Cassini, 1998.
(4) Krantz, S. G.: *A panorama of harmonic analysis*, The Mathematical Assoc. of America, 1999.

CHRONOLOGICAL TABLE

B.C.

		776	First Olympiad
		753	Founding of Rome
		743	Era of Nabonassar
		740	Works of Homer and Hesiod
		586	Babylonian Captivity
585	Thales of Miletus Deductive Geometry		
540	Pythagoras Arithmetic and Geometry (≈)* Rod numerals in China (≈) Indian Sulvasūtras (≈)		
		538	Persians took Babylon
		480	Battle of Thermopylae
		461	Beginning of the Age of Pericles
450	Spherical shape of the earth Parmenides (≈)		
430	Death of Zeno Works of Democritus Astronomy of Philolaus (≈) *Elements* of Hippocrates of Chios (≈)	430	Hippocrates of Cos (≈)
		429	Death of Pericles Plague at Athens
428	Birth of Archytas Death of Anaxagoras		
427	Birth of Plato		
420	Trisectrix of Hippias (≈) Incommensurable numbers (≈)		
		404	End of Peloponnesian War
		399	Death of Socrates *Anabasis* of Xenophon

* The symbol ≈ means approximately

369	Death of Theaetetus		
360	Eudoxus: Proportions and the method of exhaustion ($\approx$)		
350	Menaechmus: conic sections ($\approx$) Dinostratus: quadratrix ($\approx$)		
		347	Death of Plato
335	Eudemus: *History of Geometry* ($\approx$)		
		332	Founding of Alexandria
330	Autolycus: *The Moving Sphere* ($\approx$)		
		323	Death of Alexander the Great
		322	Deaths of Aristotle and Demosthenes
320	Aristaeus: *Conics* ($\approx$)		
		311	Beginning of the Seleucid Era in Mesopotamia
		306	Ptolemy I (Soter) of Egypt
300	Euclid: *Elements* ($\approx$)		
260	Aristarchus: Heliocentric system ($\approx$)		
230	Sieve of Eratosthenes ($\approx$)		
225	*Conics* of Apollonius ($\approx$)		
212	Death of Archimedes		
		210	Great Chinese Wall (beginning)
180	Cissoid of Diocles ($\approx$) Conchoid of Nicomedes ($\approx$) Hypsicles: Circle 360° ($\approx$)		
		146	Destruction of Carthage and Corinth
140	Trigonometry: Hipparchus ($\approx$)		
		121	Gaius Gracchus killed
		75	Cicero restores the tomb of Archimedes
		44	Death of Julius Caesar

A.D.

75	Works of Heron of Alexandria ($\approx$)		
100	Nicomachus: *Arithmetica* ($\approx$) Menelaus: *Spherics* ($\approx$)		
125	Theon of Smyrna and Platonic Mathematics		
150	Ptolemy: *The Almagest* ($\approx$)		
		180	Death of Marcus Aurelius
250	Diophantus: *Arithmetica* ($\approx$)		
		286	Division of the Empire by Diocletian
320	Pappus: *Mathematical Collections* ($\approx$)		
		324	Founding of Constantinople
390	Theon of Alexandria		
415	Death of Hypatia		
470	Tsu Ch'ung-chi's value of π ($\approx$)		
476	Birth of Aryabhata	476	"Fall" of Rome
485	Death of Proclus		
520	Anthemius of Tralles and Isidore of Miletus		
524	Death of Boethius		
529	Closing of the Schools of Athens		
		532	Building of "Agia Sophia" by Justinian
560	Eutocius: Commentaries on the work of Archimedes ($\approx$)		
		622	Hegira of Mohammed
		641	Library of Alexandria burned
755	Hindu works are translated into Arabic		
		814	Death of Charlemagne
830	Al-Khwarismi: *Algebra* ($\approx$)		
		1096	First Crusade
1114	Birth of Bhaskara		
1142	Adelard of Bath translated Euclid		
1202	Fibonacci: *Liber abaci*		
		1204	Crusaders sack Constantinople
		1215	Magna Carta

1260	Campanus' trisection of an angle ($\approx$)		
		1271	Travels of Marco Polo
1303	Chu Shi-kie and the Pascal triangle		
1328	Bradwardine: *Liber de proportionibus*		
		1440	Invention of typography
		1453	Fall of Constantinople
1476	Death of Regiomontamus		
1482	First printed Euclid		
1489	Use of + and − by Widman		
		1492	Discovery of America by Columbus
1494	Pacioli: *Summa*		
1527	Apian published the Pascal triangle		
1543	Tartaglia published Moerbeke's *Archimedes* Copernicus: *De revolutionibus*		
1545	Cardano: *Ars magna*		
1564	Birth of Galileo	1564	Birth of Shakespeare Deaths of Versalius and Michelangelo
1572	Bombelli: *Algebra*	1572	Massacre of Saint Bartholomew
1579	Viète: *Canon Mathematicus*		
1603	Death of Viète		
1609	Kepler: *Astronomia nova*	1609	Galileo's telescope
1614	Napier's logarithms		
		1616	Deaths of Shakespeare and Cervantes
1635	Cavalier: *Geometria indivisibilibus*		
		1636	Foundation of Harvard College
1637	Descartes: *Discours de la méthode*		
1639	Desargues: *Brouillon projet*		
1642	Birth of Newton Death of Galileo		
		1644	Torricelli's barometer

1647	Deaths of Cavalieri and Torricelli		
1655	Wallis: *Arithmetica infinitorium*		
		1662	Foundation of the Royal Society
		1666	Foundation of the Académie des Sciences
1667	Gregory: *Geometriae pars universalis*		
1668	Mercator: *Logarithmotechnia*		
1670	Barrow: *Lectiones geometriae*		
1678	Ceva's theorem		
1684	Leibniz's first paper on calculus		
1687	Newton: *Principia*		
1690	Roller: *Traité d'Algèbre*		
1696	Brachistochrone (the Bernouillis) L'Hôpital's rule		
		1699	Death of Racine
1706	Use of π by William Jones		
1715	Taylor: *Methodus incrementorum*		
1718	De Moivre: *Doctrine of Chances*	1718	Fahreneit's thermometer
1730	Stirling's formula	1730	Réaumur's thermometer
1731	Clairaut on skew curves		
1733	Saccheri: The axiom of parallelism		
1734	Berkeley: *The Analyst*		
		1738	Daniel Bernoulli: *Hydrodynamica*
1742	Maclaurin: *Treatise of Fluxions*	1742	Centigrade thermometer
1743	D'Alembert: *Traité de dynamique*		
1748	Euler: *Introduction* Agnesi: *Instituzioni*		
		1749	Volume I of Buffon's *Histoire naturelle*
1750	Cramer's rule		
1770	Hyperbolic Trigonometry		
		1776	American Declaration of Independence
1777	Buffon's needle problem		

1779	Bezout's elimination method		
		1781	Discovery of Uranus by Herschel
		1783	Composition of Water (Cavendish, Lavoisier)
1788	Lagrange: *Mécanique analytique*		
		1789	French Revolution
1794	Legendre: *Éléments de géometrie*	1794	Lavoisier guillotined
1795	Monge: *Feuilles d' analyse*	1795	École Polytechnique École Normale
1796	Laplace: *Systéme du monde*	1796	Vaccination (Jenner)
1797	Lagrange: *Fonctions Analytiques* Mascheroni: *Geometria del Compasso*		
		1799	Metric System
		1800	Volta's battery
1801	Gauss: *Disquisitiones arithmeticae*		
1803	Carnot: *Géometrie de position*		
		1804	Napoleon crowned emperor
		1815	Battle of Waterloo
1817	Bolzano: *Rein analytischer Beweis*	1817	Optical transverse vibrations (Young and Fresnel)
		1820	Dersted discovered electromagnetism
		1821	Greek Revolution
1822	Poncelet: *Traité* Fourier Series Feuerbach's Theorem		
1826	Founding of Crelle's Journal Principle of Duality (Poncelet, Plücker, Gergonne) Elliptic functions (Abel, Gauss, Jacobi)	1826	Ampère's work in electrodynamics
1827	Homogeneous coordinates (Möbius, Plücker, Feurbach) Cauchy: *Calcul de Residues*	1827	Ohm's law
1828	Green: *Electricity and Magnetism*		

1829	Lobachevski's geometry Abel's death at 26	1829	Death of Thomas Young
1830	Peacock: *Algebra*		
		1831	Faraday's electromagnetic induction
1832	Bolyai: *Absolute Science of Space* Galois' death at 20	1832	Babbage: Analytical Engine
		1836	First telegraph
		1842	Conservation of energy (Mayer and Joule)
1843	Hamilton's quaternions		
1844	Grassmann: *Ausdehnungslehre*		
		1848	Marx: *Communist Manifesto*
1852	Chasles: *Traité de géométrie supérieure*		
1854	Riemmann: *Habilitationschrift* Boole: *Laws of Thought*		
		1859	Darwin: *Origin of Species* Chemical spectroscopy (Bunsen and Kirchhoff)
1864	Weierstrass appointed at Berlin		
1865	Foundation of the London Mathematical Society		
1872	Foundation of the Mathematical Society of France Dedekind: *Stetigkeit und irrationale Zahlen* Heine: *Elemente* Méray: *Nouveau préçis* Klein's *Erlangen Programm*		
1873	Hermite proved the that e is transcendental	1873	Maxwell: Electricity and Magnetism Birth of Constantine Carathéodory
1874	Cantor: *Mengenlehre*		
		1876	Bell's telephone

1877	Sylvester appointed at John Hopkins		
1881	Gibbs: Vector Analysis		
1882	Lindemann proved the transcendence of π		
1884	Frege: *Grundlagen der Arithmetic*		
		1887	Discovery of Hertzian waves
1888	Beginnings of the American Mathematical Society	1888	Foundation of the Pasteur Institute
1889	The Peano axioms		
1895	Poincaré: *Analysis Situs*	1895	Discovery of X-rays (Roentgen)
1896	Prime Number Theorem (Hadamard and De la Vallée-Poussin)	1896	Discovery of radioactivity (Becquerel)
		1897	Discovery of electricity (J. J. Thomson)
		1898	Discovery of radium (Marie Curie)
1899	Hilbert: *Grundlagen der Geometrie*		
1900	Hilbert's problems	1900	Freud: *Die Traumdeutung*
		1901	Planck: *Quantum Theory*
1903	Lebesgue integration		
		1905	Special Relativity (Einstein)
1906	Functional Calculus (Fréchet)		
1907	Brouwer and Intuitionism		
1910	Russell and Whitehead: *Principia* (Volume I)		
1914	Hausdorff: *Grundzüge der Mengenlehre*		
		1915	Panama Canal opened
1916	Einstein's General Theory of Relativity		
1917	Hardy and Ramanujan on the theory of numbers	1917	Russian Revolution
1918	Foundation of the Greek Mathematical Society		

1923	Banach spaces		
		1926	Foundation of the Academy of Athens
		1927	Lindbergh flew the Atlantic
		1928	Fleming discovered penicillin
1931	Gödel's theorem		
		1933	Hitler became chancellor
1934	Gelfand's theorem		
1939	Volume I of Bourbaki's *Elements*		
		1945	Bombing of Hiroshima
		1946	First meeting of the United Nations
		1950	Death of Carathéodory
1955	*Homological Algebra* (Cartan and Eilemberg)		
		1957	Launching of Sputnik I
1963	Paul J. Cohen's work on the Continuum Hypothesis	1963	Assassination of President Kennedy
1966	15th International Congress of Mathematicians (Moscow)		
		1969	First man on the moon
1973	Deligne confirms the Weil conjectures		
		1975	End of Vietnam War
1976	Computational verification of the four color conjecture		
1980	Classification of finite groups		
1984	Verification of the Bieberbach conjecture ($|a_n| \leq n$)		
		1987	Death of A. N. Kolmogorov
		1990	The Collapse of the iron curtain
1994	Fermat's Last Theorem proved		

Subject Index

δ-symbol, 372
ν-body problem, 276
$\tau\epsilon\tau\rho\alpha\kappa\tau\acute{\upsilon}\varsigma$, 21
$\theta\rho\acute{\upsilon}\mu\mu\alpha$, 389
p-adic number, 374
$\aleph_0$, 301
C, 283
H, 283
N, 283
Q, 283
R, 283
Z, 283
Q$^+$, 283
R$^+$, 283
(laws of) duality, 93
écart, 325

Abelian group, 287
Abstract Field Theory, 373
abstract object, 6
actually infinite, 299
Al' muqâbala, 162
al-jabr, 162
Algebra, 208
algebra of logic, 314
algebraic number, 62, 237, 305
algorithm, 162, 405
Almagest, 135, 162
amicable numbers, 24
analytic continuation, 257
angular excess, 118
anharmonic ratio, 98
Anti-Platonism, 6
antiderivative, 254
Arithmetic in Nine Volumes, 155
Arithmetica, 141, 188
arithmetization of analysis, 256
Ars Magna, 175
associative algebras, 376
Astrolabe, 134
automorphic function, 274
Axiom of Choice, 308

Banach space, 334

Barber of Seville, 402
Beweistheorie, 319
Big Bang, 120
brachistochrone, 233
branch cut, 262
branch points, 262

calculus, 33
Calculus of Variations, 208
calculus of variations, 272
Cantor set, 387
cardinal numbers, 300, 301
categoricity, 85
category, 379
Catoptrica, 235
Cattle Problem, 129
Cauchy criterion, 251
Cauchy sequence, 288
Cauchy-Riemann equations, 263
center of curvature, 231
Chaos Theory, 386
character of a group, 373
characteristic value, 267
chromatic number, 354
closure, 329
combinatorial (or algebraic) topology, 345
commutator, 370
commutator subgroup, 370
compact, 346
complete, 289
complete continuity, 333
complete metric space, 347
complex number, 291
concrete category, 379
conformal, 232
conic sections, 101, 152
Conics, 130
conjugate space, 335
connected, 346
Consistency, 84
Consolations of Philosophy, 165
Contacts, 131
content, 360
continuous function, 251

continuum hypothesis, 302
convergent series, 252
countable, 301
criteria, 252
cryptographic codes, 17
curvature, 119
curvature of a manifold, 265
cycloid, 195

Delian Problem, 42
dense, 304
derived set, 301
developable surface, 231
differential, 254
differential calculus, 205
differential geometry, 230, 264
dimension, 347
Dioptra (Theodolite), 140
directrix, 130
Dirichlet's Theorem, 269
discrete (group), 274
distributions, 337
duplication of the cube, 59
dynamical system, 387

eccentricity, 130
eigenfunction, 328
eigenvalue, 267, 327
elementary symbols, 61
Elements, 15, 68
eliminant, 241
elliptic functions, 259
Elliptic Geometry, 77, 105
elliptic integral, 260
elliptic plane, 112
empty set ($\emptyset$), 313
entire function, 258
equivalence, 83
equivalence relation, 287
ergodic hypothesis, 397
ergodic theory, 397
essential singularity, 258
Eudoxus, 43
Euler characteristic, 349
evolute, 230
expansion, 120
exterior measure, 361
extremal, 325

Fast Fourier Transform, 385, 405
field, 287, 367
finite, 301
focus, 130
Formalists, 284, 318
founding members of the GMS, 273
four color problem, 353
Fourier series, 269
fractal, 389
free group, 370

functionals, 324
functor, 382
Fundamental Theorem of Algebra, 297

generators, 370
genus of a curve, 229
geometry of a compass, 64
golden section, 172
graph, 380
gravitation, 199
Greek era, 19
Greek Mathematical Society, 271
group, 51

harmonic tetrads, 94
Hausdorff Maximality Principle, 308
Hegira, 161
hermeneutical, 6
hertz, 426
homeomorphism, 144, 347
homomorphism, 144
Hyperbolic Geometry, 104
hypercomplex numbers, 376

idea, 284
ideal, 367
incommensurable, 22
independence, 86
index of an eigenvalue, 328
induction, 284
infimum, 309
infinitesimal, 34, 251
Infinitesimal Calculus, 205
integral, 254
integral calculus, 39, 205
integral domain, 287
integral equations, 324
integral residue, 255
interior measure, 361
Intuitionists, 284, 316
involute, 230
isomorphic, 287
isomorphic systems, 257
isoperimetric, 234

Jacobians, 259
Jordan algebra, 296

kinetic energy, 236

L. Schwartz's talk on Distributions, 411
La Dioptrique, 184
La Géométrie, 184
law of gravitation, 275
law of quadratic reciprocity, 216
Lebesgue-Stieltjes integral, 365
Les Météores, 184
Liber abaci, 167
library (of Alexandria), 67
Lie algebra, 296

SUBJECT INDEX

limit, 253
Linear Functional Analysis, 326
linguistic basis, 403
logarithm, 171
Logicians, 284
Logicism, 314
lower bound, 309

magic square, 155
Mandelbrot set, 7, 389
manifolds, 264
mathematical induction, 284
Mathematical Synagoge, 145
Mathematical Syntaxis, 135
mathematics, 21
maximal, 308
mean value theorem, 251
measurable function, 362
measurable set, 361
meniscus, 40
metamathematics, 83, 320
metaphysical project, 6
Metaphysics, 33
Method, 125
method of exhaustion, 39, 151
metric space, 323
Miletus, 150
minimal, 309
mixing matrix, 408
Mohammed, 161
monoid, 379
Monte Carlo Method, 399
Moscow's Papyrus, 11
music, 30
mutli-valued function, 258

natural numbers, 284
natural transformation, 383
nine-point circle, 280
non-Euclidean geometry, 103
normal chain, 371
normal subgroup, 371

oath, 22
oath of faith, 250
operation unary, binary, ternary, 286
operators, 324
order of a branch point, 262
order of an element a, 371
order relation, 289
ordered, 289
ordinal numbers, 300

Parabolic Geometry, 104
partial differential equations, 227
pendulum, 226
perfect numbers, 27
periodicity, 274
permutations, 52

philosophy, 21
Picard's theorem, 272
Plato's Academy, 43, 147
Platonism, 6
point at infinity, 91
pointset (or general) topology, 345
polar angle, 261
polygonal numbers, 24
positivists, 9
postulate of parallelsim, 103
potentially infinite, 299
power series, 206, 257
power set, 302
pre-symbolic language, 401
prime number, 15
Prime Number Theorem, 17
Principle of Least Action, 235
projective coordinates, 97
Projective Geometry, 89
projective plane, 352
properties of cardinal numbers, 303
Psammites, 127
Pythagorean Brotherhood, 21
pythagorean star, 24
Pythagorean Theorem, 26

quadratrix, 37
Quadrature of Parabola, 205
quaternions, 291, 294

radius of curvature, 231
rational numbers, 287
real number, 283
regular polygon with 17 sides, 249
regular polyhedra, 29
representation of groups, 372
residue, 251
resolvent, 57
Rhind's Papyrus, 11
Riemann surface, 261
ring, 286, 367, 375
Russell's paradox, 315

Sakkas, 125
self-adjoint, 331
separable space, 346
set, 300, 301
sheet, 261
simulation, 398
single-valued function, 263
singular point, 258
squaring of the circle, 59
straight line at infinity, 91
strong convergence, 329
Sulvasutras, 156
sum of a series, 252
supremum, 308
symbolic logic, 314
synthetic geometry, 279

Taylor series, 205, 253
Tetrabiblos, 162
the absolute of the plane, 109
the Fourier integral, 269
the Lebesgue integral, 361
Theory of Proportions, 23
transcendental, 62
transfinite numbers, 284
triangulation, 349
trisection of an angle, 59

upper bound, 308

Well-ordering Principle, 309

Zorn's Lemma, 309

Name Index

Πυθαγορικοί, 21

Abel, 251, 259, 327
Abraham Ben Ezra, 167
Abu Jafar al-Khazin, 144
Abu'l Jafa, 210
Achilles, 34
Ackermann, 319
Adelard of Bath, 74, 166
Ahmose (Ahmes), 11
al-Khazin, 294
Alexander the Great, 67
Alexandria, 67
Alexandrov, 347
Anaxagoras, 37
Anaximander, 20
Anaximenes, 20
Anthemius, 148, 439
Antiphon, 38
Apollonius, 67, 123, 130, 150
Appel, 354
Aquinas, 68
Archimedes, 39, 67, 80, 123, 151, 192
Archytas, 42
Aristarchus, 123, 181
Aristotle, 11, 21, 22, 49, 50, 150, 152, 221, 313
Aryabhata, 156
Arzela, 324
Ascoli, 324

Bézout, 214, 241
Bachet, 188
Baire, 317
Balaguer, 6
Banach, 333
Barrow, 183
Beltrami, 104
Berkeley, 34, 202, 221
Bernoulli, 208
Besicovitch, 390
Bhaskara, 156
Bianchi, 230
Bieberbach, 394

Biot, 280
Birkhoff, 397
Boethius, 133, 165
Bolyai, 104, 250
Bolzano, 251, 300
Bombelli, 175, 225, 291
Bonaparte, 214
Bonnet, 359
Boole, 314
Borel, 317
Bourbaki, 68, 283
Brahmagupta, 156, 161
Brianchon, 89, 280
Briggs, 179, 222
Brook Taylor, 205
Brouwer, 317
Buddha, 21
Buffon, 125
Burnside, 371

Caliph Omar I, 148
Cantor, 257, 300, 357
Carathéodory, 270, 281
Cardano, 174
Carnot, 237, 280
Cartan, 376
Cauchy, 202, 250, 300, 357
Cavalieri, 183, 281
Cayley, 295, 368
Ceres, 249
Ch'ung Chih, 156
Chasles, 90
Chen Jing-Run, 156
Clairaut, 220
Commandino, 74
Condorcet, 210
Confucius, 21
Cooley, 405
Copernicus, 180
Coxeter, 210
Cramer, 229
Crelle, 280

D'Alembert, 210

Dandelin, 281
Darboux, 359
de Branges, 394
De Gang Ma, 156
de la Place, 231
de la Vallée Poussin, 17
de Lagny, 217
de Maupertius, 235
de Moivre, 205
de Morgan, 215, 314
Dedekind, 46, 257, 288, 370
Demetrius of Phaleron, 67
Democritus, 20, 33, 35, 192
des Bertin, 220
Desargues, 92, 183
Descartes, 25, 183, 291
Dieudonné, 75
Dini, 357
Dinostratus, 46
Diocles, 152
Diogenes Laertius, 21
Diophantus, 129, 133, 167, 171, 187
Dirichlet, 15, 251, 281, 357
Dresne, 221
Dupin, 232, 280
Dyck, 370

Edison, 182
Eilenberg, 379
Einstein, 181
Empedocles, 33, 35
Eratosthenes, 16, 123
Erdös, 11
Euclid, 15, 67, 235
Eudemus, 43
Eudoxus, 23, 39, 44, 46, 288
Euler, 25, 30, 205, 208, 348
Eutocius, 148

Faltings, 189
Fermat, 17, 25, 183, 209, 235, 279
Ferrari, 175
Ferro, 174
Feuerbach, 280
Fibonacci, 144, 171, 208
Field, 6
Fischer, 326
Fourier, 252, 265
Fréchet, 324, 345
Fraenkel, 315
Frege, 314
Fresnel, 250
Frobenius, 371
Fubini, 365

Gödel, 315
Galileo, 299
Galileo Galilei, 181
Galois, 52, 55

Gauss, 15, 63, 72, 104, 217, 249, 300, 405
Gerbert, 166
Gergonne, 282
German, 129
Gherard of Cremona, 74
Gibbs, 296
Girard, 291
Goldbach, 17, 208
Grassmann, 295
Gregory, 183, 207
Grosseteste, 235
Guldin, 147
Guthrie, 353

Hölder, 370
Hachette, 229
Hadamard, 17, 317
Hahn, 333
Haken, 354
Halley, 199, 210
Halphen, 229
Hamilton, 292, 317
Hardy, 159, 397
harmonic sequence, 95
Harriot, 172
Hausdorff, 346, 390
Heawood, 353
Hegel, 33
Helioupolis, 43
Helly, 333
Helmstädt, 239
Hensel, 373
Heraclitus, 20, 33
Herbart, 264
Hermann, 208, 228
Hermite, 306
Herodotus, 11
Heron, 139, 151, 235
Hervagius of Basle, 166
Hesiod, 19
Hilbert, 6, 71, 77, 256, 319
Hipparchus, 133
Hippasus, 22
Hippias, 37, 152
Hippocrates of Chios, 40
Homer, 19
Hooke, 199
Hrotsvitha, 166
Humboldt, 244
Hurewicz, 347
Huygens, 183, 195, 227
Hypatia, 133
Hypsicles, 74

Iamblichus, 21
Isidorus, 148

Jacobi, 242, 259
Jacobo de Barbari, 172

NAME INDEX

Jones, 127
Jordan, 357, 361
Jordanus Nemorarius, 167
Justinian, 161

Körner, 409
Kant, 317
Katetov, 347
Kelvin, 269
Kempe, 353
Kepler, 192
Khayyam, 162
Klein, 90, 256, 368
Kovalevsky, 256
Krantz, 425
Kronecker, 300, 316
Kummer, 189

L'Hôpital, 234
Lagrange, 65, 205, 210, 279
Laguerre, 99
Lambert, 217
Lao Tze, 21
Laplace, 65, 205, 214
Lavoisier, 211
Le Stourgeon, 326
Lebesgue, 317
Legendre, 103, 189, 205, 216
Leibniz, 183, 205, 314
Lemaitre, 120
Lenin, 33
Leonardo da Vinci, 172
Leonardo of Pisa (See Fibonacci), 167
Libri, 251
Lie, 369
Lilavati, 158
Lindemann, 62, 270, 306
Liouville, 62
Listing, 348
Littlewood, 395
Lobachevsky, 85, 250
Lucas, 28
Lysimachus, 130

Möbius, 351
Mac Lane, 379
Maclaurin, 205, 279
Mahavita, 156, 158
Marcellus, 125
Mascheroni, 65
Matiyasevic, 16
Maxwell, 269
Mecca, 161
Medina, 161
Menaechmus, 46
Menelaus, 133, 135
Menger, 347
Mercator, 183, 221
Mersenne, 28, 183

Michelangelo, 181
Miller, 370
Minding, 242
Mittag-Leffler, 276
Mohammed ibn Musa al-Khowârismi, 162
Monge, 229
Moore, 326, 370
Morita, 347
Morley, 281
Murchiston, 178
Mydorge, 189

Napier, 178
Nasir-Eddin, 162
Nelson, 130
Netto, 368
Newton, 68, 120, 182, 183, 205, 357
Nicole Oresme, 167
Nicomachus, 133, 165
Nicomedes, 152

Olympiodorus, 235
Omar Khayyam, 174

Pacioli, 172
Paganini, 26
Pappus, 131, 133, 145, 171
Parmenides, 33
Parseval, 364
Pascal, 156, 183, 189
Pasch, 79
Peano, 257, 284, 361
Pell, 157
Peyrard, 74
Philoponus, 41
Piazzi, 249
Pingala, 156
Plato, 19, 30, 49
Plutarch, 43, 125
Poincaré, 81, 104, 274
Poncelet, 89, 90, 280
Pontius Pilate, 403
Porphyrius, 21
Pregel, 348
Proclus, 19, 103, 133, 299
Ptolemy, 151
Ptolemy (Claudius), 133
Ptolemy I, 67
Puiseux, 261
Pythagoras, 20, 21

Rademacher, 397
Radon, 365
Ramanujan, 159, 397
Recorde, 172
Regiomontanus, 172
Rhind, 11
Ricatti, 227
Richard Taylor, 188

Richardson, 391
Riemann, 105, 260, 357
Riesz, 326
Robinson, 202
Roverbal, 189
Rudolff, 172
Russell, 284

Saba, 161
Saccheri, 103
Schmidt, 326
Schumacher, 300
Schwarz, 272, 281
Servois, 280
Shi Huang-ti, 155
Shimura, 189
Simplicius, 148
Simson, 74
Socrates, 42
Spinoza, 68
Steiner, 90, 279, 281
Steinitz, 374
Stephen Dusan, 171
Stevin, 181
Stewart, 244
Stieltjes, 360
Stifel, 172
Stirling, 207
Study, 281
Syene (Aswan), 131
Sylow, 370
Sylvester II, 166

Tacitus, 165
Taniyama, 189
Tartaglia, 174
Thales, 19, 150
Theaetetus, 19, 43, 152
Theodore, 46
Theon, 74, 134
Thibault, 103
Timaeus, 30
Tonelli, 326
Tukey, 405
Tycho Brahe, 182

Urysohn, 347

Vandermonde, 239, 241, 386
Venizelos, 271
Viète, 175
Voltaire, 202
Volterra, 324

Wallis, 103, 183, 217
Wantzel, 38
wavelets, 425
Weber, 373
Wedderburn, 375

Weierstrass, 202, 256, 281
Weyl, 320
Whitehead, 315
Widman, 172
Wiener, 333
Wiles, 188, 396
Wilhelm, 249
Williams, 129
Wilson, 211
Wittgenstein, 314
Wren, 199

Zarnke, 129
Zeno, 33, 153
Zenodorus, 145, 234
Zermelo, 309, 315
Zoroaster, 21